BIOENERGY AND BIOFUEL FROM BIOWASTES AND BIOMASS

SPONSORED BY
Bioenergy and Biofuel Task Committee of the Environmental Council

Environmental and Water Resources Institute (EWRI)
of the American Society of Civil Engineers

EDITED BY
Samir K. Khanal
Rao Y. Surampalli
Tian C. Zhang
Buddhi P. Lamsal
R. D. Tyagi
C.M. Kao

Published by the American Society of Civil Engineers

Library of Congress Cataloging-in-Publication Data

Bioenergy and biofuel from biowastes and biomass / sponsored by Bioenergy and Biofuel Task Committee of the Environmental Council Environmental and Water Resources Institute of the American Society of Civil Engineers ; edited by Samir K Khanal … [et.al.].
p. cm.
Includes bibliographical references and index.
ISBN 978-0-7844-1089-9
1. Biomass energy. 2. Refuse as fuel. 3. Waste products as fuel. I. Khanal, Samir Kumar. II. Environmental and Water Resources Institute (U.S.). Bioenergy and Biofuel Task Committee.

TP339.B496 2010
662'.88--dc22 2009048102

American Society of Civil Engineers
1801 Alexander Bell Drive
Reston, Virginia, 20191-4400

www.pubs.asce.org

ISBN 978-0-7844-1089-9
Manufactured in the United States of America.

Preface

Energy demand is expected to increase by nearly 44% by 2030 due mostly to the increased demand from emerging nations such as India and China. Over 90% of the world energy demand (~500 Quadrillion Btu) is currently met by non-renewable sources such as petroleum, natural gas, coal and nuclear. In the United States, just 7% of the total energy consumption (~100 Quadrillion Btu) is currently supplied by renewable sources. Our heavy dependence on non-renewable energy sources has several irreparable consequences such as impacts on economic development, national security, and local and global environments. Thus, we must act quickly and decisively to develop a sustainable, affordable, and environmentally friendly energy sources. Biofuel and bioenergy derived from renewable feedstocks such as biowastes/residues and lignocellulosic biomass are considered to be the most promising alternatives. Significant research and technology development efforts are currently underway towards the development of the second and third generation of biofuels in the United States and other parts of the world. Many existing first generation biofuel industries such as Poet (formerly Broin), Abengoa Bioenergy, and Pacific Ethanol among others are in the process of integrating cellulosic ethanol into their existing corn-ethanol biorefineries. It is important to point out that sugar and starch-based plants will continue to be the major ethanol producers in the foreseeable future. These plants will serve as a model in the development of second generation biofuels. High strength wastewater and organic wastes are generated renewably in considerable amounts, which can be digested anaerobically to produce biomethane/biohydrogen. The generated bioenergy can contribute a significant part of energy needs in wastewater treatment plant operation. Thus, there would be multiple biofuel/bioenergy products generated from diverse feedstocks. Although there have been tremendous research efforts in microbial fuel cell, it faces several technical challenges in the process scale-up. The algal process for biodiesel production also faces a major technical barrier of low biological productivity. This book provides in-depth technical information on various aspects of biofuel/bioenergy production.

The ASCE's Technical Committee has identified biofuel/bioenergy as an important area. This 21-chapter book provides state-of-the-art reviews, and current research and technology developments with respect to the 2nd and 3rd generation of biofuels/bioenergy. The book contents are organized in such a way that each preceding chapter builds-up foundation for the following chapter(s). At the end of each chapter, the current research trends and further research needs are also outlined. The book primarily covers the biological/biochemical conversion for biofuel/bioenergy production as this option has been reported to be the most cost-effective method for biofuel/bioenergy production.

The book is divided into seven categories. Chapter 1 is the introductory chapter which gives overview of biofuel/bioenergy, advances in biofuel/bioenergy development, cost analysis of biofuel production and sustainability of biofuel industries. Chapters 2 through 5 focus on anaerobic processes for biomethane, biohydrogen and bioelectricity productions from high strength wastewaters, food wastes, organic fraction of municipal solid wastes and landfill. Chapter 6 covers microbial-based electricity production from lignocellulose-derived hydrolyzate. The second generation biofuel particular bioethanol production is covered in Chapters 7 through 12. These chapters cover lignocellulosic feedstock production, biomass preprocessing, biomass deconstruction (pretreatment), and enzyme hydrolysis of pretreated biomass. Biomass-derived syngas fermentation to ethanol is covered in Chapter 11, whereas lignin recovery and utilization are discussed in Chapter 12. The bioreactor systems, their selection, and design criteria for both gaseous and liquid biofuels (the 1st, 2nd and 3rd generation) are covered in Chapter 13. Chapters 14 to 16 cover the third generation of biofuels that focus on autotrophic and heterotrophic algal processes for biodiesel, and algal-ethanol production. Chapter 17 covers the bioconversion of residues of the 1st and 2nd generation biofuel industries into high-value biobased products. Life cycle analyses (LCA) of the 1st and 2nd generation of biofuels (from corn, soybean, Jatropha and cellulosic biomass) are discussed in Chapters 18 and 19. The last part of the book consists of two chapters (20 and 21). Chapter 20 focuses on biobutanol production from starch and hydrolyzate of agri-residues. Chapter 21 covers nanotechnology applications in biofuel production. This organization will help the readers to easily grasp the contents presented in the book.

We sincerely hope that this book will be a valuable treasure to researchers, instructors, decision-makers, practicing professionals, and others interested in the biofuel/bioenergy field. The book also will serve as a reference for senior undergraduate and graduate students, as well as for consulting engineers.

The editors gratefully acknowledge the hard work and patience of all the authors who have contributed to this book. The views or opinions expressed in each chapter of this book are those of the authors and should not be construed as opinions of the organizations they work for. Special thanks go to the graduate students at the University of Hawai'i at Mānoa (UHM): Mr. Devin Takara, Ms. Saoharit Nitayavardhana and Mr. Pradeep Munasinghe for assisting with formatting all chapters of the book. Dr. Prachand Shrestha, an SKK's former student, now at the University of California at Berkeley and Mr. Devin Takara, a current student at UHM put significant effort in designing the cover of the book. Prachand helped with the conceptual design of the cover along with SKK. Devin helped us with his excellent graphics design.

– SKK, RYS, TCZ, BPL, RDT, CMK

Contributing Authors

Bhavik R. Bakshi, *Ohio State University, Columbus, OH, USA*

S. Balasubramanian, *INRS, Universite du Quebec, Quebec, QC, Canada*

Shankha K. Banerji, *University of Missouri, Columbia, MO, USA*

Anil Baral, *International Council on Clean Transportation, Washington, DC, USA*

Puspendu Bhunia, *INRS, Universite du Quebec, Quebec, QC, Canada*

J. Brewbaker, *University of Hawaii at Manoa, Honolulu, HI, USA*

J. Carpenter, *University of Hawaii at Manoa, Honolulu, HI, USA*

Halil Ceylan, *Iowa State University, Ames, IA, USA*

Michael J. Cooney, *Hawaii Natural Energy Institute, University of Hawaii at Manoa, Honolulu, HI, USA*

Hong-Bo Ding, *Nangyang Technological University, Singapore*

Allyson Frankman, *Brigham Young University, Provo, UT, USA*

Venkataramana Gadhamshetty, *Air Force Research Laboratory (AFRL), Tyndall AFB, FL, USA*

Kasthurirangan Gopalakrishnan, *Iowa State University, Ames, IA, USA*

Christopher K.H. Guay, *Hawaii Natural Energy Institute, University of Hawaii at Manoa, Honolulu, HI, USA*

Peng Hu, *Brigham Young University, Provo, UT, USA*

P. Illukpitiya, *University of Hawaii at Manoa, Honolulu, HI, USA*

Rojan P. John, *INRS, Universite du Quebec, Quebec, QC, Canada*

Glenn R. Johnson, *New Mexico State University, Las Cruces, NM, USA*

C.M. Kao, *National Sun Yat-Sen University, Kaohsiung, Taiwan*

Samir K. Khanal, *University of Hawaii at Manoa, Honolulu, HI, USA*

Sunghwan Kim, *Iowa State University, Ames, IA, USA*

Buddhi P. Lamsal, *Iowa State University, Ames, IA, USA*

Randy S. Lewis, *Brigham Young University, Provo, UT, USA*

Hong Liu, *Oregon State University, Corvallis, OR, USA*

Xue-Yan Liu, *Nangyang Technological University, Singapore*

Saoharit Nitayavardhana, *University of Hawaii at Manoa, Honolulu, HI, USA*

Nagamany Nirmalakhandan, *New Mexico State University, Las Cruces, NM, USA*

R. Ogoshi, *University of Hawaii at Manoa, Honolulu, HI, USA*

Anup Pradhan, *University of Idaho, Moscow, ID, USA*

Marry L. Rasmussen, *Iowa State University, Ames, IA, USA*

Guo-Bin Shan, *INRS, Universite du Quebec, Quebec, QC, Canada*

Dev S. Shrestha, *University of Idaho, Moscow, ID, USA*

Prachand Shrestha, *University of California at Berkeley, Berkeley, CA, USA*

Rao Y. Surampalli, *U.S. Environmental Protection Agency, Kansas City, KS, USA*

Devin Takara, *University of Hawaii at Manoa, Honolulu, HI, USA*

Douglas R. Tree, *Brigham Young University, Provo, UT, USA*

B. Turano, *University of Hawaii at Manoa, Honolulu, HI, USA*

R.D. Tyagi, *INRS, Universite du Quebec, Quebec, QC, Canada*

G. Uehara, *University of Hawaii at Manoa, Honolulu, HI, USA*

J. (Hans) van Leeuwen, *Iowa State University, Ames, IA, USA*

C. Visvanathan, *Asian Institute of Technology, Bangkok, Thailand*

Jing-Yuan Wang, *Nangyang Technological University, Singapore*

Song Yan, *INRS, Universite du Quebec, Quebec, QC, Canada*

J. Yanagida, *University of Hawaii at Manoa, Honolulu, HI, USA*

Tian C. Zhang, *University of Nebraska-Lincoln, Omaha, NE, USA*

X.L. Zhang, *INRS, Universite du Quebec, Quebec, QC, Canada*

Contents

CHAPTER 1

Bioenergy and Biofuel Production: Some Perspectives

Samir Kumar Khanal and Buddhi P. Lamsal

1.1 Introduction

Energy is an indispensable component of society. Our modern society depends on energy for almost everything ranging from home appliances, lighting, transportation, heating/cooling, communication, to industrial processes to supply commodities for our daily needs. We currently consume around 500 Quadrillion Btu (QBtu) of energy, and about 92% of it comes from non-renewable sources such as petroleum, coal, natural gas and nuclear. Historically, the price of crude petroleum oil has been very low (in the range of $20 per barrel during 80's and 90's). From the turn of this century, the crude petroleum prices continued to rise and reached as high as $141 per barrel in early July of 2008. Dwindling reserves in the face of rapidly increasing energy consumption, combined with an increasing lack of energy security due to regional conflicts, and the environmental devastation that results from greenhouse gas (GHG) emission, clearly suggest that we must act urgently and decisively to develop sustainable, clean, affordable and renewable energy sources.

Renewable energy derived from wind, solar (photovoltaics), geothermal, ocean (tidal), hydropower, and biomass, all can equally contribute to our renewable energy portfolio. Although, only 8% of our current energy consumption comes from renewable sources, there are tremendous research and technology development efforts toward the development of various forms of renewable energy. Biofuel/bioenergy derived from biomass (lignocelluloses) has received a considerable attention lately and is considered a leading candidate for renewable energy generation, especially for liquid transportation fuel. There is also a renewed interest in autotrophic and heterotrophic algal biomass production, and nonfood-based oil seed crops such as Jatropha for biodiesel production. Moreover, anaerobic biotechnology has also received significant interest in the recent years for bioenergy production. Readers can find detailed information on anaerobic biotechnology options for bioenergy and bioelectricity production in a related publication (Khanal, 2008).

Although, 2nd and 3rd generation biofuels are the current focus of research and technology development, the 1st generation biofuel builds-up an excellent model to understand and apply the biorefinery concept to the 2nd and 3rd generation biofuels. In this opening chapter, we try to present our perspectives on biofuel and bioenergy production, rationales of moving towards a biobased economy, biochemical versus thermochemical conversions of biomass, environmental impacts, and sustainability aspects of biofuels. We also outline some cost analyses of lignocellulosic biofuels. This opening chapter helps to build a good foundation for the following chapters.

1.2 Why Biofuels?

The Organization of Economic Co-operation and Development (OECD) countries are the major consumer of energy, for example, the United States alone consumed around 99.86 QBtu of energy in 2006, which accounted for nearly 21% of the world's total energy consumption of 472.27 QBtu (International Energy Annual, 2006), and is expected to increase significantly. The International Energy Outlook (2009) projects that the world marketed energy consumption is expected to increase by 44% by 2030 primarily due to an increased demand from India and China.

Biofuels derived from renewable feedstocks are environmentally friendly fuels. The successful development of bio-based fuel is expected to provide better energy security, benefit local and national economies by contributing to agricultural sectors, and improve the local and global environments, among others. Some of these issues are elucidated in the following section.

1.2.1 Reducing Nation's Dependency on Imported Petroleum Fuels

Many OECD and non-OECD countries are dependent on imported energy, primarily petroleum fuels. Depending on the availability of biobased feedstocks, many nations can potentially develop biobased biorefineries locally to produce biofuels (such as bioethanol, biodiesel, bio-oil, jetfuel and different hydrocarbon-based fuels), and biochemicals to domestically replace their dependence on petroleums. For example, the worldwide terrestrial availability of lignocellulosic biomass is estimated to be around 200 x 10^{12} kg (220 billion ton) annually (Foust et al., 2008). If half of the biomass is converted to biofuel, it has a potential of producing as much as 7.7 trillion gallon of ethanol annually (@ 70 gallon/dry ton), which could contribute an energy equivalent of around 585 QBtu annually.

The United States is the number one petroleum oil importing country in the world. The US oil consumption in 2007 was 20,687,000 barrels/day, of which nearly 58.2% was imported (International Energy Outlook, 2009). Reducing the nation's dependency of imported petroleum fuels would greatly enhance energy security. The United States has the potential of producing 1.3 billion dry tons of plant biomass annually after accounting for food, feed and export (USDA and USDOE joint report, 2005). One billion tons of biomass could essentially substitute more than 30% of the nation's petroleum consumption thereby cutting down the oil import by as much as one-third. Thus, biomass can play an important role in the domestic bio-based economy by producing a variety of biofuels and biochemicals that are currently derived from petroleum-based feedstock.

1.2.2 Environmental Merits of Biofuels

The boom of biofuel industries in the US can be traced to the Clean Air Act Amendments of 1990, requiring the use of oxygenated gasoline in areas with unhealthy levels of air pollution. Since 1992, methyl tertiary-butyl ether (MTBE), a fuel oxygenate, octane enhancer has been used at higher concentrations in some gasoline to fulfill the oxygenated requirements, with minimum of 2 percent oxygen by weight (EPA, 2009). Oxygen in fuel helps gasoline burn more completely and cleanly, reducing harmful tailpipe emissions from motor vehicles. The oxygen dilutes or displaces gasoline components such as aromatics and sulfur while optimizing oxidation during combustion. In the past, most refiners have chosen to use MTBE over other oxygenates primarily for its blending characteristics and for economic reasons. However, concerns over ground water contamination due to the poor biodegradability of MTBE, prompted states like California to phase out MTBE use by the end of 2002 (Sissell, 1999). The ban resulted in the use of ethanol as an oxygenate to replace MTBE, and sparked ethanol boom. Ethanol is an oxygenated fuel that contains 35% oxygen, which improves fuel combustion when added to gasoline, with low tailpipe emissions of carbon monoxide, unburned hydrocarbons, and NOx. With favorable properties like higher octane number of 108, broader flammability limits, higher flame speeds and higher heats of vaporization (Balat et al., 2008), ethanol became a natural replacement of MTBE. A life-cycle analysis also showed that the use of corn-based ethanol reduces GHG emission by 18-29% compared to gasoline, and cellulosic-ethanol will achieve 85-86% GHG emission (Goettemoeller and Goettemoeller, 2007). Some discussion is presented later in the "Biofuel Sustainability" section of this chapter.

1.2.3 Economic Benefits of Biofuels

The successful creation of a biobased economy has a potential of creating local jobs and improving the rural economy. Various entities likely to be benefited from the successful development of biorefinery would be the rural farmers and the co-operatives run by the farmers, industries involved in agricultural equipment, facility design, and fabrication, the construction sector, and biotechnology industries among others. State and federal governments will also benefit from better tax revenue collection. Based on 2006 data, the economic benefits of bioethanol production in the United States are (Goettemoeller and Goettemoeller, 2007):

- $23.1 billion added to GDP
- 163, 034 new jobs
- $6.7 billion added to household income
- $2.7 billion in tax revenue for federal govt.
- $2.2 billion in tax revenue for state and local govt.

1.3 Current Status of Biofuel/Bioenergy Generation

Although the technologies for sugar and starch-derived bioethanol, and soybean, rapeseed, and palm oil-derived biodiesel are well-matured, these processes will continue to dominate in the foreseeable future for biofuel production. The 2nd and 3rd generation biofuels, which are primarily derived from non-food feedstocks such as lignocellulosic biomass, Jatropha, and microalgae, are very likely to be modeled after the 1st generation biorefinery concepts. Moreover, the integration of 1st generation biofuel plants to 2nd and 3rd generation biofuel plants could provide significant leverage in terms of resource needs, and may cut-down the production costs considerably.

As of May 2009, the United States has 161 corn-based biorefineries with an operational capacity of about 10.5 billion gallons of ethanol per year (RFA, 2009). The US produced around 650 million gallons of biodiesel in 2008 (Biodiesel 2020, 2008). In the recent years, biofuels (especially bioethanol and biodiesel) have benefited from unstable and rapidly increasing petroleum fuel prices, environmental merits and government incentives. It was, however, estimated that by year 2014, only about 15 billion gallons of corn ethanol could be produced without an adverse effect on natural resources (Sanderson, 2007). With the enactment of the Energy Independence and Security Act of 2007, which mandates 36 billion gallons of annual renewable fuel use by 2022, an alternate source of bioethanol has become a priority. Lignocellulosic biomass has been put forward as an alternate source of bioethanol feedstock. The US Department of Energy (DOE) in 2007 funded six demonstration plants for cellulosic ethanol in different regions of the country to run on different feedstocks (DOE 2007). They are:1) Abengoa Bioenergy, in Kansas, to produce 11.4 million gallons of ethanol annually using 700 tons per day of corn stover, wheat straw, sorghum stubble, and switchgrass, 2) ALICO Inc., in Florida, to produce 13.9 million gallons of ethanol a year using 770 tons per day of yard, wood, and vegetative wastes and eventually energy cane, 3) BlueFire Ethanol Inc., in California, to produce about 19 million gallons of ethanol a year using 700 tons per day of sorted green waste and wood waste from landfills, 4) POET (formely BROIN), in Iowa, to produce 125 million gallons of ethanol per year in an integrated ethanol plant of which roughly 25 percent will be from cellulosic biomass using 842 tons per day of corn fiber, cobs, and stalks, 5) Iogen Biorefinery, in Idaho, to produce 18 million gallons of ethanol annually using 700 tons per day of agricultural residues including wheat straw, barley straw, corn stover, switchgrass, and rice straw, and 6) Range Fuels, in Georgia, to produce about 40 million gallons of ethanol per year and 9 million gallons per year of methanol using 1,200 tons per day of wood residues and wood-based energy crops. Indeed, the ground-breaking of the first commercial-scale cellulosic ethanol biorefinery in the US was done in Georgia in 2007, in which Range Fuels utilized leftover wood residues from timber harvesting to produce ethanol at its facility (RFA, 2008). Table 1.1 summarizes the major cellulose-ethanol plants under development or construction. Iogen, an Ottawa-based Canadian enzyme company, has been operating a 40 ton/day pilot-scale cellulosic ethanol using wheat straw and corn stover. However, the question arises: why there isn't a single full or large-scale cellulosic-ethanol plant on the ground? The answer is probably very simple: The financial risk and costs are too high to make it profitable at this stage. The key issues are:

- Feedstock logistics (production and supply in large amounts within a reasonable distance)
- High pretreatment and enzyme costs
- Effective fermentation of mixed sugar stream
- Value of co-products and their utilization
- Long-term policy and government supports for biofuels especially ethanol

Algal processes for biofuels (biodiesel and biobased hydrocarbons) have received a significant attention in recent years. Research on algal research was carried out by the US DOE's National Renewable Energy Laboratory (NREL) under the Aquatic Species Program (ASP) from 1978 to 1996. The goal was to produce transportation fuels from algal-derived lipids cost effectively. The program was eventually abandoned due to significantly higher costs (two times higher than petroleum diesel fuel costs) at that time resulting from low biological productivity of algal cells. A detailed report can be found at NREL (1998). There has been a renewed interest in algal-research lately, primarily its potential to produce hydrocarbon-based transportation fuels. Algal technology, however, faces many engineering challenges such as good reactor design for efficient light penetration and CO_2 transfer, and algal cell recovery/harvesting and dewatering, among others. The bottle-neck for scale-up is still the low biological productivity of lipids, the precursor for biodiesel and other hydrocarbon production. Although heterotrophic algal cells showed significantly higher yield than autotrophic algae, the former requires the supplementation of organic carbon sources, primarily simple sugars. Thus, a major breakthrough is still needed in algal processes, especially in biological aspects such as metabolic engineering/genetic engineering before scale-up. A detailed discussion is given Chapters 14, 15 and 16.

Anaerobic digestion is probably the one of the earliest technologies for bioenergy production. Although, anaerobic digestion is a mature technology, there are significant research needs in both process engineering and process microbiology to improve bioenergy yield and robustness. Some of the important areas where anaerobic digestion could play a significant role in bioenergy production are covered in Chapters 2 through 5. Anaerobic digestion could also play a key part in emerging biofuel industries by supplying the heat and electricity for in-plant use. The biofuel residues/wastewaters such as stillage, vinasse, thin stillage etc. have extremely high organic content and rich in nutrients which can be anaerobically digested to produce methane. The senior author has adequately shown that the thermophilic anaerobic digestion of thin stillage from dry-grind corn ethanol plant could produce up to 1,038,425 MMBtu/year of energy equivalent from methane gas, which may replace nearly two-thirds (1,075,000 MMBtu) of boiler fuel for stream generation (Khanal, 2008).

In a subsequent study, another research group also studied methane production from thin stillage in dry-grind ethanol plants using thermophilic anaerobic sequencing batch reactors. Based on methane generation results, the researchers reported the improvement in the net energy balance ratio from 1.26 to 1.70 due to a 51% reduction in nonrenewable energy consumption (Agler et al., 2008). This option also provides an opportunity for waste remediation and water reuse. Thus anaerobic biotechnology represents an opportunity for ethanol producers to significantly improve their profitability. Such an option also needs to be explored for 2nd and 3rd generation biofuel plant residues. A detail discussion on anaerobic conversion of biofuel residues from 1st, 2nd and 3rd biofuels to bioenergy is covered in a related book (Khanal, 2008). Microbial fuel cell (MFC) or biological electricity generation also employs the principle of anaerobic biotechnology. Although there have been significant research efforts in MFC, a recent scale-up study conducted in Australia did not yield positive outcome (Keller, 2008). MFC faces significant challenges as a means for sustainable electricity generation. Some of these challenges include microbiological challenges (e.g. efficacy of exoelectrogenic bacteria, growth of methanogens in anode chamber, biofilm gradient etc.), technological challenges (e.g. membrane pH gradient, ohmic losses etc.), and economic challenges. Some of the aspects of MFC for electricity generation especially from biomass hydrolysates are presented in Chapter 6.

Table 1.1 Cellulosic-ethanol plants under development/construction as of May 2009.

Company	Location	Capacity	Technology	Feedstock
Abengoa Bioenergy LLC	Colwich, KS Hugoton, KS York, NE	11.36 mgy 11.6 mgy 11.6 mgy	Bio-chemical	Corn cobs, corn stover, switchgrass, wheat straw, milo stubble, and other biomass
AE Biofuels	Butte, MT	small scale	Ambient Temperature Cellulose Starch Hydrolysis	Switchgrass, grass seed, grass straw, and corn stalks
Bluefire Ethanol	Corona, CA Lancaster, CA	18 mgy 3.1 mgy	Arkenol Process Technology (Concentrated Acid Hydrolysis Technology Process)	Municipal solid waste
California Ethanol + Power, LLC	Brawley, CA	55 mgy		Facility powered by sugarcane bagasse

Coskata	Madison, PA	40,000 gal/yr	Biological fermentation technology; proprietary microorganisms and efficient bioreactor designs in a three-step conversion process that can turn most carbon-based feedstock into ethanol	Any carbon-based feedstock, including biomass, municipal solid waste, bagasse, and other agricultural residue
DuPont Danisco Cellulosic Ethanol LLC	Vonore, TN	250,000 gal/yr	Enzymatic hydrolysis technology	Switchgrass, corn stover, corn fiber, and corn cobs
Iogen Biorefinery Partners LLC	Shelley, ID	18.49 mgy	Bio-chemical	Wheat straw
POET	Scotland, SD Emmetsburg, IA	20,000 gal/yr 31.25 mgy	BFRAC™ separates the corn starch from the corn germ and corn fiber, the cellulosic casing that protects the corn kernel	Corn fiber, corn cobs, and corn stalks
Verenium	Jennings, LA Highlands County, FL	1.4 mgy 36 mgy	C5 and C6 fermentations	Sugarcane bagasse, specially-bred energy cane, high-fiber sugar cane
Range Fuels Inc.	Soperton, GA	20 mgy	Two-step themo-chemical process	Wood chips (mixed hardwood)
Ecofin LLC	Washington County, KY	1.3 mgy	Solid state fermentation process developed by Alltech	Corn cobs

ICM	St. Joseph, MO Shelley ID	1.51 mgy 18 mgy	Enzyme technology	Switchgrass, forage sorghum, corn stover, wheat straw, barley straw, and rice straw
Lignol Innovations	Grand Junction, CO	2.50 mgy	Biochem-organosolve	Woody biomass, agricultural residues, hardwood, and softwood
Mascoma	Monroe, TN	2.01 mgy	Bio-chemical bio-refineries	Switchgrass and hardwoods
Mascoma/New York State Energy Research and Development Authority/New York State Department of Agriculture and Markets	Rome, NY	5 mgy	"Consolidated bioprocessing" refinery would use genetically modified bacteria to break down and ferment local wood chips	Lignocellulosic biomass, including switchgrass, paper sludge, and wood chips
Mascoma/Michigan Economic Development Corporation/Michigan State University/Michigan Technological University	Chippewa County, MI	40 mgy		
Pacific Ethanol	Boardman, OR	2.7 mgy	BioGasol	Wheat straw, stover, poplar residuals
RSE Pulp	Old Town, ME	2.19 mgy	Bio-chemical bio-refineries	Woodchips (mixed hardwood)
Verenium	Jennings, LA	1.40 mgy	Bio-chemical bio-refineries	Bagasse, energy crops, agricultural residues, wood residues

Adapted from IEA Bioenergy, 2008, 2009 Ethanol Outlook Report http://www.ethanolrfa.org/industry/outlook/.

1.4 Biochemical or Thermochemical Conversion of Lignocellulosic Biomass?

The biofuels from cellulosic biomass can be produced using two major pathways as elucidated in Figure 1.1. Biochemical conversion utilizes multiple enzymes for the hydrolysis of pretreated biomass to obtain fermentable sugars. The hydrolysate obtained is then fermented to ethanol by using microbial catalysts. In thermochemical conversion, the biomass is converted into intermediate products, which are further transformed into biofuels through chemical or biological routes. Gasification converts the biomass into synthesis gas (syngas) composed primarily of hydrogen, and carbon monoxide. The syngas is then converted to liquid biofuel through chemical catalysts using a process known as Fischer-Tropsch (FT) synthesis. Pyrolysis is another thermochemical processes in which the biomass is heated in the absence of oxygen to produce liquid fuel. With fast pyrolysis, bio-oil can be produced. Although bio-oil is very unstable, corrosive and viscous, it can be reformed to transportation fuels in a petroleum refinery (Foust et al., 2008). The thermochemical route has potential to produce long chain hydrocarbons that can be reformed to synthetic diesel or aviation fuel.

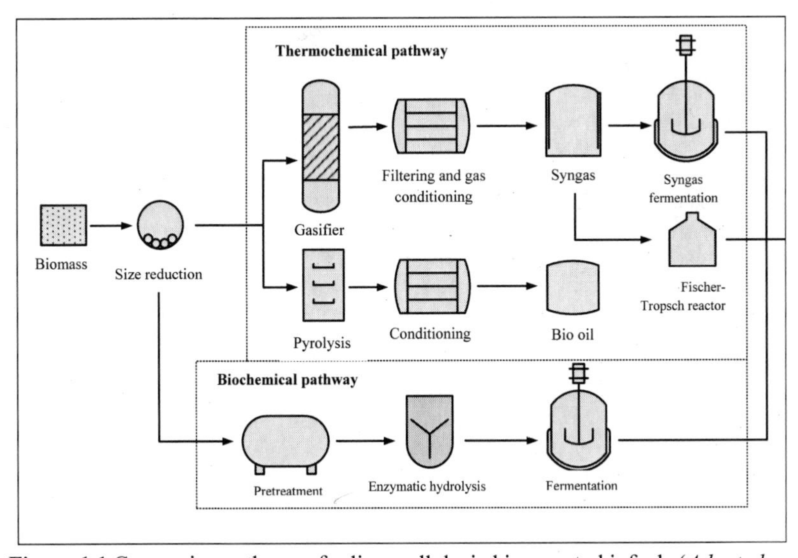

Figure 1.1 Conversion pathways for lignocellulosic biomass to biofuels (*Adapted from Munasinghe and Khanal, 2009*).

There are also hybrid processes that utilize both thermochemical and biochemical pathways to produce biofuels. For example, syngas fermentation utilizes microbial catalysts to convert syngas to biofuels (ethanol and butanol), and biomass-derived sugars into liquid alkanes of C_1 to C_{15} using aqueous-phase processing (Huber et al., 2005).

The selection of conversion technology is governed by many factors such as the local feedstock source and attributes to the local/regional markets for products/co-products, feedstock characteristics and variability, environmental attributes, government incentives and policies with respect of the products and co-products. Thermochemical conversion is more suitable for feedstocks that show considerable variability in composition, along with high lignin contents. Some of the examples include forest residues, soft woods, and mixed waste (residues) streams among others. The high lignin content facilitates efficient gasification of biomass. Agri-residues, and dedicated energy crops have more uniform chemical attributes with low lignin level. Thus, they are more suitable for biochemical conversion. It is also possible that new biorefinery may adopt a hybrid conversion technology to maximize the revenues and co-product utilization.

The International Energy Agency report suggested that biochemical conversion technology is more cost-effective for producing biofuel than the thermochemical conversion technology (IEA, 2008). The main reasons being that the research and development efforts in biochemical conversion technology are relatively recent and there are potentials of significant cost reductions in three major areas: costs of effective pretreatment and enzyme production, lower production costs based on effective process integration and additional revenue generation by value-added processing of co-products, and the development of high-yield biocatalysts that could ferment both 5 and 6-carbon sugars effectively. The thermochemical route on the other hand shows less opportunity for further cost reduction except through some novel approaches reported very recently. The costs of pretreatment and enzymes are discussed in greater details in Chapter 9 and Chapter 10, respectively.

Nonetheless, there is still uncertainty as to which pathway would be the best choice for biofuel generation since none of the technologies have been tested in an industrial-scale. There exists considerable risk for investors, technology developers and industries. This is probably the only race where the industry does not want to be the "first." The government policy may also play a significant role in terms of research and technology development priorities and the level of support for the types of biofuel/biochemical products. Cellulosic ethanol and even butanol (Chapter 20) may not be the "silver bullet" as taunted in recent years. The fuel of the next generation is still to be seen.

1.5 Cost Analysis of Cellulosic Biofuels

The estimated production costs of biomass-derived ethanol show a considerable variation ranging from \$2.27- \$4.92 per gallon. It is very likely the cost may go down to \$0.95 to \$1.32 per gallon with further research and development (IEA, 2008). The production cost of about \$1.31 per gallon ethanol was reported by Foust et al. (2008). The gasoline equivalent of \$1.91 per gallon as ethanol has energy content of two-third of gasoline. Williams et al. (2007) reported ethanol production costs of \$1.67/gallon assuming feedstock costs of \$ 44/dry Mg at a plant in California. The plant capital cost was assumed to be $2.88 per gallon-annual capacity.

Piccolo and Bezzo (2009) compared the techno-economical factors of bioethanol production through two pathways, namely biochemical conversion and gasification-syngas fermentation. The comparison was carried out by considering factors such as modeling, process optimization, heat and power generation, process sensitivity and financial analysis. The authors calculated the production cost of ethanol per gallon for the two processes, and the biochemical pathway was found to be more economically viable than the thermochemical pathway. A summary of the results of that study is shown in Table 1.2.

Table 1.2 Costs of ethanol based on the feedstock cost, and biochemical conversion and syngas- fermentation processes *(Adapted from Piccolo and Bezzo, 2009).*

		Enzymatic hydrolysis and fermentation ($/ L ethanol)	Gasification and fermentation (microbial) ($/ L ethanol)
Feedstock cost ($/dry ton of biomass)	63	1.16	1.79
	54	1.12	1.68
	45	1.08	1.67
	36	1.04	1.60
Effect of fermentation yield	Base case	1.12	1.68
	Mid-term	0.95	1.51
	Long-term	0.84	1.32

Note: all the calculations are for a 5-year payback period

Some of the critical parameters affecting the overall cellulosic ethanol production cost are:

- Feedstock cost
- Processing cost (labor, pretreatment, and other chemical/material costs)
- Enzyme cost
- Energy cost
- Revenue generation from co-products (lignin, stillage, etc.)
- Current petroleum price

The reported feedstock costs vary from $25 to $50 per dry ton. Foust et al. (2008) reported that a feedstock cost of $35/dry ton (based on 2002 dollars) could provide an economically viable lignocellulosic biorefinery, of which $10 is allocated for grower payments (to cover the actual value of biomass). The remaining amount covers the feedstock supply system costs. The expansion of cellulosic-ethanol plant may demand more feedstock, which may raise the feedstock costs, however, there is a limit that biorefineries could pay to remain competitive in the business. In the long-term, this will certainly stabilize the cost of feedstock.

The feedstock, enzymes and biomass pretreatment are three of the biggest contributors to ethanol production costs and have been subject of research for decades. The feedstock, in particular, contributes up to 35.5% of the final cost of ethanol (Piccolo and Bezzo, 2009). Improvements in technology however, have the potential to reduce expenses significantly and increase the viability of biorefineries. In recent years, the enzyme cost went down from $0.50 to $0.10-$0.30 per gallon of ethanol (Hettenhaus et al., 2002), but further cost reductions are still required. Genencor, a major enzyme developer and producer, projects that the cost of enzyme would go down to $0.30 per gallon from current cost of $1.00/gallon (Steele, 2009). For the economic viability of cellulosic ethanol, the enzyme cost should be $0.045-0.09/gal. For dry-grind corn ethanol plants, the enzyme cost is just around $0.03-0.04/gal. Biomass pretreatment-associated costs for current technologies are around $0.11 per gallon of ethanol. The National Resources Defense Council (NRDC) suggests that cost reductions of 22%, 65%, and 89% can be achieved in feedstock, pretreatment, and enzyme related costs as the industry matures (Greene and Mugica, 2005).

1.6 Sustainability of Biofuel Industries

Although sustainability has been defined in many different ways, underlying all these definitions is the common theme of meeting the needs of present and future generations, while conserving natural resources and ensuring our social and environmental well-being (USDOE/ USDA, 2009). The sustainability of biofuel from both grains and lignocellulosic biomass spans environmental, economic, and social dimensions that are intertwined with each other.

Real and/or perceived environmental benefits are important drivers for greater use of biofuels, particularly the benefits of reduced greenhouse gas emissions (GHG), and improvement in local environment. However, no fuel system is free of its environmental ramifications. Multidimensional conflicts exist in the use of land, water, energy and other environmental resources for food, feed, and biofuel production. The utilization of agricultural products for biofuel production, especially the production of first generation biofuels, has intensified the competition for resource utilization by food, feed and biofuels globally. This, among other factors, has led to rise in prices of essential staples in recent years, as cereal grains make up nearly 80% of the world's food supply (Pimentel et al., 2009). This also has had implications in clearing of forest lands for growing more grains and oilseeds intended for biofuel thereby impacting ecological balance such as biodiversity.

Apart from rising food prices, water usage has been a big concern. This is because biofuel production and processing requires a huge amount of water. Processing of lignocellulosic biomass alone is expected to consume nearly 6 gallons of water per gallon of ethanol produced (Alen, 2007). A 2007 study by National Academies of Science was quoted by Hoekman (2009), which states "currently, biofuels are a marginal additional stress on water supplies at the regional to local scale. However, significant acceleration of biofuel production could cause much greater water quantity problems depending on where the crops are grown." Concerns about both water quantity and quality are the most serious issues where increased biofuel production comes primarily from first generation biofuel feedstock like corn and soybean. The agricultural production of biomass for food and fiber was reported to require about 86% of worldwide freshwater usage (Gerbens-Leenes et al., 2009), leaving behind a huge water footprint, a term described to indicate the amount of water need to produce unit energy. Gerbens-Leenes et al. (2009) presented the water footprints of bioenergy produced from 12 crops that currently contribute the most to global agricultural production: barley, cassava, maize, potato, rapeseed, rice, rye, sorghum, soybean, sugar beet, sugarcane, and wheat. The water footprint of bioethanol was reported to be smaller than that of biodiesel. For bioethanol, sugar beet and potato were the most advantageous with water footprint at 60 and 100 m^3/giga joule (GJ) respectively, followed by sugarcane (110 m^3/GJ). Sorghum at 400 m^3/GJ) was the most unfavorable crop in terms of water footprint. Soybean and rapeseed were reported to be the most favorable crops (400 m^3/GJ) for biodiesel production, whereas biodiesel from Jatropha had a higher water requirement (600 m^3/GJ). These are some legitimate concerns of emerging biofuel industries.

Concerns regarding top soil quality and losses have been expressed with the second generation biofuel, especially with the agricultural residue, like corn stover, wheat and rice straws etc. The removal of crop residues may adversely impact soil quality, like organic matter, thus resulting in reduced agronomic productivity. It also removes the mulching effect these biomasses provide to soil in protecting nutrient-rich top soil against soil erosion. The exact amount of crop residue that can be removed to maintain a healthy top soil is a highly debated issue. Some researchers have suggested that lignocellulosic biofuel feedstock be produced through the establishment of biofuel plantations rather than by utilizing agricultural crop residues (Lal, 2006).

The economic sustainability considers the types of lignocellulosic biomass feedstock likely to be competitive in different regions, the resulting changes in spatial land use pattern and their implications for food prices. The social implications of the emergence of biofuel represent some of the most pressing and challenging sustainability issues (USDOE/ USDA, 2009). Careful consideration must be given to social structures and policies that can promote or inhibit the development of expanded biofuel production. The social processes functioning at multiple scales, from individual farms and forests to whole communities and regional ecosystems need to be taken into account.

Owing to increased feedstock demand for both first and second generation biofuels, the clearing of forests and alteration to prairies and marginal land utilization practices for expanded agricultural activity is one of the most significant human alterations of the global environment. Key criteria to assess the environmental sustainability of agricultural systems are the performance regarding soil quality, for example, organic matter content, water retention capacity, compaction, soil erosion, salination, water use, water quality, biodiversity, nitrate leaching, pesticide and herbicide loads (Muller, 2009). Measured against these criteria, many problems regarding the long-term sustainability of conventional agriculture as applied to biomass cultivation arise. One key and positive aspect of biofuel environmental sustainability, however, is the ability of biofuel to mitigate greenhouse gas emissions from the current petroleum-based transportation fuels. Biomass feedstocks usually are treated as carbon-neutral in that the carbon dioxide produced from biofuel was previously absorbed during plant growth. They could even be considered carbon negative as a result of increased carbon sequestration in the soil and root biomass. However, some studies have suggested that net GHG emission of biomass feedstock would be significantly lower if the effects of direct or indirect land use change are taken into account (Gopalakrishnan et al., 2009). Also, the greenhouse gas benefits derived from bioenergy crops may not completely offset carbon dioxide emissions from the clearing of forest land for new biomass feedstock.

The economic sustainability of biofuel, especially from biomass, should consider the possibility of spatial heterogeneity in optimal choice among different feedstock selections. Feedstock yields are the critical determinant of their economic viability. Cellulosic feedstock crops should be profitable to grow for the farmers compared to profitable alternative crops. Much of the current biofuel economic research is focused on the supply-side implications (e.g. agricultural producers, land use, and crop production). However, the demand side of the markets also needs to be considered. For example, the demand for biofuel will depend, in part, on the availability of flex-fuel vehicles, the cost and availability of biofuels and blended fuels, and market prices for gasoline (USDOE/USDA, 2009). Economic models need to incorporate the factors that determine the demand for various biofuels under various scenarios of substitution of biofuels and gasoline to assess the impact on prices and biofuel use.

Even though every person has a stake in how the global resources are being utilized for energy production, farmers and foresters, rural community, decision makers, the biofuel industries, and local, regional, and national policy makers are the direct stakeholders of emerging biofuel industries. The biofuel sector cannot develop sustainably without an understanding of, and effective response to, stakeholder values, choices, behaviors, and reactions, along with careful consideration of the social structures and policies influencing that development. The USDOE/USDA joint report on biofuel sustainability (2009) lists major social challenges of biofuel as:

- understanding stakeholders and their need;
- learning from and building on the lessons learnt from 1st generation biofuel industries and their implications;
- understanding the social effects of scale and complexity for biofuel system design; and
- understanding social dynamics, human choices, risk management, and incentives.

1.7 Summary

There have been significant research and technology development efforts in the field of second and third generation biofuels. The current focus is on lignocellulosic-derived biofuel especially ethanol. Due to economic uncertainty and risk associated with immature technologies, biofuel industries are reluctant to build full-scale plant. There also has been renewed interest in algal-biodiesel. A significant breakthrough is needed especially in process microbiology to improve biological productivity. The sustainability of biomass-derived biofuels especially with respect to social and environmental issues is yet to be examined. The second and third generation biofuel productions are still economically unfavorable. It is important to point out that the research focus may shift beyond ethanol towards alkanes and other hydrocarbons. It is still to be seen: what is the future biofuel?

1.8 References

Aden, A. (2007). *Water Usage for Current and Future Ethanol Production.* Available at http://www.swhydro.arizona.edu/archive/V6_N5/feature4.pdf (accessed May, 2009).

Agler M.T., Garcia M.L., Lee E.S., Schlicher M., and Angenent L.T. (2008). "Thermophilic anaerobic digestion to increase the net energy balance of corn grain ethanol." *Environmental Science and Technology*, 42(17), 6723-6729.

Foust, T.D., Ibsen, K.N., Dyton, D.C., Hess, J.R. and Kenney, K.E. (2008). "The biorefinery." In: Himmel, M. (Ed.) *Biomass Recalcitrance: Deconstructing the Plant Cell Wall for Bioenergy*. United Kingdom: Blackwell Publishing, pp. 7–37.

Gerbens-Leenes, W., Hoekstra, A.Y., and van der Meer, T.H. (2009). "The water footprint of bioenergy" *Proceedings of the National Academy of Sciences of the United States of America,* available at http://www.pnas.org/content/early/2009/06/03/0812619106.full.pdf+html

Gopalakrishnan, G., Negri, M.C., Wang, M., Wu, M., Snyder, S.W., and Lafreniere, L. (2009). "Biofuels, Land, and Water: A Systems Approach to Sustainability." *Environmental Science & Technology*, 43, 6094-6100.

Goettemoeller, J., and Goettemoeller, A. (2007). *Sustainable Ethanol.* Missouri, USA: Prairie Oak Publishing.

Greene, N., and Mugica, Y. (2005). *Bringing Biofuels to the Pump.* Natural Resources Defense Council. Available at http://www.nrdc.org/air/energy/pump/pump.pdf (accessed June, 2009).

Hawaii Bioenergy Master Plan Report . (2009). Conversion Technology (by Samir Kumar Khanal, Scott Turn and Charles Kinoshita). http://www.hnei.hawaii.edu/bmpp/home.asp

Hettenhaus, J., Wooley, R. and Ashworth, J. (2002). *Sugar Platform Colloquies.* NREL/SR-510-31970, available at http://permanent.access.gpo.gov/lps31319/www.ott.doe.gov/biofuels/pdfs/sugar_platform.pdf

Hoekman, S.K. (2009). "Biofuels in the U.S. – challenges and opportunities." *Renewable Energy*, 34, 14–22

Huber, G.W., Chheda, J., Barrett, C.B., and Dumesic, J.A. (2005). "Production of liquid alkanes by aqueous-phase processing of biomass-derived carbohydrates." *Science*, 308, 1446–1450.

International Energy Agency (IEA) (2008). *From 1st to 2nd generation biofuel technologies: An overview of current technologies and RD & D activities*, available at http://www.iea.org/textbase/papers/2008/2nd_Biofuel_Gen_Exec_Sum.pdf

Keller, J. (2008). "Changing paradigms: from wastewater treatment to resource recovery processes." *IWA-Croucher Foundation Advanced Study Institute (IWA-ASI)* Symposium *on Perspective of Energy and Resources Saving and Recovery in Wastewater Treatment,* The Hong Kong University of Science and Technology, Hong Kong (June 23-27, 2008).

Khanal, S.K. (2008). *Anaerobic Biotechnology for Bioenergy Production: Principles and Application.* Iowa, USA: Wiley-Blackwell Publishing.

Lal, R. (2006). "Soil and environmental implications of using crop residues as biofuel feedstock." *International Sugar Journal*, 108 (1287), 161–167.

Muller, A. (2009). "Sustainable agriculture and the production of biomass for energy use." *Climatic Change*, 94, 319–331.

Munasinghe, P., and Khanal, S. K. (2009). "Syngas fermentation to ethanol: challenges and opportunities." *Bioresource Technology* (In-review).

National Renewable Energy Laboratory (NREL) (1998). *A Look Back at the U.S. Department of Energy's Aquatic Species Program: Biodiesel from Algae.* NREL/TP-580-24190, available at http://www.nrel.gov/docs/legosti/fy98/24190.pdf. (Accessed on September 23, 2009).

Piccolo, C., and Bezzo, F. (2009). "A techno-economic comparison between two technologies for bioethanol production from lignocellulose." *Biomass and Bioenergy*, 33, 478–491.

Pimentel, D., Marklein, A., Toth, M.A., Karpoff, M.N., Paul, G.S., McCormack, R., Kyriazi, J., and Krueger, T. (2009). "Food versus biofuels: environmental and economic costs." *Human Ecology*, 37, 1–12.

Renewable Fuels Association (RFA). (2008). *Changing the Climate: Ethanol Industry Outlook 2008*, available at http://www.ethanolrfa.org/objects/pdf/outlook/RFA_Outlook_2008.pdf

Renewable Fuels Association (RFA). (2009). *2009 Ethanol Industry Outlook*, available at http://www.ethanolrfa.org/objects/pdf/outlook/RFA_Outlook_2009.pdf

Sanderson, K.W. (2007). "Are ethanol and other biofuel technologies part of the answer for energy independence?" *Cereal Foods World,* 52 (1), 5-7.

Steele, L. (2009). "Advancing enzymatic hydrolysis of lignocellulosic biomass." Extracted from technical presentation by Genencor in the *Sixth World Congress on Industrial Biotechnology and Bioprocessing*. Jul 19-22, 2009, Montreal, Canada.

United States Department of Agriculture (USDA) and United States Department of Energy (USDOE) Joint Report. (2005). *A Billion-Ton Feed Stock Supply for Bioenergy and Bioproducts Industry: Technical Feasibility of Annually Supplying 1 Billion Dry Tons of Biomass*, available at http://www1.eere.energy.gov/biomass/pdfs/final_billionton_vision_report2.pdf

United States Department of Energy and United States Department of Agriculture (USDOE and USDA). (2009). *Sustainability of Biofuels: Future research opportunities.* DOE/SC-0114.

United States Energy Information Administration (USEIA). (2006). *International Energy Annual 2006*, available at http://www.eia.doe.gov/iea/wecbtu.html

Williams, R.B., Jenkins, B.M., and Gildart, M.C. (2007). *Ethanol production potential and costs from lignocellolosic resources in California.* 15th European Biomass Conference and Exhibition, Berlin, Germany.

Appendix

Top 50 bioenergy companies as identified by Biofuels Digest (*Source: Khanal, Turn and Kinoshita, Hawaii Bioenergy Master Plan Report, 2009*) [L=lab scale, D=demonstration scale, C=commercial]

	Company	Web Site	Technology	Product	Status
1	Coskata	www.coskata.com/	Biomass gasification with microbial conversion of syngas	Ethanol	L
2	Sapphire Energy	www.sapphireenergy.com/	Photosynthetic micro-organisms	Green crude/green gasoline	L
3	Virent Energy Systems	www.virent.com	Aqueous phase reforming of sugars or fiber with catalytic conversion	Fuels, gases, chemical	L
4	POET	www.poetenergy.com	Hydrolysis and fermentation of corn	Ethanol	C
5	Range Fuels	www.rangefuels.com	Biomass gasification with catalytic conversion of syngas	Ethanol	D
6	Solazyme	www.solazyme.com	Marine microbes	Fuels, chemicals, high valued products	L
7	Amyris Biotechnologies	www.amyris.com	Conversion of plant biomass (sugars) using engineered micro-organisms	Hydrocarbon fuels, high valued products	L
8	Mascoma	www.mascoma.com	Microbially based conversion of biomass	Cellulosic ethanol	L
9	Dupont Danisco Cellulosic Ethanol	www.ddce.com	Alkaline pretreatment, enzymatic hydrolysis, fermentation	Cellulosic ethanol	L

	Company	Web Site	Technology	Product	Status
10	UOP	www.uop.com	Catalyst producer	Fuels, chemicals, high valued products	C
11	ZeaChem	www.zeachem.com	Biochemical production of ethyl acetate, lignin gasification to produce hydrogen; ethyl acetate + hydrogen to produce ethanol	Ethanol and other chemicals	L
12	Aquaflow Bionomic	www.aquaflowgroup.com	Wild strains of algae grown on waste water	Jet fuel	L
13	Bluefire Ethanol	bluefireethanol.com	Dilute acid hydrolysis of biomass with fermentation	Ethanol	D
14	Novozymes	www.novozymes.com	Enzyme, micro-organism, and protein producer	Ethanol and other high value products	C
15	Qteros	www.qteros.com	Microbial production of ethanol from fiber	Ethanol	L
16	Petrobras	www2.petrobras.com.br/ingles/	Veg. oil and animal fat and oil seed for biodiesel, fermentation of sugarcane for ethanol	Biodiesel and Ethanol	C
17	Cobalt Biofuels	www.cobaltbiofuels.com/	Fermentation of various feedstocks	Biobutanol	L
18	Iogen	www.iogen.ca	Cellulosic – from agricultural residue	Ethanol	D
19	Synthetic Genomics	www.syntheticgenomics.com/	Genomic solutions to global energy and environmental challenges	Next generation fuels and chemicals	L

	Company	Web Site	Technology	Product	Status
20	Abengoa Energy	www.abengoabioenergy.com/sites/bioenergy/en/	Hydrolysis and Fermentation of sugar cane, wheat, barley, corn and sorghum for Ethanol, soy and palm for biodiesel	Ethanol Biodiesel	C
21	KL Energy	www.klprocess.com/	Cellulosic, sugar and grain fermentation	Ethanol	D
22	INEOS	www.ineos.com/index.php	Municipal solid waste, organic commercial waste and agricultural residues are superheated and off gases fed through bacteria	Ethanol	L
23	GreenFuel	www.greenfuelonline.com/	Algae	Fuels	L
24	Vital Renewable Energy Company	www.vrec.com.br/	Sugar cane fermentation	Ethanol	C
25	LS9	www.ls9.com/	Sugar cane fermentation and cellulosic conversion of biomass	Fuels, chemicals	D
26	Raven Biofuels	www.ravenbiofuels.com/	Cellulosic two stage acid hydrolysis of bio matter	Ethanol and chemicals	L
27	Gevo	www.gevo.com/	Fermentation	Fuels and chemicals	L
28	St1 Biofuels Oy	www.st1.eu/index.php?id=2386	Fermentation of food industry side steams (dough, potatoes etc…), biowaste	Ethanol	C

	Company	Web Site	Technology	Product	Status
29	Primafuel	www.primafuel.com/	Service and technology provider	Biorefineries	C
30	Taurus Energy	www.taurusenergy.eu/EN/	Cellulosic biowaste conversion	Ethanol	L
31	Ceres	www.ceres.net/	Produce low carbon non-food energy crops	Seeds for bioenergy crops	C
32	Syngenta	www.syngenta.com/en/index.html	Agribusiness helping farmers grow more with less	Seeds and agrichemicals	C
33	Aurora Biofuels	www.aurorabiofuels.com/	Algae	Biodiesel	L
34	Bionavitas	www.bionavitas.com/	Algae	Biodiesel	L
35	Algenol	www.algenolbiofuels.com/	Algae	Ethanol	L
36	Verenium	www.verenium.com/	Cellulosic biomass	Ethanol	D
37	Simply Green	www.seacoastbiofuels.com/	Biofuels distributer	Biodiesel blends	C
38	Carbon Green	www.carbongreenllc.com/	Carbon credit management		C
39	SEKAB	www.sekab.com/	Cellulose to ethanol	Ethanol, chemicals	C
40	Osage Bioenergy	www.osagebioenergy.com/	Barley feedstock	Ethanol	L
41	Dynamotive	dynamotive.com/	Fast pyrolysis of biomass	Fuels, Biochar	C
42	Sustainable Power	www.sustainablepowercorp.us/	Hydrolysis / pyrolysis	Fuels	C
43	ETH Bioenergia	www.eth.com/website/default.asp	Sugar, fuel, energy	Ethanol	C

	Company	Web Site	Technology	Product	Status
44	Choren	www.choren.com/en/	Gasification/PO/FT	Biodiesel	D
45	OriginOil	www.originoil.com/	Quantum fracturing algae	Fuels, chemicals	D
46	Propel Fuels	www.propelfuels.com/content/	Fuel stations in CA	Biodiesel and E85	C
47	GEM Biofuels	www.gembiofuels.com/	Supply jatropha-based feedstock		C
48	Lake Erie Biofuels	www.lakeeriebiofuels.com/	Transesterification and acid esterification	Biodiesel 45 Million gallons/yr	C
49	Cavitation Technologies	www.cavitationtechnologies.com/	Produce biodiesel with flow-trough nano-cavitation technology	Sell turnkey conversion systems	C
50	Lotus/Jaguar – Omnivore	www.biofuelsdigest.com/blog2/2008/08/13/lotus-jaguar-to-debut-omnivore-multi-biofuel-engine-with-higher-mpg-than-gasoline/	Multi-Biofuels two-stroke single cylinder motor		D

CHAPTER 2

High Strength Wastewater to Bioenergy

S. K. Banerji, Rao Y. Surampalli, C. M. Kao, Tian C. Zhang, and R.D. Tyagi

2.1 Introduction

Fossil fuels have supported the economic growth of our society in the last century, but they cannot indefinitely sustain a global economy. Due to limitations of worldwide energy supply, there has been a marked increase in energy costs. Energy recovery from wastes (water) is being investigated to counter these increased costs by supplying part of the energy to local utilities in addition to meeting in-plant needs. Large volumes of high strength wastewaters are produced each year from domestic, industrial and agricultural activities. High strength wastewaters and their treatment residues (solids) can be converted into bioenergy. This chapter will mostly deal with the conversion of high strength industrial wastewater to bioenergy.

There is no set definition of high strength wastewater. Usually, high strength wastewaters have a higher concentration of biochemical oxygen demand (BOD) than typical municipal wastewater. Many municipalities have a surcharge for industrial wastewater discharges into their sewer system based on BOD, suspended solids (SS) and sometimes dissolved solids since it costs more to treat. In Baltimore County, MD, USA wastewaters having BOD and SS greater than 300 mg/L have to pay a surcharge fee as they are considered high strength wastewater. The surcharge limits for BOD and SS wastewater discharged to the sewer system in the City of Houston, TX is 350 and 375 mg/L, respectively. Thus, the definition of high strength wastewater varies from place to place. The higher the concentration of organic matter in the wastewater in terms of BOD or chemical oxygen demand (COD), the higher is the potential for energy recovery. It has been estimated that untreated domestic wastewater has an energy potential of about 280 kWh/year/person (Rogalla et al., 2008). An industrial wastewater with a higher concentration of BOD or COD than domestic wastewater would have a significantly higher energy potential.

High strength wastewaters, especially with their BOD higher than 500 mg/L, have a better energy balance if they are pretreated by an anaerobic process followed by effluent polishing using an aerobic process as compared to conventional wastewater treatment (i.e. primary, aerobic secondary and digestion of sludge). For a hypothetical wastewater with a COD of 600 mg/L, the net energy production with the anaerobic pretreatment/aerobic polishing was estimated to be 0.19 kWh/m^3, while the conventional treatment would require about 0.16 kWh/m^3 wastewater flow (Rogalla et al., 2008). Therefore, in this chapter, the emphasis will be on anaerobic processes for bioenergy recovery from high strength industrial wastewater.

2.2 Energy Recovery from Anaerobic Processes

In the past, the technical expertise required to maintain anaerobic digesters coupled with high capital costs and low process efficiencies have limited the level of their industrial application as a waste treatment technology. Recently, anaerobic digestion facilities have been recognized by the United Nations Development Programme (UNDP) as one of the most useful decentralized sources of energy supply, as they are less capital intensive than large power plants. The energy recovery from anaerobic processes could take the following alternate courses:

Alternative 1: Conversion of wastewater organic matter to mostly methane. The energy recovery from methane can have the following options:

- **Option 1a**: Methane combusted in an internal combustion engine to produce heat and electricity (cogeneration engines);
- **Option 1b**: Methane combusted in a micro-turbine to produce heat and electricity;
- **Option 1c**: Methane after processing used in a fuel cell (FC) system to directly produce electricity.

Alternative 2: Conversion of wastewater organic compounds to hydrogen using selective environmental conditions, which can be used in a fuel cell to directly produce electricity.

Alternative 3: The wastewater after pretreatment can be used in a microbial fuel cell (MFC) to directly produce electricity.

Methane is a greenhouse gas that remains in the atmosphere for approximately 9-15 years and is over 20 times more effective in trapping heat in the atmosphere than carbon dioxide (CO_2) over a 100-year period. As a result, efforts to prevent methane emission or its utilization can provide significant energy, economic and environmental benefits. In addition, the use of hydrogen as a fuel supply can greatly reduce the green house gas emission potential. Therefore there has been a lot of emphasis recently to seek low carbon footprint for all wastewater treatment processes.

2.2.1 Alternative 1–Conversion of Wastewater Organic Matter to Mostly Methane

Anaerobic Digestion Processes. For a long time, the anaerobic digestion process has been used to treat municipal and industrial wastewater residues (sludge) to form methane and CO_2. It has also been applied to treat high strength industrial wastewater from agro-food industries (e.g. starch, potato, yeast, pectin, cannery, confectionary, fruit, vegetable, dairy, and bakery), beverage industries (e.g. beer, wine, soft drinks, fruit juices and coffee), paper and pulp industries and miscellaneous industries (e.g. chemical, pharmaceutical). The biochemical, microbiological and engineering considerations of the anaerobic process have been well studied and published extensively elsewhere (Metcalf & Eddy, 2003; Speece et al., 2006; USEPA, 2006; Khanal, 2008). As such they will not be covered here.

Most anaerobic processes use the mesophilic temperatures ranging from 35 to 38°C. Temperature-phased anaerobic digestion (TPAD) has been tried to promote better destruction of high solids (e.g., sludge) and pathogen, which is needed when land application is the ultimate goal for the stable solids. In this process, the first stage digesters are operated at thermophilic temperature ranges (50–60°C) while the subsequent phase has temperatures in the mesophilic range (35–38°C). This type of operation enhances the hydrolysis and acidogenesis in the earlier phase but protects the acetogens and methanogens in the latter phase (USEPA, 2006).

The volumetric loading rates for these suspended growth anaerobic processes range from 1 to 8 kg COD/m^3·d. The solids retention time (SRT) or hydraulic retention time (HRT) for these processes varies from 15–20 days. Typically, about 0.4 m^3 CH_4 is produced per kg COD removed at 35°C. The gas phase has about 65% CH_4, and the rest is mostly CO_2 (Metcalf & Eddy, 2003). Table 2.1 summarizes the operation and performance details of several anaerobic processes used for industrial wastewater treatment (Metcalf & Eddy, 2003). It can be seen that some of these processes have significantly lower HRTs than the conventional suspended growth processes.

Anaerobic Gas Production and Energy Recovery. Depending on the quality of the wastewater, the anaerobic process gases (APGs) may contain varying amounts of CH_4, which is an important energy source. The anaerobic digestion of municipal wastewater sludge produces gases that contain about 65–70% CH_4, but for some industrial wastewater, the methane percentage in the gas phase may be different depending on the composition of the wastewater. Table 2.2 shows the methane yield from anaerobic processing of industrial wastewaters.

Table 2.1 Operational and design data for different anaerobic treatment processes.

Process	Volumetric loading rate, Kg COD/m^3•d	Hydraulic retention time (HRT)
Upflow sludge blanket reactor (UASB)	8-12*	7-9 hr*
Anaerobic baffled reactor (ABF)	0.9-4.7[1]	6-24 hr[1]
Upflow attached growth reactor	4-6[2]	1.8-2.5 days[2]
Downflow attached growth reactor	20[3]	1-2 hr[3]

*loading rate and HRT depend on the type of waste or sludge, influent COD and temperature. [1]Slaughterhouse wastewater at 25-30°C. [2]Food canning soft drink wastewater at 30 °C. [3]Brewery wastewater at 35°C.

Table 2.2 Methane percent in gas phase from anaerobic processes using industrial wastewaters.

Wastewater source	% Methane	Reference
Slaughterhouse	51	Ruiz et al., 1997
Baker's yeast	65	Britz and van der Merwe, 1997
Winery	70-74	Molina et al., 2007
Food	68-70	Xu et al., 2002

2.2.1.1 Option 1a–Methane Combusted in an Internal Combustion Engine to Produce Heat and Electricity (Cogeneration Engines)

The gases produced from different anaerobic digestion processes are first cleaned to remove impurities present. The most common impurity in the digester gas is hydrogen sulfide, which can corrode the combustion elements of a boiler or internal combustion engines. This impurity can be removed by scrubbing the gases through a column of iron sponge (hydrated ferric oxide impregnated wood chips). Moisture in the gas should also be removed. In addition, another gaseous impurity siloxane, an organic molecule that originates in cosmetics, shampoo, etc., may be present in the gases. This compound can form SiO_2 after combustion that coats the inside of the combustion chamber and the exhaust system. It can be removed by low temperature refrigeration followed by activated carbon treatment. As this treatment process is expensive, the siloxane problem often is eliminated by additional maintenance of the combustion system. If there is some ammonia in the gas, the emission of NO_x may be a problem in some places. A "lean burn" turbo charged engine could limit the NO_x emissions. Figure 2.1 shows a schematic of the treatment process with energy recovery.

As mentioned earlier, the option of converting anaerobic digester gas to energy from municipal waste sludge has been practiced for a long time. One prime example is the Point Loma Wastewater Treatment Plant in San Diego, CA, USA. This plant treats about 240 million gallons of wastewater per day. It is energy self-sufficient and sells excess electrical energy to the grid. Two internal combustion engines produce about 4.5 MW of electricity using the digester gases. Excess heat from the engines is used for heating the digesters. The electricity generated is used for process pumping, lights and other uses. In 2000, the plant saved about $3 million in energy costs and sold about $1.4 million worth of excess power to the grid (FEMP, 2004).

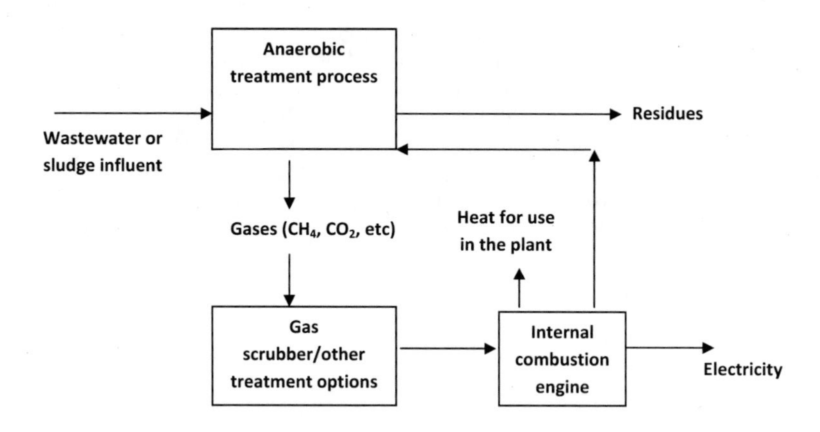

Figure 2.1 Schematic of energy production from wastewater or sludge during anaerobic treatment.

There are several examples of this technology for treating high strength industrial wastewater to produce energy. Anheuser-Busch (2007), the brewing company headquartered in St. Louis has many breweries throughout the world. They have developed a process called Bio-energy Recovery System (BERS) from wastewater anaerobic treatment processes. The aim of the company is to use the bioenergy for up to 15% of the energy needs for the manufacturing operations in all their facilities (Anhueser-Busch, 2007). In a dairy operation in New York State, an anaerobic attached growth process (mobilized film technology, MFT) was used to treat the wastewater from yogurt production that met the 30 mg/L BOD and 45 mg/L TSS effluent requirements. The gases produced from the anaerobic process were sufficient to provide about 25% of the total energy needs for the company (Spencer, 2007).

In Thailand, a palm oil factory wastewater is first pretreated and then treated by an anaerobic process at a loading rate of 4.5 kg COD/m^3·d and an SRT of 7 days. The pretreatment consists of oil recovery followed by retention in primary ponds. After the anaerobic treatment the BOD removal efficiency is about 92%. The CH_4 in the biogas was found to be about 66–67% and the CH_4 production rate was about 0.51 m^3 CH_4 /kg COD·day. The biogas is first treated to remove H_2S and moisture before being combusted in a modified diesel engine to produce electrical power. The power generated is used at the wastewater treatment plant, and some can be sold to the grid (Puetpaiboon and Chotwattanasak, 2004).

2.2.1.2 Option 1b–Methane Combusted in a Micro-Turbine to Produce Heat and Electricity

In this option, the anaerobic process gases are combusted in a microturbine system to produce electricity and heat. Microturbines are small single-stage units that use gas turbine technology to produce electricity. They are produced in the range of

60–100 kW, but some units could be as high as 250 kW. The rotating speeds may be 80,000 to 100,000 rpm. There are two types of microturbines–recuperated and non-recuperated. The recuperated microturbines have heat exchangers to recover the heat in a co-generation operation. The overall electrical efficiency of these units can be 15–30%. The gas exhaust temperatures from the microturbines are low (about 200–300°C), which can be used to generate low-pressure steam or hot water.

The microturbine system used for anaerobic digestion gas utilization has three components: a gas clean process, a gas compressor, and the microturbine unit. The digester gas cleaning system is used to remove moisture, siloxanes and H_2S (Schroedel et al., 2009). The carbon based filter media for siloxane removal can also remove H_2S. Gas cleaning is an important factor in reliable operation of the microturbines. The clean gas is compressed to about 380–450 kPa before entering into the combustion chamber. The exhaust gases turn the turbine blades to produce rotary motion that turns the electrical generator. The emissions from the microturbines are quite low requiring no emission control. Generally they have a low operating cost; however since they have not been in commercial use for a long time there may be some uncertainty about the low operating costs. Figure 2.2 shows a schematic of the microturbine process using APGs for energy recovery.

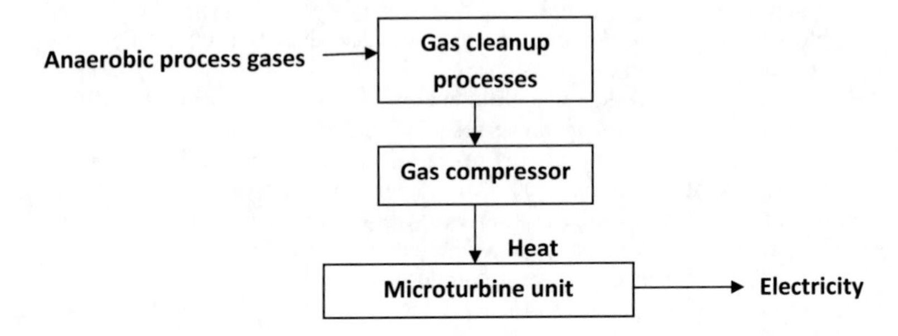

Figure 2.2 APGs combusted in a microturbine to produce heat and electricity.

The application of microturbines to cogenerate heat and electricity from APGs has been fairly recent (since 2000) and most of the applications have been for domestic wastewater sludge processing (Mignone, 2008). Thus the technology is still evolving, but it has a great promise for producing alternate power from waste sources including industrial and agricultural wastes and manures (Williams and Gould-Wells, 2004).

2.2.1.3 Option 1c–Methane after Processing Used in an FC System to Directly Produce Electricity

The APGs produced from the treatment of high strength wastewater using anaerobic processes contains mostly methane that can be used in an FC to produce electricity. An FC uses an electrochemical process where hydrogen (or hydrocarbon/biogas fuel) can combine with oxygen to produce electricity and heat. It can produce electricity with negligible air pollution (particularly NO_x, SO_x, and particulates), low greenhouse gas emissions and without much noise. The fuel cell unit has four components: (i) a gas pretreatment system; (ii) a fuel processor; (iii) cell stacks; and (iv) an inverter unit. Figure 2.3 shows an outline of an FC unit using APGs for electricity production.

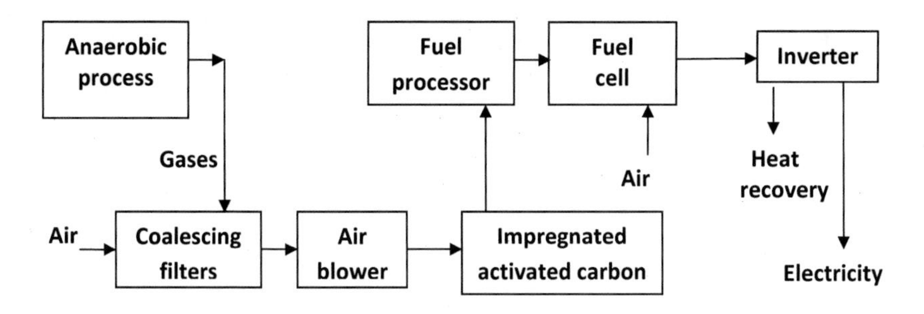

Figure 2.3 Schematic diagram of a FC using APGs.

The gas pretreatment units remove H_2S, halides, solids, and moisture. These impurities must be removed so that FC components are not damaged, fouled or poisoned. Usually a coalescing filter and impregnated activated carbon beds are used to remove water and H_2S present in the gases.

The next step is the conversion of methane to hydrogen in the fuel processor by steam at a temperature of about 800°C and a pressure of about 35 bars (34.5 atmospheres) in the fuel processor unit. A catalyst like nickel is present to facilitate the reactions. The following reactions take place:

$$CH_4 + H_2O \rightarrow CO + 3H_2 \text{ – endothermic reaction}$$
$$CO + H_2O \rightarrow CO_2 + H_2 \text{ – slightly exothermic reaction}$$

The hydrogen gas enters the anode chamber and is converted into a proton (H^+) and an electron (e^-) by a catalytic process. The electrolyte between the anode and cathode allows the proton to travel to the cathode where it reacts with O_2 in the air that is being added to form water. The electron produced in the anode travels to the cathode through an external electrical circuit. This flow of electrons produces DC electricity. A single FC only produces about 0.65–1 V but by stacking many such cells much larger voltages and power can be produced. The inverter unit converts the DC power to AC power.

There are several types of FCs that are being developed. Table 2.3 shows these various FCs being used. The main differences between these units are the electrolyte used and the operating temperature.

Table 2.3 Different types of FCs being developed (Wikipedia, 2008).

FC Types	Electrolyte	Power Output	Operating Temperature, °C
Phosphoric acid (PAFC)	Phosphoric acid soaked in a matrix	up to 100 MW	150–200
Molten carbonate (MCFC)	Carbonate solution	100 MW	600–650
Proton exchange membrane (PEMFC)	Polyperfluo-rosulfonic acid	100 W–500 kW	50-100
Alkaline (AFC)	KOH soaked in a matrix	10–100kW	< 80
Solid oxide (SOFC)	Zirconium oxide	up to 100 MW	850–1100

Figure 2.4 shows a PAFC process where the electrolyte is liquid phosphoric acid. The electrodes are made from carbon paper coated with platinum catalyst, which is not affected by CO present in the feed hydrogen stream. The operating temperature range is 150–200°C, which converts the water formed to steam. The steam can be used for heating purposes. The MCFC process is a high temperature process where the electrolyte is a mixture of molten carbonate salts suspended in a porous inert ceramic matrix. The high temperatures in the process (varying from 600 to 650°C) allow the use of non-precious metals as catalysts at anode and cathode chambers. It is also more efficient (up to 60%) than the PAFC process that is close to 40%. If the waste heat is captured for use, the overall efficiency could be about 85% (Wikipedia, 2008). In addition, due to high temperatures inside of the FC, the influent fuels do not require the external reformer to convert hydrocarbons to hydrogen as they are converted to hydrogen by internal reforming inside the FC. Durability and corrosion of components of these FCs at these temperatures is a major concern and requires more research.

There have been many applications of FC technology to produce electricity and heat from APGs from municipal wastewater sludge digesters. In 1999, Fuji Electric began developing 100 kW PAFC power units using methane fermentation gas from garbage and APGs from sewage and used the system in the Yamagata City Purification Center (Table 2.4). All of the service-power in the facility is covered by the FC power unit, and a surplus totaling 40% of the generated power can be effectively used. Energy-saving effects and environmental benefits through the project include: (i) energy saved in terms of crude oil equivalent = 458 kL/year; and (ii) reduction in gas emissions, CO_2 = 1,140 ton/year; NOx = 460 kg/year; and SOx = 412 kg/year (Kubota et al., 2003).

One of the largest digester gas FC demonstration systems was installed in 2004 at the King County's South Treatment Plant in Renton, Washington. The 1-MW FC unit uses MCFC technology operating at about 600°C. It can use either natural gas or digester gas. It has so far performed reasonably well after some adjustments, and King County is planning to continue the process in the future (Bloomquist, 2006).

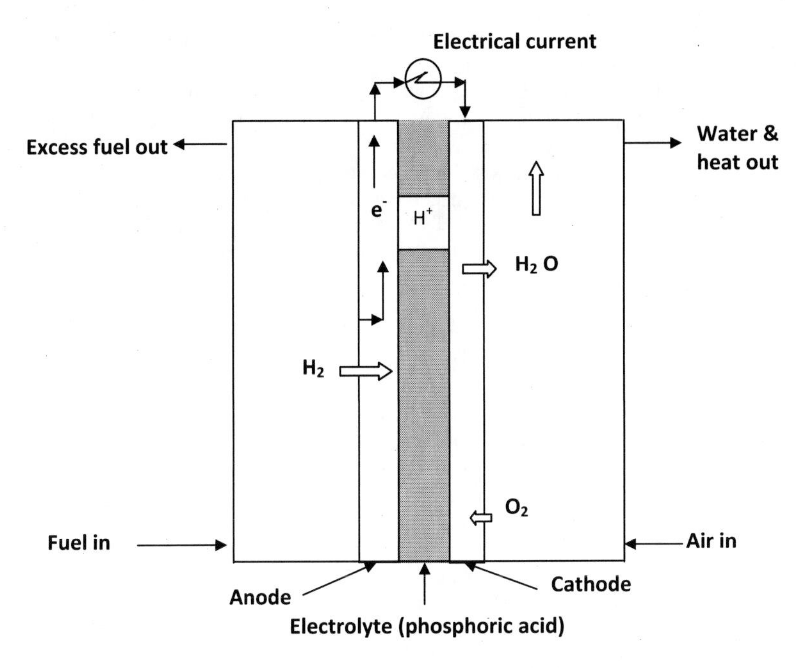

Figure 2.4 Schematics of a phosphoric acid FC unit.

Table 2.4 Specifications of the FC power unit for Yamagata City Purification Center.

Item	Specifications
Rated output	100 kW (at output terminals
Rated voltage, frequency	210 V, 50 Hz
Power generation efficiency	38% LHV (at rated operation at output terminals)
Overall efficiency	87% LHV (at rated operation at output terminals)
Fuel	Digested gas (CH_4 = 60%, CO_2 = 40%)
Fuel consumption	45 m^3/h
Operation mode	Fully automatic operation, parallel operation with mains
Thermal output	20% (90°C hot water); 29% (50°C hot water)
Exhaust gas	NOx: < 5 ppm; SOx: below the detection limit
Operating noise	65 dB(A) (average value at a distance of 1 m from the unit)
Size and mass	W x L x H = 2.2 x 4.1 x 2.5 (m^3); mass = 12 ton

Other applications of FC technology using digester gases have been in many parts of the United States including California, New York, Connecticut, Oregon and Massachusetts (Mignone, 2008). The wastewater treatment plant in Palmdale, CA, also used MCFC for converting digester gases to electricity and heat. The FC unit used about 70–80% of the digester gases to produce 225 kW of electricity that is used in the plant. The heat recovery system from the unit provided required heat for the digester operations. The overall efficiency of the FC was 73%. It has been expected that at a 90% capacity the FC units will save the city about $227,000 per year in retail electricity costs (McDannel and Wheless, 2008).

Skok (2007) reported about the installation of a 1-MW molten carbonate FC unit by Sierra Nevada Brewing Co. in CA, that uses industrial wastewater treatment digester gases as a fuel. Natural gas is used to augment the power source at times. Four MCFC units provide all the base load electrical power needed for the brewery.

At the present time, FCs for producing electricity from APGs are expensive compared to other technologies, but with more research and development it is expected that the costs will be lower in the future (Bloomquist, 2006). The reduced air pollution and greenhouse gas emissions may be a significant benefit for these units.

The cost effectiveness of the three options discussed above (combustion in an internal combustion engine, combustion in a micro-turbine or use in an FC system) for energy production from anaerobic process gases can be studied using a spreadsheet program developed by the Water Environment Research Foundation (WERF) called Life-cycle Assessment Manager for Energy Recovery (LCAMER) (Monteith et al., 2007). Using a hypothetical wastewater treatment with a wastewater inflow of 75,700 m^3/d, with two-stage conventional mesophilic anaerobic digesters with a 10-day SRT in stage 1 and a 5-day SRT in stage 2, one can calculate the performance of the three options on an equal basis. If the natural gas price is assumed to be $0.25/$m^3$ and electricity price is $0.09 kWh, the calculated net capital cost was $1,418,000, $1,909,000 and $5,749,000 for cogeneration, microturbines and PAFC, respectively. The payback time in years was 3.9, 5.6 and 14.7, respectively for the three options. The present high initial costs of the FCs make them unattractive but with more research and developments the cost will decrease significantly. The low emission from the FCs was not accounted for in the cost figures quoted above.

2.2.2 Alternative 2–Conversion of Wastewater Organic Compounds to Hydrogen Using Selective Environmental Conditions

Hydrogen is a clean source of energy but it is not available in nature. One of the methods for H_2 production is by chemical means as described in the previous section, i.e. steam reforming of H_2 from methane and CO using catalysts. There are several other methods for H_2 production as it is a feedstock for chemical synthesis and many other industries. Recently, production of hydrogen from microbial processes using wastes has received a lot of attention (Halleneck, 2005; Vijayaraghavan and Soom, 2006). It can be produced by fermentative bacteria, especially Clostridia under dark conditions as a by-product during the conversion of organic wastes to organic acids, which can then be converted to methane in the second stage. The biological hydrogen production through dark fermentation is faced with a challenge of extremely low yield (~2 mole H_2/mol glucose). More research is needed to improve biohydrogen yield (Khanal, 2008). In addition, algae can produce H_2 gas from water and CO_2 under light conditions. Some heterotrophic phototrophic bacteria can convert organic acids to produce H_2 and CO_2 (Kapdan and Kargi, 2006). The biochemical details of the fermentative hydrogen production from sugars have been thoroughly reviewed (Das and Veziroglu, 2001; Khanal, 2008). In addition, the photo-biological hydrogen production has also been reviewed by Asada and Miyake (1999).

Some algae can produce H_2 by splitting water by photosynthesis. *Chlamydomonas reinhardii* is one of the algae that can produce H_2 from water (Melis, 2002). This process can be considered to be a sustainable method for hydrogen production and elimination of the greenhouse gas CO_2. However, the strong inhibition of the hydrogenase activity by O_2 formed is a limitation for the process. In addition, the low hydrogen forming potential for the process and no wastewater utilization are the other disadvantages of the process.

Starch and cellulose containing waste materials from agricultural and food processing industries are good candidates for bio-hydrogen production (Kapdan and Kargi, 2006). In addition, wastes from the dairy industry, olive mill, baker's yeast and breweries can be used for bio-hydrogen production. Some pretreatment may be necessary before the wastes can be biologically converted to hydrogen. Wastewater sludge contains carbohydrates and proteins, which can be converted to organic acids by anaerobic bacteria, and subsequently these acids can be converted to H_2 gas by phototrophic bacteria.

Many researchers have used mixed cultures in dark fermentation reaction to convert food processing wastes to H_2 (Kapdan and Kargi, 2006). The environmental conditions for the dark fermentation process must be controlled properly to achieve desired results. The media pH affects the hydrogen yield, biogas content, type of end products formed (organic acids) and the specific hydrogen production rate. The pH range for maximum hydrogen yield varies from 5.0 to 6.0 (Fang and Liu, 2002). There are some disagreements on the optimum pH range, but the final pH after the fermentation is near 4.0–4.8. Decrease of pH affects the hydrogenase activity adversely; therefore, control of pH at the optimum level is needed. The dark fermentation media composition, temperature and type of microorganism used also affect the hydrogen yield. The end products after the dark fermentation are mostly organic acids such as acetic, butyric and propionic acids. Lactic acid and ethanol have also been reported in some situations where the substrates or the environmental conditions were different. Yu et al. (2002) studied the hydrogen production from rice winery wastewater in an upflow anaerobic reactor with mixed cultures. The biogas was mostly H_2 (53–61%) and CO_2 (37–45%) with no detectable methane. The optimum H_2 production rate at an HRT of 2 h, COD load of 34 g/L, pH 5.5 and temperature of 55°C was 9.33 L/g VSS·d. van Ginkel et al. (2005) studied H_2 production from food processing industrial wastewater (confectionary, apple and potato processor wastes). They found the highest H_2 yield of 0.21 L/g COD from potato processing wastewater. The rate of hydrogen production by the dark fermentation process is quite slow with low yields, which are the drawbacks for this process. The organic acids produced after the fermentation process can be used in a photo-heterotrophic bacterial process to produce more H_2. Thus, sequential dark and photo-fermentation processes may be used to increase the hydrogen yield from wastewaters rich in carbohydrates (Yokoi et al., 2002).

Hydrogen production by photosynthetic bacteria using industrial wastewater is possible but these organisms prefer organic acids as carbon sources. In addition, the color of the wastewater can limit the light penetration and hence the hydrogen production. The presence of high ammonia levels may also cause enzyme inhibition and some reduction in hydrogen productivity. Tofu wastewater rich in protein and carbohydrates had a hydrogen yield of 1.9 L/L wastewater at 30°C by photosynthetic bacteria, *Rhodobacter sphaeroides* RV in immobilized agar gels (Zhu et al., 1999). The process removed 41% TOC of the wastewater with no ammonia inhibition. Dilution of the wastewater by 50% increased the hydrogen yield to 4.32 L/L wastewater and improved the TOC removal efficiency to 66%. The types of photo-bioreactors used for hydrogen production are tubular, flat panel and bubble column reactors and have been reviewed by Akkerman et al. (2002). High illumination in the reactor caused higher H_2 production rates but lower light conversion yields. In addition, mixing in the photo-bioreactor is also an important parameter for obtaining better H_2 yield.

The hydrogen produced by the methods described above can be used in an FC to produce electricity (as in Option 1c) or it can be combusted in an internal combustion engine to produce power. Some automobile manufacturers are conducting research and demonstrations on cars that can use hydrogen as a fuel in place of gasoline, but because fuel hydrogen cell technology is more efficient, it is likely that the future of hydrogen utilization will be through FCs for energy production.

2.2.3 Alternative 3–The Wastewater after Pretreatment Can Be Used in an MFC to Directly Produce Electricity

Microbial fuel cell (MFC) is a bioreactor which is able to convert chemical energy stored in organic compounds to electrical energy through catalytic reactions of microorganisms under anaerobic conditions. MFCs have been studied for a long time to evaluate the possibility of converting energy stored in chemical bonds of organic compounds into electricity by microbial activities (Du et al., 2007; Logan, 2008). It is generally believed that Potter (1911) first observed the generation of an electrical current by bacteria. However, very few studies were conducted before 1990s (Allen and Bennetoo, 1993; Lewis, 1966; Logan, 2008). Before 1999, tests involved in MFCs required the use of chemical mediators or electron shuttles, which were believed to carry electrons from inside the cell to exogenous electrode. Electron transport between microorganisms (used as biocatalysts) and an electrode has been known to be possible without any additions of artificial mediators (Kim et al., 1999; Li et al., 2008; Logan, 2008). In 2004, it was demonstrated that domestic wastewater could be treated using MFCs while simultaneously generating electricity (Liu and Logan, 2004). Similar to chemical fuel cell, MFC is regarded as promising distributed power sources for mobile and stationary applications (Du et al., 2007).

Research on using MFC technology to produce energy from complex organic wastewater has been conducted in recent years (Liu and Logan, 2004; Logan, 2008). The wastewaters tested in MFCs included food, domestic, swine, chemical and other high strength wastewaters (Oh and Logan, 2005; Li et al., 2008; Min and Angelidaki, 2008). Li et al. (2008) also applied a dual-chamber MFC to treat electroplating wastewater containing Cr^{6+} with the initial concentration of 204 ppm in the MFC. Their results showed that the maximum power density of 1600 mW/m^2 was generated at a Columbic efficiency (CE) (CE measures in percent the total Coulombs calculated by integrating the current over time divided by the theoretical amount of Coulombs that can be produced by the substrate) of 12%. An MFC usually consists of an anode and a cathode separated by a proton exchange membrane. Earlier studies used specific substrates such as glucose or acetate in the anode chamber where pure culture microorganisms were present together with electron mediators (Du et al., 2007). Electron mediators accelerated the transfer of electrons to the anode. These mediators were redox compounds like dyes (e.g., methylene blue), or metallorganics (e.g., Fe(III), EDTA) (Du et al., 2007). The bacteria metabolize the organic compounds present. Electrons were transported from electron donors (carbohydrates and other substrates) to the anode electrode through the help of electron mediators. The treated wastewater and CO_2 exited the anode chamber as shown in Figure 2.5. In addition, protons were also produced during this process that moved across the proton exchange membrane to the cathode chamber, where they reacted with oxygen to form water. The electrons traveled through the external wire to the cathode to complete the electrical circuit. The separation of microorganisms in the anode chamber from oxygen or any other electron acceptor made the electrical current flow possible. The reactions in the process are as follows:

Anode reactions: Organic matter + H_2O + microorganisms $\rightarrow CO_2 + xH^+ + y\,e^-$
Cathode reactions: $zO_2 + y\,e^- + xH^+ \rightarrow a\,H_2O$

The net results of these reactions are the production of electricity and biodegradation of the organic matter in wastewater. The electrical potential developed is quite small in the range of 0.5–0.8 V, so connecting several such cells in a series can produce higher voltage (Logan, 2005, 2008).

Later it was found that there were some microorganisms that could transfer electrons directly to the anode without the help of mediators. In addition, they had a stable operation and high CE (Chaudhuri and Lovley, 2003). Although the biomass yield was not known, the highest CEs were ~85%, suggesting that about 15% of the substrate was converted by exoelectrogens into biomass (Logan, 2008). These organisms could form a biofilm on the anode surface and transfer electrons directly. Thus, the anode acts as the final electron acceptor in the dissimilatory respiratory chain of the microbes in the biofilm. Biofilm formed on a cathode surface may also play an important role in electron transfer between the microbes and the electrodes (Chaudhuri and Lovley, 2003). Further research showed that mixed microbial cultures could also transfer electrons successfully in the anode chamber with the possibility of wider substrate utilization such as domestic wastewater (Logan, 2008). Moreover, a number of publications discussed the screening and identification of microbes and the construction of a chromosome library for microorganisms that are able to generate electricity from degrading organic matters (Logan et al., 2006; Du et al., 2007; Logan, 2008). The most significant observation about the stability of an MFC is that the anode does not appear to foul over time. It is expected that within the next few year, we will gain new insight into the microbial ecology of MFC biofilms (Logan, 2008).

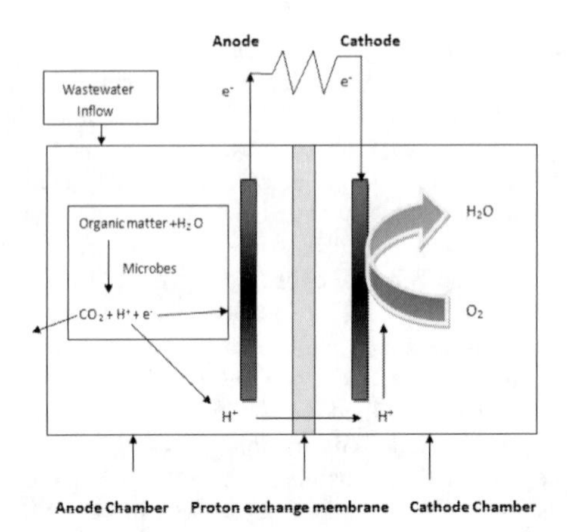

Figure 2.5 Schematic of a two-chamber microbial fuel cell.

In addition to the two-chamber MFC described in Figure 2.5, there is a simple and more efficient MFC with only one chamber, where the cathode is placed on the cation/proton exchange membrane (CEM) and thus eliminating the cathode chamber. In addition, air-cathode systems have also been developed that require no aeration and no CEM (Liu and Logan, 2004). Details of this design and other variations of this process have been discussed in literature (Logan et al., 2006; Du et al., 2007; Logan, 2008). Liu and Logan (2004) used a single chamber MFC to continuously treat domestic wastewater producing about 26 mW/m^2 (of anode area) of electricity. The BOD removal was about 80%. It was stated that increased wastewater strength could produce higher power densities. Increased power outputs up to 146 mW/m^2 with domestic wastewater influents were reported with improved reactor design. Aelterman et al. (2006) used hexacyanoferate (HCF) as catholyte in their MFCs with hospital wastewater and anaerobic digester influent wastewater as feed substrate; they reported that power densities ranged from 14 to 58 W/m^3 (based on anode compartment volume). The substrate COD removal rate was 0.5 to 2.99 kg COD/m^3 reactor·day. With artificial wastewater and platinum based open–air cathode, they obtained a power density of about 102 W/m^3 and a substrate COD removal rate of 8.9 kg/m^3·d. This indicated that higher strength wastewaters could be treated in these MFCs.

A modification of the MFC was used by Liu et al. (2005) to produce hydrogen at the cathode in an MFC from wastewater when an external power supply was applied. This process is referred as electrohydrogenesis because electrons released by bacteria are combined with protons to form hydrogen gas as a product, rather than electricity as in an MFC. This process is called a bioelectrochemically assisted microbial reactor (BEAMR), a biocatalyzed electrolysis cell (BEC), or a microbial eletrolysis cell (MEC) as it is a process dependent on the microbial production of hydrogen (via microbial-catalyzed electrolysis of organic matter). In this system no oxygen is introduced at the cathode. The recovered hydrogen can be used as a fuel for energy production. However, the capture and purification of the hydrogen gas is a complicated and energy intensive process, which may undermine the overall economics of the BEAMR process as a sustainable and cost-effective method for hydrogen production.

Although the BEAMR and MFC systems share many similar characteristics, there are some important differences between the two systems (Logan, 2008); a major difference mainly stems from the possible hydrogen loss in the BEAMR system (due to different mechanisms, e.g. diffusion through the membrane, used by microbes as a substrate in the cathode chamber). Currently, advances in MFC design that minimize internal resistance in MFC reactors are used to benefit MEC reactor designs. However, there are very few reports about using the BEAMR system for hydrogen production. Liu et al. (2005) developed the first BEAMR system and reported that, on average, 2.9 mol-H_2 were produced per mole of acetate (compared to a maximum theoretical yield of 4 mol/mol), which required an energy input equivalent to 0.5 mol of H_2 at an applied voltage of 0.25 V. Ditzig et al. (2007) used a two-cube chamber to produce hydrogen from domestic wastewater. They found that BOD removals of the BEAMR were 97 ± 2% and that of COD were 95 ± 2%, but the performance of the BEAMR as a hydrogen producer was poor, with hydrogen CE ranging from 9.6 to 26%. It is expected that the CE may be enhanced if high strength industrial wastewater is used because the CE obtained from BEAMRs with a high concentration of single compounds like acetate usually is much higher.

At the present time, no full-scale application of MFC technology has been tried on domestic or industrial wastewater. Most of the data presented in the literature is based on laboratory studies. More research needs to be conducted: to improve the design of the electrodes and reduce their costs; to design a better proton exchange membrane, to find cheaper catalysts for the cathode reactions; and to increase the power density output for the process. High nitrogen, phosphorus or sulfur compounds in the wastewater will not be removed effectively by the process. So, pretreatment or post treatment of these contaminants will be necessary. One of the advantages of MFC process is that it can treat wastewater at ambient temperatures in contrast with anaerobic processes where higher temperatures are required for better efficiency. Like all microbial processes, the presence of toxic chemical is not acceptable in the MFC process also. Another disadvantage of MFCs is the reliance on biofilms for mediator-less electron transport, while anaerobic digesters such as UASB eliminates this need by reusing the microorganisms without cell immobilization (Du et al., 2007).

2.3 Possible Future Combination of Anaerobic Technology with FCs

For high strength wastewater from food, brewery, agricultural and other industries, it may be more efficient if a combination of processes is used to maximize energy recovery and minimize energy use. For high strength industrial wastewater after pretreatment, it can be treated using high-rate anaerobic processes. The gases produced can be combusted in an internal combustion engine or micro-turbine to produce energy. The energy recovery could be 0.5–1 kW/m^3. The effluent stream may still have residual organic acids (several Kg COD/m^3). This effluent can be applied to an MFC to directly produce electricity or it can be used in a BEAMER system to produce hydrogen that can be used in the combustion process to produce energy (Aelterman et al., 2006). The effluent from the MFC/BEAMR process may still have organic contaminants (BOD/COD) that are not acceptable for discharge into the receiving water. In that case, an aerobic treatment process (activated sludge or a biofilm process) may be needed to polish the effluent to meet the discharge requirements. Figure 2.6 shows a schematic for the proposed treatment system.

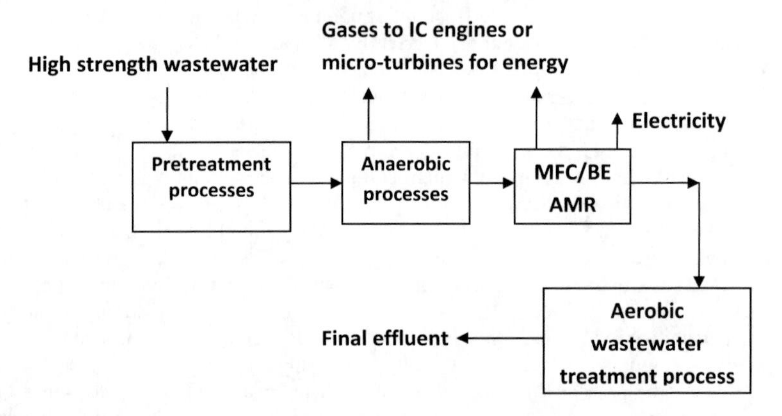

Figure 2.6 Schematic for energy recovery from high strength wastewater.

2.4 Summary

High strength wastewater from industries has a significant potential for energy recovery. Presently, many industries such as agro-food, beverage and paper and pulp industries, recover energy by utilizing anaerobic processes to generate combustible gases (mainly methane).

The methane production from the anaerobic processes can be combusted in internal combustion engines or microturbines or FCs to cogenerate energy (heat and electricity). In many instances, a major portion of energy needs for the wastewater treatment plant can be satisfied from the cogeneration process. Microturbines have been successfully tried at several locations to produce energy from APGs. FCs for energy production from APGs are still in the experimental stage and are presently not cost-effective as compared to other options.

Microbial processes to form hydrogen directly from wastewater organic matter are successful but are still in the developmental stage. More applied research needs to be conducted to determine the performance under differing operating conditions with different types of wastewater. The low environmental impact of hydrogen combustion is a big advantage for the system. Some wastewater, rich in carbohydrates (food processing), have successfully produced hydrogen in dark fermentation reactions, but the rates are slow. The wastewater effluent after the fermentation process still has a high organic content (mainly organic acids). A photofermentation process succeeding the dark fermentation process may increase the hydrogen yield. The hydrogen produced can be combusted or used in an FC to produce electricity. Perhaps in the future, wastewater treatment technology will use hydrogen production as a key goal followed by conversion of the hydrogen produced to electricity using FCs.

The use of MFC is a promising technology for treating wastewater to produce electricity, but it is still in its infancy and needs more research and development. It can be used in conjunction with an anaerobic treatment process for some industrial wastewater to cogenerate energy and treat the wastewater. It appears that lot of exciting research work needs to be done to manage industrial high strength wastewater in a sustainable way and to make the process energy sufficient.

2.5 References

Aelterman, P., Rabaey, K., Clauwaert, P., and Verstaete, W. (2006). "Microbial fuel cells for wastewater treatment." *Water Sci. Technol.*, 54(8), 9–15.

Akkerman, I., Jenssen, M., Rocha, J., and Wijffels, R.H. (2002). "Photobiological hydrogen production: photochemical efficiency and bioreactor design." *Int. J. Hydrogen Energy*, 27, 1195–1208.

Allen, R.M., and Bennetto, H.P. (1993). "Microbial fuel cells: electricity production from carbohydrates." *Appl. Biochem. Biotechnol.*, 39(2), 27–40.

Anheuser-Busch (2007). "Bio-energy Recovery Systems." Company report. Available at http://www.anheuser-busch.com.

Asada, Y., and Miyake, J. (1999). "Photobiological hydrogen production." *J. Biosci. Biotechnol.*, 88, 1–6.

Bloomquist, R.G. (2006). *Case Study: King County South Treatment Plant Renton, Washington: Combined heat and power using a molten carbonate fuel cell–1.0 Megawatt of electrical output (MWe).* Washington State University Extension Energy Program, July 2006.

Britz, T.J., and van der Merwe, M. (1997). "Anaerobic treatment of baker's yeast effluent using a hybrid digester and polyurethane as support material." *Biotechnol. Letters*, 15(7), 755–760.

Chaudhuri, S.K., and Lovley, D.R. (2003). "Electricity generation by direct oxidation of glucose in mediatorless microbial fuel cells." *Nat. Biotechnol.*, 21, 1229–1232.

Das, D., and Veziroglu, T.N. (2001). "Hydrogen production by biological processes: a survey of literature." *Int. J. Hydrogen Energy*, 26, 13–28.

Ditzig, J., Liu, H., and Logan, B.E. (2007). "Production of hydrogen from domestic wastewater using a bioelectrochemically assisted microbial reactor (BEAMR)." *Int. J. Hydrogen Energy*, 32(13), 2296–2304.

Du, Z., Li, H., and GU, T. (2007). "A state of the art review on microbial fuel cells: a promising technology for wastewater treatment and bioenergy." *Biotechnol. Advances*, 25(5), 464–482.

Fang, H.H.P, and Liu, H. (2002). "Effect of pH on hydrogen production from glucose by mixed culture." *Bioresource Technol.*, 82(1), 87–93.

FEMP (Federal Energy Management Program) (2004). "Wastewater Treatment gas to energy for federal facilities." Fact sheet, ORNL 2004-0594/abh, 2004.

Halleneck, P.C. (2005). "Fundamentals of the fermentative production of hydrogen." *Water Sci. Technol.*, 52(1–2), 21–29.

Kapdan, I.K., and Kargi, F. (2006). "Bio-hydrogen production from waste materials." *Enzyme and Microbial Technol.,* 38(5), 569–582. Kim, J.R., Jung, S.H., Regan, J.M., and Logan, B.E. (1999). "Electricity generation and microbial community analysis of ethanol powered microbial fuel cells." *Bioresource Technol.*, 98(13), 2568–2577.

Khanal, S.K. (2008). *Anaerobic Biotechnology for Bioenergy Production: Principles and Applications.* Wiley-Blackwell Publishing, Ames, IA, USA.

Kubota, K., Kuroda, K., and Akiyama, K. (2003). Present status and future prospects of biogas powered fuel cell power units." *Fuji Electric Review*, 49(2), 68-72.

Kim, B.H., Kim, H.J., Hyun, M.S., and Park, D.H.(1999). "Direct electrode reaction of Fe(II)-reducing bacterium *Shewanella putrefaciens*." *J. Microbiol. Biotechn.*, 9, 127-131.

Lewis, K. (1966). "Symposium on bioelectrochemistry of microorganisms: IV. Biochemical fuel cells." *Bacteriol. Rev.*, 30(1), 101-113.

Lettinga, G., Vanvelsen, A.F.M., Hobma, S.W., Dezeeuw, W., and Klapwijk, A. (1980). "Use of the upflow sludge blanket (USB) reactor concept for biological wastewater treatment, especially for anaerobic treatment." *Biotechnol. Bioengr.*, 22, 699–734.

Li, Z., Zhang, X., and Lei, L. (2008). "Electricity production during the treatment of real electroplating wastewater containing Cr^{6+} using microbial fuel cell." *Process Biochemistry*, 43, 1352–1358.

Liu, G., and Shen, J. (2004). "Effects of culture medium and medium conditions on hydrogen production from starch using anaerobic bacteria." *J. Biosci. Bioengr.,* 98, 251-256.

Liu, H., and Logan, B.E. (2004) "Electricity generation using an air-cathode single chamber microbial fuel cell in the presence and absence of a proton exchange membrane." *Environ. Sci. Technol.*, 38, 4040-4046.

Liu, H., Grot, S., and Logan, B.E. (2005). "Electrochemically assisted microbial production of hydrogen from acetate." *Environ. Sci. Technol.,* 39(11), 4317–4320.

Liu, H., Ramnarayanan, P., and Logan, B.E. (2004). "Production of electricity during wastewater treatment using a single chamber microbial fuel cell." *Environ. Sci. Technol.,* 28, 2281–2285.

Logan, B.E. (2005) "Simultaneous wastewater treatment and biological electricity generation." *Water Sci. Technol.*, 52(1–2), 31–37.

Logan, B.E., Hamelers, B., Rozendal, R., Schroder, U., Keller, J., Freguia, S., Aelterman, P., Verstraete, W., and Rabaey, K. (2006). "Microbial fuel cells: methodology and Technology." *Environ. Sci Technol.,* 40(17), 5181–5192.

Logan, B.E. (2008). *Microbial Fuel Cells*. John Wiley & Sons, Inc., Hoboken, NJ.

Melis, A. (2002). "Green alga hydrogen production: progress, challenges and prospects," *Int. J. Hydrogen Energy*, 27, 1217–1228.

McDannel, M., and Wheless, E. (2008). "The power of digester gas," *Water Environ. Technol.*, 20(6), 36–41.

Metcalf & Eddy (2003). *Wastewater Engineering–Treatment and Reuse*, 4th ed., McGraw-Hill, New York.

Mignone, N.A. (2008). "Fad or Future?" *Water Environ. Technol.*, 20(11), 42–47.

Min, B., and Angelidaki, I. (2008). "Innovative microbial fuel cell for electricity production from anaerobic reactors." *J. of Power Sources,* 180, 641–647.

Molina, F., Ruiz-Filppi, G., Garcia, C., Roca, E., and Lema, J.M. (2007). "Winery effluent treatment at an anaerobic hybrid USBF pilot plant under normal and abnormal operation." *Water Sci. Technol.,* 56(2), 25–31.

Monteith, H., Bagley, D., Kalogo, Y., Ramani, R., and Fillmore, L. (2007). "Power (recovery) tool." *Water Environ. Technol.*, 19(12), 66–69.

Oh, S., and Logan, B.E. (2005). "Hydrogen and electricity production from a food processing wastewater using fermentation and microbial fuel cell technologies." *Water Research*, 39, 4673–4682.

Potter, M.C. (1911). "Electrical effects accompanying the decomposition of organic compounds." *Proc. Roy. Soc. Lond Ser. B*, 84, 260–276.

Puetpaiboon, U., and Chotwattanasak, J. (2008). "Anaerobic treatment of palm oil mill wastewater under mesophilic condition." Available at http://web.eng.psu.ac.th/ enghome/web _Guideline/Treatmentweb/PuetpaiboonU.pdf.

Rogalla, F., Tarallo, S., Scanlan, P., and Wallis-Lage, C. (2008). "Sustainable solutions." *Water Environ. Technol.,* 20(4), 30–33.

Ruiz, I., Viega, M.C., deSantiago, P., and Blazquez, R. (1997) "Treatment of slaughter house wastewater in a UASB reactor and an anaerobic filter." *Bioresource Technol.*, 60, 251-258.

Schroedel Jr., R.B., Cavagnaro, P.V., and Peterson, J.W. (2009). "Siloxanex the hidden threat to biogas systems." *Water Environ. Technol.*, 21(8), 59-64.

Speece, R.E., Boonyakitsombut, S., Kim, M., Azbar, N., and Ursillo, P. (2006). "Overview of anaerobic treatment: thermophilic and propionate implications." *Water Environ. Res.*, 78(5), 460–473.

Spencer, R. (2007). "Anaerobic treatment of high-strength wastewater." *Biocycle*, pp. 41–43, December 2007.

Skok, A. (2007). "Pristine power, premium beer." *Sustainable Facility*, October 2007.

USEPA (2006). "Milti-stage Anaerobic Digestion," Biosolids Technology Fact Sheet, Office of Water, EPA 832-F-06-031, 2006.

van Ginkel, S., Oh, S.E., and Logan, B.E. (2005). "Biohydrogen production from food processing band domestic wastewater." *Int. J. Hydrogen Energy*, 30(15), 1535–1542.

Vijayaraghavan, K., and Soom, M.A.M. (2006). "Trends in biological hydrogen production–a review." *Environ. Sciences*, 3(4), 255–271.

Williams, D.W., and Gould-Wells, D. (2004). "Biogas production from a covered lagoon digester and utilization in a microturbine." *Resources: Engineering and Technology for Sustainable world*, April 2004.

Wikipedia (2008). http://en.wikipedia.org/wiki/fuels_cell. 2008.

Xu, H.L., Wang, J.Y., and Tay, J.H. (2002) "A hybrid anaerobic solid-liquid bioreactor for food waste digestion." *Biotechnol. Letters*, 24(10), 757-761.

Yokoi, H., Mahi, R., Hirose, J., and Hayashi, S. (2002). "Microbial production of hydrogen from starch manufacturing wastes." *Biomass Bioenergy*, 22, 389–395.

Yu, H., Zhu, Z., Hu, W., and Zhang, H. (2002). "Hydrogen production from rice winery wastewater in an upflow anaerobic reactor by using mixed anaerobic culture." *Int. J. Hydrogen Energy*, 27(11–12), 1357–136.

Zhu, H., Suzuki, T., Tsygankov, A.A., Asada, Y., and Miyake, J. (1999). "Hydrogen production from tofu wastewater by *Rhodobacter sphaeroides* immobilized agar gels." *Int. J. Hydrogen Energy*, 24, 305–310.

CHAPTER 3

Food Waste to Bioenergy

Xue-Yan Liu, Hong-Bo Ding, and Jing-Yuan Wang

3.1 Introduction

Bioenergy is a renewable energy that can be generated from feedstock of biological origin. The use of bioenergy is key to mitigate the problem of excessive greenhouse gas (GHG) emission, to minimize the use fossil fuels and to provide national energy security. Bioenergy will be the significant source of renewable energy for the next several decades until the solar or wind power becomes an economically attractive for large-scale production (Nallathambi Gunaseelan, 1997).

Food wastes include food preparation waste, leftover/residual foods from households and commercial establishments such as restaurants and market, institutional sources like school cafeterias, and industrial sources like food processing factories and factory lunchrooms, etc. Food waste is a highly biodegradable organic fraction of municipal solid waste (OFMSW) and is the major source of odor emanation, leachate, toxic gas emission and groundwater contamination (Kim et al., 2004).

In light of rapidly rising costs associated with energy supply and waste disposal, and increasing public concern with environmental quality, successful application of anaerobic digestion (AD) technology could provide an economical and an environmentally friendly means for bioenergy recovery from food wastes with simultaneous remediation of wastes. Thermochemical conversion technologies, such as combustion and gasification are not suitable due to relatively high moisture content of food waste (Cecchi et al., 1992; Edelmann et al., 2000). A high-solid digestion known as dry-digestion is discussed in Chapter 4.

This chapter discusses the biochemical principle of AD and key affecting factors in AD design and operation, especially food wastes. Different AD processes and the current technology available are outlined. Some experience with advanced digestion processes in the US, Europe, Singapore and Korea are also shared. Future research direction is also outlined.

3.2 Anaerobic Digestion

Anaerobic digestion (AD) is a biological process in which complex organic materials are broken down into simpler compounds by microbes using exo-enzymes in absence of oxygen. The polymers of carbohydrates, lipids, and proteins are converted into simpler monomers (e.g. sugars, glycerin and fatty acids, and amino-acids), and subsequently into the biogas primarily methane and carbon dioxide by methanogenic bacteria.

3.2.1 Biochemical Principles of AD

AD consists of a series of biochemical steps mediated by distinct groups of microorganisms, mostly bacteria and archaea. Generally, there are three main groups of bacteria: the fermentative bacteria, the acetogenic bacteria and the methanogens. Each of them has a specific action in the overall process, described by the following four steps: (1) hydrolysis (liquefaction/solubilization), (2) acidogenesis, (3) acetogenesis, and (4) methanogenesis. An overall pathway of AD is given in Figure 3.1.

Hydrolysis is the first and necessary step in AD of high solid waste in which large, complex and insoluble organics are broken down into their component units: proteins to amino acids, carbohydrates to soluble sugars, and lipids to long chain fatty acids and glycerin. These compounds can only be transported, metabolized and assimilated into microbial cells in water soluble state (Schieder et al., 2000). The hydrolysis process is mediated by extra-cellular hydrolytic enzymes such as amylase, cellulase, protease, and lipase secreted by bacteria. It is often a rate-limiting step in AD of high solid wastes and governs the overall AD process.

Acidogenesis is the acid-forming fermentation of water-soluble compounds such as sugars and amino acids. Acidogenic bacteria convert them into organic acids (mainly volatile fatty acids (VFAs)), hydrogen, and carbon dioxide. These products vary highly with the types of substrate, bacteria, and environmental conditions such as pH and hydrogen partial pressure.

In *Acetogenesis,* bacteria such as *Syntrobacter wolini* and *Syntrphomonas wolfei* convert higher VFAs (e.g. propionate and butyrate) and alcohol into acetate, hydrogen, and carbon dioxide, which are the main substrates for methane fermentation. In addition, there is also the other group of bacteria called homoacetogens which form acetate from hydrogen and carbon dioxide.

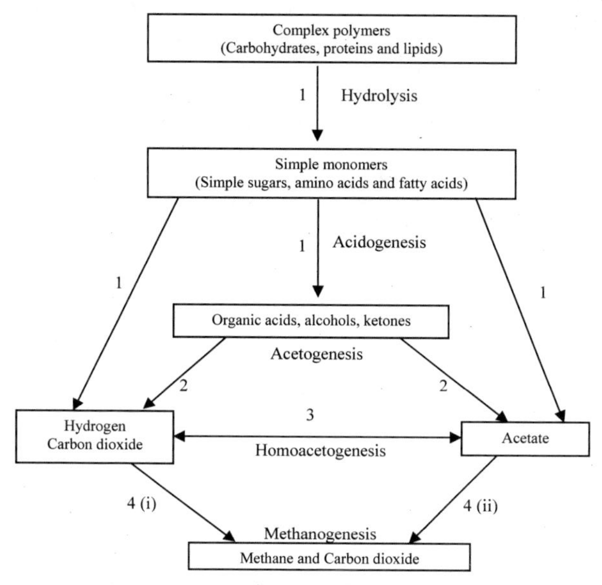

Figure 3.1 Bioconversion of complex organic material in anaerobic digestion (Adapted from Khanal, (2008)).

Methanogenesis or the formation of methane occurs by two major routes: One from acetate splitting (aceticlastic methanogensis) and the other one from the reduction of carbon dioxide by hydrogen (hydrogenotrophic methanogenesis). About 60-75% of methane is produced from the former and the remaining from the latter (Mackie and Bryant, 1981).

3.2.2 Two-Phase AD

In conventional AD, both acidogenesis and methanogenesis occur in a single-reactor system. Acidogenesis normally causes a consequential decrease in pH due to the production of VFAs. Because the acid-forming and the methane-forming microorganisms differ widely in terms of physiology, nutritional requirement, growth kinetics and sensitivity to environmental conditions, it takes quite a long time to get a balance between them in a single reactor.

In order to overcome this problem, Pohland and Ghosh (1971) first proposed the physical separation of the acid-former and the methane-former in two separate reactors where optimum environmental conditions for each group of microorganisms were provided to enhance overall process stability and control. Over the last 30 years, an improved two-phase AD became more popular, in which the first phase is responsible for acid formation and the second phase is optimized for methane fermentation.

3.2.2.1 Acid-phase Digestion

Han et al. (2004) operated an innovative two-stage process, BIOCELL, at a temperature of 37°C in lab-scale, that was developed to produce hydrogen (H_2) and methane (CH_4) from food waste in Korea on the basis of phase separation, reactor rotation mode, and sequential batch technique.

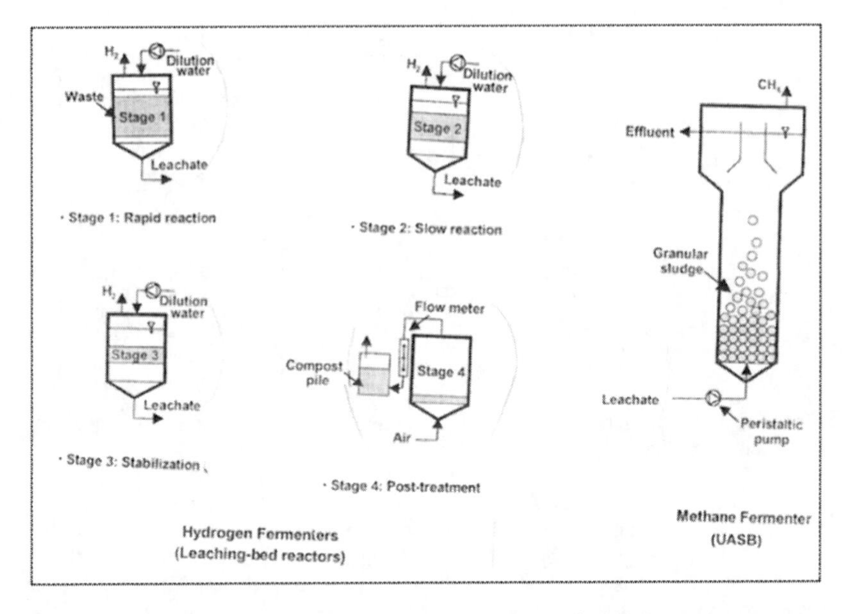

Figure 3.2 The BIOCELL process (Adapted from Han et al., 2004).

The BIOCELL process shown in Figure 3.2 consisted of four leaching-bed reactors for H_2 recovery and post-treatment and a UASB reactor for CH_4 recovery. The leaching-bed reactors were operated in a rotation mode with a 2-day interval between degradation stages. The sequential batch technique was useful to optimize environmental conditions during H_2 fermentation.

The BIOCELL process demonstrated that, at the high volatile solids (VS) loading rate of 11.9 kg $m^{-3}d^{-1}$, it could remove 72.5% of VS and convert $VS_{removed}$ to H_2 (28.2%) and CH_4 (69.9%) on a chemical oxygen demand (COD) basis in 8 days. H_2 gas production rate was 3.63 m^3 m^{-3} d^{-1}, while CH_4 gas production rate was 1.75 m^3 m^{-3} d^{-1}. The yield values of H_2 and CH_4 were 0.31 and 0.21 m^3 kg^{-1} VS_{added}, respectively. Moreover, the output from the post-treatment could be used as a soil amendment. The principal advantages of this process are (1) the recovery of H_2 as well as CH_4 that can be used as a fuel for the production of energy; (2) the production of compost that can be used as a soil amendment; (3) the stability of performance by supplying preferred environments for acidogenic hydrogenesis and methanogenesis in two separate spaces; (4) the ease of operation by employing reactor rotation mode and sequential batch technique; (5) no need of agitation by using leaching-bed reactors; and (6) the convenience of post-treatment by treating residues in the same reactor without troublesome moving. In summary, the BIOCELL process proved stable, reliable, and effective in treating food waste.

3.2.2.2 Hybrid Anaerobic Solid-liquid (HASL) System

The hybrid anaerobic solid-liquid (HASL) system is a modified two-phase anaerobic digester and has been successfully developed for food waste digestion in Singapore. It includes an acidogenic reactor (Ra) to hydrolyze high solids food waste and a modified upflow anaerobic sludge blanket (UASB) as the methanogenic reactor (Rm) to convert the leachate from Ra into biomethane. The leachate from Ra is 5 times diluted with the effluent from Rm. This mixture is fed into Rm as a new influent. The effluent from Rm is divided into two streams, S1 and S2, and the initial flow rate ratio of S1 and S2 is kept at 1:4. S1 flows into Ra, while S2 is used to dilute the leachate from Ra and pump it back into Rm. After the HASL set up, the flow rates of S1 and S2 may change according to the chemical oxygen demand (COD) concentrations in the leachate from Ra and the effluent from Rm in order to keep a COD concentration of less than 5,000 mg L^{-1} and an organic loading rate (OLR) of less than 10 g COD L^{-1} d^{-1} as the optimal conditions in Rm (Wang et al., 2002; Xu et al., 2002). The schematic diagram of the HASL system is shown in Figure 3.3.

The high methane content (68-70%) from the Rm demonstrated that the HASL system is effective and efficient for the conversion of food waste to bioenergy. Besides higher efficiency of methane production, another advantage of the HASL system is small volume of discharge from the system due to recirculation. The salient features of the HASL system include: (1) effective acidification by leachate recirculation; (2) washout of VFAs by circulating the effluent from Rm into Ra; (3) efficient methanogenesis in a modified UASB reactor; (4) built-in mechanism to prevent system instability by efficient VFA removal in an active Rm; (5) no pH control requirement due to the leachate dilution prior to entering into Rm; and (6) no mixing, less excess fresh water addition and nearly no wastewater production. All of these make the HASL system highly efficient, flexible and stable.

The high methane content (68-70%) from the Rm demonstrated that the HASL system is effective and efficient for the conversion of food waste to bioenergy. Besides higher efficiency of methane production, another advantage of the HASL system is small volume of discharge from the system due to recirculation. The salient features of the HASL system include: (1) effective acidification by leachate recirculation; (2) washout of VFAs by circulating the effluent from Rm into Ra; (3) efficient methanogenesis in a modified UASB reactor; (4) built-in mechanism to prevent system instability by efficient VFA removal in an active Rm; (5) no pH control requirement due to the leachate dilution prior to entering into Rm; and (6) no mixing, less excess fresh water addition and nearly no wastewater production. All of these make the HASL system highly efficient, flexible and stable.

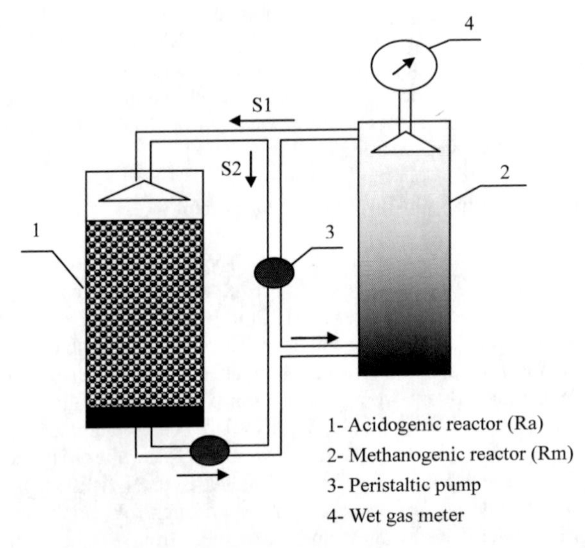

Figure 3.3 Schematic diagram of the HASL system (Adapted from Wang et al., 2002; Xu et al., 2002).

3.2.2.3 Temperature-phase Anaerobic Digestion (TPAD)

Two-phase digestion may be the best process option for those seeking increased VS reduction and gas production without the complications of temperature staging. But a general consensus has emerged that one of the two phases will need to be operated at thermophilic temperatures if Class A pathogens reduction is an objective.

Temperature-phased anaerobic digestion (TPAD), a patented process developed at Iowa State University (ISU) by Dague and coworkers, is one of the innovative sludge digestion processes that operate in the temperature-phased mode. Figure 3.4 shows a schematic of the TPAD system. It consists of a two-stage system, which operates at high thermophilic temperatures (typically 55°C) in the first stage and lower mesophilic temperatures (typically 35°C) in the second stage. The arrangement utilizes the mesophilic reactor as a polishing stage alleviating the drawbacks of thermophilic process. It has been shown to be a reliable and effective means of sludge stabilization that achieves bioconversion and methane production rates higher than the existing mesophilic anaerobic systems. A series of bench-scale studies conducted at ISU have also demonstrated that the system capability of producing class A biosolids, the highest ranked stabilized sludge for surface disposal regulated by USEPA (Han et al., 1997a&b). It has been shown to achieve significant higher organic removals than is possible for single-stage systems operated at either 55 or 35°C. In addition, the ability to treat higher solids and organic loadings at relatively shorter hydraulic retention times (HRTs) indicates the fact that TPAD would be less than one half the size of conventional systems currently in use.

The completely mixed TPAD system is known to handle high-solid concentrations, but has never been evaluated for animal wastes until Sung et al. reported the performance characteristics of a completely mixed TPAD system over an entire range of total solid (TS) concentrations and organic loading rates (OLR).

Anaerobic digestion of cattle wastes using the TPAD technology not only recovers the energy by-product methane, but also retains the nutrient value of manure and provides near pathogen-free biosolids. The arrangement of two reactors in series, with the thermophilic unit as the first stage followed by the mesophilic unit, can take advantage of both thermophilic and mesophilic conditions. The thermophilic first stage enhances the hydrolysis of some of the recalcitrant organics in cattle wastes that makes it available for acidogenic and methanogenic bacteria in the mesophilic stage. The thermophilic unit operated at a higher temperature and VS loading, achieves higher VS destruction rate. The second mesophilic stage completes the digestion process converting the partially digested organics to methane and carbon dioxide thus fully recovering the energy by-product from cattle wastes.

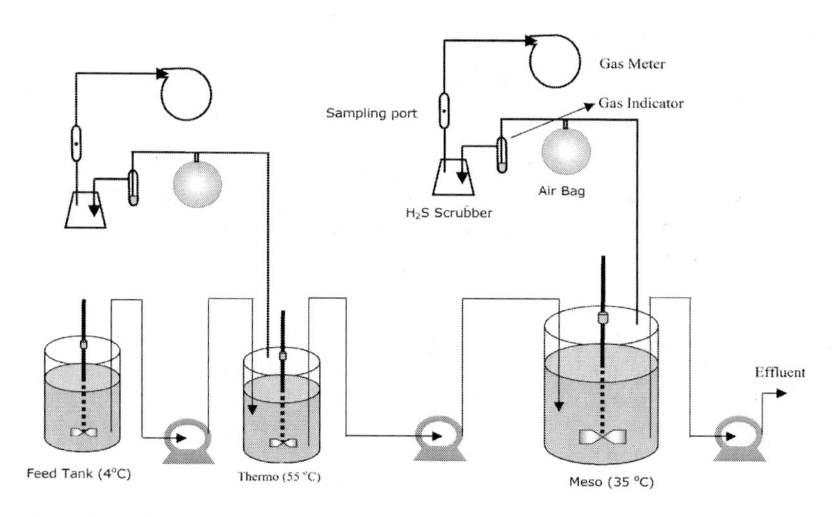

Figure 3.4 Schematic of the TPAD system (Adapted from Sung et al., 2003).

Bouallagui et al. (2005) studied AD of fruit and vegetable wastes (FVW) under different operating conditions using different types of bioreactors for material recovery and energy production. The study reported conversion of 70-95% of organic matter into methane, with an OLR of 1-6.8 g VS L^{-1} d^{-1}. Using a two-stage system involving a thermophilic liquefaction reactor and a mesophilic anaerobic filter, over 95% VS were converted to methane at a volumetric loading rate of 5.65 g VS L^{-1} d^{-1}. The average methane yield was about 0.42 L g^{-1} added VS.

Conventional mesophilic systems could be modified to two-stage systems by upgrading one of the mesophilic for operation at thermophilic temperatures. In practice, it would also be advisable to place an effluent heat exchanger on the first-stage thermophilic digester. This approach could reduce the temperature of thermophilic effluent to the optimum mesophilic level and recover a portion of the energy used in raising the temperature of the incoming waste stream to the thermophilic level. The TPAD process provides sufficient energy to keep the digesters at operating temperature and still provide an additional amount of net energy. King County (Seattle) recently completed an extended pilot plant investigation of TPAD for its West Point treatment plant to reduce the number of digesters. A full-scale TPAD system treating dairy cattle waste at Tinedale Farms, Wrightstown, Wisconsin is generating excess energy to power 250 houses (Sung et al., 2003). However, it would be unwise to associate TPAD technology with generation of electricity alone. The economic value of pathogen-free residual solids and liquid end products has to be identified as well to make the process economically attractive.

3.3 Key Affecting Factors in Food Waste Digestion

AD has been widely applied for food waste digestion. Due to the difficulty in isolating anaerobes and the complexity of the bioconversion processes, many factors affect the design and performance of the AD processes. Some of the key factors are: feedstock characteristics, reactor design and operational conditions (Hansen and Cheong, 2007). The following section discusses the effects of food waste characteristics, co-digestion, pH, temperature, hydraulic retention time, pretreatment of food waste and process configurations on food waste digestion.

3.3.1 Characteristics of Food Waste

The physical and chemical characteristics of food waste are important for designing and operating an anaerobic digester and significantly affect the biogas production and process stability. They include, but not limited to, moisture content (MC), volatile solids (VS), nutrient contents, particle size, composition, biodegradability, carbon/nitrogen (C/N) ratio, etc (Zhang et al., 2007). Typical characteristics of some food wastes are summarized in Table 3.1.

Table 3.1 Characteristics of food wastes.

Source	*Characteristics*			*Country*	*References*
	MC (%)	VS/TS (%)	C/N		
Mixed municipal sources	90	80	NA*	Germany	Nordberg and Edstrom (1997)
Mixed municipal sources	74	90-97	NA	Australia	Steffen et al. (1998)
Fruit and vegetable waste	88-92	86-92	15.2-30.4	Tunisia	Bouallagui et al. (2003)
Fruit and vegetable waste	85	89	36.4	India	Rao and Singh (2004)
University canteen	92	92	18.3	Singapore	Wang et al. (2004)
A dining hall	80	95	15.4	Korea	Han and Shin (2004)
University cafeteria	80	94	NA	Korea	Kwon and Lee (2004)
A dining hall	93	94	18.3	Korea	Shin et al. (2004)
A dining hall	84	96	NA	Korea	Kim et al. (2004)

*NA-not available.

Biodegradability. Food waste is a typical biodegradable organic waste and mainly composed of carbohydrates, lipids, cellulose and proteins. Its anaerobic biodegradability depends on the relative amount of each component. Biodegradability of a feedstock is indicated by biogas or methane yield and percentage of solids (TS or VS) destroyed. Cho and Park (1995) determined that the methane yields were 482, 294, 277 and 472 mL g^{-1} VS for cooked meat, boiled rice, fresh cabbage and mixed food wastes respectively at 37°C and with 28 days of digestion time, which corresponded to 82%, 72%, 73% and 86% of the stoichiometric methane yield respectively, based on elemental composition of raw materials. Heo et al. (2004) reported that a methane yield of 489 mL g^{-1} VS at 35°C after 40 days of digestion for traditional Korean food consisting of boiled rice (10-15%), vegetables (65-70%), and meat and eggs (15-20%). Stabnikova et al. (2008) reported a methane yield of 300 mL g^{-1} VS removed at 35°C after 12 days of digestion for a typical Singapore meal consisting of vegetable roots (50%), fruits peel (20%), boiled rice (15%) and noodles (15%).

Lipids content. The composition of food waste may vary depending on location, season, economical conditions, business type and food type. For example, the waste from meat processing plants contain high fat and protein content which are difficult to degrade, while waste from the canning industry contains high concentrations of sugars and starches which are easily degradable organic matter. In general lipids (characterized as oil, grease and fat), which are present in high quantity in Asian food, are most problematic due to the toxicity of the long chain fatty acids (LCFA) produced during hydrolysis of lipids (Forster et al., 1992). Stabnikova et al. (2005) reported two important effects of lipids content on the performance of the HASL system: low degradability and high energetic value. The positive effect of lipids on food waste digestion dominated when the lipids content was in the range of 20-30%, indicated by an increase in COD removal rate of 13% and methane production of 7-15%. However, when lipids content was greater than 40%, it was detrimental to HASL performance since the growth of methanogens was inhibited by LCFA. In this case, COD removal rate dropped from 89% to 78% and the maximum growth rate of methanogens reduced from 0.2 h^{-1} to 0.03 h^{-1}. Cirne et al. (2007) also observed a stronger inhibition at higher lipids contents (31%, 40% and 47%). When evaluating the effect of addition of lipase on enzymatic hydrolysis of lipids, the enzyme appeared to enhance the hydrolysis, but the intermediates produced resulted inhibition of the later steps in the degradation process. Since the VFA profiles presented similar trends for the different concentrations of lipid tested, the major obstacle to methane production was believed to be due to LCFA formation.

C/N ratio. C/N ratio is used to measure the relative amounts of organic carbon and nitrogen present in the feedstock. A C/N ratio of 20-30 is considered to be optimum for an efficient digestion, based on biodegradable organic carbon. If the C/N ratio of OFMSW is very high, the waste used as the single substrate will be deficient of nitrogen, which is needed for the build-up of bacterial communities. As a result, the biogas production will decrease. If the C/N ratio is very low, nitrogen will be liberated and accumulate in the form of ammonia there by resulting in pH increase beyond 8.5, which shows toxicity effect on the methanogenic communities (Hartmann and Ahring, 2006).

The C/N ratio varies significantly depending on material composition. For example, the average C/N ratio of food and yard waste is below 20, while it is 24 for animal manure, nearly 70 for rice straw, over 100 for mixed paper and more than 200 for sawdust. Proteins such as meats are high in nitrogen while paper products and plant materials contain a high percentage of carbon (Kayhanian and Rich, 1995). To maintain optimal C/N ratio, materials with high C/N ratio are generally mixed with those with a low C/N ratio, i.e. organic solid waste can be mixed with municipal sewage, biosolids or animal manure. Co-digestion with nutrient-rich organic wastes like manure would be another option to overcome nutrient deficiency conditions (Hartmann and Ahring, 2005).

3.3.2 Co-digestion

Co-digestion is an effective option to improve biogas yield in AD of food waste due to the positive synergistic effect established in the digestion medium and the supply of missing nutrients by the co-substrates. In addition, economic advantages from the sharing of equipment are quite significant. Sometimes, the use of a co-substrate can also help establish the required moisture contents of the feedstock. Other advantages are the easier handling of mixed wastes, the use of common access facilities and the known effect of economies of scale (Mata-Alvarez et al., 2000). Mata-Alvarez and Cecchi (1990) referred to the co-digestion of OFMSW with sewage sludge in existing digesters because both wastes are produced in large quantities and in many places. Much research has been focused on these examples.

Other innovative processes dealing with the co-digestion of different wastes are currently being examined. For example, Kujawa-Roeleveld et al. (2002) reported the co-digestion of black water and kitchen refuse. Kaparaju and Rintala (2005) investigated the possible use of potato tuber and its industrial by-products (potato stillage and potato peels) on farm-scale co-digestion with pig manure in a laboratory study. The methane yields (m^3 kg^{-1} VS added) were 0.13-0.15 at a feed ratio of 100:0 (VS% pig manure to VS% potato co-substrate), 0.21-0.24 at 85:15 and 0.30-0.33 at 80:20 on semi-continuous co-digestion at loading rate of 2 kg VS m^{-3} d^{-1} in continuously-stirred tank reactors (CSTR) at 35°C. Methane yields of 0.28-0.30 m^3 kg^{-1} VS added were achieved when the loading rate increased from 2 to 3 kg VS m^{-3}

d^{-1} at a feed ratio of 80:20. The results suggested that co-digestion of potatoes and/or its industrial by-products with manures on a farm-scale would generate renewable energy and provide a means of waste treatment for industry.

Parawira et al. (2004) studied anaerobic batch biodegradation of potato waste alone and co-digested with sugar beet leaves. The effects of increasing concentration of potato waste expressed as percentage of TS and the initial inoculum-to-substrate ratio (ISR) on methane yield and productivity were investigated. The ISRs studied were in the range of 9.0-0.25 and increasing proportions of potato waste from 10% to 80% of TS. A maximum methane yield of 0.32 L g^{-1} VS degraded was obtained at 40% of TS and an ISR of 1.5. A methane content of up to 84% (v/v) was obtained at this proportion of potato waste and ISR. Higher ISRs led to faster onset of biogas production and higher methane productivity. Furthermore, co-digestion of potato waste and sugar beet leaves in varying proportions was investigated at constant TS. Co-digestion improved the accumulated methane production and improved the methane yield by 31-62% compared to digestion of potato waste alone.

Carucci1 et al. (2005) assessed the anaerobic treatability of fresh vegetable waste, precooked food waste and their co-digestion with sludge in batch tests at different solids contents both on the single wastes and on appropriate mixtures of these wastes (also in order to simulate the seasonal factory production). Both fresh vegetable and precooked food wastes strongly inhibited methanogenesis from unacclimated inoculum at 10% solids content (undiluted waste) and 5% solids content (eight- to nine-fold diluted waste), respectively. Co-digestion of the fresh vegetable waste and sludge (60 and 40% on wet basis) was more effective both in terms of rate and methane yield with respect to the single wastes due to dilution and synergistic effects.

3.3.3 pH

The pH varies in response to biological conversions during different processes in AD. At low total alkalinity of waste, a stable pH indicates system equilibrium and digester stability. A falling pH can point toward acid accumulation and digester instability. In general, the acid-producing bacteria tolerate a low pH and have an optimal pH of 5.0 to 6.0. However, most methane-producing bacteria can only function optimally in a very narrow pH range of 6.7 to 7.4 (Elefsiniotis and Oldham, 1994).

In a conventional single anaerobic stage system, the pH is typically maintained within methanogenic limitations to prevent the predominance of acid-forming bacteria, which will cause acid accumulation and lower pH to below 5.0, a level lethal to methanogens. This potentially causes system failure. Many studies reported that the pH required in an anaerobic digester for good performance and stability is in the range of 6.5-7.5, although stable operation has been observed beyond this range. AD can operate over a wide range of VFA concentrations if proper control is maintained (Parkin and Owen, 1986; Anderson and Yang, 1992). On the other hand, prolific methanogenesis may result in a higher concentration of ammonia, increasing the pH to above 8.0, and will impede acidogenesis (Lusk, 1999). Under normal conditions, this pH reduction is buffered by the bicarbonate produced by methanogens. Under adverse environmental conditions, the buffering capacity within the system can be upset, eventually stopping the methane production. Therefore, excess alkalinity or buffering ability to control pH must be available to avoid acid accumulation.

Maintaining optimal pH is especially delicate in the start-up stage because organic wastes undergo acidogenesis before methanogenesis. To raise the pH during the early stages or when the pH drops during operation, a buffer must be added to keep bicarbonate alkalinity as high as possible in order to maintain pH high enough for methanogens to survive.

The common materials used as buffer to increase the alkalinity are lime, sodium hydroxide, ammonia, ammonium bicarbonate, soda ash and sodium bicarbonate. The first three are the least expensive (Parkin and Owen, 1986; Anderson and Yang, 1992). An advantage of adding these alkalis is that they induce swelling of particulate organics, making the cellular substances more susceptible to enzymatic attack (Vlyssides and Karlis, 2004). In some cases, automatic pH control is considered more economical than adding alkalis in a random manner because lesser chemicals are consumed.

3.3.4 Temperature

AD is strongly affected by temperature and can be classified based on temperatures: psychrophilic (0-20°C), mesophilic (20-42°C) and thermophilic (42-75°C). Anaerobic bacteria resist temperature changes. In the mesophilic range, the activity and growth rate of bacteria decrease by 50% for each 10°C drop, while the overall process kinetics doubled for every 10°C increase until it reaches optimal value. Despite high hydrolytic activity at higher temperatures, the activity of other groups of bacteria, such as acetate degrading bacteria, decrease when temperature increases to more than 60°C due to the increased VFA concentration which causes a rapid drop in microbial activity.

The optimum temperature for mesophilic digestion is 35°C at which bacteria are more robust and can tolerate greater changes in the environmental conditions. This stability makes the mesophilic AD more popular even though longer retention time is needed. Thermophilic digestion allows higher loading rates and achieves a higher rate of pathogen destruction as well as a higher degradation of the substrate and smaller digester size at lower capital cost (Mackie and Bryant, 1995). It is, however, more sensitive to toxins and small fluctuations in the environmental conditions and it take longer time to establish thermophilic population. Another disadvantage is more energy is required to heat the feedstock to reactor temperature which makes it less attractive from an energy point of view (Parkin and Owen, 1986; van Lier et al., 1996). Over the past 15 years, however, more and more biogas plants have been established and operated under thermophilic conditions since it offers a higher reaction rate and a more profitable process with a lower retention time.

When compared with the high-solids digestion of OFMSW (see also Chapter 4 on dry-digestion) and the co-digestion of OFMSW with sewage sludge, it was found that biogas production at thermophilic conditions, with a hydraulic retention time (HRT) of 12 days, was almost double that at mesophilic conditions with a HRT of 15 days. This surplus in biogas production was enough to compensate for the additional energy required to heat the digester. The change from mesophilic to thermophilic conditions was achieved over 2 month period without digester instability (Cecchi et al., 1992; Hansen et al., 2006). The effect of temperature on acidogenesis at an HRT of 30 hour was also tested at 22°C, 30°C and 35°C using mixed industrial wastewater and primary sludge as feedstock. The net VFA production went up by 15% when the temperature increased from 22°C to 30°C but decreased by 23% at 35°C. As the VS was fairly constant, it can be suggested that there was an optimum at around 30°C for this particular biomass and wastewater composition (Banerjee et al., 1998). However, other studies showed 45°C was the optimum temperature for pre-acidification of coffee waste with acetic acid and n-butyric acid production (Kozuchowska and Evison, 1995). Rapid temperature drops also affected starch degradation significantly at a temperature level between 15°C and 30°C (Cha and Noike, 1997).

3.3.5 Hydraulic Retention Time (HRT)

HRT (or SRT in case of CSTR) is determined by the average time it takes for organic material to digest and is used to measure the rate of substrate flow into and out of a reactor. HRT is in the range of 14-30 days for most dry (solids content is above 20%) anaerobic processes and can be as low as 3 days for wet (solids content is below 20%) anaerobic processes. The optimal values vary significantly with temperature, characteristics of solid waste and the specific technology used. Therefore, for a specific anaerobic digester, the HRT may change from day to day or from season to season.

A shorter HRT will reduce the size of the digester, leading to a higher production rate per volume unit, but a lower overall degradation. These two effects have to be traded in the design of a full-scale anaerobic digester. One method generally accepted for minimizing HRT is mixing in the digester. The other method is the recirculation of water and biogas in the digester to keep materials moving. This will ensure that microbial populations have rapid access to as many digestible surfaces as possible and that environmental characteristics are consistent throughout the digester (Lin et al., 1997; Vlyssides and Karlis, 2004).

A new study on reducing HRT focuses on the following areas: (1) separate the acidogenesis and methanogenesis into individual reactors so that the microbial population in each reactor can be independently optimized, i.e. two-phase AD; (2) alternate the mixing flow patterns to improve circulation within the digester; (3) introduce surfaces/support media (for attachment of microbes) or combination of surfaces to the reactor on which the anaerobic bacteria can grow, reducing microbial population wash-out and (4) employ pretreatment of the organic wastes to increase digestion efficiency (Marjoleine et al., 1998; Lissens et al., 2001; van Lier et al., 2001; Libanio et al., 2003; Mata-Alvarez et al., 2000; Nguyen et al., 2007).

3.3.6 Pretreatment of Food Waste

Various pretreatment methods are used to break down the complex structure, increase biodegradability and improve the performance of AD based on the waste characteristics and the effects they have on AD. There is an obvious link between successful pretreatment and improved biogas yield. By means of efficient pretreatment, the suspended solids may become more accessible to the anaerobic microbial consortium, thereby improving hydrolysis and liquefaction which are often the rate-limiting steps in AD (Shin et al., 2001).

The most promising pretreatment method is source separation which provides an immediate clean waste stream ready to digest. Other pretreatment methods can enhance biodegradability of a particular substrate mainly by improving the accessibility of the substrate to enzymes (Vavilin et al., 2002; Vavilin and Angelidaki, 2005). These pretreatment methods can be grouped as mechanical, biological and physico-chemical methods (Mata-Alvarez et al., 2000; van Lier et al., 2001) and are described as following:

Mechanical methods. The representative mechanical pretreatment methods are disintegration and grinding of solid particles which reduce particle size and solid content of the feedstock. As a result, the amount of soluble organics increases in the influent. Shredding, pulping or crushing of the waste allows bacteria to access a greater surface area and therefore reduces the retention time required for the treatment. Diluting the waste with water also allows the bacteria to move more freely inside the digester (Mata-Alvarez et al., 2000).

Mechanical reduction of particle size increases the specific surface area of organic waste, which results in speeding up of the rate of AD and biogas production (Sharma et al., 1988; Mata-Alvarez et al., 2000). However, when particle size is small enough, further shredding is not necessary as there is no significant hydrolysis enhancement; to obtain appreciable positive effects on biogas production from agricultural and forest residuals, they must be shredded to particles of sizes ranging from 0.088 to 0.40 mm (Sharma et al., 1988). Meanwhile, mechanical mixing can only be used at low solid content. When solid content is more than 30%, the blender is easily damaged and can be difficult to repair during operation.

Engelhart et al. (1999) studied the effects of mechanical disintegration by a high-pressure homogenizer on anaerobic biodegradability of sewage sludge. The degradation of soluble proteins and carbohydrates showed that a slowly degradable fraction of carbohydrates was released through disintegration. Meanwhile, a 25% increase in VS reduction was achieved. Hartmann et al. (1999) also found an increase of up to 25% in biogas from fibers in manure feedstock after pretreatment in a macerator before digestion.

In mechanical pretreatment methods, the energy necessary for the destruction of cell membranes is supplied in the form of pressure, translational or rotational energy. The cells are broken up, such that the substances contained within are set free without any chemical changes or denaturing. Mechanical treatment methods like maceration and ultrasonication have also shown promising results in overcoming the hydrolysis stage. During ultrasonication, extreme conditions are created by hydrodynamic shear forces that cause cell disruption or lysis. When cells are disrupted, intracellular matter is released, thus enhancing the digestion process.

Biological methods. Enzymatic and microbial pretreatments are commonly occurred in nature. Various extracellular enzymes generated by microbes facilitate breakdown of proteins, carbohydrates and lipids into simpler compounds. Enzymatic hydrolysis of cellulose with immobilized cellulase was reported by Kumakura (1997). Microbial solubilization and liquefaction of food waste at 35°C in an aerated reactor with supplied water and activated sludge added as inoculum was also proposed to reduce organic loading for further biological treatment (Gonzales et al., 2005).

Physical-chemical methods. Chemical pretreatments such as ozonation method have been effective in reducing COD and making the compounds more biodegradable. Another form of chemical treatment is the use of an alkaline agent such as NaOH at room temperature. Lin et al. (1997) reported an increase of 20% COD degradation an anaerobic digester with an HRT of 20 days using NaOH solubilization of waste activated sludge. The alkaline wet oxidation for pretreatment of wheat straw was also reported by Klinke et al. (2002). Acid hydrolysis of cellulose and hemicellulose under high temperature and pressurewas proposed as a promising pretreatment method by Mosier et al. (2005).

Thermal pretreatment was also used for the solubilization of food waste. It was reported that pyrolysis of cellulose at temperatures around 400°C can produce pyrolysate containing levoglucosan (Zhuang et al., 2001). The treatment of food wastes with temperatures from 160 to 200°C and pressure of 4 MPa for 60 minutes was recommended to transform organic solid materials of food waste to liquid hydrolyzate, which could be used for AD directly or after separation of the solid fraction (Tsukahara et al., 1999; Schieder et al., 2000). Use of this hydrolyzate increased the rates of the anaerobic process and biogas production in comparison to the digestion rate of crushed food waste. Biogas productions was 500 and 470 L kg^{-1} with added COD, and 850 and 780 L kg^{-1} with dry organic matter for pretreated and raw food waste respectively (Schieder et al., 2000). Wang et al. (2006) also investigated the effect of thermal pretreatment on the performance of the HASL system. It was found that preheating of food waste at 150°C for 60 minutes and at 70°C for 120 minutes enhanced AD of food waste in terms of higher and faster SCOD, VFA and methane production, and diminished the time to produce the same quantity of methane by 48% and 25% respectively, compared to AD of fresh food waste without any pretreatment.

Recently, freeze/thaw pretreatment has also been used in food waste digestion based on the food cell wall disruption mechanism in the process of freezing and thawing. It was reported the total methane production increased by 8.4 % and the time needed to produce the same quantity of methane was reduced by 42%. Meanwhile, freeze/thaw pretreatment of food waste would improve the quality of residue with less toxicity and higher stability (Stabnikova et al., 2008).

All these methods result in particle size reduction and enhance solubilization of solid waste because biopolymers are split up into shorter chains. Therefore, pretreatment of food waste is essential to enhance digestibility and bioenergy production. However, biological solubilization of food waste requires a special aerated reactor equipped with a mixing system (Gonzales et al., 2005). Both chemical and biological methods require a long time to reach steady-state conditions. Amongst the pretreatment methods, heating of food waste appears to be the simplest pretreatment method (Schieder et al., 2000). The main limitation for thermal pretreatment of food waste is the energy cost. To cover these expenses, biogas produced should compensate the cost of thermal pretreatment.

3.3.7 Process Configurations

Wet or dry process. Depending on the moisture content or TS concentration of the feedstock, AD is termed as wet process with TS of 10-15%, dry process with TS of 25-40%, and semi-dry process with TS of 16-22%. Since food waste contains a high moisture content of about 80-90%, the wet AD process is widely used to treat food waste. The wet AD of food waste can be performed in conventional reactor systems where process homogeneity is obtained by mechanical mixing or a combination of mechanical mixing and biogas injections. In order to lower the TS concentration, addition of liquid is necessary, by recirculation of the liquid effluent fraction (Hartaman and Ahring, 2006).

Batch or continuous process. In batch process, bioreactor is fed with food waste and is sealed to allow degradation. When the process is completed, the reactor is opened and the content is discharged. There are distinct stages of digestion throughout the batch process. When food waste is first loaded, hydrolysis takes place and gas production is low, forming primarily carbon dioxide. Methane production starts during the acid forming stages, reaching a maximum halfway through the degradation period, when methanogenesis dominates the processes. Toward the end of the degradation period, when the digestible material ceases, the biogas production drops (Lissens et al., 2001; Bolzonella et al., 2006). In a continuous AD process, fresh organic substrate is added and an equal amount of digested material is removed perpetually. With consistent feedstock input, all reactions occur at a fairly steady rate and equilibrium is achieved, resulting in approximately constant biogas production (Lissesn et al., 2001; Bolzonella et al., 2006). Because of the constant movement, the substrate inside the digester is mixed and does not become stratified. This allows more optimal use of the tank volume. The disadvantage of the continuous process is that the removed reactor content is a combination of completely digested and partially digested material.

The semi-continuous process falls between the batch and continuous processes, in which food waste is added at a certain internal and the digested content is discharged periodically. Wang et al. (2003) operated a lab-scale HASL system on a semi-continuous mode for 36 days. This study showed (1) the effective acidification in a semi-liquid recycle reactor and the high-efficiency of methanogenesis in a UASB reactor; (2) during the first 23 days of operation, the leachate from Ra with a total VFA (TVFA) of 9,500-11,500 mg L^{-1} and a COD of 8,000-11,800 mg L^{-1} was efficiently treated in the UASB reactor, with about 88.5% of TVFA removal and 85.7% of COD removal; (3) 99% of the total methane was generated from the Rm with average methane content of 71% and (4) at the end of operation, a 78% VS removal rate and a methane yield of 0.29 L g^{-1} VS added were achieved.

Single stage or multi-stage process. In a single-stage process digestion occurs in one reactor, while the multi-stage process consists of several reactors in series. Two-phase AD is the most common multi-stage process, in which the acid-forming stage (acidogenesis) is separated from the methane-forming stage (methanogenesis), resulting in increased efficiency as the acidogenic and methanogenic microorganisms are separated in terms of nutrient needs, growth capacity and ability to cope with environmental conditions (Babel et al., 2004). Some multi-stage systems also use a preliminary aerobic stage to raise the temperature and increase the degradation of the organic material (Raynal et al., 1998). In other systems the digesters are separated into a mesophilic stage and a thermophilic stage as the temperature-phase anaerobic digestion (TPAD).

3.4 Case Study of Biomethane Production

Methane, the main end product in AD, has been promoted as a part of the solution to energy problems. Besides waste minimization and stabilization, the main purpose of food waste digestion is to produce methane gas.

3.4.1 A Demonstration HASL System Plant for Food Waste Bioconversion in Singapore

Food wastes in Singapore amounted to approximately 542,700 tonnes in 2006. Almost all food wastes incinerated. As waste collection is highly regulated in Singapore, it is manageable to collect food waste separately for further bioconversion. Anaerobic digestion appears to be the most promising method for food waste management. It reduces the volume of food wastes, generates bioenergy in the form of methane (55 to 75% (v/v)), and produces nutrient-rich organic residues that can be used as soil conditioner or fertilizer.

A pilot plant capable of treating 3 tonnes of food waste per day was built and operated on the campus of Nanyang Technological University (NTU) from November 2005 to February 2008 (photo in Figure 3.5 and schematic diagram in Figure 3.6). The HASL pilot-scale study was a collaboration project among governmental agencies, industrial partner and NTU. The main goal of the project was to test and gain experience from building and operating a home grown technology, HASL food waste digestion system, developed by NTU. The ultimate goal is to build an industrial-scale HASL system for commercialization to help minimize the amount of food waste for disposal.

Figure 3.5 Pilot-scale HASL plant in NTU for food waste bioconversion (Adapted from Wang et al., 2008).

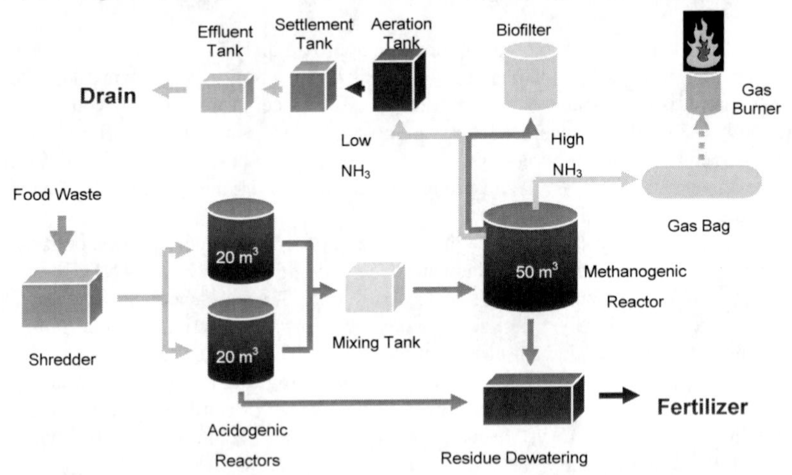

Figure 3.6 Schematic diagram of the pilot-scale HASL plant (Adapted from Wang et al., 2008).

The food waste was collected from a local shopping mall by a licensed waste collector. After source separation of foreign substances like paper, plastic and metallic pieces, the food waste was transported in 120-litre plastic containers to the HASL pilot plant site. The containers were lifted, tilted into a hopper, and then transferred to a shredder (AZ05F, MOCO, Germany) via a screw conveyor system. The shredded food waste with particles size less than 10 mm along with an equal amount of water was mixed in a storage tank and then transferred to two acidogenic reactors. The seeding of the acidogenic reactor was carried out by adding 4 m^3 of anaerobically digested sewage sludge collected from a local wastewater treatment plant with 1% total solids (TS) and VS/TS ratio of 0.7. The acidogenic reactor was a 20 m^3 cylindrical tank with a vertical side leachate chamber separated by a perforated steel plate. This design permitted separation of digested solids and leachate inside the tank. The acidogenic fermentation provided hydrolyzed and acidified leachate with pH of 4.2-5.2 and high COD of 15-40 g L^{-1}. Leachate was then diluted using the effluent from the methanogenic reactor inside a mixing tank and used for feeding the methanogenic reactor. The methanogenic reactor with an effective volume of 50 m^3 was designed as a modified UASB reactor with a gas hopper installed at the top of the tank to trap biogas and transfer it to a gas bag. The operation of the methanogenic reactor was started-up by seeding 25 m^3 of anaerobically digested sewage sludge. The biogas produced in the methanogenic reactor was temporarily stored in a gas bag and subsequently flared. The remaining part of the effluent from the methanogenic reactor, not used for dilution of leachate, was then treated sequentially in the aeration and sedimentation tanks. The stream of the effluent used for the dilution of leachate was monitored for the presence of ammonia.

Operational parameters such as pH and flow rates were monitored at each process step with the aid of control panels. It also allowed control over pumps, valves, mixers, blowers and other mechanical equipment in the plant. The ambient temperature varied between 23 and 25°C at night and rainy days and was in the range from 31 to 33°C during sunny days. Leachate from the acidogenic reactors and effluent from the methanogenic reactor were collected at least once in three days for analyses.

The food waste collected had high moisture content in the range of 75-85% (82.8%), and it contained high concentrations of organic material with a volatile solid content in the range of 85-90% (88.1%) of the total solids. This indicated that the food waste supplied was a favourable feedstock for the pilot plant as volatile solids content is an estimation of the amount of available food for various microorganisms in anaerobic digestion.

An approximate chemical formula representing the organic part of the raw food waste can therefore be expressed as $C_{23}H_4O_{16}N_4$ S based on the elemental analysis. Due to high meat residues, C/N ratio in the mixed food waste was much lower than the desirable C/N ratio of feedstock for anaerobic digestion, which ranges from 20:1 to 30:1 (Yadvika et al., 2004). So, adjustment of C/N ratio of the food waste should be done before anaerobic treatment. It can be adjusted by adding vegetable and fruit wastes to achieve higher C/N ratio. The high content of protein in food waste could cause high ammonia concentration, which might negatively affect anaerobic digestion of food waste (Lay et al., 1997). Although a two-phase system is more tolerant of the inhibiting effect of ammonia than a single-stage system (Lissens et al., 2001), to ensure stable operation of the pilot HASL system, the recycled effluent from the methanogenic reactor was treated for ammonia removal in the submerged biofilter in case the ammonia concentration was higher than 1000 mg

NH_4^+-N L^{-1}. Process parameters of the pilot HASL system batch operation are shown in Table 3.2.

The concentration of soluble COD (SCOD) in leachate from the acidogenic reactor increased from 15.1 g L^{-1} to 42.6 g L^{-1} in a period of 20 days. VS removal was 60% during 20 days of food waste anaerobic digestion. A similar VS removal rate of 63%, was observed for anaerobic treatment of food waste with high content of lipids, 30% (w/w of TS in food waste), in a lab-scale HASL system (Stabnikova et al., 2005). The food waste treated in the HASL pilot plant contained a lot of meat residuals, which resulted in relatively low VS removal. So, adding vegetable/fruit food waste and/or other organic wastes with high carbon content and low lipids content are recommended.

Table 3.2 Performance of HASL pilot plant.

Process parameters	
OLR into the acidogenic reactor, kg VS m^{-3} d^{-1}	3.3
OLR into the methanogenic reactor, g SCOD m^{-3} d^{-1}	70-1200
SCOD in leachate from the acidogenic reactor, g L^{-1}	15.0-42.6
SCOD in effluent from the methanogenic reactor, g L^{-1}	0.12-0.44
VFA in leachate form the acidogenic reactor, g L^{-1}	33.3
Average SCOD removal in the methanogenic reactor, %	95.5
Average methane production, L g^{-1} SCOD removed	0.594
VS removal, %	60.1
Average methane content in biogas, %	74.0

Operation of the HASL pilot plant provided a good learning experience to understand process capabilities, inadequacies and challenges to be taken care of towards commercialization of the HASL system. During commissioning, some unexpected mechanical problems were observed in the acidogenic reactor with respect to solid-liquid separation. After a series of mechanical modification, much better performance of the acidogenic reactor for solid-liquid separation was achieved. Besides solid-liquid separation, pump and pipe clogging was also experienced in different parts of the plant, especially for the acidogenic reactor.

3.4.2 A Full-scale Plant Using the BTA Process for Co-digestion of OFMSW and Sewage Sludge or Manure in Germany

The BTA (Biotechnische Abfallverwertung) process was developed to transform the OFMSW from households, commercial, and agricultural waste into biogas and compost. The full-scale plant was installed in Germany for the co-digestion of the OFMSW and sewage sludge or manure. The system consists of three major processes: mechanical wet pre-treatment in a pulper for size reduction, anaerobic hydrolization, and biomethanation. The full scale installations are illustrated in Figure 3.7.

The whole treatment scheme for BTA application is shown in Figure 3.8. After passing over a scale, the delivered waste is unloaded into a flat bunker in a receiving hall. It is then fed by a front loader into two screw mills that coarsely chop the organic material, which is fed into two dissolution tanks (pulpers). The core element of the BTA process is the hydro-pulper where the preshredded feedstock is diluted to 8-10% TS (maximum 12% TS) and chopped. Contaminants such as plastics, textiles, stones, and metals are separated by gravity. Sand and stones sink and can be later removed from the bottom; plastic materials tend to float to the surface and are removed by a rake. An essential component of the process is the grit removal system, which separates the residual fine matter such as sand, small stones, and glass splinters by passing the pulp through a hydrocyclone that is designed to withstand the abrasion these materials can cause. The mechanical treatment is followed by a heating step for hydrolysis enhancement step (30 minutes at 70°C) before the pulp is processed by the biological degradation step.

Figure 3.7 BTA wet two-stage system at Kirchstockach, Germany (Adapted from Zaher et al., 2007).

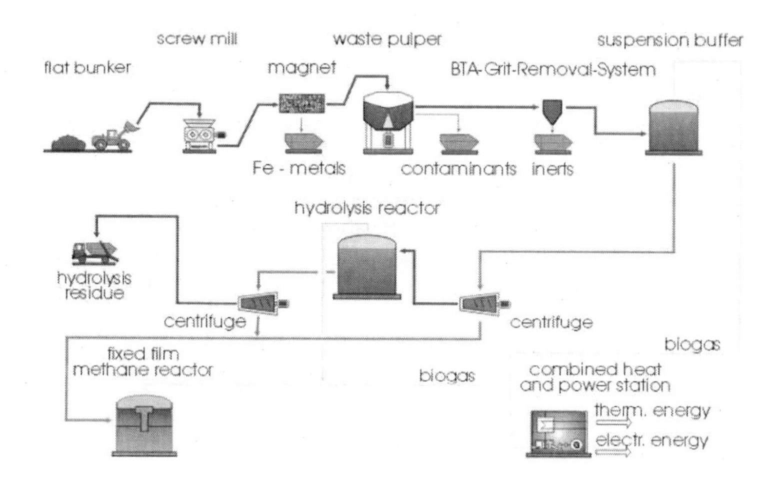

Figure 3.8 The treatment scheme for the BTA system (Adapted from Zaher et al., 2007).

The biological degradation step is divided into a hydrolysis step and a biomethanization step that occurs in a fixed film reactor. Before the hydrolysis step, the suspended materials are dewatered and separated into liquid and solid fractions. The liquid contains a high volume of previously dissolved organics, and is pumped directly into the AD reactor. The dewatered solids are re-mixed with process water and fed into the hydrolysis reactor to dissolve the remaining organic solids. After 2-4 days the hydrolyzed suspension is dewatered and the hydrolysis-liquid is also fed into the AD reactor. The fiber that remains after hydrolysis is a high quality material: it is free of pathogens with a low-salt concentration. Post-digestion composting is generally not needed. The liquid fraction is treated by a cleaning system that consists of sedimentation steps and a biological nitrification/denitrification step to remove some of the nutrients. Most of the cleaned liquid is reused as process water by the pulpers for the treatment of further waste. A small amount of the liquid is discharged as mechanical-biological pre-cleaned surplus water and is fed into the public sewer for final handling by a municipal waste water treatment plant.

3.5 Future Research Direction

AD has been successfully operating in many parts of the world. Some modern plants have been developed with few technical barriers. The future research in AD, therefore should concentrate on improving the efficiency and economics of a facility. Research and development are needed in both process engineering and process microbiology for efficient conversion of food waste to bioenergy.

The first subgroup includes strategies of proper waste management, such as source separation, waste composition, storage, handling and collection. All of these should take into account issues such as odor, sanitation, pest control and economics. The waste stream should also be analyzed for the following characteristics: moisture content, salinity, density, calorific value, chemical composition including C/N ratio and VS content, which determine whether AD is a viable technology. Research should be expanded to optimize the biogas production with a variable waste stream.

An important area of research involves system design, development and modification that will greatly improve digestibility, at short retention time and thereby reducing the footprint of the digester. Waste preprocessing including mechanical, thermal, chemical and thermochemical processes, demand further research for better hydrolysis. Another area for future research is better understanding of process microbiology for efficient digestion. Identifying, isolating and enriching microbes that can efficiently catalyze biological reaction at shorter duration for various food wastes.

For practical and policy reasons, the following areas of research can also be investigated: more experience in scale-up from lab-scale to pilot-scale and full-scale; co-digestion of MSW with sewage waste, agriculture waste, dairy manure etc.; separation of the wet and dry stream of MSW to improve resource recovery; government incentives in the form of tax credit for bioenergy production.

Considering the energy and environmental benefits, it is apparent that the future for AD technology is very promising in both developing and developed nations.

3.6 Summary

AD has the advantage to be an integral part of the solution to two of the most pressing environmental concerns of urban centers: waste management and renewable energy. Through AD, organics are decomposed and biogas and a stable solid can be produced. Each of these products can be used for beneficial purposes to close the loop in organic waste management. The biogas, which consists of up to 65% methane, can be combusted in a cogeneration unit and produce green energy. The solid digestate can be used as an organic soil amendment. As a waste management strategy employed in over 20 countries, AD has been successful in reducing the volume of waste going to landfill, decreasing emissions of greenhouse gases and creating organic fertilizer, all at a profit. Future research is needed that focuses on both process engineering and microbiology of anaerobic digestion especially for food wastes.

3.7 References

Anderson, G. K., and Yang, G. (1992). "pH control in anaerobic treatment of industrial wastewater." *J. Environ. Eng.*, 118(4), 551-567.

Babel, S., fukushi, K., and Sitanarassamee, B. (2004). "Effect of acid speciation on solid waste liquefaction in an anaerobic acid digester."*Water Res.*, 38(9), 2417-2423.

Banerjee, A., Elefsiniotis, P., and Tuhtar, D. (1998). "Effect of HRT and temperature on the acidogenesis of municipal primary sludge and industrial wastewater." *Water Sci. Technol.*, 38(8-9), 417-423.

Bolzonella, D., Paven, P., Mace, S., and Cecchi, F. (2006). "Dry anaerobic digestion of differently sorted organic municipal solid waste: a full-scale experience." *Water Sci. Technol.*, 53 (8), 23-32.

Bouallagui, H., Cheikh, R.B., Marouani, L., and Hamdi, M. (2003). "Mesophilic biogas production from fruit and vegetable waste in a tubular digester." *Bioresour. Technol.*, 86(1), 85-89.

Bouallagui, H., Touhami, Y., Cheikh, R.B., and Hamdi, M. (2005) "Bioreactor performance in anaerobic digestion of fruit and vegetable wastes." *Process Biochem.*, 40(3-4), 989-995.

Carucci1, G., Carrasco, F., Trifoni, K., Majone, M., and Beccari, M. (2005). "Anaerobic Digestion of Food Industry Wastes: Effect of Codigestion on Methane Yield." *J. Envir. Engrg.*, 131(7), 1037-1045.

Cecchi, F., Mata-Alvarez, J., Pavan, P., Sans, C., and Merli, C. (1992). "Semidry anaerobic-digestion of MSW - Influence of process parameters on the substrate utilization model." *Water Sci. Technol.*, 25(7), 83-92.

Cha, G.C., and Noike, T. (1997) "Effect of rapid temperature change and HRT on anaerobic acidogenesis." *Water Sci. Technol.*, 36(6), 247-253.

Cho, J.K., and Park, S.C. (1995). "Biochemical methane potential and solid state anaerobic digestion of Korean food wastes." *Bioresour. Technol.*, 52 (3), 245-253.

Cirne, D.G., Paloumet, X., BjÖrnsson, L., Alvesb, M.M., and Mattiassona, B. (2007). "Anaerobic digestion of lipid-rich waste-Effects of lipid concentration." *Renew. Energ.*, 32(6), 965-975.

Edelmann, W., Schleiss, K., and Joss, A. (2000). "Ecological, energetic and economic comparison of anaerobic digestion with different competing technologies to treat biogenic wastes." *Water Sci. Technol.*, 41(3), 263-273.

Elefsiniotis, P., and Oldham, W.K. (1994) "Influence of pH on the Acid-Phase Anaerobic-Digestion of Primary Sludge." *J. Chem. Technol. Biotechnol.*, 60(1), 89-96.

Engelhart, M., Kruger, M., Kopp, J., and Dicht, N. (1999). "Effects of disintegration on anaerobic degradation of sewage excess sludge in downflow stationary fixed film digesters." In: Mata-Alvarez, J., A. Tilche, and F. Cecchi, F. (Eds.), *Proceedings of the Second International Symposium on Anaerobic Digestion of Solid Wastes, Barcelona*, Vol. 1. Grafiques 92, 15-18, June, pp. 153-160.

Forster, C.F. (1992). "Oils, fats and greases in wastewater treatment." *J. Chem. Technol. Biotechnol.*, 55(4), 402-404.

Gonzales, H.B., Takyu, K., Sakashita, H., Nakano, Y., Nishijima, W., and Okada, M. (2005). "Biological solubilization and mineralization as novel approach for the pretreatment of food waste." *Chemosphere*, 58(1), 57-63.

Han, S. K., and Shin, H. S. (2004). "Performance of an innovative two-stage process converting food waste to hydrogen and methane." *J. Air Waste Manage. Assoc.*, 54(2), 242-249.

Han, Y., Sung, S., Dague, R.R. (1997a). "Temperature-phased anaerobic digestion of wastewater sludges. " *Water Sci. Technol.*, 36(6-7), 367-374.

Han. Y., Dague, R.R. (1997b). "Laboratory studies on the temperature-phased anaerobic digestion of domestic wastewater sludge." *Water Environ. Res.*, 69(6), 1139-1143.

Hansen, C.L., and Cheong, D.Y. (2007). Agricultural waste management in food processing. In *Food Machinery Hand Book*. William Andrew Publishing, Norwich, N.Y., pp. 609-661.

Hansen, T.L., Sommer, S.G., Soren, G., and Christensen, T.H. (2006). "Methane production during storage of anaerobically digested municipal organic waste." *J. Environ. Qual.*, 35 (3), 830-836.

Hartmann, H., and Ahring, B.K. (2005). "Anaerobic digestion of the organic fraction of municipal solid waste: influence of co-digestion with manure." *Water Res.*, 39(8), 1543-1552.

Hartmann, H., and Ahring, B.K. (2006). "Strategies for the anaerobic digestion of the organic fraction of municipal solid waste: an overview." *Wat. Sci. Technol.*, 53 (8), 7-22.

Hartmann, H., Angelidaki, I., and Ahring, B.K. (1999). "Increase of anaerobic degradation of particulate organic matter in full-scale biogas plants by mechanical maceration." In: Mata-Alvarez, J., A. Tilche, A., and F.Cecchi (Eds.), *Proceedings of the Second International Symposium on Anaerobic Digestion of Solid Wastes, Barcelona*, Vol. 1. Grafiques 92, 15-18 June, pp. 129-136.

Heo, N.H., Park, S.C., and Kang, H. (2004). "Effects of mixture ratio and hydraulic retention time on single-stage anaerobic co-digestion of food waste and waste activated sludge." *J. Environ. Sci. Health.*, A39 (7), 1739-1756.

Kaparaju, P., and Rintala, J. (2005). "Anaerobic co-digestion of potato tuber and its industrial by-products with pig manure." *Resour. Conservat. Recycl.*, 43(2), 175–188.

Kayhanian, M., and Rich, D. (1995). "Pilot-scale high-solids thermpohilic anaerobic digestion of municipal of municipal solid waste with an emphsis on nutrient requirements." *Biomass Bioenerg.*, 8 (6), 123-136.

Khanal, S.K. (2008). *Anaerobic Biotechnology for Bioenergy Production: Principles and Applications*. Wiley-Blackwell Publishing, Ames, IA, USA.

Kim, S.H., Han, S.K., and Shin, H.S. (2004). "Feasibility of biohydrogen production by anaerobic co-digestion of food waste and sewage sludge." *Int. J. Hydrogen Energy*, 29(15), 1607-1616.

Klinke, H.B., Ahring B.K., Schmidt, A.S., and Thomsen, A.B. (2002). "Characterization of degradation products from alkaline wet oxidation of wheat straw." *Bioresour. Technol.*, 82(1), 15-26.

Kozuchowska, J., and Evison, L.M. (1995) "VFA production in pre-acidification systems without pH Control." *Environ. Technol.*, 16(7), 667-675.

Kujawa-Roeleveld, K., Elmitwalli, T., Gaillard, A., van Leeuwen, M., and Zeeman, G. (2002). "Codigestion of concentrated black water and kitchen refuse in an accumulated system-within a DESAR concept." In *Proc. 3rd Int. Symp. on Anaerobic Digestion of Solid Wastes*, Munich, Germany, September 18-20.

Kumakura, M. (1997) "Preparation of immobilized cellulase beads and their application to hydrolysis of cellulosic materials." *Process Biochem.*, 32(7), 555-559.

Kwon, S.H., and Lee, D.H. (2004). "Evaluation of Korean food waste composting with fed-batch operations I: using water extractable total organic carbon content (TOCw)." *Process Biochem.*, 39(1), 1183-1194.

Lay, J. J., Li, Y. Y., Noike, T., Endo, J., and Ishimoto, S. (1997). "Analysis of environmental factors affecting methane production from high-solids organic waste." *Water Sci. Tech.*, 36(6), 493-500.

Libanio, P.A.C., Costa, B.M.P., Cintra, I.S., and Chernicharo, C.A.L. (2003). "Evaluation of the start-up of an integrated municipal; solid waste and leachate treatment system." *Water Sci. Technol.*, 48 (6), 241-247.

Lin, J.G., Chang, C.N. and Chang, S.C. (1997) "Enhancement of anaerobic digestion of waste activated sludge by alkaline solubilization." *Bioresour. Technol.*, 62(3), 85-90.

Lissens, G., Vandevivere, P., De. Baere, L., Biey, E.M., and Verstraete, W. (2001). "Solid waste digestors: process performance and practice for municipal solid waste digestion." *Water Sci. Technol.*, 44 (8), 91-102.

Lusk, P. (1999). Latest progress in anaerobic digestion. *Biocycle*, 40 (7), 52-56.

Mackie, R.I., and Bryant, M.P. (1981) "Metabolic activity of acetate, propionate, butyrate and CO2 to Methanogenesis in cattle waste at 40 and 60°C." *Appl. Environ. Microbiol.*, 41(6), 1363-1373.

Mackie, R. I., and Bryant, M. P. (1995). "Anaerobic digestion of cattle waste at mesophilic and thermophilic temperatures." *Appl. Microbiol. Biotechnol.*, 43(2), 346-350.

Marjoleine, P., Weemaes, J., and Verstraete, W.H. (1998). "Evaluation of current wet sludge disintegration techniques." *J. Chem. Technol. Biotechnol.*, 73(2), 83-92.

Mata-Alvarez, J., and Cecchi, F. (1990) "A review of kinetic models applied to the anaerobic bio-degradation of complex organic matter. Kinetics of the biometahnization of organic fractions of municipal solid waste." In: Kamely, D., Chackrobordy, A., Ommen, G.S. (Eds) *Biotechnology and Biodegradation*, Gulf Publishing Company, Houston, Texas, pp 27-54.

Mata-Alvarez, J., Mace, S., and Llabres, P. (2000). "Anaerobic digestion of organic solid wastes. An overview of research achievements and perspectives." *Bioresour. Technol.*, 74(1), 3-16.

Mosier, N., Wyman, C., Dale, B., Elander, R., Lee, Y.Y., Holtzapple, M., and Ladisch, M. (2005). "Features of promising technologies for pretreatment of lignocellulosic biomass." *Bioresour. Technol.*, 96(6), 673-686.

Nallathambi Gunaseelan, V. (1997). "Anaerobic digestion of biomass for methane production: a review." *Biomass Bioenerg.*, 13(1), 83-114.

Nguyen, P.H.L., Kuruparan, P., and Visvanathan, C. (2007). "Anaerobic digestion of municipal solid waste as a treatment prior to landfill." *Bioresour. Technol.*, 98(2), 380-387.

Nordberg, A., and Edstrom, M. (1997). "Co-digestion of ley crop silage, source sorted municipal solid waste, and municipal sewage sludge." In: *Proceedings of 5th FAO/SREN Workshop, Anaerobic Conversion for Environmental Protection, Sanitation and Re-Use of Residuals*". March 24-27, Gent, Germany.

Parawira, W., Murto, M., Zvauya, R., and Mattiasson, B. (2004). "Anaerobic batch digestion of solid potato waste alone and in combination with sugar beet leaves." *Renew. Energ.*, 29(11), 1811-1823.

Parkin, G.F., and Owen, W.F. (1986). "Fundamentals of anaerobic digestion of wastewater sludges." *J. Environ. Eng.*, 112 (5), 867–920.

Pohland, F.G. and Ghosh, S. (1971) "Developments in anaerobic stabilization of organic wastes-the two-phase concept." *Environ. Lett.*, 1(4), 255-266.

Rao, M.S., and Singh, S.P. (2004). "Bioenergy conversion studies of organic fraction of MSW: kinetic studies and gas yield-organic loading relationships for process optimization." *Bioresour. Technol.*, 95(2), 173-185.

Raynal, J., Delgenes, J.P., and Moletta, R. (1998). "Two-phase anaerobic digestion of solid wastes by a multiple liquefaction reactors process." *Bioresour. Technol.*, 65(1-2), 97-103.

Schieder, D., Schneider, R., and Bischof, F. (2000). "Thermal hydrolysis (TDH) as a pretreatment method for the digestion of organic waste." *Water Sci. Technol.*, 41(3), 181-187.

Sharma, S.K., Mishra, I.M., Sharma, M.P., and Saini, J.S. (1988) "Effect of particle size on biogas generation from biomass residues." *Biomass*, 17(4), 251-263.

Shin, H.S., Han, S.K., Song, Y.C., and Lee, C.Y. (2001). "Multi-step sequential batch two-phase anaerobic composting of food waste." *Environ. Technol.*, 22(33), 271-279.

Shin, H.S., Youn, J.H., and Kim, S.H. (2004). "Hydrogen production from food waste in anaerobic mesophilic and thermophilic acidogenesis." *Int. J. Hydrogen Energ.*, 29(13), 355-1363.

Stabnikova, O., Ang, S.S., Liu, X.Y., Ivanov, V., Tay, J.H., and Wang, J.Y. (2005). "The use of hybrid anaerobic solid-liquid (HASL) system for the treatment of lipid-containing food waste." *J. Chem. Technol. Biotechnol.*, 80(4), 455-461.

Stabnikova, O., Liu, X.Y., and Wang, J.Y. (2008). "Digestion of frozen/thawed food waste in the hybrid anaerobic solid-liquid system." *Waste Manage.*, 28 (9), 1654-1659.

Steffen, R., Szolar, O., and Braun, R. (1998). *Feedstocks for Anaerobic Digestion.* Institute of Agrobiotechnology Tulin, University of Agricultural Sciences, Vienna.

Sung, S., and Santha, H. (2003). "Performance of temperature-phased anaerobic digestion (TPAD) system treating dairy cattle wastes." *Water Res.*, 37 (7), 1628-1636.

Tsukahara, K., Yagishita, T., Ogi, T., and Sawayama, S. (1999) "Treatment of liquid fraction separated from liquidized food waste in an upflow anaerobic sludge blanket reactor." *J. Biosci. Bioeng.*, 87 (4), 554-556.

van Lier, J.B., Martin, J.L.S., and Lettinga, G. (1996). "Effect of temperature on the anaerobic thermophilic conversion of volatile fatty acids by dispersed and granular sludge." *Water Res.*, 30(1), 199-207.

van Lier, J.B., Tilche, A., Ahring, B.K., Macrie, H., Moletta, R., Dohanyos, M., Hushoff Pol, L.W., Lens, P., and Verstraete, W. (2001). "New perspectives in anaerobic digestion." *Water Sci. Technol.*, 43 (1), 1-18.

Vavilin, V. A., and I. Angelidaki. (2005). "Anaerobic degradation of solid material: importance of initiation centers for methanogenesis, mixing intensity, and 2D distributed model. " *Biotechnol. Bioeng.*, 89 (1): 113-122.

Vavilin, V.A., Schelkanov, M.Y., Lokshina, L.Y., Rytov, S.V., Jokela, J., and Salminen, E. (2002). "A comparative analysis of a balance between the rates of polymer hydrolysis and acetoclastic methanogenesis during anaerobic digestion of solid waste." *Water Sci. Technol.*, 45 (10), 249-254.

Vlyssides, A.G., and Karlis, P.K. (2004). "Thermal-alkaline solubilization of waste activated sludge as a pretreatment stage for anaerobic digestion." *Bioresour. Technol.*, 91(2), 201-206.

Wang J.Y., Liu X.Y., Ding H.B., Sreeramachandran S. and Jiao Y. (2008). “Demonstration plant using the hybrid anaerobic solid-liquid system for waste bioconversion.” In *the Proceedings of ISWA/WMRAS World Congress*, 2008, Singapore.

Wang, J.Y., Liu, X.Y., Kao, J.C.M., and Stabnikova, O. (2006). "Digestion of pretreated food waste in hybrid anaerobic solid-liquid (HASL) system." *J. Chem. Technol. Biotechnol.*, 81(3), 345-351.

Wang, J.Y., Xu, H.L., and Tay, J.H. (2002) "A hybrid two-phase system for anaerobic digestion of food waste." *Water Sci. Technol.*, 45(12), 159-165.

Wang, J.Y., Xu, H.L., Zhang, H., and Tay, J.H. (2003). "Semi-continuous anaerobic digestion of food waste using a hybrid anaerobic solid-liquid bioreactor." *Water Sci. Technol.*, 48(4), 169-174.

Wang, J.Y., Stabnikova, O., Tay, S.T.L., Ivanov, V., and Tay, J.H. (2004). "Biotechnology of intensive aerobic conversion of sewage sludge and food waste into fertilizer." *Water Sci. Technol.*, 49(10), 147-154.

Xu, H.L., Wang, J.Y., and Tay, J.H. (2002) "A hybrid anaerobic solid-liquid bioreactor for food waste digestion." *Biotechnol. Lett.*, 24(10), 757-761.

Yadvika, Santosh, Sreekrishnan, T. R., Kohli, S., and Rana, V. (2004). "Enhancement of biogas production from solid substrates using different techniques--a review." *Bioresour. Technol.*, 95(1), 1-10.

Zaher U., Cheong D.Y., Wu B.X., and Chen S.L. (2007). "Producing Energy and Fertilizer from Organic Municipal Solid Waste." Washington State University, Project Deliverable # 1. June 26, 48-51.

Zhang, R.H., El-Mashad, H.M., Hartman, K., Wang, F.Y., Liu, G.Q., Choate, C., and Gamble, P. (2007). "Characterization of food waste as feedstock for anaerobic digestion." *Bioresour. Technol.*, 98(4), 929-935.

Zhuang, X.L., Zhang, H.X., Yang, J.Z. and Qi, H.Y. (2001). "Preparation of levoglucosan by pyrolysis of cellulose and its citric acid fermentation." *Bioresour. Technol.*, 79(1), 63-66.

CHAPTER 4

Bioenergy Production from Organic Fraction of Municipal Solid Waste (OFMSW) through Dry Anaerobic Digestion

C.Visvanathan

4.1 Introduction

Disposal of municipal solid wastes (MSW) is a major solicitude in most urban centers of developing and developed countries, and it has reached an alarming stage. According to Tanaka et al. (MOEJ, 2006), the total amount of waste generated was 12.7 billion tonnes across the world in 2000 and is expected to grow approximately 19 billion tonnes in 2025. Hence, the cost effective disposal of MSW has become a major challenge. One approach of handling the MSW is to promote the concept of source separation, where the organic fraction can be effectively converted to biomethane through anaerobic digestion.

There are two different methods of biological processing of the organic fraction of municipal solid waste (OFMSW), namely aerobic composting and anaerobic digestion, where the natural degradation of biodegradable materials takes place. Composting refers to a microbial degradation process that consumes oxygen and produces humus-rich material. On the other hand, anaerobic digestion is a microbial degradation process that occurs in the absence of oxygen. The final product of anaerobic digestion is a methane-rich biogas, which is a renewable energy source, and the digested residues.

Anaerobic digestion can be classified as wet and dry processes based on the water content. The wet process is similar to the conventional anaerobic sludge digestion. The basic components of a dry anaerobic digestion system include substrate handling, digestion, and biogas utilization. Generally, the substrate from the storage facility is transported to the digester, by a pump or through gravity flow for digestion. During anaerobic digestion, biogas is produced as a result of the bacterial decomposition of organic matter. The produced biogas is collected and supplied to a generator where it undergoes combustion to generate electricity. The digestion process also decreases the amount of organic matter and as a result, reduces the volume of waste to be disposed off. The residual digested material, known as digestate, can be directly applied to the land as a soil conditioner or pressed to increase the solid contents for use as compost.

The biomethanization of organic wastes is accomplished by a series of biochemical transformations which can be separated into four major steps. The first step consists of the transformation of complex compounds into simple soluble substrates such as amino acids, simple sugars and fatty acids. In the second step, these soluble compounds are converted into volatile fatty acids (VFA) ($C > 2$) such as propionic and butyric acids, and ethanol. In the third step, these long chain VFAs are converted into acetate, hydrogen and carbon dioxide. In the final step, acetate and hydrogen + carbon dioxide are converted into methane and carbon dioxide (Figure 4.1). During hydrolysis, the fermentative bacteria convert the insoluble complex organic matter, such as cellulose, into soluble substrates. The hydrolytic activity is of significant importance in wastes with high organic content and may become a rate-limiting state in overall digestion.

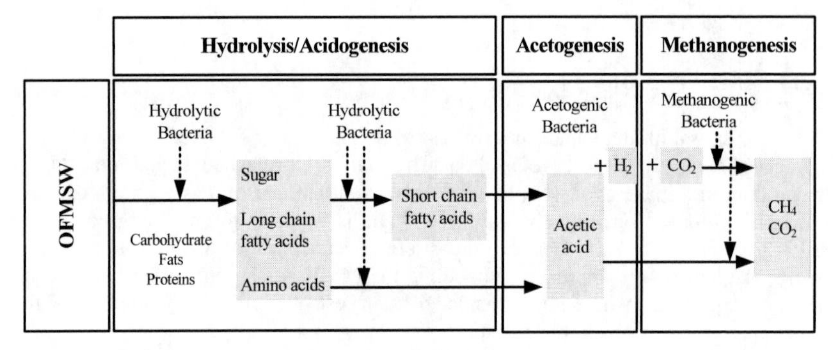

Figure 4.1 Metabolic stages of anaerobic digestion.

Depending on the stage of operation, dry anaerobic digestion can be classified into i) a single-stage anaerobic digestion or ii) two-stage anaerobic digestion. In a single-stage system, all microbial activities take place in a single reactor, while in a two-stage system, the activities take place sequentially in two reactors. The amount of methane gas produced during digestion is dependent on characteristics of the feedstock. The biogas production in a dry anaerobic digestion process is two-fold higher than that of a wet anaerobic digestion process because higher organic loading rates can be applied in the former one.

Anaerobic digesters can be operated under mesophilic or thermophilic conditions. In the mesophilic digestion, the temperature range is between 35–40°C. Because of low temperatures, the process takes a longer time for waste stabilization. In contrast, the thermophilic digestion, with a temperature range of 50-55°C, produces biogas at a higher rate along with more sanitized digestate (Table 4.1).

Table 4.1 Comparison of mesophilic and thermophilic digestions.

Parameters	Mesophilic	Thermophilic
Temperature	35-40° C	>45 °C
Residence time	15-30 days	10-20 days
Biogas yield	110 Nm^3 CH_4/tonne VS	275-410 Nm^3 CH_4/tonne VS
Merits	- More robust and tolerant process - Less sensitive to the temperature changes (within 2°C) - Less energy consumption for digester heating	- High gas production - Faster throughput - Short residence time - Small digester volume - High organic loading rate
Limitations	- Low gas production - Require relatively larger volume - Long residence time	- Need meticulous process control - Very sensitive to change in environmental factors - High energy consumption - Higher decay rate

The challenge of operating dry anaerobic digestion is not only keeping the biochemical reactions occurring at high total solids (TS) levels, but also on handling, pumping and disposing of solid streams. Most industrial-scale units built prior to the 80's relied mainly on wet processes. However, a majority of new plants built during the past decade use the dry digestion process. Table 4.2 compares both techniques for the handling of the OFMSW.

The major focus of this chapter is to present the dry digestion of OFMSW, factors affecting dry digestion, design considerations, case studies and future perspectives of dry anaerobic digestion with particular emphasis on bioenergy production.

4.2 Dry Anaerobic Digestion of OFMSW

In recent years, a new trend has developed in which dry anaerobic digestion is applied increasingly in OFMSW processing. In dry systems, the fermenting organic waste within the reactor is kept at a solids content in the range 20 - 40 % TS, so that only

Table 4.2 Comparison of dry and wet anaerobic digestion technology.

Parameters	Dry Anaerobic Digestion (DAD)	Wet Anaerobic Digestion (WAD)
TS	22-40 %	< 15 %
Water requirement	Little additional water is added to the reactor	Water must be added to maintain the required TS
Pretreatment	Not necessary to remove inert (soil and grits) and plastic hence, no need for extensive pretreatment	Inert and plastic need to be removed hence, demands extensive and costly pretreatment operations

Abrasion of reactor	Not prone to abrasion from sand and grit within the reactor	Abrasion from sand and grit, and frequent clogging of nozzles. Deposition at the reactor bottom, create extensive operational problems
Loss of volatile solids	Minimal loss during pretreatment	High loss with inert and plastic removal during pre-treatment
Organic loading rate	High 5- 17 kg VS/m^3.d	Low 2- 12 kg VS/m^3.d
Volume and heating requirement	Smaller volume, less heating is required	Larger volume, high heating is required
Phases involved	Single phase of feedstock	Two phases: sink and float
Dispersion of inhibitors	Less dispersion because of less mixing and less solubilization of N	More dispersion as sensitive to shock loads
Wet waste treatment	Can not be treated alone	Can be treated alone
Operational problem	Few moving parts within the reactor which limit short circuiting and operational problems	Mixers are needed, often leading to short circuiting and operational problems
Dewatering of digestate	Solid digestate does not require extensive dewatering	Dilute digestate requires costly dewatering processes
Wastewater and compost	Less wastewater and more compost formed as the final product	More wastewater and less compost formed as the digestate
Digestate characteristics	More stable than wet process	Less stable with high VS
Commercial suppliers	DRANCO, Kompogas, Valorga, Biocel, BRV and SUBBOR.	BTA, KCA, BIOSTAB, WAASA

very dry substrates (> 50 % TS) need be diluted with process water (Oleszkiewicz and Varaldo, 1997). The physical characteristics of the wastes at such high solids content impose technical challenges in terms of handling, mixing and pretreatment, which are fundamentally different from those of wet systems.

The process of dry anaerobic digestion for OFMSW can be divided into four stages:

(1) Pretreatment

The level of pretreatment depends on the type of feedstock. For example, animal manures simply need mixing, whereas MSW needs to be sorted-out and shredded.

(2) Digestion

The digestion stage takes place in the digester. There are different types of dry digesters with different configurations and mixing. To enhance the digestion process, the liquid refreshment is done by recirculation of the leachate.

(3) Gas upgrading
The biogas produced during the digestion has to be upgraded to remove the impurities (hydrogen sulfide, water vapor, carbon dioxide, and siloxane) that can damage boilers or engines.

(4) Post-treatment
The digestate from the digester after partial degradation is subjected to post treatment.

Most of the larger scale industrial systems process OFMSW alone. However simpler and smaller-scale systems are more successful when co-digestion with animal manure is used. The animal manure improves the C/N ratio of the feedstock and aids the anaerobic digestion process for optimal microbial growth in the digester.

In the dry anaerobic digestion system, the fermenting wastes move via plug flow in the reactors in contrast to wet systems where complete mix reactors are usually used. The use of plug flow reactor offers the advantage of simplicity as no mechanical devices need to be installed within the reactor. However, it does not address the problem of mixing the incoming wastes with the fermenting waste and also the occurrence of localized overloading and acidification. Figure 4.2 presents different dry anaerobic digestion designs used in pilot or commercial-scale plants.

In the DRANCO (Figure 4.2a), the organic material is converted into biogas in an enclosed vertical digester capable of treating a wide range of wastes with a dry matter content of 15 to 40%. The digestion takes place in the thermophilic range of 50 – 55°C. Incoming wastes are fed through the top of the digester once per day. Mixing of the digester contents occurs via-recirculation of the materials extracted at the bottom of the digester, mixing with fresh wastes (usually one part fresh wastes and six parts digested wastes), and pumping the material to the top of the digester. There is no internal mixing equipment within the DRANCO digester. This simple process design ensures that there is little wear and tear on the digester and has proven effective for the treatment of wastes ranging from 20 to 50% TS.

The reactor retention time is between 20 to 30 days and the biogas yield ranges between 80-120 m^3/tonne of wastes. Sometimes the collected biogas is stored temporarily for further purification before it is sent to combined heat and power (CHP) facilities. The extracted material from the bottom of the digester is then dewatered to a total solids (TS) content of about 50% and is stabilized aerobically for a period of about two weeks. The composted product is used as a soil conditioner.

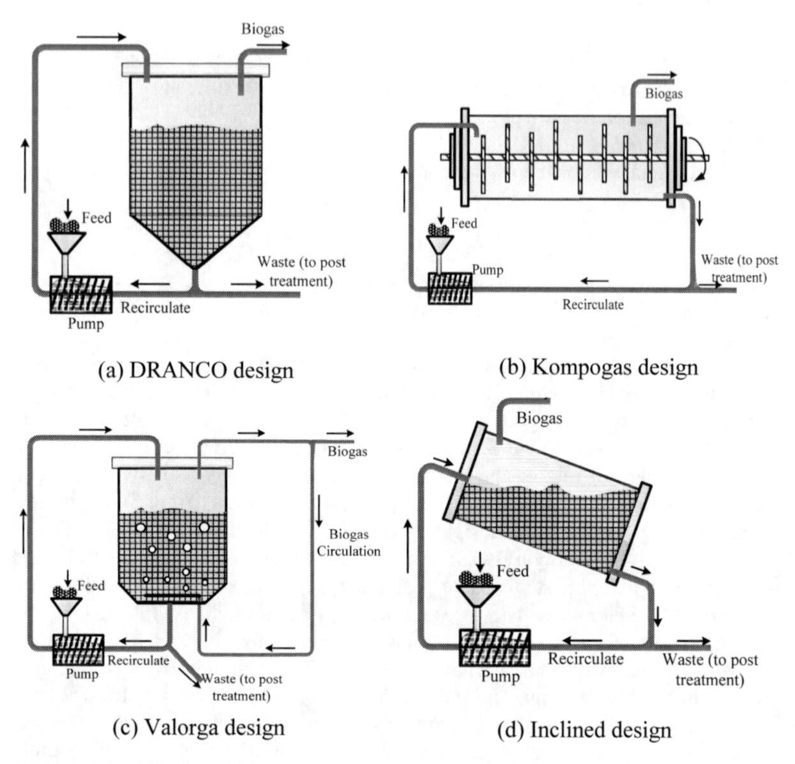

Figure 4.2 Different digester designs used in dry digestion systems.

The Kompogas design (Figure 4.2b) uses a plug flow digester with a 15 to 20 day retention time. It is a thermophilic process. For facilitating the digestion process, the incoming waste is shredded and then sorted to remove the contaminants such as plastics, glass, etc. A magnetic separator is used to remove any ferrous materials ahead of the digester.

For mixing and regulating the flow to the digester, the shredded wastes are stored in the storage tank for 2 days, where it is heated prior to entering the digester. Due to mechanical constraints, the volume of the Kompogas reactor is fixed, and the capacity of the plant is adjusted by constructing several reactors in parallel (Thurm and Schmid, 1999). The digester is usually of concrete or steel tank and with a piston pump used for feeding the waste. To maintain the thermophilic temperature, a heat exchanger is used. A screw press dewaters the digestate to 50% solids content.

The Valorga process (Figure 4.2c) is a single-stage, dry processing technology that was developed in France and has been used extensively throughout the Europe for the past 25 years. The process is designed to treat organic solid wastes but some plants also handle mixed, unsorted MSW. In the Valorga process, about a 30% solid content is maintained by diluting with water obtained from the dewatering system.

Conveyor belt, screws and powerful pumps designed especially for highly viscous materials are used for transporting and handling of materials. The equipment used in this system is very robust so only the courser impurities with a size greater than 40 mm are removed during pretreatment.

The Valorga digestion process is a semi-continuous, high solid, single step, plug flow type process. The reactors are vertical cylinders without any mechanical device inside the reactor. This allows the process to operate in dry conditions without any hindrance. Mixing is accomplished via biogas injection at high pressure at the bottom of the reactor every 15 min. One disadvantage of this design is that the gas injection ports often become clogged and maintenance is also tedious. The range of retention time for all Valorga plants is 18 to 25 days. The solid-liquid separation of the treated material is carried out by a screw press.

Figure 4.2d represents the inclined design for the dry anaerobic digestion. The purpose of this design is to facilitate the feeding and withdrawal operation especially during continuous digestion process. Such a type of design has been proven to be a viable system for decentralized municipal solid waste management. Since the system is energy efficient with more than 80% surplus energy, and is a completely closed system, the implementation of this technology is recommended for the locations where there is lack of available space, energy and odor problems. Tables 4.3 and 4.4 represent the operating conditions for various dry anaerobic digestion facilities and the energy production from anaerobic digestion plants.

Table 4.3 Performance comparison of different dry anaerobic digestion systems.

System	Substrate	T (°C)	OLR ($kgVS/m^3.d$)	HRT (d)	VS removal (%)	Methane yield ($Nm^3/kgVS$ removed)	Reference
DRANCO	SS-OFMSW	50-55	10-15	20	40–70	0.21–0.30	De Gioannis et al., 2008
Valorga Dry	SS-OFMSW	37–55	10-15	20	60	0.21–0.30	De Gioannis et al., 2008
Kompogas	OFMSW	55	4.3	29	-	0.58	Illmer and Gstraunthaler, 2009
Inclined*	SS-OFMSW	55	2.5	25	59.39	0.28	Chaudhary, 2008

* For pilot scale digester; SS- OFMSW – Source Separated OFMSW

Table 4.4 Energy production from dry digestion plants.

Plants	System	Waste (tonnes/year)	Biogas production (m^3/tonne)	Electric energy production (million kWh)
Passau, Hellersberg, Germany	Kompogas	39,000	115	9.1
Kaiserslautern, Germany	DRANCO	20,000	158	5.2
Rumlang, Switzerland	Kompogas	5,000	94.9	2.85
Baar, Switzerland	DRANCO	18,000	85	0.64
Tilburg, Netherland	Valorga	52,000	82	18
Lemgo, Germany	DRANCO	40,000	102	6.0
Roppen, Austria	Kompogas	10,000	111.3	1.43
Brecht-II , Belgium	DRANCO	51,525	126.6	8.0
Kahlenberg, Germany	Biopercolat	20,000	70-80	5.0
Buchen, Germany	ISKA	30,000	40-50	7.6

Source: RIS (2005).

4.3 Factors Affecting Dry Anaerobic Digestion

The rate at which the microorganisms grow is of vital importance in the dry anaerobic digestion process. The operating parameters of the digester must be controlled so as to enhance the microbial activity and thus increase the anaerobic degradation efficiency of the system. Some of these parameters are discussed in the following section.

4.3.1 Substrate Characteristics

In general MSW can be characterized into biodegradable organic, combustible and inert fractions.

a. Biodegradable organic fraction - Kitchen scraps, food residues, and trimmed grass.
b. Combustible fraction - Slow degrading lignocelluloses e.g. coarser wood, paper, grass, and cardboard. Since these lignocellulosic materials do not readily degrade under anaerobic conditions, they are better suited for incineration-based energy generation plants.
c. Inert fraction - Stone, glass, sand, metal, etc and biologically nondegradable material and needs to be removed prior to digestion to reduce wear and tear of the equipment.

The volatile solids consist of the biodegradable volatile solids (BVS) fraction and the refractory volatile solids (RVS). Kayhanian and Rich (1995) reported that knowledge of the BVS fraction in MSW helps in better evaluating the biodegradability of waste, biogas generation, organic loading rate and C/N ratio. Waste characterized by high VS and low non-biodegradable matter is best suited to anaerobic digestion. The quantity and quality of wastes affect the yield and biogas composition as well as the compost quality. In a dry anaerobic digestion, a smaller size of the particles promotes faster decomposition of waste and shortens the overall digestion time. Moreover, Fernandez et al. (2008) reported that the OFMSW with 20% total solids content showed higher performance in terms of dissolved organic carbon (DOC) removal and biogas yield than at 30%TS content.

4.3.2 Alkalinity and pH

Since the buffering capacity can be provided by a wide range of substances, the total alkalinity is interpreted as a determination of the carbonate, bicarbonate and digestion processes. The alkalinity results from the release of amino groups and production of ammonia from the degradation of protein-rich wastes. High-solids anaerobic digestion may be especially sensitive to ammonia overproduction. Anaerobic bacteria, especially the methanogens, are sensitive to the acid built-up in the digester, and their growth can be inhibited by acidic conditions. The optimum pH for anaerobic digestion is normally in the range of 7-8; pH levels that deviate significantly from this range can indicate potential toxicity and digester failure.

During digestion, the two processes of acidification and methanogenesis require different pH levels for optimal process control.

Acidogenesis can lead to accumulation of large amounts of organic acids resulting in decline in pH below 5.0. Excessive generation of acid can inhibit methanogens. Optimal pH can be maintained by addition of lime. As digestion reaches the methanogenesis stage, the concentration of ammonia increases and the pH increases to above 8.0. Once methane production is stabilized, the pH level stays between 7.2 and 8.2.

4.3.3 Volatile Fatty Acids (VFA) Concentration

VFA is an important intermediate compounds in the metabolic pathway of methane fermentation and causes microbial stress if present at high concentrations. The intermediates produced during the anaerobic degradation of organic compounds are mainly acetic, propionic, butyric and valeric acids (Buyukkamaci and Filibeli, 2004). High propionate concentrations inhibit hydrolysis, a rate-limiting step in dry anaerobic digestion.

4.3.4 Temperature

Due to the strong dependence of temperature on digestion rate, temperature is the most critical parameter for effective digestion. There are two temperature ranges that provide optimum digestion conditions for the production of methane: the mesophilic and thermophilic ranges. The optimum temperature for mesophilic digestion is 35°C, and the digester must be kept between 30-35°C for most favorable functioning. The operation in mesophilic range is more stable and requires a smaller energy expense (Fernandez et al., 2008). The thermophilic temperature is in the range of 50-65°C.

Dry anaerobic digestion at the higher temperatures in the thermophilic range has been found to lower the required retention time. In addition, thermophilic digestion allows higher loading rates and achieves a higher rate of pathogen destruction, as well as a higher degradation efficiency of the substrate. Thermophilic digestion, however, is more susceptible to toxic compounds and smaller changes in the environmental conditions. A decrease in operating temperature from 60°C to 37°C in anaerobic digesters reduced the inhibition caused by free ammonia as indicated by increase in biogas yield (Chen et al., 2008).

4.3.5 C/N Ratio

The relationship between the amount of carbon and nitrogen present in organic materials is represented by the C/N ratio. Microorganisms need nitrogen for the production of new cells. A nutrient ratio C:N:P:S of 600:15:5:3 is sufficient for methanization. Optimum C/N ratios in anaerobic digesters should be between 20–30 in order to provide sufficient nitrogen supply for the cell synthesis and the degradation of organic carbon present in the wastes (Fricke et al., 2007). As the reduced nitrogen compounds are not eliminated in the process, the C/N ratio in the feed material plays a crucial role.

A high C/N ratio is an indication of rapid consumption of nitrogen by methanogens and results in lower gas production. On the other hand, a lower C/N ratio results in ammonia accumulation and pH exceeds 8.5, which is toxic to methanogenic bacteria (Kadam and Boone, 1996). An optimum C/N ratio in the digester can be achieved by mixing materials of high and low C/N ratios such as organic solid waste and sewage sludge or animal manure.

4.3.6 Retention Time (RT)

The required retention time for completion of the dry anaerobic digestion reactions varies with different technologies, process temperature, and waste composition. The retention time for wastes treated in a mesophilic digester range from 10 to 40 days. A high solids reactor operating in the thermophilic range has a retention time of 14 days (Verma, 2002).

4.3.7 Organic Loading Rate (OLR)

The OLR is a measure of the biological conversion capacity of the AD system. Feeding the system above the allowable OLR results in a low biogas yield. This is mainly due to the accumulation of inhibiting substances such as VFAs in the digester slurry. In such a case, the feeding rate to the system must be reduced. Many plants have reported system failure due to overloading. Table 4.2 compares the typical OLR for wet and dry anaerobic digestions.

4.3.8 Mixing

Mixing is important in anaerobic digestion to ensure that conditions are consistent throughout the digester, providing favorable conditions for bacteria to rapidly access the substrate. Mixing in dry digestion is usually performed by recirculating the leachate, injecting the biogas produced or slowly rotating the mechanical devices. It is highly dependent on the configuration of the digester and the total solids contents.

4.4 Case Studies

4.4.1 EcoTechnology JVV Oy System – Finland/Germany

Since 1995, EcoTec has had a bio-waste facility operating in Bottrop, Germany, which processes 6,500 tonnes/yr (Figure 4.3). This company also has several other semi-dry anaerobic digestion plants with capacities ranging from 17,000 tonnes/yr in Shilou, China to a 30,000 tonnes/yr in Berlin, Germany.

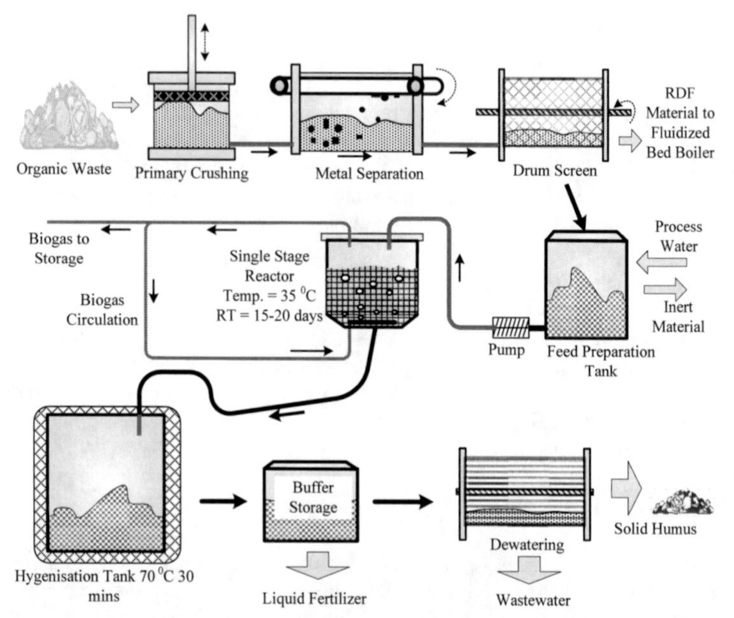

Figure 4.3 Eco Technology semi- dry anaerobic digestion process flow diagram.

The source separated waste is delivered to the plant and passed through a primary crushing stage and to a magnetic separator before entering a drum screen. Here, combustible materials are separated from the waste and are conveyed to a fluidized bed boiler. The remaining organic material is conveyed to the feed preparation tank, where it is mixed with process water to adjust the total solids content to 15%. Solid impurities are removed, and the feedstock is then pumped into the digester.

The digestion process occurs at mesophilic temperatures in single stage semi-dry anaerobic digestion system. The retention time for this plant is between 15 to 20 days. The plant is also good in post treatment of the digestate, which produces a good quality of compost. Since the feed TS content is maintained at 15%, this plant is classified as a semi-solid anaerobic digestion process and demands extensive pretreatment. The liquid digestate also needs further expensive post treatment to meet the local effluent discharge standards.

The amount of biogas produced varied between 75 and 90 Nm^3 per ton of waste loaded. The percentage of methane content was around 55% in biogas samples and the total methane yield by the plant ranged between 49 and 63 Nm^3 per ton of organic waste loaded.

4.4.2 *Organic Waste Systems – DRANCO Process, Brecht, Belgium (Dry Continuous Single Stage System)*

Organic Waste Systems (OWS) developed the DRANCO process of dry anaerobic digestion, and their plant at Brecht in Northern Belgium has a capacity of 42,500 tonnes/year (Figure 4.4). The incoming feedstock composed of source separated organics, such as garden, kitchen, food waste and shredded prior to passing through a screen to separate the large particles. A magnetic separator removes metallic particles, and the feedstock is then mixed with water recycled from the process. The feedstock is then pumped into the top of the reactor which has a capacity of 808 m^3.

The DRANCO process consists of a single stage, thermophilic reactor (operating between 50 and 58 °C). The retention time in the reactor is 20 days. 5% of the contents are removed each day and dewatered to 55% solids using a screw press. The filtrate is pretreated in on-site aeration ponds prior to discharge to the local municipal wastewater treatment plant. The digested residues are extracted from the reactor, dewatered to a TS content of 50%, and then stabilized aerobically for a period of approximately 2 weeks in a facility with in-floor air ducts. The final product is called Humotex and is a hygienic, stabilized soil conditioning product. About 7% of the gas produced is used to heat the digester. The total solids content of the feedstock varies between 15-40% depending on the input material.

Around 5.8 to 7.4 million m^3 of biogas were produced and consumed in two gas engines with an electrical power output of 625 kW each between the year 2005-2008. The electrical production amounted to 9.1 million kWh in the year 2008, enough to provide power to 2000 to 2500 homes. The gas engines operated during 97% of the time during the year.

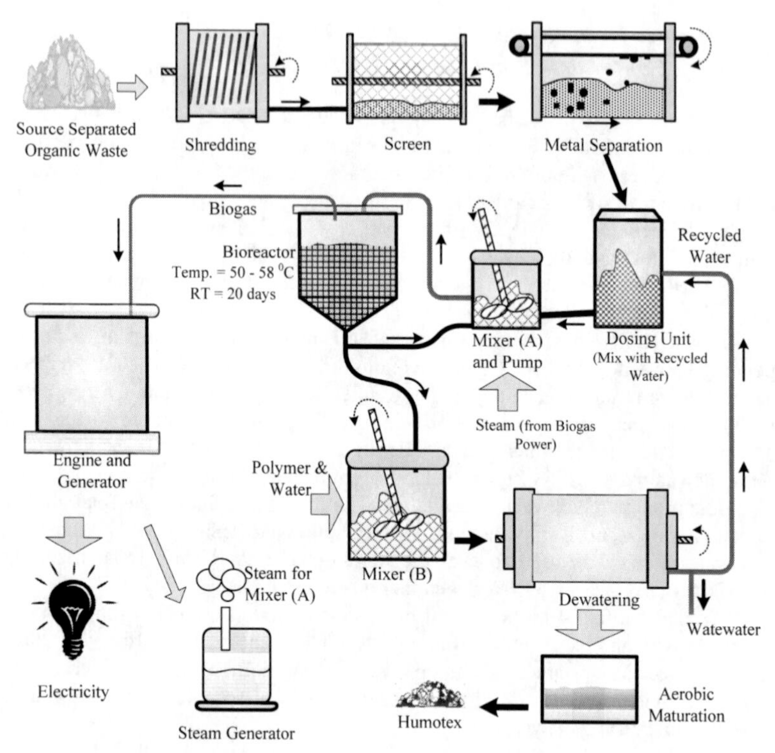

Figure 4.4 DRANCO dry anaerobic digestion process flow diagram.

4.5 Future Perspective of Dry Anaerobic Digestion

During the past two decades, the implementation of dry anaerobic digestion for the treatment of biowaste and recovery of resources from the organic fraction of municipal solid waste has been considered as the major development in the field of solid waste management in Europe. Nowadays, anaerobic digestion has captured a significant part of the European market. There are a number of environmental benefits of using anaerobic digestion to stabilize either a source separated organics or a mixed waste stream. These include production of green energy to supplement the use of fossil fuels for electricity and heat generation. In New York, the total cost of anaerobic digestion facilities is between \$47 and \$58 per tonne of waste. This cost includes all the cost involved for the construction of digestion facility as well as plant operation and maintenance costs. Moreover, the economic viability of anaerobic digestion depends on the control on landfilling or dumping, market for both compost and biogas, and the organic contents of the waste stream.

Dry anaerobic digestion will continue to expand in Europe and elsewhere if the economics of the waste management and energy markets provide significant incentives. There is increasing interest in anaerobic digestion in the United States and Canada, as green energy and energy from biomass are getting more attraction. Various dry anaerobic digestion systems are gaining escalating markets due to their high efficiency in bioenergy production.

4.5.1 *Mesophilic Versus Thermophilic Digestion*

Until the early 90's, all plants were operating at mesophilic conditions, mainly because this was considered as a more stable process with lesser energy need for heating. The first thermophilic dry digestion plant was constructed in the early 1990s to digest OFMSW. It was noted, that unlike the wet process, the thermophilic dry process could be operated easily with significantly high biogas generation. The current trend in dry digestion, especially for OFMSW is to use the thermophilic digester.

4.5.2 *Wet Versus Dry Digestion*

Although applications of both dry and wet anaerobic digestion have continued to increase in total capacity, dry digestion has been dominant since the beginning of the nineties. An increase of wet systems was observed between 2000 and 2005 as a number of large-scale wet plants were put into operation, while more dry fermentation plants have been being installed since 2005. Dry anaerobic digestion currently provides almost 54% of the capacity for treatment of OFMSW in Europe.

4.5.3 *Single-Phase Versus Two-Phase Digestion*

Two-phase anaerobic digestion systems which are very difficult to operate in dry digestion of MSW are expected to be only 2% of total capacity at the end of 2010 and most of the systems will be derived from batch single-phase digestion. For most organic wastes, the biological performance of one-phase digestion is as high as that of two-phase digestion if the digester is well designed and the operating conditions are carefully chosen (Weiland, 1992).

The high prices of fossil fuels together with a diminishing capacity of landfills will keep stimulating the growth of dry anaerobic digestion for bioenergy production from organic fraction of municipal solid wastes.

Also, the research trend is increasing towards maximizing the biogas recovery potential from the waste through the decoupling of anaerobic digestion in a single reactor to provide shorter HRTs for acidified liquid and higher SRTs for solid substrate. Different reactor configurations, with various operating conditions at the lab- and pilot- scale studies are in progress. Furthermore the management of the digestate from anaerobic reactors through aerobic processing facility is becoming a more important topic presently.

4.6 Summary

Dry anaerobic digestion systems for digesting OFMSW are being widely used worldwide. A majority of plants are large scale, processing over 2,500 tonnes of waste per day and involves complex plant design. Germany and Denmark lead in terms of technology development with numerous successful full-scale plants in operation. The basic technology is well implemented, but improvements are still under way in terms of the optimization and pre- and post-treatments of the feed and end-products, respectively. The current EU policy to attain 20% of the primary energy demand from renewable sources by 2020 is the major driving force to promote this technology in the urban waste treatment sector. Whereas, decentralized biological treatment have been adopted in developing Asian countries like China, India, Thailand, Vietnam, Nepal, and Sri Lanka to treat OFMSW. Compared to Europe, Asian countries are characterized with having high amounts of biodegradable components (40 – 80 %) in the waste stream, with a high moisture content, which can be potentially converted into energy through anaerobic degradation to meet the global energy crises.

4.7 References

Buyukkamaci, N., Filibeli, A. (2004). "Volatile fatty acid formation in an anaerobic hybrid reactor." *Process Biochemistry*, 39(11), 1491–1494.

Chaudhary, B.K. (2008). "Dry continuous anaerobic digestion of municipal solid waste in thermophilic conditions." *Master AIT Thesis No. EV-08-4*. Asian Institute of Technology, Thailand.

Chen, Y., Cheng, J.J., and Creamer, K.S. (2008). "Inhibition of anaerobic digestion process: a review." *Bioresource Technology*, 99(10), 4044–4064.

De Gioannis, G., Diaz, L. F., Muntoni, A., and Pisanu, A. (2008). "Two-phase anaerobic digestion within a solid waste/wastewater integrated management system." *Waste Management*, 28(10), 1801–1808.

Fernandez, J., Perez, M., and Romero, L.I. (2008). "Effect of substrate concentration on dry mesophilic anaerobic digestion of organic fraction of municipal solid waste." *Bioresource Technology*, 99, 6075–6080.

Fricke, K., Santen, H., Wallmann, R., Huttner, A., and Dichtl, N. (2007). "Operating problems in anaerobic digestion plants from nitrogen in MSW." *Waste Management*, 27, 30–43.

Illmer, P., and Gstraunthaler, G. (2009). "Effect of seasonal changes in quantities of biowaste on full scale anaerobic digester performance." *Waste Management*, 29(1), 162–167.

Kadam, P.C., and Boone, D.R. (1996). "Influence of pH on ammonia accumulation and toxicity in Halophilic, Methylotrophic methanogens." *Applied and Environmental Microbiology*, 62, 4486–4492.

Kayhanian, M., and Rich, D. (1995). "Pilot-scale high solids thermophilic anaerobic digestion of municipal solid waste with an emphasis on nutrient requirements." *Biomass and Bioenergy*, 8, 433–444.

MOEJ, (2006). "Sweeping policy reform towards a sound material-cycle society starting from Japan and spreading over the Entire globe, the 3R loop connecting Japan with other countries." *Ministry of Environment, Japan*, 36.

Oleszkiewicz, J.A., and Varaldo, P. (1997). "High solids anaerobic digestion of mixed municipal and industrial wastes." *Journal of Environmental Engineering*, 123, 1087–1092.

RIS International Ltd. (2005). *Feasibility of generating green power through anaerobic digestion of garden refuse from the Sacramento area.* Available at http://www.nerc.org/documents/sacramento_feasibility_study.pdf (accessed September 2, 2009).

Thurm, F., and Schmid, W. (1999). "Renewable energy by fermentation of organic waste with the Kompogas process." *In II International Symposium on Anaerobic digestion of solid waste*, held in Barcelona, 15-17 June, 1999, 342–345.

Verma, S. (2002). *Anaerobic digestion of biodegradable organics in municipal solid wastes.* Available at http://www.seas.columbia.edu/earth/vermathesis.pdf (accessed January 15, 2009).

Weiland, P. (1992). "One and Two-step anaerobic digestion of solid agroindustrial residues." In: *International Symposium on Anaerobic digestion of solid waste*, Venice, 14-17 April, 1992, 193–199.

CHAPTER 5

Bioenergy from Landfills

Guo-Bin Shan, Tian C. Zhang, and Rao Y. Surampalli

5.1. Introduction

Each year, approximately hundreds of millions of tons of municipal solid waste (MSW) is produced (e.g., 254 million tons of MSW was generated in 2007 in the U.S., USEPA, 2008), with a similar amount of general industrial wastes around the world. There are several general methods that can be used to dispose of MSW (Fig. 5.1); land-filling is the least preferred tier of the hierarchy of waste management options because of the advantages of waste minimization, reuse and recycling, and incineration with energy recovery. In spite of this the use of landfills currently represents the most common practice for waste disposal and about 84–90% of MSW is disposed of in landfills (Porteous, 1993; USEPA, 1994).

The land-filling process involves compacting and covering waste with soil or clay to minimize MSW. The wastes are covered and compressed mechanically by the weight of the material that is placed above the wastes. In landfills, landfill gas (LFG) is produced during anaerobic decomposition of organic waste. Due to minimal exposure to atmosphere, anaerobic microbes thrive. LFG builds up and is slowly released into the atmosphere if the landfill site is not engineered to capture the produced gas (Gendebien et al., 1992). LFG is of great concerns because it is hazardous for three key reasons. First, LFG becomes explosive when it escapes from the landfill and mixes with oxygen. The lower explosive limit is 5% methane and the upper explosive limit is 15% methane. Second, the methane is 20 times more potent as a greenhouse gas than carbon dioxide. The methane generated by landfills accounts for 12% of global methane emissions or nearly 750 million metric tons of CO_2 equivalent ($MMTCO_2E$), USEPA, 2008) and makes up about 25% or even more of anthropogenic contribution to global warming (IPCC, 1992). Therefore, if uncontainable LFG escapes into the atmosphere, it may significantly contribute to global warming. In addition, methane oxidation (45 g/m^2-d) to CO_2 has been observed in a soil-covering landfill in the presence of methanotrophic bacteria (Whalen et al., 1990). Third, LFG contains volatile organic compounds (VOCs) that can contribute to the formation of photochemical smog.

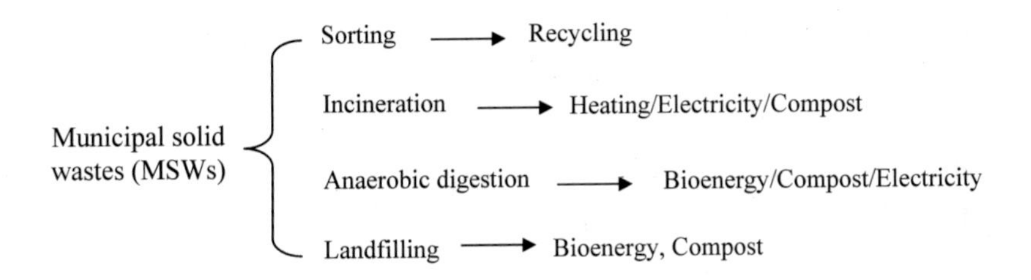

Figure 5.1 Management methods of municipal solid wastes and possible ways for their beneficial utilization.

LFG results from the biological decomposition of organic fraction of municipal waste and is a flammable and odorous gaseous mixture, consisting mostly of methane (CH_4) and carbon dioxide (CO_2) together with a few parts per million (ppm) of hydrogen sulfide (H_2S), nitrogen (N_2) and volatile organic compounds (VOCs) (Qin et al., 2001; Liamsanguan and Gheewala, 2008). Using LFG to generate energy not only encourages more efficient collection thereby reducing greenhouse emissions into the atmosphere but also generates revenues for operators and local governments. In a landfill gas-to-energy system (Figure 5.2), a series of wells are drilled into the landfill and a piping system connects the wells and collects the produced gas. Dryers remove moisture from the gas, and filters remove impurities. The LFG can be used for a variety of purposes, e.g. (i) the LFG can be used to produce electricity with an engine-generator set, gas turbines, microturbines, and other technologies, (ii) LFG can be used as an alternative energy to local industrial customers or other organizations that need a constant energy supply; the direct use of LFG is reliable and requires limited processing and modifications to existing combustion equipment and (iii) further gas cleanup improves biogas to pipeline quality (the equivalent of natural gas), which can be used to replace natural gas or used as alternative vehicle fuel. In addition, reforming the gas to hydrogen would make it possible to use it as fuel cell technology for electricity production.

LFG production and utilization is a complex subject, covering from microbiology to combustion technology, and involving gas migration, leachate production and treatment, the safe and effective utilization of gas, and health risks associated with toxic constituents present in the gas. At the same time, many studies have investigated environmental pollution resulting from the land-filling process (Gronow 1990a, 1990b). This chapter introduces the background and current status of LFG technology. LFG technology involves: (i) estimation of ultimate yield and generation rates of LFG on the basis of waste composition; (ii) design of an LFG abstraction system appropriate to site conditions and landfilling practices; and (iii) cost-effective gas utilization schemes. This chapter focuses on principles and benefits of LFG recovery/utilization as well as issues for technology or project development.

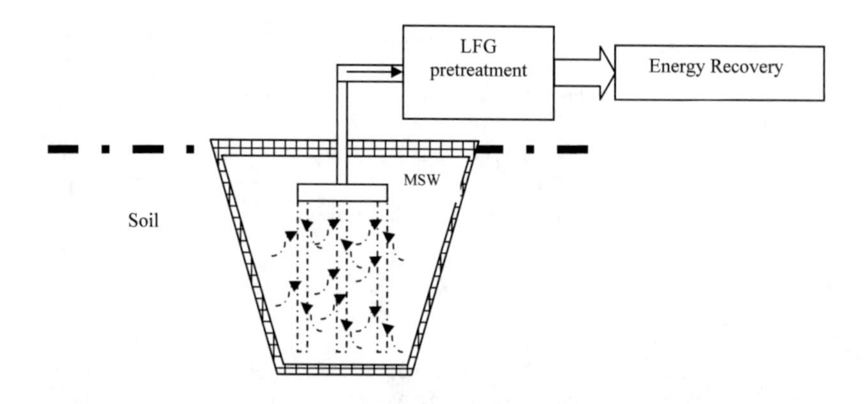

Figure 5.2 Schematic diagram of a landfill gas-to-energy system.

5.2 Development and Utilization of Bioenergy from Landfills

5.2.1 Landfill Gas (LFG) Production and Related Research Activities

LFG in landfills is produced via a number of simultaneous and interrelated biological, chemical and physical changes by utilization of MSW as substrates. Among these, the most important biological reactions occurring are those involving the organic material in MSWs that lead to the evolution of LFG. In aerobic reactions, CO_2 is the principal gas produced. In anaerobic reactions, organic matter is converted CH_4 and CO_2 with trace amounts of ammonia and hydrogen sulfide. According to the compositions of the MSW, the total volume of potentially produced gas can be estimated by representing the organic matter with a generalized formula of the form of $C_aH_bO_cN_dS_e$. The anaerobic decomposition of solid waste is shown below for the biodegradable part of the organic matter (Tchobanoglous et al., 1993):

$$C_aH_bO_cN_dS_e + \left(\frac{4a-b-2c+3d+2e}{4}\right)H_2O \rightarrow \left(\frac{4a+b-2c-3d-2e}{8}\right)CH_4 + \left(\frac{4a-b+2c+3d+2e}{8}\right)CO_2 + dNH_3 + eH_2S. \quad (5.1)$$

A typical composition of raw landfill gas is shown in Table 5.1. Due to the varying nature of the contents of landfill sites and age, the constituents of landfill gas vary widely. However, CH_4 and CO_2 make up nearly 90% of the gas volume produced.

Table 5.1 Typical composition of raw landfill gas*.

Component	Chemical formula	Content
Methane	CH_4	40–60% (by vol.)
Carbon dioxide	CO_2	20–40% (by vol.)
Nitrogen	N_2	2–20% (by vol.)
Hydrogen sulfide	H_2S	40-100 ppm
Heavier hydrocarbons	C_nH_{2n+2}	< 1% (by vol.)
Oxygen	O_2	< 1% (by vol.)
Ammonia	NH_3	0.1-1% (by vol.)
Complex organics	-	1000–2000 ppm
Siloxane, chlorinated organics	-	at ppb level

* Demirbas, 2006; DHHS, 2001.

It is not easy to evaluate landfill performance with respect to LFG (and leachate) generation. Sometimes, landfill test cells (small landfills or declined portions of landfills) are used as physical models for evaluation purposes and/or landfill research. The amounts of the wastes these test cells treated vary from a couple of hundred tons up to about a million tons. Test cell is a very important research tool for the development of landfill technology and the test cell tools have been very interest for the LFG activity, because it is difficult to represent landfill processes in a laboratory while considering the number of influencing and site specific factors. However, not many landfill test cells have been built because they are fairly large and expensive. Nonetheless, LFG activity data on test cell experiments was collected and analyzed (Lagerkvist, 1995). During 1992-1994, a so called Brogborough test cell was built in UK and Swedish; the "Integrated Landfill Gas"-program with test cells were operated in the cities of Malmö, Helsingborg and nearby Stockholm (Lawson et al., 1992; Karlsson et al., 1993; McLeod, 1994). These test cells experiments involved waste compositions (e.g., waste categories, sludge, and ashes), filling parameters (such as waste crushing, compaction degree, and lift thickness), and operation techniques (such as water additions, leachate recycle and air blowing). Results obtained from these tests were used for improving the design and operation of landfill processes as well as monitoring and experimental methods. One of the important findings was that the rate of gas generation is variable but generally much greater than what is achieved at existing LFG schemes. Most of results show the methane yield is ~10 m^3 of CH_4 per ton of waste (on TS basis) (Lagerkvist, 1995).

In order to produce LFG and treat waste more efficiently, bioreactor landfill technology has been developed because of the advantages of the increased moisture content, enhanced microbial activities with accelerated waste stabilization, and the reduced content of organic pollutants in landfills. Since the late 1960s, semi-aerobic landfilling has been developed. In a semi-aerobic landfilling site, air entered the leachate collecting pipe, and thus, some aerobic zones in the landfill pile existed. Yasushi and Masataka (1997) reported that about 90% organic substances in waste were converted into gas phase in a semi-aerobic landfilling treatment, while 90% organic substances were converted into the leachate phase in an anaerobic landfilling treatment. Onay and Pohland (1998) showed that the use of leachate recirculation in simulated landfill bioreactors was feasible for the *in-situ* removal of ammonia nitrogen. Huang et al. (2008) constructed a large-scale simulated semi-aerobic landfill and studied the variation of LFG component contents and temperature in the landfill.

Presently, the existing landfill techniques have been enhanced by combining with other techniques or technologies, such as a combination with aerobic pre-treatment (e.g., through the use of a composted bottom layer of waste) and is a very efficient method to promote the onset of methane production. Adding sludge is beneficial to enhance microbial growth. . In order to make biogas production more profitable, many plants use co-fermentation of waste products to increase CH_4 yields (Sebastian et al., 2006). Igoni et al. (2008) reported the key issues concerning the design of a high-performance anaerobic digester for producing biogas from MSWs and process optimization in the development of anaerobic digesters aimed at creating useful commodities from the ever-abundant MSW; the results of the study show such anaerobic digestion systems may be used to design better landfill systems as renewable energy sources.

LFG can be used to produce bio-methane by removing the CO_2 from the LFG. LFG can also be used to produce: (a) liquid hydrogen (LH_2) by steam reforming (SMR), partial oxidation, pyrolysis, autothermal pyrolysis, and autothermal SMR followed by cryogenic separation or pressure swing adsorption (PSA) purification; (b) synthetic natural gas via methanation; (c) methanol via methanol synthesis; and (d) synthetic hydrocarbons via Fischer-Tropsch (FT) synthesis (Muradov and Smith, 2008). For H_2 separation and purification, the cryogenic process is more efficient than the PSA and membrane process. The cryogenic H_2 separation utilizes partial condensation to separate H_2 from impurities with higher boiling points, such as H_2O, CO, CO_2, CH_4, and hydrocarbons (Muradov et al., 2006). However, design and efficiency determination of the cryogenic process is difficult and time consuming, which results in the limitation of development and widespread application of the cryogenic process. On the basis of these processes, Huang and T-Raissi (2007) combined the cryogenic process and H_2 liquefaction into a single process to produce LH_2 with zero carbon emission and a minimal H_2 loss using CH_4 or LFG to increase the efficiency and reduce costs due to no PSA step involvement.

5.2.2 USEPA Landfill Methane Outreach Program (LMOP)

Methane emissions from municipal landfills represent 3% of the total US greenhouse gas emissions that contribute to climate change. These methane emissions can be released into the air, flared, and collected for generating electricity. In 1994, the US Environmental Protection Agency (USEPA) created the LMOP, with the objective of reducing landfill greenhouse gas emissions by promoting the development of landfill-gas to-energy projects (Jaramillo and Matthews, 2005).

Since the 1996 enactment of the New Source Performance Standard and Emission Guidelines for Municipal Solid Waste Landfills, the LMOP has become a tool to help landfills meet the new regulations. As a result of these efforts, methane emissions from landfills have decreased. In 1995, methane emissions from landfills were estimated to be 216.1 Teragrams (Tg) of CO_2 equivalents. In 1997, these emissions were 207.5 Tg of CO_2 equivalents, and by 2001 they had been reduced to

202.9 Tg of CO_2 equivalents. The USEPA estimated that, between 1995 and 2001, 253 Tg of CO_2 equivalents were used in landfill-gas-to-energy projects, while 350.3 Tg of CO_2 equivalents was flared (USEPA, 2004). In these projects the LFG is used as a direct fuel in industrial processes or as the fuel that runs electricity-generating equipment. It is important to note that the LMOP offers a wide array of free technical, promotional and informative tools, plus support services to assist with the development of LFG energy projects. Information can be found from the USEPA's website (http://www.epa.gov/lmop) on these technical documents, toolkit, brochures, fact sheets, case studies, software, media/press and corporate users.

5.2.3 LFG Management

According to Directive 31/1999/CE "LFG shall be collected from all landfills receiving biodegradable waste and the LFG must be treated and used. If the gas collected cannot be used to produce energy, it must be flared." In 1996, the USEPA established the New Source Performance Standards and Emission Guidelines for Municipal Solid Waste Landfills. Under these standards, large landfills (i.e., those with the potential to emit more than 50 Mg/year of non-methane volatile organic compounds) have to collect and combust the LFG.

For the LFG management, during the five years of operation and the 30 years of the post-closure phase, the following program has been proposed (Lombardia et al., 2006): no LFG collection during the first year; 50% LFG collection efficiency from the second to the fifth year (operation) and energy recovery; 80% LFG collection efficiency from the sixth to the 35th year (post-closure) and energy recovery or flaring. A detailed landfill rule can be found via the USEPA's LMOP website. For example, the landfill rule guidance document (SCS Engineers, 2008) is very useful for landfill owners to understand how federal LFG regulations apply to their landfill and explain how energy recovery can be a cost-effective compliance option.

5.3 Benefits of LFG Recovery and Utilization

5.3.1 Major Benefits

Environmental Benefits. Use of LFG has an additional benefit in that the methane generated in landfills that would otherwise escape into the atmosphere is instead combusted thereby resulting only in carbon dioxide release. Methane is a potent greenhouse gas, with 20 times more heat-trapping potential than carbon dioxide. Like any gaseous fuel, LFG can be burned to generate electricity, heat, or steam (Figure 5.2). LFG can also be fed into a stationary or vehicle fuel cell after being transformed into a hydrogen-rich gas (Lombardi et al., 2006). In addition, utilization of LFG can upgrade the environment and living quality by reduction of odor problems, mitigation of VOCs hazards and prevention of landfill fires; it also can enhance land utilization by acceleration of landfill remediation.

Economic Benefits. Utilization of LFG can decrease environmental protection stress from the United Nations Framework Convention on Climate Change

(UNFCCC); it can treat solid waste without long-term follow-up costs that are usually due to soil and water pollution; and it also reduces foreign exchange needs through production of compost to reduce fertilizer, chemical herbicides and pesticides demand and through direct utilization of energy produced (biogas/electricity/heat) in the treatment process to reduce fossil energy demand. Utilization of LFG can generate income through compost and energy sales (biogas/electricity/heat) to the public; it can improve agricultural productivity through long-term effects on soil structure and fertility through compost utilization; and it can recover materials to be recycled or sold to the recycling industry (Taleghani and Kia, 2005). Utilization of LFG results in CH_4 reduction, electricity generation, and CO_2 mitigation (Tsai, 2007).

Energy Benefits. Incineration of LFG and other conversion technologies allow energy recovery to displace electricity generated from fossil fuel. This is because a large quantity of organic waste can be almost completely converted into energy (for electricity production, heating, or truck and automobile fuel) by these methods. The recovered energy can reduce the amount of energy consumption. It is helpful for coordinating with energy policy by the promotion of diversification of primary energy and fulfillment of energy-related technology development; it also upgrades social awareness and national impression by the promotion of energy and environmental education of the public.

LFGE Benefits Calculator. The LFGE Benefits Calculator can be found at http://www.epa.gov/lmop/res/calc.htm. Reductions of methane and carbon dioxide are expressed in equivalent environmental and energy benefits within the *LFGE Benefits Calculator*. As shown in Table 7.2, the calculator can be used to estimate direct, avoided, and total greenhouse gas reductions, as well as environmental and energy benefits for the current year of a LFG energy project. For both electricity generation and direct-use projects, reductions of GHG emissions are derived from capturing and destroying landfill methane. GHGs are also reduced by the offset of CO_2 emissions. Electricity generation projects displace CO_2 that would have otherwise been generated from fossil fuels burned at conventional power plants. For direct-use projects, the methane in landfill gas displaces fossil fuels and avoids CO_2 that would have otherwise been released. For example, the total 2008 benefits for a typical 3-megawatt (MW) electricity generation project are approximately equal to powering 1,900 homes or heating 3,200 homes (as energy benefits) or equal to any one of the following environmental benefits: (a) annual greenhouse gas emissions from 23,580 passenger vehicles; (b) carbon sequestered annually by 29,261 acres of pine or fir forests, (c) carbon dioxide emissions from burning 672 railcars' worth of coal, and (d) carbon dioxide emissions from 14.6 million gallons of consumed gasoline.

5.3.2 LFG Utilization for Beneficial Purposes and Related Issues

Energy recovery in LFG can be accomplished by (a) flaring or combustion in reciprocating engines, (b) directly feeding LFG to a fuel cell (FC), and (c) producing a hydrogen-rich gas to feed a FC (Lombardia et al., 2006). Possible uses of LFG include the supplementary or primary fuel, pipeline quality gas and vehicle fuel to produce electric power and heat.

Historically, exploitation of LFG as a replacement fuel in kilns, boilers and furnaces began in the early 1970's. The first LFG gathering site was built in Palos Verds in South California, US, in 1977. Later, schemes were established for the use of LFG as a fuel for engines for power generation. Sims and Richards (1990) identified and reviewed 242 sites in 20 nations where LFG was used as a fuel with total energy contribution > 2.037 million tons of coal equivalent per annum (mtcepa). The three biggest users of LFG were U.S.A., U.K., and Germany. The end users of such systems were power generation units accounting for 55% of the total number of projects, followed by use in boilers (23%), kilns and furnaces (13%) and gas cleaning/pipeline distribution (9%).

To date, it is increasingly becoming an attractive source of energy in many nations of the world. In 2004 alone, there were more than 375 operational LFG energy projects in 38 states in the U.S., which supplied 9 billion kilowatt hours (KWh) of electricity, and 74 billion cubic feet of LFG to end users. The estimated annual environmental benefits and energy savings associated with currently operational projects are equivalent to planting 19 million acres of forest, or preventing the use of 150 million barrels of oil, or removing the CO_2 emissions equivalent to 14 million cars, or offsetting the use of 325,000 railcars of coal. To date, the landfills that capture LFG in the US collect about 2.6 million tones of methane annually, 70% of which is used to generate heat and/or electricity.

International utilization of LFG in place of a conventional fuel such as natural gas, fuel oil, or coal in boilers is an established practice with a track record of more than 25 years of success. Upgrading LFG to replace fossil vehicle fuels has been practiced in some countries (EASEPA, 2004). The Finnish magazine Suomen Iuonto reported that the gas is used to fuel a car owned by a farm owner. Sweden has a similar story as many city-buses are powered by biogas, and some gas stations there offer biogas in addition to other fuels. In fact, all over the world, biogas has been used for heating purposes and/or electricity generation. For example, in U.K., Xuereb (1997) reported that, although the use of biogas for electricity generation was still at an experimental stage, it already accounted for about 0.5% of the total electricity output. There are more than 45 schemes in the U.K. where LFG is used. Direct use of the gas still is the most popular option because it requires minimum gas pretreatment and is so far the most cost effective and economical. However, in the U.K., power generation is a preferred option by a price premium that is given to the producer.

Table 5.2 Benefit calculation of a LFG energy project with megawatt capacity of 3.0 MW[a].

For electricity generation projects, enter megawatt (MW) capacity: 3.00

- OR -

For direct-use projects, enter landfill gas utilized by project: [ ] million standard ft^3/d (mmscfd) or [ ] standard cubic feet per minute (scfm)

Direct Equivalent Emissions Reduced [Reduction of CH_4 emitted directly from the landfill]		Avoided Equivalent Emissions Reduced [Offset of from CO_2 avoiding the use of fossil fuels]		Total Equivalent Emissions Reduced [Total = Direct + Avoided]		
$MMTCO_2E/yr$ million metric tons of carbon dioxide equivalents per year	**tons CH_4/yr** tons of methane per year	**$MMTCO_2E/yr$** million metric tons of carbon dioxide equivalents per year	**tons CO_2/yr** tons of carbon dioxide per year	**$MMTCO_2E/yr$** million metric tons of carbon dioxide equivalents per year	**tons CH_4/yr** tons of methane per year	**tons CO_2/yr** tons of carbon dioxide per year
0.1139	5,976	0.0149	16,418	0.1287	5,976	16,418
Equivalent to any one of the annual benefits below:		**Equivalent to any one of the annual benefits below:**		**Equivalent to any one of the annual benefits below:**		
Environmental Benefits		Environmental Benefits		Environmental Benefits		
• Annual greenhouse gas emissions from passenger vehicles:	20,852	• Annual greenhouse gas emissions from passenger vehicles:	2,728	• Annual greenhouse gas emissions from passenger vehicles:		23,580
• Carbon sequestered annually by acres of pine or fir forests:	25,876	• Carbon sequestered annually by acres of pine or fir forests:	3,385	• Carbon sequestered annually by acres of pine or fir forests:		29,261
• CO_2 emissions from burning railcars' worth of coal:	595	• CO_2 emissions from burning railcars' worth of coal:	78	• CO_2 emissions from burning railcars' worth of coal:		672
• CO_2 emissions from gallons of gasoline consumed:	12,923,212	• CO_2 emissions from gallons of gasoline consumed:	1,690,670	• CO_2 emissions from gallons of gasoline consumed:		14,613,881

To date, LFG is only marginally exploited. In the developed countries LFG systems are established, and a sizeable proportion of renewable energy is generated and utilized from landfills. Biogas fuels represent about 1% of US electricity generation, which is equivalent to reducing CO_2 emissions in the electricity sector by more than 10%. Biogas is also presently used in India, China, Brazil, Singapore, and many other countries. In developing countries, the labor-oriented solid waste management systems focus more on the collection and transportation stages. One important issue for landfill project development in the developing countries is that open dumps and unmanaged landfills are the dominant disposal options there. These sites can only generate lesser amounts of methane. Fortunately, many developing countries are currently transforming to landfills from more uncontrolled systems.

Another important issue is energy price structure. Government policies on energy and solid waste management can promote or inhibit the beneficial use of LFG; a lack of regulations governing landfills and LFG energy projects (i.e., no incentive to collect and combust LFG) hinder project development. Garg et al. (2007) analyzed the policy framework for solid recovered fuel (including LFG) in Europe; they found that the European Commission (EC) is seeking to clarify the legal situation on waste used as fuel and is collating case studies on industry byproducts that become waste. For example, in order to encourage the use of clean energy in Taiwan, the regulations, known as "Regulation of Tax Deduction for Investment in the Procurement of Equipments and/or Technologies by Energy conservation, or emerging/Clean Energy Organizations" have been revised for a few times (Tsai et al., 2004).

5.4 Gas Quantities, Characteristics, and Monitoring

5.4.1 Estimating Gas Flow and Monitoring Methods

LFG produced during anaerobic digestion of MSWs is mainly composed of methane and carbon dioxide. A wide range of gas production rates of between 0.187 and 0.424 $m^3\ kg^{-1}$ wet wastes or between 0.009 and 0.02 $m^3\ kg^{-1}$ wet wastes per year has been reported (Rees, 1985). Theoretical and experimental studies indicate that complete anaerobic biodegradation of MSWs will generate about 200 m^3 of methane per dry ton of contained biomass. However, the reported rate of methane production in industrial anaerobic digestion reactors is about 40–100 m^3 of methane per dry ton of MSWs. Therefore, a conservative estimate is about 50 m^3 of methane produced per ton of MSWs landfilled, and annually the corresponding rate of methane generation at landfills is 75 billion m^3 for the estimated global landfilling of 1.5 billion tones of MSWs. At present, less than 10% of this estimated amount is collected and utilized (Themelis and Ulloa, 2007).

Most LFG emissions can be measured using chamber techniques, subsurface gradient techniques, micro-meteorological techniques, and tracer techniques. Gas (e.g., CH_4) concentrations generally are determined by gas chromatography with a flame ionization detector (Whalen and Reeburgh, 1988). These existing techniques can measure at a variety of sites where landfill methane emissions are in the range of 0.003 to 3000 $g/m^2 \cdot d$. The gas emission rate through cover soil can be estimated by the close flux chamber method (Reinhart et al., 1992).

5.4.2 Gas Characteristics

LFG is a complex gaseous mixture consisting primarily of CH_4 and CO_2 and small amounts of N_2 and O_2 (typically, less than 10% (v/v)) and trace quantities (in most cases, < 0.1%) of H_2S and sulfur-organics, nonmethane organic compounds (NMOC), and other impurities (see Table 5.1). The composition of LFG varies from one source to another. As a common feature, however, CH_4 is a major component in all these gases with the CH_4 concentration typically ranging from 50 to 75% (v/v) (the balance being, primarily, CO_2). In addition, the following properties of LFG always need to be considered:

- It is flammable, potentially explosive and a readily controllable source of energy.
- Its use helps to reduce the amount that would otherwise be released naturally into the atmosphere, and so reduces the excessive greenhouse effect.
- When LFG is burned, CO_2 is released. It is not considered as a net contributor to the global CO_2 level because it originated from plants, which have absorbed the CO_2 from the atmosphere.
- The harnessing of LFG helps to minimize the unpleasant decomposition smells produced in landfill sites because, otherwise, these gases would be released directly into the atmosphere. Hence, especially where landfills are situated close to inhabited areas, harnessing LFG makes landfills more socially acceptable.

5.5 Factors Affecting Biogas Production

For proper planning and design of LFG systems, the gas generation in a landfill must be estimated. As one of the most important processes for LFG production, anaerobic digestion is a versatile biochemical process. It is capable of converting almost all types of biodegradable organic matter under anaerobic conditions into an energy-rich biogas consisting of methane and carbon dioxide. In the anaerobic process, consortia of microbial populations grow harmoniously and produce reduced end products. The production rate, amount and composition of LFG typically and mainly depend on the conditions that affect the anaerobic digestion process (e.g., temperature, refuse concentration, moisture, and pH), on the transport processes (diffusion and convection) occurring in the landfill, and on the MSW composition and particle size. Decreasing particle size of fibrous plant materials can enhance anaerobic digestion (Mshandete et al., 2006).

In landfills, polysaccharide breakdown is a limiting factor in the anaerobic degradation of waste in landfills primarily involving the decomposition of complex polymers, including cellulose, hemi-cellulose, and lignin. The anaerobic degradation in landfills is attributed primarily to bacteria. Culture-based techniques and molecular biological techniques have been used to detect, identify, and monitor bacteria in landfills (Barlaz et al., 1989). Van Dyke and McCarthy (2002) have designed primer sets specific for 16S rRNA genes for four phylogenetic groups of clostridia known to contain mesophilic cellulolytic species; the primers have been authenticated for use in the rapid identification of clostridia in anaerobic environments.

Methane is generated by anaerobic digestion of methanogenic bacteria in landfills. Methanogenic bacteria isolated from landfill sites mainly include *Methanobacterium formicicum*, *Methanosarcina barkeri*, *Methanobacterium bryantii*, and a coccoid methanogen (Fielding et al., 1988). Maximum methane production requires an optimal pH, temperature and suitable sites deeply capped with an effective impermeable layer (like clay). Maximum methane formation requires a moisture content of around 40%, but too much water in a site can cause difficulties in gas collection. Thus, leachate control is an important factor for gas yield.

5.6 Gas Collection, Treatment and Energy Recovery

The technology used in a LFG energy recovery project mainly includes two aspects: (a) gas collection technology including gas wells and collection pipe work, gas blowers, condensate management and monitoring; and (b) gas end-use technology, including gas pre-treatment, gas compression, combustion plant and further monitoring (Sims and Richards, 1990; Brown and Maunder, 1994).

5.6.1 Gas Collection Systems

LFG generated by landfills should be collected to avoid the potential risks to the environment (Martín et al., 1997). Extraction and flare of the LFG are the main methods to solve the problem of the escape of LFG. In this way the pressure of the LFG within the landfill is decreased which reduces the escape of LFG from the landfill. The flare of the LFG also reduces the problem of odor. The main products of flare of LFG are CO_2 and H_2O.

LFG can be collected by gas well systems that are installed throughout the landfill. In order to collect LFG efficiently and cost-effectively, two main types of gas well systems have been developed: vertical wells and horizontal collection systems. Choice is determined mainly by site specific factors such as site status (planned, filling, or closed), fill depth, leachate level, site filling techniques, and by carrying out pumping trials on areas of a landfill site in order to assess the optimal balance between suction pressure, gas production rates, domains of well influence, and potential for the ingress of air (Brown and Maunder, 1994). An LFG collection system can be classified as a passive or active system. In a passive collection system, the pressure of the gas generated within the landfill serves as the driving force for the movement of the gas. In an active system, a blower must be used to generate a vacuum to force the gas flowing along the pipe. The active system is commonly used in LFG-to-energy systems (Hao et al., 2008). In order to capture rapidly and effectively biogas in the sanitary landfill, an active collection system usually is designed by using blower equipment. The check valve is installed at the suction inlet and output of the blower to prevent the recycle from the pipeline while extracting. Under the induced extraction, the well will become a low-pressure vacuum zone, resulting in the biogas inflow through the vertically percolated pipe. Each extraction well is also regulated by the control valve, which may reduce a considerable amount of the biogas flow rate.

Popov (2005) developed a landfill design and system for LFG containment and air ingress prevention. The system offered more control over the processes of anaerobic digestion and biogas extraction than conventional landfills. Rodríguez-Iglesias et al. (2006) developed a general method (and a model) to characterize and evaluate the optimum potential of LFG extraction at each vertical well in landfills. They found that the biogas extraction yield in the wells under study varied between ~26 and 97%, with a mean recovery value of 82%. The low yields found in certain cases were generally caused by a sealing defect that leads to excessive incorporation of air into the landfill gas through the surrounding soil or through the extraction shaft, which make LFG's subsequent utilization difficult.

Gas transport in the landfill has been modeled by assuming that there is a convective flow of gases towards the well. The extraction system is designed such that laminar flow conditions can be maintained in the well on the basis of Darcy's Law, with the Reynolds number as a check for laminar flow. The concept of a radius of influence helps to decide the well spacing. However, these models have met with only varying successes, as they did not account for site-to-site and/or zone-to-zone variability as well as the effect of other influencing factors. A hydraulic circuit model also has been proposed to understand the working of biogas wells based on fluid mechanics basic equations, continuity equation and mechanic energy balance. The mathematical model has been used to simulate the biogas extraction well in which the gas generation rate in the landfill increases and decrease after a constant rate period, respectively (Martín et al., 1997).

5.6.2 Gas Treatment Methods

Gas is drawn under suction from the extraction wells and fed to either the utilization plant or a flare stack. Suction is controlled to prevent gas migration, without drawing excessive amounts of air into the waste. Fresh LFG is warm and saturated with water vapor. High grade LFG is mostly composed of CH_4 and CO_2, with some smaller amounts of oxygen and nitrogen. The large quantity of CO_2 in LFG (typically 40–50%) presents a severe problem with its utilization for energy production due to its negative impacts on combustion efficiency and stability. The Tsotsis group has been actively involved in the development of LFG clean-up technology (He et al., 1997).

LFG contains trace compounds such as aliphatic and aromatic hydrocarbons, halogenated compounds and silicon-containing compounds. For example, LFG contains up to 20,000 ppmv of hydrogen sulfide (H_2S), a corrosive substance that attacks power engines and can affect the health of the industrial staff. Removing H_2S from LFG is a prerequisite step for LFG utilization. Redondo et al. (2008) have designed and constructed an automated H_2S on-line analyzer to assess the composition of the liquid and gas phases of gas-phase bioreactors.

To prevent liquid from damaging compressors and other equipment, pipelines are laid at an incline position, and condensate is removed by means of baffled expansion chambers (knock out pots). Regarding the prevention of corrosion associated with the interaction between H_2S and moisture, the moisture is first condensed and further separated from the biogas by compression and cooling.

Condensate drains are located at all low spots and proper intervals along biogas collection pipelines. The condensed water was percolated back into the landfill. In order to further purify the dehumidified biogas, the filtration device was equipped to remove particulates (> 4 mm) from the dehumidified stream. Particulate filters (often incorporated as part of flame attesters) are also included in the gas supply lines to protect gas extraction and energy recovery plants.

Additionally, more extensive gas cleanup may be needed. Various techniques have been used, including high pressure water scrubbing, chemical processing or semi-permeable membranes. The decision to incorporate such gas pre-treatment equipment depends largely on the balance of additional capital cost against predicted lower operating costs and higher potential availabilities.

The countercurrent water wash process is commonly used to produce biogas in sewage plants or biogas plants treating manure and other organic wastes. The potential for a high-pressure countercurrent water absorption system to enrich LFG has been demonstrated (Rasi et al., 2008). It is possible to use the countercurrent water wash process to produce high quality methane gas (Air Liquide, 2006). Halldorsson (2006) used the water scrubber technology to upgrade LFG; the methane content in the product gas has been increased to around 95%. A landfill can provide fuel for 3000-4000 vehicles. Currently, water absorption has become an effective upgrading process for LFG, as CO_2, sulfur compounds and halogenated compounds are more soluble in water than methane. The countercurrent water wash process is a simple, efficient and versatile method capable of treating corrosive compounds at a relatively low cost (Hunter and Oyama, 2000).

If LFG reforming is needed, the potentially harmful impurities (e.g., sulfurous and silicon- and halogen-containing compounds) could easily deactivate catalysts used in the reforming process (Muradov and Smith, 2008). Some of these impurities are not present in natural gas or other industrial gases, so the pretreatment of LFG could be very different from that of the conventional feedstocks for reforming processes. Nevertheless, these contaminants could efficiently be removed from LFG before the reforming stage using conventional technologies (e.g., adsorption, absorption, etc.). Other minor components that are present in LFG such as N_2 and nonmethane organic compounds may not adversely affect the catalyst activity during the LFG reforming reaction.

5.6.3 Energy Recovery of LFG

Upon removal of most of the trace organic compounds from LFG, it can be used as fuel in ICEs or gas turbines for generation of heat and electricity (Bove and Lunghi, 2006). The most common technologies used at the present time for energy production with LFG are discussed as follows:

Internal Combustion Engines. Reciprocating internal combustion engines (ICEs) is one of the most widely used technologies for generating electricity at landfills. More than two-thirds of the operational landfills use this type of equipment to generate electricity because of the compatibility of the power with the economic feasibility of the system. Another advantage is that ICEs are compact and easy to transport. Thus, they can be moved from one well to another in case of LFG shortage in one well. The disadvantage of the ICEs is the high air pollution generation. The amounts of NO_x and CO, compared to the other technologies, are significantly high.

ICEs burn the landfill gas in the presence of oxygen. The engine is connected to a crankshaft that turns a generator and produces electricity. In ICEs, the purified biogas is used to generate electricity. The capacity of each power generation module is set at 1.0–1.4 MW which is generated by the engine generator. In order to control the noise from ICEs, each module is also installed in a soundproof housing, which contains an independent electricity operation system. A programmable logic controller unit controls the entire generation system, including startup, shutdown, loading control and alarm. The voltage of electricity is transformed to 11.4 kV or a proper voltage, and the electricity (the green energy) is transferred to the end user. It is reported that ICEs have higher emission costs than flaring the landfill gas.

Micro-Turbines. Gas turbines (GTs) and steam turbines can also be used in landfill-gas-to-energy projects, and represent the second most used technologies for LFG energy conversion. GTs combust LFG to heat compressed air, making it to expand to power a turbine, which in turn drives a generator. Like ICEs, gas turbines have a lifetime of over 25 years. In steam turbines, LFG is used to heat up water and produce steam that spins the generator. Steam turbines have a lifetime of up to 50 years. Steam turbines have the lowest emission costs, followed by gas turbines.

Molten Carbonate Fuel Cell. Molten carbonate fuel cells (MCFC) operate at relatively high temperatures (650 °C) without using noble metals as catalysts for the electrochemical reaction. This property enables MCFCs to operate with higher impurity concentrations than low temperature fuel cells. In the MCFC, no combustion occurs, and thus, the NO_x and CO concentrations are almost zero. Pollutants are produced during the fuel treatment, in particular for the steam reforming of the methane embedded in the LFG. A life cycle assessment (LCA) study for MCFCs fuelled with LFG shows that the CO, CO_2 and NO_x are drastically reduced when LFG is used instead of natural gas. The US Department of Energy (USDOE) has shown that the MCFCs using LFG as the fuel are very promising, compatibility of the gas composition with the fuel cell requirements, and demonstrated the high energy conversion efficiency obtainable through MCFCs. The disadvantage of MCFCs is their high capital cost.

Spiegel et al. (1997) tested fuel-cell (FC) energy recovery and control of LFG emissions. Their results show the FC-LFG energy-conversion system provides net reduction in total emissions while simultaneously removing the methane from the LFG, and encourage the use of LFG in FCs, one of the cleanest energy-conversion technologies available. The fuel cell system is more attractive due to its high efficiency. Spiegel et al. (1999) reported a test successfully demonstrated the operation of a commercial phosphoric acid fuel cell on LFG at the Penrose Power Station in Sun Valley, California. The output of the fuel cell was up to 137 kW; 37.1% efficiency at 120 kW; exceptionally low secondary emissions of 0.77 ppmV

CO, 0.12 ppmV NO_x, and undetectable SO_x. The LFG pretreatment unit operated for a total of 2297 h with the fuel cell.

Solid Oxide Fuel Cell. Solid Oxide Fuel Cell (SOFC) co-generators could substantially upgrade the value of LFG, due to their advantages of their modularity and relative insensitivity to micro-contaminants in the LFG. SOFC units in the 5–500 kW range can be installed on farms, landfill sites, sewage treatment plants, organic solid waste digesters or organic liquid effluent treatment sites (van Herle and Membrez, 2002). Typically, operating temperatures of SOFCs are in the range of 800–1000 °C. The high temperature makes LFG and biomass derived gas to be internally reformed. It is possible to realize combined heat and power plant solutions (CHP) or hybrid systems.

van Herle et al. (2003) analyzed energy balance by building a process flow model and an electrochemical model in a SOFC-biogas combined small heat and power system in order to assess conversion efficiencies of biogas by SOFCs, as a function of the variation of adjustable operating parameters such as reforming conditions, air excess rate, SOFC stack temperature, imposed stack current, etc. In western Finland, a planar SOFC unit from Wärtsilä (Wärtsilä Corporation, 2008) will produce 20 kW of electric power and 14–17 kW of thermal output with very low emissions from a biogas-fueled fuel cell. It will be the first in the world fueled by biogas or methane originating from a nearby landfill. The performance of SOFCs and MCFCs are very close.

Organic Rankine Cycle. The Organic Rankine Cycle (ORC) is based on the same thermodynamic cycle as the classical traditional Rankine cycle, with the exception of the working fluid (typically, iso-butane or propane, rather than water) is used. The ORC is generally used for geothermal energy conversion. Since the ORC is an external combustion engine, if LFG is used to replace geothermal energy as the energy source, both plant layout and operation would be similar. Therefore, the data collected from the ORC systems that use the geothermal energy as the energy source can be considered as a reference for an ORC operating on LFG. However, some deviations might occur for the emissions, when operating on LFG.

Others. LFG can also be used to produce bio-fuels, including compressed natural gas (CNG), liquid natural gas (LNG) and hydrogen. CNG is estimated to be similar to pipeline quality LFG recovery. LNG produced from LFG requires steps beyond CNG to further purify and liquefy the LFG. Hydrogen is relatively easy to measure and is a critical intermediate that may be used to monitor the status of landfill fermentations. The relationship between H_2 level and methanogenesis in landfills has been investigated (Mormile et al., 1996). Hydrogen production in landfills results from a thermodynamic uncoupling of fatty acid oxidation. Inhibitory effects of low pH and accumulation of volatile fatty acids result in low growth rates of methanogenesis. According to the results of current studies, it may be more technically and economically feasible to use hydrogen as an enrichment to LFG to improve the combustion process for lowering criteria pollutant emissions from engines using the LFG to produce electricity.

In addition, LFG as a fuel for a spark ignition engine to produce power will be an effective way. However, the performance of the LFG fueled engine deteriorated in comparison with methane operation. Increases in the compression ratio and advancing spark timing can improve the performance of the LFG operation. Additions of hydrogen in the LFG can improve the performance, extend the operational limits, improve the combustion characteristics, and reduce cyclic variations of LFG operations, especially at the lean and rich mixtures (Shrestha and Narayanan, 2008).

Recent developments in Fischer–Tropsch (FT) technology point to a possibility of producing FT hydrocarbons from a CO_2-rich syngas and even from H_2-CO_2 mixtures. This technological option would be particularly advantageous for the production of synthetic liquid hydrocarbons (SLH) from LFG, biogas, and biomass that generate syngas with a relatively low H_2:CO ratio and a high CO_2 content. It will also significantly simplify the process by eliminating gas conditioning and CO_2 removal steps. Muradov and Smith (2008) assess the technical feasibility of direct catalytic reforming (i.e., without preliminary methane recovery from these gases) of LFG and biogas to synthesis gas (syngas) that could further be processed to synthetic liquid hydrocarbon fuels via the FT synthesis. They evaluated the catalytic activity and selectivity of a number of noble metal (Ru, Ir, Pt, Rh, Pd) and Ni-based catalysts for reforming of a model CH_4-CO_2 mixture mimicking a typical LFG into a synthesis gas. They also investigated issues related to the catalysts stability and process sustainability under the conditions that are favorable for carbon deposition. Their results have shown that the syngas produced by Ni-catalyzed steam-assisted reforming of a model LFG is suitable for production of liquid hydrocarbons via FT synthesis (Muradov and Smith, 2008).

5.7 Tools and Models

Landfill sites can be assessed by modeling and physical measurements. In order to quantify the available information and to plan and design the system, more precise mathematical models have been developed to predict gas production and flow characteristics under varying conditions. Models of LFG formation have been produced which attempt to simulate the processes and kinetics that are believed to control the gas formation in landfills. For example, Young (1992) reported that gas emissions are sensitive to the rate of change in barometric pressure, which can also affect the methane content of the gas released. These models basically assume reasonably homogeneous condition and primarily aim at the prediction of total gas yield or gas production rate. Besides, a large number of factors and/or processes influence the gas production rate in MSW landfills. To simulate the effects of these various interacting processes, mechanistic models are needed (Haarstrick et al., 2001; Batstone et al., 2002; White et al., 2004). Mechanistic models require input in terms of rate constants and effects of environmental conditions on these rate constants.

Many researchers have used Monod kinetics to describe biodegradation reactions. Simultaneous effects of 10 environmental factors on the rate of methane production from the Fresh Kills landfill, which is the world's largest landfill, have been evaluated by multiple-regression analysis (Gurijala et al., 1997). Among these factors, volatile solids, moisture content, sulfate, and the cellulose-to-lignin ratio were significantly associated with methane production from refuse. A generic spatially distributed numerical model has been developed to link sub-models of landfill processes in order to simulate solid waste degradation and gas generation in landfills and provide the basis of the modeling of the chemistry and microbiology of solid waste degradation. The model includes the simulation of the transport of leachate and gases, and the consolidation of the solid waste (White et al., 2004).

In order to improve the economics of LFG utilization for energy production, it is important to develop a fundamental knowledge base about LFG burning characteristics. A simple radiation model (SRM) and a detailed radiation model (DRM) are two common models of radiating heat transfer. The SRM includes the use of the Planck's mean absorption coefficient, and the assumption of an optically thin gas (Egolfopoulos, 1994). The DRM includes the use of the subroutine RADCAL that predicts the spectral structure of various combustion products over a wide range of temperature, pressure, and path-length (Grosshandler, 1993) The RADCAL provides the spectral absorption co-efficiencies for the radiating species of CO_2, H_2O, CO, and CH_4. Experimentally and numerically quantified combustion characteristics of LFG in a stagnation flow configuration have made it possible to simulate the experiments using a complete description of molecular transport and the detailed GRI 2.11 chemical kinetic mechanism. The SRM and DRM have been implemented for both the stagnation flow configuration and the freely propagating flame. Both the experimental and the simulation results have proved that the presence of CO_2 in LFG significantly decreases the laminar flame speeds and extinction strain rates; the results also indicate that an increase in CO_2 concentrations would increase the amount of NO emissions per gram of consumed CH_4. Considering the DRM and SRM, the optically thick (DRM) model results in higher laminar flame speeds, extinction strain rates, and maximum flame temperatures as compared to the optically thin (SRM) model. The effect of CO_2 on the flame response is of thermal rather than kinetic nature, as the mixture is diluted by the "inert" CO_2 that is also an efficient radiator. By calculating fundamental flammability limits, it has been found that as the CO_2 content in the fuel feed increases, the equivalence ratio that is required to achieve the same rate of CH_4 consumption increases, the NO starts to be produced and, finally, the flammable range noticeably decreases. Thus, the use of LFG may not be advantageous in terms of NO emissions, and its combined use with natural gas may be an optimal solution in terms of efficient energy utilization and emissions.

Making estimates of LFG potentials is a critical step to establish a LFG utilization facility. Multi-phase models are the preferred method, where the degradation rates of different waste fractions are considered (Lagerkvis, 1995). On the basis of fundamental principles governing the physical, chemical and microbiological processes, El-Fadel et al. (1996a, b, 1997) reported formulation, application and sensitivity analysis of a mathematical model for describing the dynamics of the microbial landfill ecosystem into multi-layer, time-dependent transport and generation of gas and heat models on the Mountain View Controlled Landfill Project (MVCLP) in California, USA. The Sahimi group (Hashemi et al., 2002; Sanchez et al., 2006, 2007) has developed a comprehensive three-dimensional (3-D) model of gas generation and transport in landfills. The model considers the heterogeneity in the distributions of the permeability and porosity in the landfill, as well as that of the soil that surrounds it; it also considers an arbitrary number of gas extraction wells with an arbitrary spatial distribution. The optimal spatial distributions of the porosity, permeability, the tortuosity factor and the total potential of various types of wastes for producing the gases in a landfill are determined. The model not only reproduces accurately the data, but also provides accurate predictions for the future behavior of the landfill's properties.

"Life-cycle-assessment (LCA) is a process to evaluate the environmental burdens associated with a product, process, or activity by identifying and quantifying energy and materials used and wastes released to the environment; and to identify and evaluate opportunities to affect environmental improvements" (SETAC, 1993). The LCA includes every process, every product and raw material connected to the system to be analyzed, including raw material processing, production, recycling and disposal. LCA has been conducted for evaluation of environmental consequences of fuel-cells-based landfill-gas energy conversion technologies and to provide a guide for further environmental impact reduction (Lunghi et al., 2004).

5.8 Economic Evaluation

It should be pointed out that many opportunities exist to reduce greenhouse gas emissions, in MSW management systems (Figure 5.1). Incineration of MSWs is the most effective technology, yet the investment is the highest, and some air pollution problems are generated. Energy conversion of MSWs by biological actions (mainly including anaerobic digestion and alcoholic fermentation) that convert semi-solid or liquid biomass into LFG or liquid fuel has been widely utilized (Tsai et al., 2004). Systems containing energy recovery devices are credited for selling energy. For example, selling electricity at $0.05/kWh will reduce the tipping fee (TF) at the gate of an incineration plant by $9.4/ton of waste; at an anaerobic digestion plant by $3.5/ton, and at a landfill that recovers energy from LFG by $1.6/ton (USDOE, 1993). The investment required to reduce a ton of CO_2 equivalent by aerobic composting, using the windrow technology, is the lowest, but the efficiency of this method to reduce GHG emission is high. Therefore, the lowest cost alternative to mitigate GHG emissions from the waste sector is to construct composting plants. In this option, all organic waste will be processed, some of the materials (i.e., paper, plastics, etc.) will be recycled, and only the non-recyclable waste that does not produce GHG will be landfilled. By adopting an integrated waste management approach, a reduction of 8–9% of total GHG emissions could be achieved at reasonable costs. Moreover, soil conditioner recovered from both systems and energy recovered from the anaerobic digestion system can avoid the production of chemical soil conditioners and generation of energy, respectively, and the consequent reduction in GHG emissions.

The investment to reduce one ton of CO_2 equivalent by collecting and burning the LFG in landfills is rather low, less than $20. The gas pretreatment unit flare safely disposed of the removed LFG contaminants by achieving destruction efficiencies greater than 99%. The cost of a system containing energy recovery would be more than twice of that collecting and burning the LFG (USEPA, 1997a, b). Yet, these systems can recover only 40–90% of the emitted gas. Hao et al. (2008)

evaluated the feasibility of trigeneration using LFG as fuel in terms of energy saving, environment protection and economic benefit, especially in Hong Kong, and showed that LFG for trigeneration can achieve not only a higher energy saving and a lower GHG emission but also a more impressive advantage in economy than the other scenarios.

Table 5.3 lists characteristics of different ways for energy recovery of FLG. The ICE, the most used LFG energy recovery systems presents the worst environmental performance of all. The emissions of SCE and high temperature fuel cells (both MCFC and SOFC) are very low. The table also lists an economic analysis for evaluating the favorable conditions for cleaner and higher efficiency technologies.

In order to evaluate the abatement costs of GHG emissions, the investment cost of each alternative has been evaluated. Costs of a LFG power generation scheme mainly include capital costs and running costs. The capital costs involve prime mover and generator civil works, electrical and control equipment; the running costs involve operation and maintenance of the plant (O&M), rates, refurbishment and management. Among these, O&M costs are mainly affected by site specific variables that are a function of factors such as salaries, transportation, insurance rates, taxes and prices obtained for recyclables or energy sales, etc. It should be noted that there is a rough correlation between investment costs and O&M costs. As a rule of thumb, O&M represent ca. 40% of the costs of landfilling and 35%–40% of the incineration cost (excluding revenues from energy sales) (USEPA, 1997a). In UK, methane typically begins to be produced within about two years of waste emplacement, builds up to a maximum within about five years, and thereafter declines slowly. Gas production may be measurable for several decades, but usable amounts (containing at least 28% (v/v) methane) are produced for about 10 to 15 years (Brown and Maunder, 1994).

Table 5.3 Characteristics of different ways for energy recovery of FLG (Adapted from Bove and Lunghi, 2006).

Method for energy recovery	Electric efficiency	Fuel consumption (MJ/kWh)	Emission (μg/kJ)		Installed power (MW)	Investment cost (M$)	O & M ($/kW-y)
			NO_x	CO			
ICE	33%	11	56.6	56.6	1	1.2	115.2
GT	28%	12.9	15	19	0.84	1.26	99.84
ORC (Geothermal)	18%	19.2	16	18.9	0.54	0.81	61.44
MCFC	50%	7.2	Trace	1.4	1.5	4.2	96
SCE	38.6%	9.4	3.11	15	-	-	-
SOFC	-	-	-	-	1.5	5.25	84

Note: the hypotheses include availability of the systems (320 d/y), systems life (10 years) or fuel cell stack life (5 years), and fuel cell stack cost equal to one third of the fuel cell system cost.

In addition, capital and O&M costs for the gas extraction plant or the cost of LFG supply to the prime mover may also be included. The inclusion of these latter costs, or a proportion of them, in the overall figure depends on a judgment of how much of the gas collection system would be required for environmental protection. For example, it may be decided that a blower to extract the LFG would be required for migration control, and its cost should therefore not be borne by the energy recovery scheme.

Although energy recovery from LFG is one of the most promising renewable energy technologies, LFG energy recovery projects are not always successful. LFG energy recovery projects face non-technical barriers, which include awareness of LFG energy recovery and risk perception by potential end-users and financiers, project and energy supply economics, market access, finance, planning, permitting, licensing, and bureaucracy (He et al., 1997). In order to promote the biomass energy utilization from waste landfills, there is a need to increase the public awareness of LFG utilization as a valuable, free and renewable energy resource, make related policy, regulations or laws to ensure efficient utilization of LFG, and encourage and stimulate the operators of landfills and the potential building owners to use the energies produced by the LFG trigeneration plants off-site the landfills. In particular, the government could tie the gap between the LFG operators and the potential building owners.

5.9 Future Work

The valorization and application of LFG as a source of energy not only eliminate the possibility of methane combustion and an increased greenhouse effect, but also decrease the use of fossil fuel. However, the valorization is affected by landfill systems, LFG collection, LFG concentration and purity, LFG online monitoring, policies, and so on. To alleviate the contradiction between rapid economic growth and the shortage of raw materials/energy and to establish a synergy between environmental protection and energy production, future work on landfill technology may focus on:

- *LFG collection and purification.* It is important to collect and control LFG from a landfill site or system. This might be realized by optimization of well systems based on improved understanding of gas and leachate flows in landfills. Usually, LFG contains a trace amount of organic contaminants. For commercial applications of LFG, it is necessary to develop cheaper purification processes.
- *Improvement of methodologies for in-situ and post-closure monitoring.* Current landfill management has to take into account many impacts or harmful effects to the local environment. Monitoring technologies are always needed to help landfill managers with solving the problem, reducing the volume of lost LFG, and thus, optimising the LFG management system. For example, modern landfill technologies, such as an upper liner over completed landfill cells and the LFG extraction system, contribute to drastically reduce greenhouse gas emissions. But the integrity of landfill capping remains the weakest element because LFG could escape through leakages. Therefore, development of surface monitoring of LFG emission (i.e., portable analyser devices for leakage detection) is needed.
- *Development and optimization of large-scale capacity engineered landfills (or bioreactors).* Currently, landfilling is done in engineered landfills equipped with leachate and LFG collection and treatment systems. However, this technique presents two major shortcomings, i.e., it requires a large area, and generates emissions over several years. As a new landfill technology, the bioreactor addresses these problems by accelerating the degradation of the organic fraction, and therefore, reduces the time necessary to stabilize the landfilled waste and increase the site's capacity. However, systematic information is not available on

how to develop and optimize large-scale capacity engineered landfills and bioreactors.

- *Life-cycle assessment (LCA).* Although landfill system design, LFG production and collection system, theories, models and case projects have been well studied (e.g., Huber-Humer et al., 2009), the influence of other parameters (e.g., bottom layer, final cover, leachate/LFG collection systems' efficiencies, CO_2 sequestered, temporal boundary corresponding to waste stabilization) on landfill performance and potential environmental impacts has not been fully evaluated. Improvement of evaluation approaches both in the planning stage and throughout operation of the landfill system is needed. It's helpful to estimate the contribution to the greenhouse effect and improve the confidences of the investigators.
- *Sustainability, broader visions and non-technical barriers.* Clean production was the first ecological economy. Later, eco-industrial came into being, and it was not until the 1990s that the economic cycle caught the attention of every country. At the beginning of the 21st century, circular economy (CE) is emerging as an economic strategy rather than a purely environmental strategy (Yuan et al., 2006). Currently, CE has been used as an economic development model, which is at the core of resource efficiency and recycling, following the principle of reduction, reuse and resource (the "3R" principle), with the feature of lower-input, lower consumption, lower emissions and high efficiency and in line with the concept of sustainable development (Chen et al., 2009). We need to have broader visions for future development of landfill technology, such as developing landfill technology within the CE framework. We also need to expand LCA of landfill technology by including other MSW management activities (source reduction, recycling, composting and incineration) and non-technical barriers (e.g., related government policy, landfill operations and securing permission to construct the plant from the local planning).

5.10 Summary

Application of LFG means a synergy between environmental protection and energy production. LFG is one of the important bioenergies from landfills. Utilization of LFG has received much attention lately. The recovered energy can reduce the amount of energy consumption and also emissions that would otherwise be produced by other energy systems such as fossil-fired power plants.

Gas collection and control from a greater proportion of landfill sites are important factors for strengthening environmental protection standards governing waste management. Many studies are focusing on the improvement of gas collection systems and *in-situ* assessment methodology. Pretreatment of waste can ensure better mixing and recirculation of leachate to maintain optimal moisture content for decomposition. Meanwhile, non-technical barriers to LFG exploitation need to be considered including negotiation of rights to the gas, securing permission to construct the plant from the local planning authority, obtaining finance and gaining access to the end-user market. The ultimate goal is to make LFG energy recovery be self-sustaining on the open market.

Although ICEs present the poorest environmental performance, these are the most widely used technology due to economical reasons. Fuel cells, on the contrary, are the cleanest energy conversion systems, but the investment cost is still too high to compete with traditional energy systems. However, fuel cells can become economically competitive because of their high energy conversion efficiency. Catalytic reforming of LFG to a syngas has been actively researched over the years and may have a very bright future.

5.11 References

Air Liquide (2006). *Gas Encyclopedia.* http://encyclopedia.airliquide.com/Encyclopedia.asp?languageid=11.

Shrestha, B, S.O., and Narayanan, G. (2008). "Landfill gas with hydrogen addition–a fuel for SI engines." *Fuel*, 87, 3616–3626.

Barlaz, M.A., Schaefer, D.M., and Ham, R.K. (1989). "Bacterial population development and chemical characteristics of refuse decomposition in a simulated sanitary landfill." *Appl. Environ. Microbiol.*, 55, 55–65.

Batstone, D.J., Keller, J., Angelidaki, I., Kalyuzhnyi, S.V., Pavlostathis, S.G., Rozzi, A., Sanders, W.T.M., Siegrist, H., and Vavilin, V.A. (2002). "The IWA anaerobic digestion model no 1 (ADM1)." *Water Sci. Technol.*, 45(10), 65–73.

Bove, R., and Lunghi, P. (2006). "Electric power generation from landfill gas using traditional and innovative technologies." *Energy Conversion and Management*, 47, 1391–1401.

Brown, K.A., and Maunder, D.H. (1994). "Using landfill gas: a uk perspective." *Renewable Energy*, 5, 774–781.

Chen, W.-Y., Ma, Y.-M., Li, D.-H., Yuan, B., and Gao, S. (2009). "The circular economy mode and research on sustainable development strategy of chemical enterprises." Bulletin of Wuhan Institute of Technology, 2009, 1-7.

Demirbas, A. (2006). "Biogas production from the organic fraction of municipal solid waste." *Energy Socurces, Part A*, 28, 1127–1134.

DHHS (Department of Health and Human Services) (2001). *Landfill Gas Primer – An Overview for Environmental Health Professionals,* available at http://www.atsdr.cdc.gov/HAC/landfill/html/toc.html.

Egolfopoulos, F.N. (1994). "Geometric and radiation effects on steady and unsteady strained laminar flames." *Proc. Combust. Inst.*, 25(1), 1375–1381.

El-Fadel, M., Findikakis, A.N., and Leckie, J.O. (1996a) "Numerical modeling of generation and heat in landfills I. model formulation" *Waste Management & Research*, 14, 483–504.

El-Fadel, M., Findikakis, A.N., and Leckie, J.O. (1996b). "Numerical modeling of generation and heat in landfills II. model application." *Waste Management & Research*, 14, 537–551.

El-Fadel, M., Findikakis, A.N., and Leckie, J.O. (1997). "Numerical modeling of generation and heat in landfills III. sensitivity analysis." *Waste Management & Research*, 15(1), 87–102.

EASEPA (Environment Agency and Scottish Environment Protection Agency). (2004). *Guidance on Gas Treatment Technologies for Landfill Gas Engines, Bristol.*http://www.sepa.org.uk/pdf/guidance/landfill_directive/gas_treatment_tech.pdf.

Fielding, E.R., Archer, D.B., de Macario, E.C., and Macario, A.J.L. (1988). "Isolation and Characterization of Methanogenic Bacteria from Landfills" *Applied and environmental microbiology*, 54(3), 835–836.

Garg, A., Smith, R., Hill, D., Simms, N., and Pollard, S. (2007). "Wastes as co-fuels: the policy framework for solid recovered fuel (SRF) in Europe, with UK Implications." Environ. Sci. Technol., 41, 4868–4874.

Gendebien, A., Pauwels, M., and Constant, M. (1992). *Landfill Gas -from Environment to Energy.* Commission of the European Communities Directorate General Energy Final Report. Contract No 88-B-7030-11-3-17.

Glebs, R.T. (1989). "Subtitle D: how will it affect landfills?" *Waste Alternatives*, 1(3), 56–64.

Gronow, J. (1990a). "Waste management research in the U.K.–progress and priorities." Paper from *Proc. 1990 Harwell Waste Management Symp on "The 1980s–a Decade of Progress and Achievements in Waste Management Research,"* held at Harwell, Oxon, U.K., May 1990. AEA Environment and Energy, 1-11.

Gronow, J. (1990b). "The dept. of environment landfill gas programme." *Proceedings of International Conference on Landfill Gas: Energy and Environment 90,* Bournemouth, U.K., 16–20 October 1990, pp. 178–189.

Grosshandler W.L. (1993). "RADCAL: a narrow-band model for radiation calculations in a combustion environment, NIST Technical Note, 1402.

Gurijala, K.R., Sa, P., and Robinson, J.A. (1997). "Statistical modeling of methane production from landfill samples." *Applied and Environmental Microbiology*, 63(10), 3797–3803.

Haarstrick, A., Hempel, D.C., Ostermann, L., Ahrens, H., and Dinkler, D. (2001). "Modelling of the biodegradation of organic matter in municipal landfills." *Waste Manage. Res.*, 19(4), 320-331.

Halldorsson, B. (2006). "Experience of landfill gas upgrading to vehicle fuel." In: *Proceedings of Nordic Biogas Conference*, Feb. 8–9, 2006, Helsinki, Finland.

Hao, X., Yang, H., and Zhang, G. (2008). Trigeneration: a new way for landfill gas utilization and its feasibility in Hong Kong." *Energy Policy*, 36, 3662–3673.

Hashemi, M., Kavak, H.I., Tsotsis, T.T., and Sahimi, M. (2002). "Computer simulation of gas generation and transport in landfills–I: quasi-steady-state condition." *Chemical Engineering Science*, 57, 2475–2501.

He, C., Herman, D., Minet, R.G., and Tsotsis, T.T. (1997). "A catalytic/sorption hybrid process for landfill gas cleanup." *Ind. Eng. Chem. Res.*, 36(10), 4100–4107.

Hessami, M., Christensten, S., and Gani, R. (1996). "Anaerobic digestion of household organic waste to produce biogas." *WREC*, 9(1-4), 954–957.

Huang, C., and Raissi, T.A. (2007). "Analyses of one-step liquid hydrogen production from methane and landfill gas." *Journal of Power Sources*, 173, 950–958.

Huang, Q., Yang, Y., Pang, X., and Wang, Q. (2008). "Evolution on qualities of leachate and landfill gas in the semi-aerobic landfill." *Journal of Environmental Sciences*, 20, 499–504.

Huber-Humer, M., Röder, S., and Lechner, P. (2009). "Approaches to assess biocover performance on landfills." *Waste Management*, 29, 2092–2104.

Hunter, P., and Oyama, S.T. (2000). "Control of volatile organic compound emissions." In: *Conventional and Emerging Technologies*. John Wiley & Sons Inc., New York, USA.

Igoni, A.H., Ayotamuno, M.J., Eze, C.L., Ogaji, S.O.T., and Probert, S.D. (2008). "Designs of anaerobic digesters for producing biogas from municipal solid-waste." *Applied Energy*, 85(6), 430-438.

IPCC (1992). *Climate Change. The IPPC Scientific Assessment.* Houghton, E.J.T., Jenkins, G.J., and Ephramus, J.L. (eds.). Cambridge University Press, Cambridge.

Jaramillo, P., and Matthews, H.S. (2005). "Landfill-gas-to-energy projects: analysis of net private and social benefits." *Environmental Science and technology*, 39(19), 7365–7373.

Karlsson, H., Lagerkvist, A., Meijer, J.E., and Nilsson, P. (1993). "Optimized biogas generation from landfill cells." Proceedings of ISWA-RVF Conference: Better Waste Management-a global challenge, Jonkoping, Sweden. 1993.

Lagerkvist, A. (1995). "The landfill gas activity of the IEA bio-energy agreement." *Biomass & Bioenergy*, 9(1-5), 399–413.

Lawson, P., Campbell, D.J.V., Lagerkvist, A., and Meijer, J,E. (1992). *Landfill Gas Test Cell Data Exchange*. IEA-report, ETSU, Harwell, UK, 1992.

Liamsanguan, C., and Gheewala, S.H. (2008). "The holistic impact of integrated solid waste management on greenhouse gas emissions in Phuket." *Journal of Cleaner Production*, 16, 1865–1871.

Lombardia, L., Carnevalea, E., and Corti, A. (2006). "Greenhouse effect reduction and energy recovery from waste landfill." *Energy*, 31, 3208–3219.

Lunghi, P., Bove, R., and Desideri, U. (2004). "Life-cycle-assessment of fuel-cells-based landfill-gas energy conversion technologies." *Journal of Power Sources*, 131, 120–126.

Martín, S., Maranon, E., and Sastre, H. (1997). "Landfill gas extraction technology: study, simulation and manually controlled extraction." *Bioresource Technology*, 62, 47–54.

McLeod, S. (1994). "Landfill gas enhancement - large scale R & D–the brogborough test cells." Proceeding from "Landfill Gas Seminar, Energy from Landfill Gas - Making it Work," 17/3 at Solihull, UK, ETSU, UK. 1994.

Mormile, M.R., Gurijala, K.R., Robinson, J.A., Mcinerney, M.J., and Suflita, J. (1996). "The importance of hydrogen in landfill fermentations." *Applied and Environmental Microbiology*, 62(5), 1583–1588.

Mshandete, A., Björnsson, L., Kivaisi, A.K., Rubindamayugi, M.S.T., Mattiasson, B. (2006). "Effect of particle size on biogas yield from sisal fiber waste, *Renewable Energy*, 31(14), 2385–2392.

Muradov, N., Smith, F., Huang, C., T-Raissi, A. (2006). "Autothermal catalytic pyrolysis of methane as a new route to hydrogen production with reduced CO_2

emissions." *Catal. Today*, 116(3), 281–288.

Muradov, N., and Smith, F. (2008). "Thermocatalytic conversion of landfill gas and biogas to alternative transportation fuels." *Energy & Fuels*, 22, 2053-2060.

Onay, T.T., and Pohland, F.G. (1998). In situ nitrogen management in controlled bioreactor landfills." *Water Research*, 32(5), 1383–1392.

Popov, V. (2005). "A new landfill system for cheaper landfill gas purification." *Renewable Energy*, 30, 1021–1029.

Porteous, A. (1993). "Developments in, and environmental impacts of, electricity generation from municipal solid waste and landfill gas combustion." *IEE Proceedings-A*, 140(1), 86–93.

Qin, W., Egolfopoulos, F.N., and Tsotsis, T.T. (2001). "Fundamental and environmental aspects of landfill gas utilization for power generation." *Chem. Eng. J.*, 82, 157–172.

Rasi, S., Läntelä, J., Veijanen, A., and Rintala, J. (2008). "Landfill gas upgrading with countercurrent water wash." *Waste Management*, 28, 1528–1534.

Rees, J.F. (1985). "Landfill for treatment of solid wastes." In: *The Principles, Applications, Regulations of Biotechnology in Industry, Agricultural and Medicine. Comprehensive Biotechnology*, M. Moo-Yong (Ed.), vol. 4, U.K., Pergamon Press.

Redondo, R., Machado, V.C., Baeza, M., Lafuente, J., and Gabriel, D. (2008). "On-line monitoring of gas-phase bioreactors for biogas treatment: hydrogen sulfide and sulfide analysis by automated flow systems." *Anal Bioanal Chem*, 391, 789–798.

Reinhart, D.R., Cooper, D.C., and Walker, B. (1992). "Flux chamber design and operation for the measurement of municipal solid waste landfill gas." *J. Air Waste Manage. Assoc.*, 42, 1067–1070.

Rodríguez-Iglesias, J., Vázquez, I., Marañón, E., Castrillón, L., and Sastre, H. (2006). "Extraction wells and biogas recovery modeling in sanitary landfills." *Air & Waste Manage. Assoc.*, 55, 173-180.

Sanchez, R., Hashemi, M., Tsotsis, T.T., and Sahimi, M. (2006). "Computer simulation of gas generation and transport in landfills II: Dynamic conditions." *Chemical Engineering Science*, 61(14), 4750–4761.

Sanchez, R., Tsotsis, T.T., and Sahimi, M. "Computer simulation of gas generation and transport in landfills. III: development of landfills' optimal model." *Chemical Engineering Science*, 62, 6378–6390.

SCS Engineers (2008). *Helping Landfill Owners Achieve Effective, Low-Cost Compliance with Federal Landfill Gas Regulations.* Work assignment 1–2, Task 6, Submitted to: USEPA, Landfill Methane Outreach Program, by SCS Engineers. Available at http://www.epa.gov/lmop (accessed Oct. 2008).

Sebastian, W., Jager, P., and Dohler, H. (2006). "Balancing of greenhouse gas emissions and economic efficiency for biogas-production through anaerobic co-fermentation of slurry with organic waste." *Agriculture, Ecosystems and Environment* ,112, 178–185.

Sims, R.E.H., and Richards, K.M. (1990). "Anaerobic digestion of crops and farm wastes in the United Kingdom." *Agriculture, Ecosystems & Environment*, 30(1–2), 89–95.

SETAC (Society of Environmental Toxicology and Chemistry). (1993). *Guidelines for Life-Cycle Assessment: A "Code of Practice"*, SETAC, 1993.

Spiegel, R.J., Trocciola, J.C., Preston, J.L. (1997). "The results for fuel-cell operation on landfill gas." *Energy*, 22(8), 777–786.

Spiegel, R.J., Preston, J.L., and Trocciola, J.C. (1999). "Fuel cell operation on landfill gas at Penrose Power Station." *Energy*, 24(8), 723–742.

Taleghani, G., and Kia, A.S. (2005). "Technical–economical analysis of the Saveh biogas power plant." *Renewable Energy*, 30, 441–446.

Tchobanoglous, G., Theisen, H., and Vigil, S. (1993). *Integrated Solid Waste Management. Engineering, Principles and Management Issues*. McGraw-Hill, New York, 1993.

Themelis, N.J., and Ulloa, P.A. (2007). "Methane generation in landfills, *Renewable Energy*, 32(7), 1243 –1257.

Tsai, W.T., Chou, Y.H., and Chang, Y.M. (2004). "Progress in energy utilization from agrowastes in Taiwan." *Renewable and Sustainable Energy Reviews*, 8, 461–481.

Tsai, W.T. (2007). "Bioenergy from landfill gas (LFG) in Taiwan." *Renewable and Sustainable Energy Reviews*, 11, 331–344.

USDOE. (1993). *Externalities from Landfill and Incineration. a Study by CSERGE and EFTEL*. HMSO, Edinburgh, UK.

USEPA (1994). *Characterization of Potential of Municipal Solid Waste (MSW) Components. Municipal Solid Waste in the United States: 1992 Update.* EPA/530-R-94-042, NTS #PB 95–147690. Solid Waste and Emergency Response (5305), Washington, D.C.

USEPA. (1997a). *Energy Project LFG Utilization Software (EPLUS)*. EPA 430-B97-006, www.epa.gov/globalwarming.

USEPA. (1997b) *Feasibility Assessment for Gas-to-Energy at Selected Landfills in Sao Paulo, Brazil*. EPA 68-W6-0004.

USEPA (2004). *Inventory of U.S. Greenhouse Gas Emissions and Sinks: 1990–2002*; Office of Atmospheric Programs, U.S. Environmental Protection Agency, Government Printing Office, Washington, DC, 2004.

USEPA (2008). Landfill Methane Recovery and Use Opportunities. Available at www.methanetomarkets.org.

USEPA (2008). Municipal Solid Waste Generation, Recycling, and Disposal in the United States: Facts and Figures for 2007. EPA-530-F-08-018, USEPA Solid Waste and Emergency Response (5306P), Washington, DC 20460, Nov. 2008.

van Dyke, M.I., and McCarthy, A.J. (2002). "Molecular biological detection and characterization of clostridium populations in municipal landfill sites." *Applied & Environmental Microbiology*, 68(4), 2049–2053.

van Herle, J., and Membrez, Y. (2002). "Biogas exploitation in SOFC." In: U. Bossel (Ed.), *Proceedings of the Fifth European Solid Oxide Fuel cell Forum*, Lucerne, Switzerland, European Forum Secretariat, CH 5442-Oberrohrdorf, Switzerland, July 2002, pp. 1003–1010.

van Herle, J., Maréchal, F., Leuenberger, S., and Favrat, D. (2003). "Energy balance model of a SOFC cogenerator operated with biogas." *Journal of Power Sources*, 118, 375–383.

Wärtsilä Corporation. (2008). "Wärtsilä to deliver SOFC system using landfill gas." *Fuel Cells Bulletin*, March 2008, page 1.

Whalen, S.C., Reeburgh, W.S., and Sandbeck, K.A. (1990). "Rapid methane oxidation in a landfill cover soilt." *Applied and environmental microbiology*, 56(11), 3405–3411.

Whalen, S.C., and Reeburgh,W.S. (1988). "A methane flux time series for tundra environments." *Global Biogeochem. Cycles*, 2, 399–409.

White, J., Robinson, J., and Ren, Q. (2004). "Modelling the biochemical degradation of solid waste in landfills." *Waste Management*, 24, 227–240.

Xuereb, P. (1997). Biogas–a fuel produced from waste. http://www.synapse.net.mt/mirin/newsletter/3836.asp/1997.

Yasushi, M., and Masataka, H. (1997). "Characteristic and mechanism of semi-aerobic landfill on stabilization of solid waste." *Proceedings of the first Korea-Japan Society of Solid Waste Management*, 87–94.

Young, A. (1992). "The effects of fluctuations in atmospheric pressure on landfill gas migration and composition." *Water, Air and Soil Pollution*, 64, 601–616.

Yuan, Z., Bi, J., and Moriguichi, Y. (2006). "The circular economy: A new development strategy in China." *J. Industrial Ecology*, 10(2), 4-8.

CHAPTER 6

Microbial Electricity Generation from Cellulosic Biomass

Hong Liu

6.1 Introduction

Biomass is drawing global interest as a renewable feedstock for bioenergy production due to its potential to reduce CO_2 emissions and to reduce dependency on fossil fuels. Cellulosic biomass, such as agricultural residues and woody biomass, is available in abundance, and does not compete with food and feed. Extensive research is currently underway to convert cellulosic biomass into ethanol. This chapter discusses power generation from biomass-derived sugars using a microbial fuel cell (MFC) technology. MFCs are devices that directly convert chemical energy stored in organic matter into electricity through catalytic activities of microorganisms. Compared to ethanol production from biomass, microorganisms in MFCs can directly utilize a mixture of sugars and sugar derivatives for energy generation. Separation of glucose from other sugars and byproducts is not necessary. In addition, the final products of an MFC process are electricity, carbon dioxide and water. No extra distillation step is needed. In this chapter, we will discuss basics of the MFC system followed by a section on electricity generation from biomass hydrolysates using MFCs. The effects of furan derivatives and phenolic compounds derived from cellulosic biomass on electricity generation, direct electricity generation from biomass, and engineering and economic considerations are also covered.

6.2 Microbial Fuel Cell (MFC)

Figure 6.1 shows the essential physical components of the MFC that consists of an anode, a cathode, an electrolyte and electrochemically active microbes. Microbes catalyze the oxidation of reduced substrates releasing electrons and protons during cell respiration. The electrons are transferred to the anode electrode and flow to the cathode electrode, which generates current. The protons released from the cells are transferred to the cathode through the electrolyte to sustain the current. Typically, the electrons and protons react with oxygen on the cathode, aided by a catalyst such as platinum, to form water. The reactions occurring at the anode and cathode with glucose as a substrate are given by:

$$\text{Anode: } C_6H_{12}O_6 + 6H_2O \rightarrow 6CO_2 + 24H^+ + 24e^- \quad (6.1)$$

$$\text{Cathode: } O_2 + 4H^+ + 4e^- \rightarrow 2H_2O \quad (6.2)$$

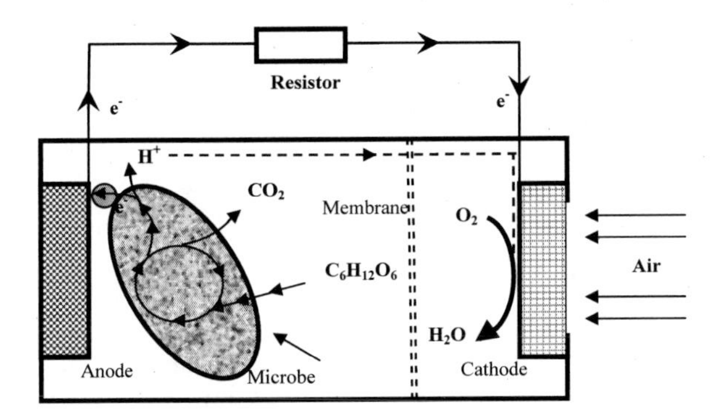

Figure 6.1 A schematic of a two-chamber MFC, showing the anode, where microbes form a biofilm on the surface, a single microbe in yellow (for illustration), a membrane, and an air cathode. The membrane can be removed to form a single-chamber MFC.

6.2.1 Electrochemically Active Microorganisms

Electricity generation in an MFC is directly linked to the ability of electrochemically active microbes on the anode to facilitate the transfer of electrons from substrate to the anode. These microorganisms have been referred to as exoelectrogens (Logan and Regan, 2006), electrogens (Lovley, 2006), or anode-respiring bacteria (Torres et al., 2007) in previous publications. Electrochemically active microbes can be enriched from various environments, such as domestic wastewater (Liu et al., 2004), ocean sediments (Reimers et al., 2001), and anaerobic sewage sludge (Kim et al., 2005). Different genetic groups of bacteria have shown electron transfer activity to anodes, including β-*Proteobacteria* (Chaudhuri and Lovley, 2003), γ-*Proteobacteria* (Kim et al., 1999; Rabaey et al., 2004), δ-*Proteobacteria* (Bond et al., 2002; Pham et al., 2003; Holmes et al., 2004a,b; Bond and Lovley, 2005), *Firmicutes* (Park et al., 2001), and *Acidobacteria* (Bond & Lovley, 2005). The mechanisms used for extra-cellular transport of electrons by these microbes are still under investigation. Some exoelectrogens, such as *Pseudomonas aeruginosa* (Myers and Myers, 1992) and *Geothrix fermentans* (Bond and Lovley, 2005) can excrete mediators to shuttle electrons to anode surfaces. Other exoelectrogens, such as *Geobacter* and *Shewanella species* may rely on cell-bound outer membrane cytochromes and conductive pili for electron transfer (Myers and Myers., 1992; Reguera et al., 2005; Gorby et al., 2006).

6.2.2 Electrode Materials

Carbon-based materials are commonly used as electrodes for MFCs due to their good conductivity, biocompatibility, and chemical stability. Carbon can be fabricated with different morphologies, such as plate, rod, fiber, paper, cloth and felt. The simplest materials for anode electrodes are graphite plates or rods as they are relatively inexpensive, easy to handle, and have a defined surface area. Much larger surface areas can be obtained with carbon cloth, carbon fiber and graphite felt electrodes, which can enhance electron transfer efficiency from microbes to anode electrode. A power density of 6.8 W/m^2 at a current density of 26 A/m^2 was achieved using a microfiber carbon cloth anode, which is the highest MFC power density ever reported (at the time of preparing this write-up) (Fan et al., 2008).

While various carbon based materials have been used as cathode electrode in the two-chamber (aqueous cathode) MFC system (see 6.2.3), carbon cloth with diffusion and poly(tetrafluoroethylene) (PTFE) layers was commonly used in an air cathode MFC system (Liu et al., 2005a; Cheng et al., 2006a; Catal et al., 2008a; Fan et al., 2008). The presence of PTFE layers can facilitate air diffusion to the reaction sites and enhance the water holding capability of cathode. A catalyst is also required for the cathode of an MFC to efficiently catalyze the oxygen reduction reaction. Platinum is commonly used in MFCs (Liu et al., 2004). The high cost of platinum may, however, limit the commercial application of MFCs. Some non-precious metal macrocycles, such as cobalt tetramethoxyphenylporphyrin (CoTMPP) and iron phthalocyanine (FePC) have been developed as low cost cathode catalysts for oxygen reduction in MFCs (Cheng et al., 2006b; Zhao et al., 2005; 2006; Yu et al., 2007).

6.2.3 MFC Configuration

MFCs are traditionally designed as a two-chamber system with the bacteria in the anode chamber separated from the cathode chamber by an ion exchange membrane (Figure 6.1). Proton exchange membrane, such as Nafion, is commonly used in MFC systems due to its popularity in chemical fuel cells for excellent proton conductivity, and thermal and mechanical stability. However, protons cannot be effectively transferred through the Nafion membrane to maintain high current generation due to the neutral pH of the medium solution (Kim et al., 2007; Chae et al., 2008; Fan et al., 2007a). The applications of anion exchange membranes (AEMs), such as AMI-7001 (Kim et al., 2007), have been explored in MFCs systems to enhance proton transfer through the facilitation of anions, such as bicarbonate and phosphate in solution (Fan et al., 2007a).

Single-chamber MFC systems are developed through the use of air cathode electrodes and the removal of expensive and complicated membranes (Liu and Logan, 2004). Single chamber MFCs can achieve much better performance than two-chamber MFCs due to the higher mass transfer coefficient of oxygen in air compared to water and a decrease in internal resistance due to the removal of membrane (Liu and Logan, 2004; Liu et al., 2005b). For both single- and two-chamber systems, various flat plate and tubular designs have been developed (Logan et al., 2006). A flat plate single chamber MFC with a cloth electrode assembly structure (Fan et al., 2007b) showed the highest reported volumetric power density (1550 W/m^3) (at the time of this writing) due to an increased surface/volume ratio and reduced internal resistance (Fan et al., 2007a).

6.3 Electricity Generation from Cellulosic Biomass Hydrolysates

Similar to direct conversion of biomass to ethanol, the structural complexity of cellulosic biomass limits its direct conversion into electricity in MFCs by microbes. Pretreatment and subsequent hydrolysis are often essential to obtain fermentable sugars. While the composition of the products from pretreatment and hydrolysis depends on the biomass sources as well as the pretreatment methods, cellulosic biomass hydrolysates mainly contain monosaccharides and some sugar derivatives.

6.3.1 Monosaccharides

Electricity can be directly generated from nearly all monosaccharides commonly found in the hydrolysate of biomass in single chamber MFCs with a mixed bacterial culture (Catal et al., 2008a). The monosaccharides tested include six hexoses (glucose, galactose, fructose, mannose, fucose, and rhamnose) and three pentoses (xylose, arabinose, and ribose). The mixed bacterial culture enriched from domestic wastewater with acetate as initial carbon source can adapt well to all monosaccharides tested in MFCs with the adaptation times varied from 1 to 70 hours (Catal et al., 2008a). The maximum power density obtained from these carbon sources ranged from 1240 mW/m^2 to 2330 mW/m^2. Mannose produced the lowest maximum power density, whereas xylose generated the highest one. The Coulombic efficiency, which is used to evaluate the electron recovery as current from the organic matter, ranged from 21 to 37% based on the total substrate concentration. The chemical oxygen demand (COD) removal at the end of the batch operation was 80-95% for all monosaccharides tested. Table 6.1 summarizes the performance of MFCs with various monosaccharides as carbon sources.

Table 6.1 The performances of MFCs using various monosaccharides (modified from Catal et al., 2008a).

Carbon sources		Power density (mW/m^2)	Coulombic efficiency[a] (%)	COD removal (%)
	Glucose	2160 ± 10	28	93 ± 2
	Galactose	2090 ± 10	23	93 ± 2
Hexoses	Fructose	1810 ± 10	23	88 ± 2
	Fucose	1760 ± 10	34	84 ± 4
	Rhamnose	1320 ± 110	30	90 ± 2

	Mannose	1240 ± 10	25	88 ± 4
	Xylose	2330 ± 60	31	95 ± 2
Pentoses	Arabinose	2030 ± 20	27	93 ± 2
	Ribose	1520 ± 40	30	86 ± 3

[a] Coulombic efficiencies at 120 Ω external resistance.

The relationship between the maximum voltage output and the monosaccharide concentration appears to follow the Monod type kinetics (equation 6.3) (Liu et al., 2005a; Catal et al., 2008a).

$$V = \frac{V_{max} S}{K_s + S} \tag{6.3}$$

where S the monosaccharide concentration, V the voltage output at concentration S, V_{max} the maximum voltage, and K_s (V) the half-saturation constant.

When single chamber MFCs were operated at an external resistance of 120 Ω, the estimated maximum voltage output ranged between 0.26-0.44 V and half-saturation kinetic constants ranged from 111 to 725 mg L^{-1} (Catal et al., 2008a).

6.3.2 Sugar Derivatives

Polyalcohols can be found in water soluble materials of biomass such as corn stover, which account for 14-27% of the total biomass on a dry weight basis (Chen et al. 2007). Polyalcohols are also by-products of ethanol fermentation (Catal et al., 2008b). Electricity can be directly generated from three pentitols (xylitol, arabitol, and ribitol) and three hexitols (galactitol, mannitol, and sorbitol) (Table 6.2) and the power densities were comparable to those generated from monosaccharides using the same type of MFCs and microbial sources (section 6.3.1). Among the three pentitols tested, ribitol generated a maximum power density of 2350 mW/m^2 at a current density of 0.73 mA/cm^2 followed by xylitol and arabitol. Among the three hexitols, sorbitol generated the highest power density of 2650 mW/m^2 at a current density of 0.78 mA/cm^2 followed by galactitol. Mannitol generated the lowest power density of 1490 mW/m^2 at a current density of 0.55 mA/cm^2. Although same mixed bacterial cultures were used initially in this study, denaturing gradient gel electrophoresis (DGGE) of polymerized chain reaction (PCR) amplification of 16S rRNA gene segments of the anode biofilms showed that more bacterial species were present in the MFC anode biofilms fed with hexitols than those fed with pentitols (Catal et al., 2008b).

Uronic acids, such as galacturonic and glucuronic acids can be released from the hydrolysis of hemicelluloses. Two uronic acids (D-galacturonic acid and D-glucuronic acid) and one aldonic acid (D-gluconic acid) have been examined for electricity generation in MFCs (Catal et al., 2008a). Glucuronic acid generated a maximum power density of 2770 mW m^{-2} at a current density of 1.18 mA cm^{-2}, which was 35% higher than that from gluconic acid and 86% higher than that from galacturonic acid (Table 6.2). The maximum power density generated from glucuronic acid was even higher than those from glucose and xylose, indicating that glucuronic acid can be a good substrate for electricity generation in MFCs.

The COD removal rates for these sugar derivatives were in the range of 71-93%. The Coulombic efficiency was in the range of 10-30% at 120 Ω (Table 6.2). The significant electron loss was possible due to the electron transfer from substrate to other electron acceptors in solution, such as oxygen diffused through the air cathode, and the substrate utilization for bacterial growth (Liu et al., 2005a).

6.3.3 Biomass Hydrolysates

Hydrolysates from dilute-acid and steam explosion pretreated corn stover were also examined for electricity generation in MFCs (Zuo et al., 2006). The hydrolysates were mainly composed of glucose, xylose, galactose, arabinose, and mannose following the removal of lignin-derived phenolic compounds through lime ($Ca(OH)_2$) addition. Maximum power densities of 810 - 860 mW/m^2 have been achieved in fed-batch tests using single chamber MFCs with Coulombic efficiencies of 20 to 30%. Almost all the biodegradable organic matters in the hydrolysates were utilized as evident from the overall biochemical oxygen demand (BOD) removal efficiencies of 93- 94%.

Table 6.2 The performances of MFCs using polyalcohols and other sugar derivatives (modified from Catal et al., 2008b).

Compound		Power density (mW/m^2)	Coulombic Efficiency[a] (%)	COD removal (%)
5-carbon polyalcohols	Xylitol	2107 ± 84	21	91 ± 2
	Arabitol	2027 ± 117	25	91 ± 1
	Ribitol	2347 ± 141	28	92 ± 1
6-carbon polyalcohols	Galactitol	2650 ± 12	13	90 ± 5
	Mannitol	1488 ± 160	19	91 ± 1
	Sorbitol	1691 ± 84	10	71 ± 1
Sugar derivatives	Galacturonic acid	1480 ± 150	22	80 ± 2
	Glucuronic acid	2770 ± 30	24	89 ± 1
	Gluconic acid	2050 ± 30	30	93 ± 6

[a] Coulombic efficiencies at 120 Ω external resistance.

6.4 Effects of Furan Derivatives and Phenolic Compounds on Electricity Generation

The pretreatment and the subsequent acid hydrolysis of biomass not only produce monosaccharides and sugar derivatives, but also generate a number of byproducts, such as furan derivatives (2-furaldehyde and 5-hydroxymethyl-2-furaldehyde), phenolic compounds (cinnamic acids, syringaldeyde, vanillin, cetophenone and dimethoxybenzyl alcohol) and carboxylic acids (acetic, formic, and levulinic acids) (Almeida et al., 2007; Cantarella, et al., 2004). It has been demonstrated that these byproducts negatively affect the cell membrane function, the growth, and the glycolysis by ethanol-producing yeast and bacteria (Taherzadeh et al., 1999; Larsson et al., 2001; Klinke et al., 2002). Some of these byproducts such as acetic acid, however, are good substrates for electrochemically active microbes, and have been used as model substrates in many MFC studies (Liu et al., 2005a). Recently, the effects of furan derivative and phenolic compounds on electricity generation in MFCs were systematically investigated by Catal et al. (2008c). Table 6.3 summaries the minimum inhibitory concentrations (MICs) of these compounds on power output. The MIC here is defined as the lowest concentration that will significantly affect the voltage output of an MFC to less than 0.1 V at 1000 Ω.

Table 6.3 The minimum inhibitory concentrations of furan derivatives and phenolic compounds for electricity generation in MFCs (Tunc, 2008c).

	Compounds	MIC (mM)
Group 1	Trans-cinnamic acid	> 40
	3-5-dimethoxy-4- hydroxycinnamic acid	> 40
	5-hydroxymethyl-2-furaldehyde	> 40
	Vanillin	> 40
Group 2	Syringaldehyde	20
	Trans-4-Hydroxy-3-methoxy cinnamic acid	20
	4-Hydroxycinnamic acid	20
Group 3	2-Furaldehyde	0.2*
	Acetophenone	0.1
	3-4-dimethoxybenzyl alcohol	0.2

*In another study, no inhibition was observed at a 2-formaldehyde concentration up to 20.8 mM (Borole et al. 2009)

6.4.1 5-hydroxymethyl-2-furaldehyde

5-hydroxymethyl-2-furaldehyde (5-HMF), a furan derivative, is formed by dehydration of hexoses, such as glucose, during the pretreatment of biomass

(Palmqvist and Hahn-Hagerdal, 2000). It can inhibit ethanologenic microorganisms such as yeasts at a concentration of 8-120 mM depending on the strain used (Almeida et al., 2007). While the concentration of 5-HMF is generally low (from 0.47 mM to 4.7mM) in the hydrolysate from steam explosion pretreatment of sugarcane bagasse, corn stover and poplar (Martin and Jonsson, 2003; Öhgren et al., 2006), higher concentrations of 5-HMF (16-46 mM) were detected in the hydrolysate of spruce when one- or two-step dilute acid hydrolysis was performed (Almeida et al., 2007; Oliva et al., 2003). Although electricity can be produced from 5-HMF without the addition of other carbon sources, the voltage output could be much lower than that using glucose (Catal et al., 2008a). In a single- chamber MFC with a mixed bacterial culture enriched from wastewater, the power generation from glucose was not affected by 5-HMF at a concentration of up to 10 mM. But a decrease in voltage output was observed when 5-HMF concentration increased to 40 mM (Catal et al., 2008c). Power generation quickly recovered after replacing the solution with a glucose-containing medium without 5-HMF.

6.4.2 Furaldehyde

2-Furaldehyde, another furan derivative and a well known inhibitor for ethanologenic microorganisms, can be generated from the acid hydrolysis of hemicellulose (Rao et al., 2006). The concentration of 2-furaldehyde was as high as 10 mM in the hydrolysate of spruce when two stage dilute acid hydrolysis pretreatment was used, and it decreased to 4.5 mM when one stage dilute acid pretreatment was used (Larsson et al., 1999). Catal et al. (2008c) reported that the MFC performance was greatly affected by the presence of 2-furaldehyde using a mixed culture enriched from municipal wastewater. Although the power output was not affected by the addition of 0.01 mM 2-furaldehyde, a nearly 17% decrease in electricity production was observed when the 2-furaldehyde concentration was increased to 0.05 mM. Voltage output was completely inhibited at 0.2 mM. In another study reported by Borole et al. (2009), electricity was produced from furaldehyde at a concentration of 0.1 g/l to 2 g/l (20.8 mM) using a mixed culture enriched from anaerobic sludge. No decrease in the voltage output was observed with increasing concentrations. The different results in these studies were possible due to the varied microbial consortia enriched from different sources.

6.4.3 Cinnamic Acids

Cinnamic acids have been found in the hydrolysates of cellulosic biomass (Klinke et al., 2002). A concentration of about 0.007 mM in the hydrolysate of spruce with dilute acid pretreatment was reported (Klinke et al., 2002). Trans-cinnamic acid and 3,5-dimethoxy-4-hydroxy-cinnamic acid did not inhibit power generation in MFCs with a mixed culture when their concentrations were lower than 10 mM and demonstrated about 20-25% decrease in voltage output when their concentrations were increased to 40 mM (Catal et al., 2008c). Trans-4-hydroxy-3-methoxy cinnamic acid and 4-hydroxy cinnamic acid, however, can severely inhibit power generation at a concentration of 20 mM although the addition of up to 5 mM did not affect the maximum voltage output at an external resistance of 1000 Ω. Following the MFC operation with 20 mM of these two compounds, voltage production was not recovered by subsequent replacement of medium solutions lacking these compounds.

6.4.4 Syringaldeyhde and Vanillin

Syringaldeyde and vanillin can be formed in the degradation of syringyl propane units and guaiacylpropane units of lignin (Jonsson et al., 1998). It was reported that vanillin and syringaldehyde concentrations were around 0.6-0.8 mM in the hydrolysate of spruce with dilute acid pretreatment (Larsson et al., 2001), 0.1-0.2 mM in the hydrolysate of wheat straw with wet oxidation hydrolysis (Klinke et al., 2002), and 0.1-0.3 mM in the steam exploded poplar hydrolysates (Jonsson et al., 1998). Their concentrations in poplar wood hydrolysate could be further decreased to 0.08-17 μM using enzymatic treatment (Cantarella et al., 2004). Although syrinaldehyde can severely inhibit the voltage output at a concentration of 20 mM, electricity generation was not affected at concentrations up to 10 mM in single chamber MFCs inoculated with a mixed culture. No decrease in power generation was observed when the vanillin concentration was higher than 40 mM. Due to the much lower concentrations obtained using various pretreatments and hydrolysis methods, these two compounds may not be a concern when the hydrolysates of lignocellulosic biomass were directly used in MFCs for power generation.

6.4.5 Dimethoxybenzyl Alcohol and Cetophenone

Benzyl alcohol (6 mM) was reported in the hydrolysate of poplar treated with steam explosion (Oliva et al., 2003). Acetophenone of 0.03 mM was found in the hydrolysate of wheat straw treated with alkaline wet oxidation. MFC performance was affected significantly by the presence of acetophenone and 3,4-dimethoxybenzyl alcohol at a concentration less than 0.1 mM (Tunc et al., 2008c). Power generation was not recovered after single chamber MFCs were operated with 0.2 mM acetophenone or dimethoxybenzyl alcohol solutions (Tunc et al., 2008c).

For all the compounds, e.g. acetophenone and 3,4-dimethoxybenzyl alcohol, when their concentrations in biomass hydrolysates are similar or higher than that at which inhibitory effects have been observed, alternative or additional approaches should be employed to increase the efficiency of electricity generation from the biomass hydrolysates. These approaches include use of appropriate pretreatment methods to yield a lower concentration of inhibitory compounds, removal of strong inhibitors prior to the MFC process, and increasing the tolerance of bacteria towards these compounds through the enrichment of new bacterial cultures or genetic modification of the bacterial strains.

6.5 Direct Electricity Generation from Cellulosic Biomass

The direct electricity generation from cellulosic biomass can also be achieved without enzymatic pretreatment or an exogenous catalyst. Ren et al. (2007) reported a direct conversion of cellulose into electricity using a binary culture of *Clostridium cellulolyticum* and Geobacter *sulfurreducens*. In the coculture, *C. cellulolyticum* can ferment cellulose into small molecules, such as acetate, ethanol, hydrogen, and carbon dioxide, while electrochemically active *G. sulfurreducens* transfer electrons from some of these fermentation products to the anode. Maximum power densities of 143 mW/m^2 and 59.2 mW/m^2 were obtained from carboxymethyl cellulose (CMC) and MN301 cellulose, respectively using two chamber MFCs, while neither pure culture alone produced electricity from these substrates. In comparison with a pure *C. cellulolyticum* culture, the co-culture also increased the CMC degradation rate from 42% to 64%. However, the COD removal rates were low. Less than 40% COD was removed by the end of the test (Ren et al., 2007). Directly converting cellulose into current has also been demonstrated in microbial electrolysis cells using uncharacterized mixed cultures enriched from soil (Cheng and Logan, 2007). However, much lower current density was generated from cellulose compared to acetate and glucose using the same cell configuration. The lower COD removal and current density in these studies indicate that the metabolism of the insoluble cellulose limits the MFC processes, and pretreatment and hydrolysis may be needed for effective utilization of cellulosic biomass.

6.6 Engineering and Economic Considerations

Although electrical current generation in half cell using microorganism was first demonstrated in 1910, the MFC technology had not received much attention until recently. Many studies focused on energy generation from wastewater to offset the cost of wastewater treatment. However, the low energy content in wastewater limits its potential for energy generation. The abundant and renewable cellulosic biomass could be an excellent feedstock for MFCs. One of the major advantages of using MFC systems for energy generation is that nearly all biomass constituents and many pretreatment/hydrolysis byproducts, such as sugar derivatives and carboxylic acids can be used for power generation. Efficient utilization of all these components is indispensable for economic conversion of cellulosic biomass to energy. The other advantage is that the MFC process is relatively simple. No energy extensive separation and purification processes are required for feedstocks and final products, which could improve the economical feasibility of renewable energy production from biomass.

The realization of practical application of generating energy from cellulosic biomass using the MFC technology, however, will rely on the cost effectiveness of pretreatment and hydrolysis of biomass and the efficiency of MFC systems. It is expected that the costs of pretreatment and enzymes will continue to decrease with the tremendous research efforts in cellulosic ethanol. Significant increase in power density of MFC systems has also been demonstrated in the past several years.

Although a pilot-scale study using brewery wastewater has been demonstrated in Queensland, Australia, most of the MFCs studies are being conducted in bench-scale. Commercialization of this technology will require additional breakthroughs in the development of highly efficient microbes, low-cost but high-performance cathode materials, and scalable reactor designs (Liu et al., 2008).

6.7 References

Almeida, J., Modig, T., Petersson, A., Hahn-Hägerdal, B., Lidén, G., and Gorwa-Grauslund, M.F. (2007). "Increased tolerance and conversion of inhibitors in lignocellulosic hydrolysates by *Saccharomyces cerevisiae*." *J. Chem. Technol. Biotechnol.,* 82, 340-349.

Bond, D.R., and Lovley, D.R. (2005). "Evidence for involvement of an electron shuttle in electricity generation by *Geothrix fermentans*." *Appl. Environ. Microbiol*., 71, 2186–2189.

Bond, D.R., Holmes, D.E., Tender, L.M., and Lovley D.R. (2002). "Electrodereducing microorganisms that harvest energy from marine sediments." *Science,* 295, 483–485.

Borole, A.P., Mielenz, J.R., Vishnivetskaya, T.A., and Hamilton C.Y. (2009). "Controlling accumulation of fermentation inhibitors in biorefinery recycle water using microbial fuel cells." *Biotechnol. Biofuels*, doi: 10.1186/1754-6834-2-7.

Cantarella, M., Cantarella, L., Gallifuoco, A., Spera, A., and Alfani, F. (2004). "Effect of inhibitors released during steam-explosion treatment of poplar wood on subsequent enzymatic hydrolysis and SSF." *Biotechnol. Prog.,* 20, 200-206.

Catal, T., Li, K., Bermek H., and Liu, H. (2008a). "Electricity Production from Twelve Monosaccharides Using Microbial Fuel Cells." *J. Power Sources*, 175, 196-200.

Catal, T., Xu, S., Li, K., Bermek, H., and Liu, H. (2008b). "Electricity generation from polyalcohols in single-chamber microbial fuel cells." *Biosensors Bioelectronics,* 24, 855-860.

Catal, T., Fan, Y., Li, K., Bermek, H., and Liu, H. (2008c). "Effects of furan derivatives and phenolic compounds on electricity generation in microbial fuel cells." *J. Power Sources*, 180, 162-166.

Chae, K.J., Choi, M., Ajayi, F.F., Park, W., Chang, I.S., and Kim, I.S. (2008). "Mass transport through a proton exchange membrane (Nafion) in microbial fuel cells.". *Energy Fuels,* 22, 169-176.

Chaudhuri, S.K., and Lovley, D.R. (2003). "Electricity generation by direct oxidation of glucose in mediatorless microbial fuel cells." *Nat. Biotechnol.* 21, 1229–1232.

Chen, S.F., Mowery, R.A., Scarlata, C.J., and Chambliss, C.K. (2007). "Compositional analysis of water-soluble materials in corn stover." *J. Agric. Food Chem.,* 55, 5912-5918.

Cheng, S., Liu H., and Logan B.E. (2006a). "Increased power generation in a continuous flow MFC with advective flow through the porous anode and reduced electrode spacing." *Environ. Sci. Technol.,* 40(7), 2426-2432.

Cheng, S., Liu H., and Logan B.E. (2006b). "Power densities using different cathode catalysts (Pt and CoTMPP) and polymer binders (Nafion and PTFE) in single chamber microbial fuel cells." *Environ. Sci. Technol.,* 40(1), 364–369

Cheng, S., and Logan B.E. (2007). "Sustainable and efficient biohydrogen production via electrohydrogenesis." *PNAS*, 104(47), 18871–18873.

Fan, Y., Hu, H., and Liu, H. (2007a). "Sustainable power generation in microbial fuel cells using bicarbonate buffer and proton transfer mechanisms." *Environ. Sci. Technol.*, 41(23), 8154-8158.

Fan, Y., Hu, H., Liu, H. (2007b). "Enhanced Coulombic efficiency and power density of air-cathode microbial fuel cells with an improved cell configuration." *J. Power Sources*, 171(2), 348-354.

Fan, Y., Sharbrough, E., Liu, H. (2008). "Quantification of the internal resistance distribution of microbial fuel cells." *Environ. Sci. Technol.*, 42 (21), 8101–8107.

Gorby, Y.A., Yanina, S., and Malean, J.S. (2006). "Electrocally conductive bacterial nanowires produced by *Shewanella oneidensis* strain MR-1 and other microorganisms." *PNAS,* 103, 11358-11363.

Holmes, D.E., Bond, D.R., and Lovley, D.R. (2004a). "Electron transfer by *Desulfobulbus propionicus* to Fe(III) and graphite electrodes." *Appl. Environ. Microbiol.,* 70, 1234–1237.

Holmes, D.E., Nicoll, J.S., Bond, D.R., and Lovley, D.R. (2004b). "Potential role of a novel psychrotolerant member of the family Geobacteraceae, *Geopsychrobacter electrodiphilus* gen. nov., sp. nov., in electricity production by a marine sediment fuel cell." *Appl. Environ. Microbiol.,* 70, 6023–6030.

Jonsson, L.J., Palmqvist, E., Nilvebrant, N.O., and Hahn-Hagerdal B. (1998). "Detoxification of wood hydrolysates with laccase and peroxidase from the white-rot fungus Trametes versicolor." *Appl. Microbiol. Biotechnol.,* 49(6), 691-697.

Kim, H.J., Hyun, M.S., Chang, I.S., and Kim, B.H. (1999). "A fuel cell type lactate biosensor using a metal reducing bacterium *Shewanella putrefaciens." J. Microbiol. Biotechnol.,* 9, 365-367.

Kim, J.R., Min, B., and Logan, B.E. (2005). "Evaluation of procedures to acclimate a microbial fuel cell for electricity production." *Appl. Microbiol. Biotechnol.* 68(1):23-30.

Kim, J. R., Oh, S.E., Cheng, S., and Logan. B.E. (2007). "Power generation using different cation, anion and ultrafiltration membranes in microbial fuel cells." *Environ. Sci. Technol.,* 41(3), 1004-1009.

Klinke, H.B., Ahring, B.K., Schmidt, A.S., and Thomsen, A.B. (2002). "Characterization of degradation products from alkaline wet oxidation of wheat straw." *Bioresource Technol.*, 82, 15-26.

Larsson, S., Reimann, A., Nilvebrant, N.O., and Jonsson, L.J. (1999). "Comparison of different methods for the detoxification of lignocellulose hydrolyzates of spruce." *Appl. Biochem. Biotechnol.,* 77, 91-103.

Larsson, S., Cassland, P., and Jonsson, L.J. (2001). "Development of a *Saccharomyces cerevisiae* strain with enhanced resistance to phenolic fermentation inhibitors in lignocellulosic hydrolysates by heterologous expression of laccase." *Appl. Environ. Microbiol.* 67, 1163-1170.

Liu, H., Cheng, S., Huang, L., and Logan, B.E. (2008). "Scale-up of membrane-free single-chamber microbial fuel cells." *J. Power Sources*, 179, 274–279.

Liu, H., Cheng, S., and Logan, B.E. (2005a). "Production of electricity from acetate or butyrate using a single-chamber microbial fuel cell." *Environ Sci Technol.*, 39(2), 658-662.

Liu, H., Cheng, S., and Logan, B.E. (2005b). "Power generation in fed-batch microbial fuel cells as a function of ionic strength, temperature, and reactor configuration." *Environ Sci Technol.*, 39(14), 5488-5493.

Liu, H., and Logan, B.E. (2004). "Electricity generation using an air-cathode single chamber microbial fuel cell in the presence and absence of a proton exchange membrane." *Environ Sci Technol.*, 38, 4040-4046.

Liu, H., Ramnarayanan, R., and Logan, B.E. (2004). "Production of electricity during wastewater treatment using a single chamber microbial fuel cell." *Environ. Sci. Technol.* 38, 2281-2285.

Logan, B.E., Hamelers, B., Rozendal, R., Schröder, U., Keller, J., Freguia, S., Aelterman, P., Verstraete, W., and Rabaey. K. (2006). "Microbial fuel cells: methodology and technology." *Environ. Sci. Technol.*, 40, 5181-5192.

Logan, B.E., and Regan, J.M. (2006). "Microbial fuel cells: challenges and applications. " *Environ. Sci. Technol.*, 40(17), 5172-5180.

Lovley, D.R. (2006). "Bug juice: harvesting electricity with microorganisms." *Nat. Rev. Microbiol.*, 4, 497–508.

Martin, C., and Jonsson, L.J. (2003). "Comparison of the resistance of industrial and laboratory strains of *Saccharomyces* and *Zygosaccharomyces* to lignocellulose-derived fermentation inhibitors." *Enzyme Microb. Technol.*, 32, 386–395.

Myers, C.R., and Myers, J.M. (1992). "Localization of cytochromes to the outer membrane of anaerobically grown *Shewanella putrefacians* MR-1." *J. Bacteriol.*, 174, 3429–3438.

Öhgren, K., Rudolf, A., Galbe, M., and Zacchi, G. (2006). "Fuel ethanol production from steam-pretreated corn stover using SSF at higher dry matter content." *Biomass Bioenergy*, 30, 863–869.

Oliva, J.M., Saez, F., Ballesteros, I., Gonzalez, A., Negro, M.J., Manzanares, P., and Ballesteros, M. (2003). "Effect of lignocellulosic degradation compounds from steam explosion pretreatment on ethanol fermentation by thermotolerant yeast *Kluyveromyces marxianus*." *Appl. Biochem. Biotechnol.*, 105, 141-154.

Palmqvist, E., Hahn-Hagerdal, B. (2000). "Fermentation of lignocellulosic hydrolysates: I: inhibition and detoxification." *Bioresource Technol.*, 74, 17-24.

Park, H.S., Kim, B.H., Kim, H.S., Kim, H.J., Kim, G.T., Kim, M., Chang, I.S., Park, Y.K., and Chang, H.I. (2001). "A novel electrochemically active and Fe(III)-reducing bacterium phylogenetically related to *Clostridium butyricum* isolated from a microbial fuel cell." *Anaerobe,* 7, 297-306

Pham, C.A., Jung, S.J., Phung, N.T., Lee, J., Chang, I.S., Kim, B.H., Yi, H., and Chun, J. (2003). "A novel electrochemically active and Fe(III)-reducing bacterium phylogenetically related to *Aeromonas hydrophila,* isolated from a microbial fuel cell." *FEMS Microbiol. Lett.,* 223, 129–134.

Rabaey, K., Boon, N., Siciliano, S.D., Verhaege, M., and Verstraete, W. (2004). "Biofuel cells select for microbial consortia that self-mediate electron transfer." *App. Environ. Microbiol.,* 70, 5373-5382.

Rao, R.S., Jyothi, C.P., Prakasham, R.S., Sarma, P.N., Rao, L.V. (2006). "Xylitol production from corn fiber and sugarcane bagasse hydrolysates by *Candida tropicalis.*" *Bioresource Technol.*, 97, 1974-1978.

Reguera, G., McCarthy, K.D., Mehta, T., Nicoll, J.S., Tuominen, M.T., and Lovley, D.R. (2005). "Extracellular electron transfer via microbial nanowires." *Nature,* 435, 1098–1101.

Reimers, C.E., Tender, L.M., Fertig, S., Wang, W. (2001). "Harvesting energy from the marine sediment-water interface." *Environ. Sci. Technol.,* 35, 192-195.

Ren, Z., Ward, T.W., and Regan, J.M. (2007). "Electricity production from cellulose in a microbial fuel cell using a defined binary culture." *Environ. Sci. Technol.,* 41(13), 4781- 4786.

Taherzadeh, M., Gustafsson, L., Niklasson, C., Liden, G. (1999). "Conversion of furfural in aerobic and anaerobic batch fermentation of glucose by *Saccharomyces cerevisiae.*" *J. Biosci. Bioeng.,* 87, 169-174.

Torres, C.I., Marcus, A.K., and Rittmann, B.E. (2007). Kinetics of consumption of fermentation products by anode-respiring bacteria. *Appl. Microbiol. Biotechnol.,* 77, 689-697.

Yu, E.H., Cheng, S., Scott, K., and Logan, B. (2007). "Microbial fuel cell performance with non-Pt cathode catalysts." *J. Power Sources,* 171(2), 275–281.

Zhao, F., Harnisch, F., Schroder, U., Scholz, F., Bogdanoff, P., Herrmann, I. (2005). "Application of pyrolysed iron(II) phthalocyanine and CoTMPP based oxygen reduction catalysts as cathode materials in microbial fuel cells." *Electrochem. Commun.,* 7(12), 1405–1410

Zhao, F., Harnisch, F., Schroder, U., Scholz, F., Bogdanoff, P., Herrmann, I. (2006). "Constraints and challenges of using oxygen cathodes in microbial fuel cells." *Environ. Sci. Technol.,* 40(17), 5193–5199.

Zuo, Y., Maness, P.C., and Logan, B.E. (2006). "Electricity production from steam exploded corn stover biomass." *Energy Fuels,* 20(4), 1716-1721.

CHAPTER 7

Evaluation of Cellulosic Feedstocks for Biofuel Production

R. Ogoshi, B. Turano, G. Uehara, J. Yanagida, P. Illukpitiya, J. Brewbaker, and J. Carpenter

7.1 Introduction

Dedicated cellulosic crops are expected to provide significant quantities of feedstock for biofuel production. In the National Biofuels Action Plan (BRDB, 2008), biofuel production is based on three generations of feedstock crops. First generation feedstocks are starch, sugar, and oil produced from food and feed crops. These feedstocks are currently converted to ethanol and biodiesel. Second generation feedstocks are lignocellulosic residues from agricultural and forest production. These feedstocks are expected to produce biofuels in the near-term. Third generation feedstocks are lignocellulose derived from dedicated crops such as grasses and trees, or oil from algae. Further development is required before commercialization, but third generation feedstocks are expected to be important contributors to biofuel production. An assessment of the potential quantities of lignocellulose across the U.S. showed that third generation feedstocks (perennial crops and trees) could provide up to 342 million dry Mg per year out of 1.2 billion Mg total biomass that is potentially available for bioenergy production (Perlack et al., 2005). The 1.2 billion Mg biomass is estimated to produce liquid fuel equivalent to one-third of the U.S. transportation fuel consumption.

7.2 Criteria for Third Generation Feedstock Crops

Grass and tree species have been selected based on criteria that evolved over time. The Herbaceous Energy Crop Program and the Biofuel Feedstock Development Program at Oak Ridge National Laboratory (ORNL) were initiated in the 1980s and 1990s. Out of these programs came criteria for species selection (Wright, 2007). These criteria may be categorized around four issues:

1) Energy production,
2) Economics,

3) Environmental impacts, and
4) Maintenance of food supply and food price.

While it is recognized that a single criteria may address more than one issue, the following is presented as a starting point:

Energy production refers to measure of energy produced, energy consumption during production that affects net energy, or expansion of feedstock production systems. The criteria are:

a) **Net energy production**. Net energy is the energy in the final product minus the energy required to produce it. This parameter may be used to compare energy consumed in a production system from planting to harvest, or planting to point of use.
b) **Species compositional analysis matched to conversion technology.** Gains in yield or efficiency are realized when feedstock composition is matched to conversion technology. For example, high ash content reduced the fuel yield in thermochemical conversion, and high cellulose content increased the biofuel yield in biochemical conversion.
c) **Wide environmental adaptation**. Wide adaptation implies that a crop has high efficiencies of water- or nutrient-use, and/or tolerance to cold, wet, or dry soils. The reduction or elimination of water or nutrient application to a crop reduces the energy input for its production.
d) **Potential to expand feedstock production systems**. The large demand for fuel necessitates rapid expansion of feedstock production. Feedstock crops that are easily propagated and established are needed to meet the demand.

Environmental impacts include beneficial and/or harmful effects that feedstock crops may have.

a) **Low soil erosion**. Crops and cropping systems that minimize soil erosion benefit the environment. Crops such as perennials and cropping systems such as cover cropping and intercropping can reduce soil loss.
b) **Carbon sequestration in soil**. Land use change has the potential to alter (increase or decrease) soil carbon content. Any change would affect the net carbon emission of biofuels. Deep rooted crops were thought to sequester more carbon in soils.
c) **Low chemical inputs**. Crops that require less chemical inputs may garner more environmental support for biofuel production.
d) **Low invasive potential or sterile genotypes**. Feedstock crops share characteristics with invasive species such as fast growth. Concern is raised that feedstock crops could become invasive weeds. Sterility may reduce the invasive potential of feedstock crops.

Maintenance of food supply and food price was an important criterion since the early 1980s. Grass and tree crops, and marginal lands were emphasized, in part, to avoid competition with food crops. However, during the early 1990s when agricultural surpluses increased, the food criterion was de-emphasized.

Economics refers to profit as well as the components of revenue and cost. Revenue is composed of price and yield.

a) **Profitable on productive lands**. Producing feedstock must be a profitable enterprise. Producing feedstock on productive lands was considered in the 1990s. However, because productive lands are recognized as important for food production, calls for the use of marginal lands to produce biofuels are being made (Tilman et al., 2009).
b) **High yield and yield stability**. Feedstock crops that produced high and stable yields across years and locations were considered favorable traits.
c) **Low pest and disease susceptibility**. Feedstock crops that were less susceptible to pests and diseases reduced the costs for control and increased yield stability.
d) **Operations and management**. The type and number of operations to produce a crop has a large effect on costs. For example, harvesting grasses four times a year or once a year affects harvesting costs and biomass quality. Other management considerations include timing of harvest, storage and transportation of feedstock.
e) **Alternative usage**. Crops with uses other than energy reduce the risk should the energy price fall. Crops could then be sold in other markets such as forage or mulch.

The list of criteria developed by Wright (2007) is relatively comprehensive. Others have reported criteria for feedstock evaluation (McKendry, 2002; Gomez et al., 2008), but these are included in the former list (Wright, 2007).

7.3 Additional Criteria to Evaluate Feedstock

Competition for water will be an important issue both in the growing and processing of biofuel feedstock. Since most energy crops grown for their fiber will be perennial in nature, their water requirements will not be as critical compared to annual crops that are managed to produce uniform, blemish-free products. That said, biomass yield will depend heavily on water availability, particularly in cases where marginal lands in semi-arid environments, not suited for annual crops, are planted to perennial energy crops. Given the pressures to keep prime agricultural lands for food crops, it may very well be that the world's semi-arid zone will be the land area where much of our biofuel feedstock will be produced. Using a combination of deep-rooted, perennial crops and rain harvesting technology (Kablan et al., 2008), it may be possible to transform this now underutilized area into a productive zone for energy production. Rain harvesting not only increases biomass yield in this zone, but can increase soil carbon sequestration, recharge ground water (Kablan et al., 2008) and reduce soil erosion, and in the long term have a lasting beneficial impact on the cultivated land and downstream environment. However, the consequence of cultivating new land for energy crop production is still uncertain and Robertson et al. (2008) warns of the unintended consequences and the need to take a systems approach, focus on ecosystems services and to attempt to understand the implications of policy and management practices at different spatial scales from farm to the planet as a whole.

Water consumption for processing ranges from 2 - 4 liters of water to produce a liter of ethanol (http://kswarps.sproutsofware.net/files/active/2/Challenges%20for%20 Future-Biofuels%20and%20Water.pdf), but no commercial scale cellulosic biorefinery is operational so this number is still uncertain. Irrigating energy crops with wastewater from biorefineries is possible, but competition for high quality water for other uses will pose a problem in water short areas.

In addition to net energy produced, the ratio of energy required to produce a unit of biofuel to the energy in that unit biofuel is equally critical. For example, one-unit energy of gasoline required 1.25 units of fossil fuel energy to produce it resulting in an energy ratio of 0.8 (http://www/utbioenergy.org/TNBiofuelsIntiative /FAQs/). The expectation is that biofuels would result in ratios greater than 1.0, but Pimentel and Patzel (2005) report a ratio for corn grain ethanol of 0.84, and Goldemberg (2007) report a ratio of 10 for ethanol produced from sugarcane in Brazil. It turns out, however, that this ratio varies not only with feedstock, but with how the ratio is defined and the variables and boundary conditions included in its computation. This issue is addressed by Hammerschlag (2006) by reviewing 6 papers reporting energy ratios for corn ethanol and 4 papers reporting ratios for cellulosic ethanol. In the paper Hammerschlag refers to r(E) as the energy return on investment and defines it as a ratio of energy in a liter of ethanol to the nonrenewable energy required to make. Specifically

$$r(E) = E(out)/E(in), \text{(nonrenewable)} \tag{7.1}$$

where, E(out) is the energy in a certain amount of ethanol output, and E(in), (nonrenewable) is nonrenewable energy input to the manufacturing process for the same amount of ethanol. By using identical units in the numerator and denominator, the ratio remains unitless. According to Hammerschlag, the ratio tells us how well an energy technology leverages its nonrenewable energy inputs to deliver renewable energy. In the end, the surveyed article shows $r(E)_s$ ranging from a low 0.84 reported by Pimentel and Patzek (2005) to 1.65 (Lorenz and Morris, 1995), and higher values for cellulosic ethanol ranging from 4.40 (Sheehan et al., 2004) to 6.61 (Tyson et al., 1993). A low value for 0.69 for cellulosic ethanol reported by Pimentel and Patzek (2005) was treated as an outlier by the author and not included in the range. In this analysis, the energy used for producing corn starch ethanol is partitioned between ethanol and co-products, where as in the case of cellulosic ethanol, excess electricity generated from using lignin as an energy source is taken into account. The ratio would clearly be different if feedstock were to be converted into biofuels by thermal conversion processes since lignin would be converted to biofuels and no longer be available for power generation, but given that commercial scale processing of cellulosic ethanol is still in its infancy, there is still much to learn about how energy ratios for cellulosic feedstock will change over time.

7.4 Evaluation of Three Feedstock Crops

The choices for cellulosic feedstock crops are extensive. Researchers have investigated crops, grasses and trees that grow well in their particular locations such as China (Cuiping et al., 2004), New Zealand (Senelwa and Sims, 1999), and Eastern Europe (Fischer et al., 2005). In the U.S., the list of feedstock crops included the tree species: poplar, hybrid poplar, willow, sycamore, black locust, silver maple, sweetgum, and eucalyptus; and the grasses: switchgrass, reed canary grass, sorghum, sorghum-sudangrass, sweet sorghum, johnsongrass, pearl millet, giant reed, energy cane, giant miscanthus, bermudagrass, napiergrass, bahiagrass, tall fescue, and *Sericea lespedeza* (Wright, 2007). Of these, switchgrass has received much attention. Among the tropical species, banagrass (*Pennisetum purpureum*, a tropical grass) and leucaena (*Leucaena spp.*, tropical, nitrogen-fixing tree) have been considered for feedstock production (Osgood and Dudley, 1993; Shleser, 1994; Kinoshita, 1999; Keffer et al., 2006), particularly in Hawaii.

A summary evaluation was prepared for three of these feedstock crops, namely switchgrass, banagrass, and leucaena. Data were collected from literature to address criteria within the issues of energy production and economics. The specific criteria addressed were wide adaptation, feedstock composition, net energy, and profitability. The first three criteria will be presented by crop in the proceeding sections followed by an economic analysis.

The evaluation of feedstock based on its composition differs according to the conversion process. Ash content reduces the amount of energy on a mass basis, but the components of ash are important as well for thermochemical conversion. Obernberger (1998) set guidelines for feedstock composition to reduce harmful emissions and fouling of equipment in thermochemical conversion. To prevent NOx emissions, a powerful greenhouse gas, feedstock N should be $<0.6\%$ of dry weight. To lower corrosion rate of the conversion equipment, Cl and S in the feedstock should be $<0.1\%$ dry weight basis for each element. To prevent slagging and fouling of internal components in the conversion unit, feedstock Ca should be 15-35% and K $<7\%$ dry weight basis. In the biochemical conversion process, the structure of cell walls determines its efficiency of conversion. Cellulose is a polymer of glucan molecules that forms hydrogen bonds with adjacent glucan polymers giving it a semi-crystalline form. A number of hydrogen bonded chains form a microfibril that is interwined with hemicellulose and lignin. Crystallinity affects the rate of hydrolysis. However, lignin content governs the extent of hydrolysis (Laureano-Perez et al., 2005).

7.4.1 Switchgrass

Range of Habitation. Switchgrass (*Panicum virgatum*) is a native perennial warm-season grass with a north-south distribution from Southern Canada to Central America and east-west distribution from the East Coast to Arizona and Nevada (Hitchcock, 1935). Switchgrass comprises two ecotypes, a lowland type that grows best in the warmer, higher rainfall of its southern range and an upland type that is adapted to its mid and northern range. Lowland ecotypes are not as cold tolerant as the upland ecotypes, but generally have higher dry matter (DM) yields, are taller, wider leafed, and thicker stemmed (Anon, 1954; Porter, 1966; Vogel et al., 1985). Lowland cultivars are found in the southern latitudes of the south central U.S., while the upland varieties are found in the mid-central and northern latitudes up to 51°N (Jefferson et al., 2002).

Traditionally, switchgrass research focused on its usefulness as forage. Its adaptation to diverse environments, low water and nutrient requirements, and high production potential on marginal lands are reasons for its selection as a dedicated energy crop model by the U.S. Department of Energy in the early 1990s (McLaughlin, 1992).

Water Requirement and Water Use Efficiency. Switchgrass like all C-4 grasses has highly efficient water use mechanisms. McLaughlin and Kszos (2005) reported a water use efficiency of 2.08-3.77 mmol mol^{-1} for 25 switchgrass accessions. Parrish and Fike (2005) stated that it should be possible to develop switchgrass cultivars that can sustainably produce 15 Mg ha^{-1} in areas with greater than 700 mm annual precipitation.

Biomass Properties. Schmer et al. (2008) conducted a life cycle analysis of switchgrass to ethanol on 10 farms of 3-9 ha of marginal lands each and reported switchgrass yields of 5.2-11.1 Mg ha^{-1}, a net energy yield of 60 $GJha^{-1}$ yr^{-1}, and a net energy ratio of 5.4. The report also stated that emissions from cellulosic ethanol from switchgrass would produce 94% less greenhouse gases than gasoline.

Effect of Plant Development, Genotype, Environmental Stress, and Management on Biomass Properties. Above ground plant biomass comprises stems, leaves, and reproductive structures. Stems contain more cell wall components, as compared to leaves, whereas leaves contain more soluble sugars and starch and less lignin. The ratio of these organs and their chemical composition depends on genotype, plant maturity, and time of harvest (Sanderson and Wolf, 1995; Jenkins et al., 1998; Madakadze, et al., 1999; Vogel et al., 2002; Casler and Boe, 2003; Adler et al., 2006; Dien et al., 2006; Sarath et al., 2007). Plants harvested at inflorescence had higher efficiency for ethanol production than older plants, but have lower yields (Dien et al., 2006). The concentration of inorganic elements and ash content of switchgrass decreases with maturity (Sanderson and Wolf, 1995; Madakadze et al., 1999; Adler et al., 2006). Time of harvest also affects biomass quality. Adler et al. (2006) showed that plants harvested in the spring instead of the fall exhibited decreased yield (40%), moisture (80%), mineral content, and ash (30%), but increased cellulose, hemicellulose, and lignin content.

Growth and Yield. Quality seeds and weed management are critical for establishing switchgrass stands (Vogel, 2004; Schmer et al., 2006). Switchgrass forms extensive root systems that have been linked to its drought tolerance, low input requirements, and high productivity (McLaughlin and Kszos, 2005). The expenditure of energy to produce the roots system results in the production of about half of its yield potential by the end of its first year (McLaughlin and Kszos, 2005; Schmer et al., 2006). It has been suggested that selection of cultivars for breeding programs should not be based solely on first year yield performance, due to the need for plants to form extensive root systems. Once established, switchgrass stands require minimal weed management.

Fuentes and Taliaferro (2002) reported an overall dry matter mean yield of 13 Mg ha^{-1} for 10 switchgrass cultivars (lowland and upland) over seven years at two locations in Oklahoma. Switchgrass is a short day plant and its yield is greatly affected by genotype by environment interactions. Sanderson et al. (1996) showed that planting lowland cultivars in northern latitudes increased yields due to shorter days occurring later in the season, while growing upland cultivars in Southern latitudes decreased yields.

Optimal production of switchgrass depends on cultivar, site location, and crop management. An important consideration in any crop is the application of fertilizer containing nitrogen, phosphorous, and potassium. Nitrogen management recommendations have varied greatly in the literature due, in part, to research based on production of switchgrass for forage as opposed to biofuel feedstock (reviewed in McLaughlin and Kszoz, 2005; Parrish and Fike, 2005; Sarath et al., 2008). Current recommendations are 10-12 kg ha-1 $year^{-1}$ of applied nitrogen for each 1 Mg ha^{-1} biomass produced (Vogel et al., 2002). Switchgrass, like other warm-season grasses, appears to show little or no response to added phosphorus or potassium (Hall et al., 1982; Brejda, 2000). Switchgrass has moderate tolerance to acidic soils, but can benefit from liming (Brejda, 2000).

Single harvest systems are optimal under most conditions for sustainable biofuel production (Parrish and Fike, 2005; Sanderson and Adler, 2008; Sarath, et al., 2008). Nevertheless, the timing of harvest varies by region (Sanderson et al., 1999; Vogel et al. 2002; Casler and Boe, 2003) and affects yield, feedstock quality, and stand productivity (Sanderson and Wolf, 1995; McLaughlin et al., 1996; Jenkins et al., 1998; Christian et al., 2002; Vogel et al., 2002; Casler and Boe, 2003; Adler et al., 2006).

7.4.2 Banagrass/Napiergrass

Range of Habitation. *Pennisetum purpureum* (Schumach), commonly known as Banagrass, Napiergrass, and elephantgrass, is classified in the tribe Paniceae of the Gramineae (Panicoideae) family (Bogdan, 1977). Banagrass was thought to be an interspecies hybrid (Boonman, 1993), but recent molecular data suggests it is not (Lowe, et al., 2003). Alcantara et al. (1980) list numerous Banagrass cultivars, although the pedigrees and origins of cultivars are difficult to establish due to the ease of transporting vegetative cuttings, seeds, and the rhizomes. It is common for a single cultivar to receive different names in different countries and regions (Teacenco and Botrel, 1995).

Banagrass is indigenous to equatorial Africa where annual rainfall exceeds 1000 mm, but has been introduced to all tropical areas of the world and has become naturalized throughout Southeast Asia (Mannetje, 1992). It typically grows as a perennial in tropical areas of South America and Asia. Currently, it is grown in some parts of the southern United States and Hawaii (Duke, 1983).

The grass has substantial drought tolerance due to its deep fibrous root system, but it responds to irrigation. In the wild, Banagrass is normally, only, found in areas with rainfall >1000 mm, and on river banks in low rainfall areas. Banagrass does not tolerate prolonged flooding or water logging.

Biomass Properties. The chemical composition of Banagrass varies according to maturity at time of harvest (Hana et al, 2004; Otieno et al. 1992; Johnson et al. 1973). Seasonal variations have also been reported (Johnson et al., 1973). Direct comparisons of chemical composition analysis by different groups are not always possible. Johnson et al. (1973) reported a crude protein content ranging from over 20 to 55%; cellulose ranging from 20 to 36%; and lignin ranging from 1 to 9 % in immature leaves and mature leaves, respectively. Bain et al. (2003) collected data that showed banagrass was composed of 0.44-0.84% N, 0.61-0.84% Cl, 0.03% S, 0.25% Ca, 3.5% K, and 9.88% ash.

Otieno et al. (1992) reported that Banagrass is comprised of 27.7% hemicellulose, 34.3% cellulose, 4.0% lignin, 10.1% crude protein, and 4.4% lignin. Turn et al. (2003) reported a 12.0 % moisture content, 27.3% fixed carbon, 67.7% volatile matter, and 4.9% ash, and higher heating value of 18.53 MJ kg^{-1} (dry basis). Recently, Strezov (2008) reported similar values.

In general, the nutritive value of Banagrass becomes poorer with maturity (Hana et al., 2004). Johnson et al. (1973) reported that crude protein decreased with age, while cellulose and lignin increased. As stated previously, seasonal changes in chemical composition have been reported (Johnson et al., 1973).

Growth and Yield. Banagrass is a robust creeping rhizomatous perennial grass found in the tropics and subtropics. It generally grows in dense bamboo-like clumps that are 3.0-3.7 m tall, but it can grow over 6.7 m in height. The plant produces numerous tillers and the base (width) of the plant can be 0.9 m or more. Stems are coarse and hairy, can have 20 or more internodes and are about 2.5 cm thick near the base. The leaf blades have smooth upper and lower surfaces, a prominent midrib on the lower side, and the edges of the leaves are razor-sharp (Bogdan, 1977; Skerman and Riveros, 1990). The ligules are thin with hairy rims. Inflorescence is long, cylindrical, bottle-brush-like. Banagrass has an extensive root system penetrating to 4.5 m.

The plant is adapted to various soil conditions: low fertility poorly drained soils, both acid to slightly alkaline, but grows best in deep, well-drained, fertile soils

(Hanna *et al.*, 2004). It does not grow well on heavy clay soils and cannot tolerate soils that remain waterlogged for long periods of time.

Once established, Banagrass requires 100-300 kg N ha^{-1} yr^{-1} , together with other nutrients as indicated by soil tests. Responses at much higher levels of applied N have been obtained. Yields decline rapidly if fertility is not maintained. Weeds may invade if fertilizer regime is relaxed. Other nutrients must also be provided to sustain high levels of productivity. Banagrass is a luxury consumer of potassium (K). Approximately 30-35 kg ha^{-1} yr^{-1} phosphorus (P) was required to replenish P removed in Banagrass herbage (Boonman, 1993). Normally not sown with legumes, but will grow with vigorous twining legumes such as *Pueraria phaseoloides* , *Neonotonia wightii* and *Centrosema molle* (*pubescens*), or with the shrub/tree legume, *Leucaena leucocephala*.

Banagrass is an obligate quantitative short-day plant, with a critical photoperiod of 12-13 hours, flowering under a relatively wide range of photoperiods (e.g. flowers January to June in South Africa). There is some variation among ecotypes in flowering time. Seed set is usually poor, possibly due to low pollen viability. Forage stand density, plant maturity, soil moisture, and soil fertility are all factors positively correlated to yield. Banagrass does not tolerate much frost, so culture is limited to warmest parts of mainland United States and Hawaii.

The work of Kinoshita, Osgood and collaborators in Hawaii have shown that 40 and 49 Mg ha^{-1} yr^{-1} of Banagrass yields could be achieved on irrigated arid and rain fed lands, respectively. Moreover, yield from ratoon crop was 13% higher than the planted crop (Osgood et al., 1996). Banagrass grows from sea level to 2,000 m altitude and has moderate shade tolerance. Banagrass grown in Florida was persistent and productive when cut at a 10 cm stubble height, if the defoliation interval was 9 wk or longer (Charparro et al., 1995), and persistence was also excellent if stubble height was increased to 34 to 46 cm for a 3-wk cutting frequency.

Banagrass is usually propagated vegetatively. Sollenberger et al. (1990) reported that an end-to-end arrangement of cuttings was adequate for "Mott" Banagrass to obtain good stands. Overlapping the stems was recommended if the stems were not mature or lacked vigor because of inadequate fertilization or drought. Grof (1969) reported that stem cuttings of tall plant should be obtained from hard stems with well-developed leaf buds or "eyes." Kipnis and Bnei-Moshe (1988) showed that aerial tillers, those that arise from auxiliary buds along the stem, are excellent for propagation (Hanna et al., 2004). Banagrass is planted in the same way as sugar cane (*Saccharum officinarum* L.): the culms are cut into pieces, each with three nodes, and are buried in the soil just deep enough to cover the second node and to leave the third above the ground.

To control weeds at the first stages of establishment, inter-row cultivation and herbicides can be used. Fertilizer of 100–150 kg N ha^{-1} after each harvest gives best uniform production, with total of 900 kg N ha^{-1} for six harvests. Potential cattle carrying capacity of banagrass is very high with application of 50 kg ha^{-1} N after each harvest, thus maintaining about 27 head ha^{-1} (Duke, 1983). banagrass should not be cut closer than 10-15 cm from the ground.

Poor persistence of Banagrass can be a problem under some defoliation regimes, especially multiple harvests to low stubble. In Taiwan, Banagrass cultivars were harvested throughout the year when the uppermost leaf collar was a 1.0, 1.5, and 2.0 m above soil surface (Hsu et al., 1990). Cutting intervals were 58, 68 and 88 d for the 3 treatments. Averaged across cultivars and years, DM yield was 34, 41, and 52 Mg ha^{-1} yr^{-1} for the 3 treatments, respectively. In Kenya, Banagrass fertilized with 200 kg ha^{-1} N produced 8, 13, 18 Mg ha^{-1} yr^{-1} DM when harvested every 6, 9, ad 12 wk, respectively (Boonman, 1993). Yields increased to 12, 17, and 25 Mg ha^{-1} yr^{-1} DM when fertilization was increased to 400 kg N ha^{-1}. Similar responses with ages of re-growth and fertilization have also been reported in Puerto Rico (Velez-Santiago and Arroyo-Aguilu, 1981) and Florida (Calhoun and Prine, 1985). Their data also suggest that Banagrass has a long linear growth phase. The production potential at these locations ranged from a low of 15 to a high of 49 Mg ha^{-1} with lower yields occurring with short re-growth periods of 30-45 days and higher yields at longer regrowth periods of 60-120 days. Under normal management, stands are invaded by weeds and "run out" after two or three years, so that they have to be ploughed up and replanted.

7.4.3 *Leucaena spp.*

Range of Habitation. Leucaena is an American genus of mimosoid legume trees that consists of 22 species distributed from Texas to Peru (van den Beldt and Brewbaker, 1975; Anon, 1977). Noted as multipurpose crops, the leucaenas are best known by the one species that moved worldwide in 1600 through Spanish galleons, *L. leucocephala* (Shelton and Brewbaker, 1994). The 21 other species are now more widely evaluated and at least 15 are involved in hybrids made in Hawaii (Sorensson and Brewbaker, 1994), Australia and India. The leucaenas grow rapidly, are highly drought tolerant and provide hard wood and highly digestible forage. They require mid-range pH similar to corn and sugarcane.

Leucaena grows well where annual rainfall is 600 to 2000 mm (Parrotta, 1992). Leucaena survives with rainfall as low as 300 mm. Three varieties bred in Hawaii dominate world production; these are K8 (Brewbaker, 1975), K636 (marketed in Australia as "Tarramba") and KX2. K8 and K636 are grown both as fodder and fuelwood worldwide, and are self-fertile and thus seedy. "KX2-Hawaii" (Brewbaker, 2008) is a newly released hybrid that is self-sterile, bee-pollinated and thus very poorly seedy. KX2 was bred with cold tolerance, while K8 and K636 excel only in the lowlands. A new hybrid ("LXL" in Hawaii, "Wondergraze" in Australia) promises to dominate forage production in the future. Generally the leucaenas have no pests or diseases of significance, and KX2 was bred to be resistant to a psyllid insect, now worldwide, that has come under heavy predation. Also promising are seedless triploid hybrids (KX4, KX5) with high wood yields and valuable hardwood quality, but they must be vegetatively cloned.

Biomass Properties. Keffer et al. (2008) reported a ranges of cellulose, hemi-cellulose, and lignin content of 41- 43%, 15-17%, and 25-27%, respectively. Wood from seven year old leucaena trees had 74% volatile matter, 25% fixed carbon, and 0.93% ash (Fuwape and Akindele, 1997). Gross CV was 22.03 MJ kg^{-1}. Ultimate analysis showed wood was 42% C, 6% H, 52% O, 0.6% N, 0.04% S and 0.93% ash.

Growth and Yield. Four-year old trees of varieties K8, K636 and KX2 yielded around 112 Mg ha^{-1} DM, or 28 Mg ha^{-1} yr^{-1}, with specific gravity ~0.6 and heat of combustion of 19.3 GJ Mg^{-1} (Brewbaker, 1980). Limited studies of regrowth after coppicing gave similar yields. Net energy produced was 323 GJ ha^{-1} yr^{-1} (Brewbaker, 1980).

7.5 Economic Analysis of Three Feedstock Crops

7.5.1 Nomenclature

Annual Equivalent Cost (AEC): Cost per year averaged over the life of the project. AEC is the present value of all costs of a project divided by the annuity factor for the life of the project, used to evaluate projects with different timelines.

Annual Equivalent Revenue (AER): Revenue per year averaged over the life of the project. AER is the present value of all revenues of a project divided by the annuity factor for the life of the project, used to evaluate projects with different timelines.

Discount Rate (*i*): The interest rate used in discounting future cash flows.

Net Present Value (NPV): Difference between the present values (PVs) of the future cash flows from an investment and the cost of the investment. Net Present Value on an annual basis is computed by discounting the annual net returns by the appropriate discount rate.

7.5.2 Framework for Economic Analysis

A net economic returns model was used to analyze the economic costs and returns of selected crops for biofuel production. The cost of producing each crop includes commonly used cost components such as land preparation, planting, fertilization, weed control, harvesting, etc. The analysis assumes that the producer has the necessary capital except for farm machinery and irrigation equipment for feedstock production.

7.5.3 Data

The majority of the investigated crops are not widely grown in Hawaii, hence, finding appropriate field data for economic analysis is challenging. For banagrass and leucaena, data from previous Hawaii studies were used. For switchgrass, data from previous North American studies were used.

7.5.4 Economic Analysis

Certain field operations are not performed regularly and uniformly year after year. Therefore, annual costs may differ over the crop's life. From an economic point of view, the overall approach is to estimate average annual costs and returns over the entire economic life of the crop, which allows for direct comparison among different crops. To calculate costs and revenues in annual equivalent terms, the present values of all costs and revenues over the useful life of the crop is transformed into an equivalent annuity. The following procedure was adopted in estimating annual equivalent cost and revenue (Monti et al., 2007).

1. Present value of the total investment over useful life (25 years) is estimated as:

$$PV = \sum_{t=0}^{n} FV_n x(1+i)^{-n} \quad (7.2)$$

where, PV and FV are present value and future values respectively and i denotes the appropriate discount rate.

2. Annual Equivalent Cost (AEC) and Annual Equivalent Revenue (AER) were estimated as:

$$\text{AEC} = \frac{PV_c xi}{1-(1+i)^{-n}} \quad (7.3)$$

$$\text{AER} = \frac{PV_R xi}{1-(1+i)^{-n}} \quad (7.4)$$

In this analysis, n was assumed equal to 25 years and i was 4.5% (average historical discount rate during 1986 – 2006 from the Federal Reserve System). PV_C and PV_R are the present value of cost and revenue respectively over the 25 year period. The analysis for each crop was based on assumptions, modifications and adjustments which are explained in the technical notes for each crop (see Appendix A at the end of this chapter).

Feedstock cost of ethanol per MJ was estimated by dividing the cost per ha of producing each feedstock by the corresponding crop's total per ha energy production. Total energy per liter of ethanol is 21.3 MJ (Jaeger et al., 2007). Feedstock cost per liter of ethanol was estimated by dividing the per ha cost of producing the feedstock by the total liters (per ha) of ethanol for each crop.

Break-even price of feedstock is that price of feedstock such that net revenue equals zero. Break-even price of ethanol is the price of energy such that net returns in terms of energy equals zero. The break-even price is calculated as cost divided by yield where yield is either in terms of feedstock or the appropriate conversion to energy.

7.5.5 Results of the Net Returns Model

Table 7.1 provides a comparison of crop yields, ethanol yields and feedstock costs per liter for the selected crops. Note that crop yields are the average crops yields. Banagrass has the highest ethanol production (2,207 liters ha^{-1} yr^{-1}). Banagrass, switchgrass, and Leucaena have feedstock costs less than \$0.26 $liter^{-1}$.

Table 7.1 Comparison of biofuel yields, feedstock cost $Liter^{-1}$ for selected bio-fuel crops.

Crop	Yield Mg ha^{-1} yr^{-1} (dry matter)	Conversion factor	Ethanol liter ha^{-1} yr^{-1}	Feedstock cost $liter^{-1}$
1. Switch grass	13.5	279 liter Mg^{-1}	3,767	\$0.21
2. Banagrass	48.2	279 liter Mg^{-1}	13,448	\$ 0.23
3. Leucaena	22.5	279 liter Mg^{-1}	6,278	\$ 0.21

Table 7.2 summarizes the major components of the economic analysis including the analysis involving feedstock and conversion of the feedstock to ethanol. The major findings are as follows:

- Net return (based on feedstock price) is not available for Leucaena because of the absence of feedstock price data. Banagrass and switchgrass show a positive net return ha^{-1}.
- For these bio-fuel crops, high production costs are primarily due to field operation costs (fertilizer, pesticide and other chemical applications) and harvesting costs. With improved yields, the cost component can be reduced and net returns improved.
- Net returns after conversion to ethanol show banagrass and switchgrass production as having positive net returns from ethanol production. This is due to the crop's high energy yield (conversion to ethanol). However, it should be noted that conversion costs to ethanol from cellulosic feedstock is still under investigation and hence the results should be interpreted with caution.

Table 7.2 Summary of feedstock and ethanol data (production, costs, gross revenues and net revenues): cellulosic feedstock.

Cost items	**Unit**	**Annual Equivalent Costs and Revenues**		
		Switchgrass	**Banagrass**	**Leucaena**
1. Land preparation	\$ ha^{-1}	\$3.63	\$86.40	\$20.50
Machinery				
Labor				
2. Planting	\$ ha^{-1}			\$10.55
Seeds/plants	\$ ha^{-1}	\$7.71	\$4.12	
Machinery				
Labor				

Chemicals		$11.56		
Other	$\$\ \text{ha}^{-1}$			
3. Field operations				
Machinery	$\$\ \text{ha}^{-1}$	$46.34		
Labor	$\$\ \text{ha}^{-1}$			
Fertilizer	$\$\ \text{ha}^{-1}$	$103.12	$464.31	$104.23
Herbicides	$\$\ \text{ha}^{-1}$	$73.68	$347.06	$120.04
Pesticides				
Irrigation	$\$\ \text{ha}^{-1}$			
Other	$\$\ \text{ha}^{-1}$			
4. Harvesting	$\$\ \text{ha}^{-1}$	$298.43	$684.76	$663.32
Machinery	$\$\ \text{ha}^{-1}$			
Labor	$\$\ \text{ha}^{-1}$			
Other	$\$\ \text{ha}^{-1}$		$553.43	$101.07
5. Other operations	$\$\ \text{ha}^{-1}$	$143.88	$698.81	$135.83
6. Operating overhead	$\$\ \text{ha}^{-1}$	$68.84	$283.90	$101.96
Total variable costs	$\$\ \text{ha}^{-1}$	$757.15	$3,122.80	$1,314.09
Fixed cost	$\$\ \text{ha}^{-1}$	$33.59	$138.52	$125.77
Total Cost	$\$\ \text{ha}^{-1}$	$790.75	$2,984.28	$1,131.75
A. Production				
Primary Production (Units)		Mg	Mg	Mg
1. Primary production		13.5	48.2	22.5
Average price per unit	$/unit		$92.39	-
3. Gross revenue	$\$\ \text{ha}^{-1}$	$1,123.28	$4,453.39	-
B. Net revenue		$332.53		-
Net revenue (ha^{-1})	$\$\ \text{ha}^{-1}$	$498.84	$1995.91	-
C. Production of ethanol				
Table 7.2 Cont'd				
2. Total processing cost	$\$\ \text{ha}^{-1}$	$1,302.53	$4,838.93	$2,600.91
3. Total production cost	$\$\ \text{ha}^{-1}$	$2,093.28	$7,961.72	$3,858.44
Gross revenue	$\$\ \text{ha}^{-1}$	$2,311.77	$8,558.95	$3,862.07
C. Net revenue from ethanol	$\$\ \text{ha}^{-1}$	$218.52	$597.22	$3.63
Feedstock cost of ethanol	$\$\ \text{liter}^{-1}$	$0.21	$0.23	$0.21
Break even price of feedstock	$\$\ \text{Mg}^{-1}$	$58.61	$64.68	-
Break even price of ethanol	$\$\ \text{liter}^{-1}$	$0.56	$0.59	$0.63

The net returns analysis shows that Banagrass, switchgrass, and Leucaena have a positive net return (i.e., when the price of Banagrass, switchgrass, and Leucaena are measured as a feedstock or in terms of ethanol).

The lowest break-even price of ethanol comes from switchgrass. However, for Banagrass and switchgrass caution should be used since the technology for cellulosic feedstock conversion to ethanol is still developmental. Cost and revenue figures should also be treated with caution due to geographical variations of the crops considered for the analysis (i.e., Hawaii vs North America).

7.5.6 Evaluation of Feedstock

Based on the criteria of wide adaptation, crop properties, net energy, and economics, switchgrass is a suitable feedstock crop for biofuel production (Table 7.3). The low N, Cl, S, Ca, and K indicates that switchgrass should not cause excessive NOx emissions, corrosion or fouling during thermochemical conversion. The fairly high lignin content may make biochemical conversion difficult without pretreatment. With its high lignin and low ash, N, and K contents, switchgrass has properties more like a tree than a grass.

Banagrass has grass-like properties. High N, K, and ash concentration, and low lignin are typical of grasses (Table 7.3). Banagrass has fairly wide adaptation and positive net revenue from ethanol that makes for a suitable biofuel feedstock.

Leucaena has fuel properties suitable for thermochemical conversion processes (Table 7.3). Low N, S, and ash concentration, but high lignin would make it more suitable for thermochemical conversion. The net revenue from ethanol is negative which does not make it suitable as a feedstock. However, the analysis was based on available yield data from older varieties. With the recent release of the new varieties "KX2-Hawaii" and "LXL" or "Wondergraze", the higher yields may change the economic analysis.

Table 7.3 Summary of criteria and data to evaluate three cellulosic feedstock crops. Fuel property and chemical data are from Adler et al. (2006), Bain et al. (2003), Keffer et al. (2008), and Otieno et al. (1992).

Criteria	Value	Switchgrass	Banagrass	Leucaena
Wide adaptation	Good-fair-poor	Good	Fair	Fair
Match crop to conversion	N < 0.6%	0.3 – 0.6%	0.4 – 0.8%	0.6%
	Cl < 0.1%	0.02 – 0.07%	0.6 – 0.8%	-
	S < 0.1%	0.05 – 0.06%	0.03%	0.04%
	Ca 15-35%	0.3 – 0.5%	0.3%	-
	K < 7%	0.06 – 0.35%	3.5%	-
	Lignin	13.0 – 18.2%	4%	25 – 27%
	Ash	2.3 – 3.5%	9.9%	0.9%
Net energy	GJ ha^{-1} yr^{-1}	60	-	323
Economics	Net revenue from ethanol	\$219 ha^{-1}	\$597 ha^{-1}	\$4 ha^{-1}

7.6 Summary

Several of the proposed criteria (Wright, 2007) were not included in the summary evaluation such as carbon sequestration and food supply. While these criteria are critical, they have no well-defined method for evaluation. Soil carbon accounting systems are currently being proposed (SM CRSP, 2008) indicating that no single method to determine soil carbon sequestration has been widely accepted. The food vs. fuel issue gained prominence after the rapid rise of food prices in 2008. In place of well defined methods, surrogates are used such as selecting deep rooted species or using marginal lands. Yet, well defined methods for assessment are needed.

Evaluations of feedstock crops are location specific. The environment, cost/price, and management data are affected by local conditions. The difference in costs between the mainland U.S. and Hawaii is large and may be an extreme example. Yield difference is also large as Hawaii has a year round growing season.

Improvements in plants are being made at a rapid pace. Efforts are underway to improve biomass and conversion yields, lengthen the growing season by shortening dormancy, reduce lignin content of biomass, and increase drought tolerance (Han et al., 2007). Feedstock evaluations will need to be done frequently to account for improvements in germplasm.

What is needed is a method to do evaluations rapidly and accurately. The assessment should be flexible to incorporate several criteria, site-specific to account for local conditions, and rapid to keep pace with the ever changing biophysical, genetic, and economic landscape.

7.7 References

Alcantara, P.B., Alcantara, V.D.B.G., and Almeida, J.E.D. (1980). "Studies on 25 prospective varieties of elephantgrass (*Pennisetum purpureum* Schum.)." *Bol. Ind. Anim.*, 37, 279-302.

Adler, P.R., Sanderson, M.A., Boateng, A.A., Weimer, P.J., and Jung, H. (2006). "Biomass yield and biofuel quality of switchgrass harvested in fall or spring." *Agron. J.,* 98, 1518-1525.

Anon. (1954). "Indian grass (*Sorghastrum nutans* L. Nash) and switchgrass (*Panicum virgatum* L.)." *Oklahoma Agr. Res. Sta. Forage Crops Leaflet 17*. Stillwater, Oklahoma.

Anon. (1977). "Leucaena, Promising forage and tree crop for the tropics." *Natl. Acad. Sci.,* pp. 155.

Bain, R.L., Amos, W.A., Downing, M., and Perlack, R.L. (2003). *Biopower technical assessment: state of the industry and technology*. Technical report NREL/TP-510-33123, National Renewable Energy Laboratory, Golden, CO.

Biomass Research and Development Board (BRDB). (2008). *National biofuels action plan*. Biomass Research and Development Initiative.

Bogdan, A.V. (1977). "Tropical pasture and fodder crops." *Tropical agriculture series*. Longman Group, London.

Boonman, J.G. (1993). *East Africa's grasses and fodders: Their ecology and husbandry.* Kluwer Academic Publ., London.

Brewbaker, J.L. (1975). "'Hawaiian Giant' koa haole." *Hawaii Agric. Exp. Sta. Misc. Publ.*, 125, 1-4.

Brewbaker, J.L. (1980). "Giant Leucaena energy tree farm; An economic feasibility analysis for the Island of Molokai, Hawaii." *Hawaii Natural Energy Inst. Publ.*, 60-6, pp. 90.

Brewbaker, J.L. (2008). "Registration of KX2-Hawaii, interspecific-hybrid leucaena." *J. Plant Registrations*, 2, 190-193.

Cuiping, L., Chuangzhi, W., Yanyongjie and Haitao, H. (2004). "Chemical elemental characteristics of biomass fuels in China." *Biomass and Bioenergy*, 27, 119-130.

Brejda, J.J. (2000). "Fertilization of native warm-season grasses." In: Anderson, B.E., and Moore, K.J. (Ed.) *Native warm season grasses: research trends and issues.* Madison, Wisconsin: CSSA Special Pub. No. 30. Crop Science Society of America, pp. 177-200.

Calhoun, D.S., and Prine, G.M. (1985). "Response of elephantgrass to harvest interval and method of fertilization in the cooler subtropics." *Soil Sci. Soc. Fla. Proc.,* 44, 112-115.

Casler, M.D., and Boe, A.R. (2003). "Cultivar x environment interactions in switchgrass." *Crop Sci.,* 43, 2226-2233.

Chaparro, C.J., Sollenberger, L.E., and Jones, C.J. (1995). "Defoliation effects on "Mott" elephantgrass productivity and lead percentage." *Agron. J.*, 87, 981-985.

Christian, D.G., Richie, A.B., and Yates, N.E. (2002). "The yield and composition of switchgrass and coastal panic grass grown as a biofuel in southern England." *Bioresource Tech.*, 83, 115-124.

Dien, B.S., Jung, H.G., Vogel, K.P., Casler, M.D., Lamb, J.F.S., Weimer, P.J., Iten, L, Mitchell, R.B., and Sarath, G. (2006). "Chemical composition and response to dilute-acid pretreatment and enzymatic saccharification of alfalfa, reed canary grass, and switchgrass." *Biomass Bioenergy*, 30, 880-891.

Duke, J.A. (1983). *Handbook of Energy Crops.* Available at http://www.hort.purdue.edu/newcrop/duke_energy/pennisetum_purpureum.html (accessed August, 2009).

Fischer, G., Prieler, S., and van Velthuizen, H. (2005). "Biomass potentials of miscanthus, willow and poplar: results and policy implications for Eastern Europe, Northern and Central Asia." *Biomass and Bioenergy*, 28, 119-132.

Fuentes, R.G., and Taliaferro, C.M. (2002). "Biomass yield stability of switchgrass cultivars." In: Janick, J., and Whipkey, A. (Ed.) *Trends in new crops and new uses*. Alexandria, VA: ASHS Press.

Fuwape, J.A., and Akindele, S.O. (1997). "Biomass yield and energy value of some fast-growing multipurpose trees in Nigeria." *Biomass and Bioenergy*, 12, 101-106.

Goldemberg, J. (2007). "Ethanol for sustainable energy future." *Science*, 315, 808 – 810.

Gomez, L.D., Steele-King, C.G., and McQueen-Mason, S.J. (2008). "Sustainable liquid biofuels from biomass: the writing's on the walls." *New Phytologist*, 178, 473-485.

Grof, B. (1969). "Elephant grass for warmer and wette lands." *Queensl. Agric. J.*, 94, 227-234.

Hall, K.E., George, J.R., and Riedl, R.R. (1982). "Herbage dry matter yields of switchgrass, big bluestem and indiangrass with N fertilization." *Agron. J.*, 74, 47-51.

Hammerschlag, R. (2006). "Ethanol energy return on investment: A survey of the li terature 1990 - present." *Environ. Sci. Technol.*, 40, 1744 - 1750.

Han, K.H., Ko, J.H., and Yang, S.H. (2007). "Optimizing lignocellulose feedstock for improved biofuel productivity and processing." *Biofpr*, 1, 135-146.

Hanna, W.W., Chaparro, C.J., Mathews, B.W., Burns, J.C., Sollenberger, L.E., and Carpenter, J.R. (2004). "Perennial Pennisetums." In: Moser, L.E., and Solenberger, L.E. (Ed.) *Warm-season (C4) Grasses*. Madison, WI: Agronomy Monograph. No. 45. ASA, pp. 503-535.

Hitchcock, A.S. (1935). *Manual of grasses of the United States*. U.S. Department of Agriculture, Washington, D.C.

Hsu, F.K., Hong, K.Y., Lee, M.C and Lee, K.C. (1990). "Effects of cutting height on forage yield, forage and silage quality of napier grass." *J. Agric. Assoc.* (China), 151, 77-89.

Jefferson, P.G., McCaughey, W.P., May, K., Woosaree, J., McFarlane, L., and Wright, S.M. (2002). "Performance of American native grass cultivars in the Canadian prairie provinces." *Native Plants J.*, 3, 24-33.

Jenkins,B.M., Baxter, L.L., Miles Jr., T.R., and Miles, T.R. (1998). "Combustion properties of biomass." *Fuel Processing Technol.*, 54, 17-46.

Johnson, W.L., Guerrero, J., and Pezo, D. (1973). "Cell wall constituents and in vitro digestibility of napier grass (Pennisetum purpureum)." *J. Anim. Sci.*, 37, 1255-1261.

Kablan, R., Yost, R.S., Brannan, K., Doumbia, M., Traore, K., Yorote, A., Toloba, Y. , Sissoo, S., Samake, O., Vaksman, M., Dioni, L., and Sissoko, M. (2008). "'Amengagement en courbes de niveau', increasing r ainfall capture, storage, and drainage in soils of Mali." *Arid Lands Research and Management*, 22, 62 - 80.

Keffer, V.I., Evans, D.E., Turn, S.Q., and Kinoshita, C.M. (2006). *Potential for ethanol production in Hawaii*. Prepared for State of Hawaii, Department of Business, Economic Development and Tourism, Honolulu, Hawaii.

Keffer, V.I., Turn, S.Q., Kinoshita, C. M., and Evans, D.E. (2008). "Ethanol technical potential in Hawaii based on sugarcane, banagrass, Eucalyptus, and Leucaena." *Biomass and Bioenergy*, doi:10.1016/j.biombioe.2008.05.018.

Kinoshita, C. (1999). "Siting evaluation for biomass-ethanol production in Hawaii." *College of Tropical Agriculture and Human Resources*, University of Hawaii at Manoa, Honolulu, Hawaii.

Kipnis, T., and Bnei-Moshe, S. (1988). "Improved vegetative propagation of napiergrass and pearl millet x napiergrass interspecific hybrids." *Trop. Agric.* (Trinidad), 65, 158-160.

Laureano-Perez, L., Teymouri, F., Alizadeh, H., and Dale, B.E. (2005). "Understanding factors that limit enzymatic hydrolysis of biomass." *Applied Biochemistry and Biotechnology*, 124, 1081-1099.

Lorenz, D.,
and Morris, D. (1995). "How much energy does it take to make a gallon of ethanol?" *Institute for local self-reliance*: Washington, DC.

Lowe, A.J., Thorpe, W., Teale, A., and Hanson, J. (2003). "Characterisation of germplasm accessions of Napier grass (*pennisetum purpureum* and *P.purpureum x P. glaucum* Hybrids) and comparison with farm clones using RAPD." *Genetic Resources and Crop Evolution*, 50, 121-132.

Madakadze, I.C., Stewart, K., Peterson, P.R., Coulman, B.E., and Smith, D.L. (1999). *Switchgrass biomass and chemical composition for biofuel in Eastern Canada. Agron. J.*, 91, 696-701.

Mannetje, L. T. (1992). "*Pennisetum purpureum* Schumach." In: Mannetje, L. T., and Jones, R.M. (Ed.) *Plant resources of Southeast Asia, forages*. Netherlands: Pudoc Sci. Publ., pp. 191-192.

McKendry, P. (2002) "Energy production from biomass (part 1): overview of biomass." *Bioresource Technology*, 83, 37-46.

McLaughlin, S.B. (1992). "New switchgrass biofuels research program for the Southeast." In: *Proc. Ann. Auto. Technol. Dev. Contractors Coordinating Meeting*, Dearborn, MI, Nov. 2-5, 1992. pp. 111-115.

McLaughlin, S.B., and Kszos, L.A. (2005). "Development of switchgrass (Panicum virgatum) as a bioenergy feedstock in the United States." *Biomass & Bioenergy*, 28, 515-535.

McLaughlin, S.B., Samson, R., Bransby, D.I., and Weislogel, A. (1996). "Evaluating physical, chemical, and energetic properties of perennial grasses as biofuels." In: *Bioenergy '96. Proc. Seventh National Bioenergy Conference: Partnerships to Develop and Apply Biomass Technologies*, Nashville, TN. September 15-20,1996, pp.1-8.

Obernberger, I. (1998). "Decentralized biomass combustion: state of the art and future development." *Biomass and Bioenergy*, 14, 33-56.

Osgood, R.V., and Dudley, N.S. (1993). *Comparative study of biomass yields for tree and grass crops grown for conversion to energy, Final report*. Prepared for State of Hawaii, Department of Business, Economic Development and Tourism, Honolulu, Hawaii.

Osgood, R.V., Dudley, N.S., and Jakeway, L.A. (1996). *A demonstration of grass biomass production on Molokai*. Diversified Crops Report 16, from Hawaii Agriculture Research Center.

Otieno, K., Onim, J. F. M., and Semenye, P. P. (1992). "Feed production and utilisation by dual-purpose goats in smallholder production systems of western Kenya." In: Stares, J.E.S., Said A.N., and Kategile J.A. (Ed.) *The complementarity of feed resources for animal production in Africa*. Proceedings of the joint feed resources networks workshop held in Gaborone, Botswana 4-8 March 1991. African Feeds Research Network. ILCA (International Livestock Centre for Africa), Addis Ababa. Ethiopia, pp. 430.

Parrotta, J.A. (1992). *Leucaena leucocephala (Lam.) de Wit: leucaena, tan tan.* Available at http://www.fs.fed.us/global/iitf/pubs/sm_iitf052%20%20(8).pdf (accessed August 2009).

Perlack, R.D., Wright, L.L., Turhollow, A.F., Graham, R.L., Stokes, B.J., and Erbach, D.C. (2005). *Biomass as feedstock for a bioenergy and bioproducts industry: The technical feasibility of a billion-ton annual supply*. ORNL/TM-2005/66. Oak Ridge National Laboratory, Oak Ridge, TN.

Pimentel, D., and Patzek, T. (2005). "Ethanol production using corn, switchgrass, and wood; biodiesel production using soybean and sunflower." *Nat. Resour. Res.*, 14 (1), 65 -76.

Porter, C.L. (1966). "An analysis of variation between upland and lowland switchgrass, *Panicum virgatum* L. in central Oklahoma." *Ecology*, 47, 980-992.

Robertson, G.P., Dale, V.H., Doering, O.C., Hamburg, S.P., Melillo, J.M., Wander, M.M, Parton, W.J., Adler, P.R., Barney, J.N., Cruse, R.M., Duke, C.S., Fearns ide, P.M, Follet, R.F., Gibbs, H.K., Goldemberg, J., Mladenoff, D.J., Ojima, D., Palmer, M.W., Sharpley, A., Wallace, L., Weathers, K.C., Wiens, J.A., Wi lhelm, W.W. (2008). "Sustainable biofuels redux." *Science,* 322, 49 -50.

Sanderson, M.A., Reed, R.L., McLaughlin, S.B., Wullschleger, S.D., Conger, B.V., Parrish, D.J., Wolf, D.D., Taliaferro, C., Hopkins, A.A., Ocumpaugh, W.R., Hussey, M.A., Read, J.C., and Tischler, C.R. (1996). "Switchgrass as a sustainable bioenergy crop." *Bioresource Tech.*, 67, 209-219.

Sanderson, M.A., Reed, R.L., Ocumpaugh, W.R., Hussey, M.A., Van Esbroeck, G., Read, J.C., Tischler, C.R., and Hons, F.M. (1999). "Switchgrass cultivars and germplasm for biomass feedstock production in Texas." *Bioresource Technology*, 67, 209-219.

Sanderson, M.A., and Wolf, D.D. (1995). "Switchgrass biomass composition during morphological development in diverse environments." *Crop Science*, 35, 1432-1438.

Sarath, G., Baird, L.M., Vogel, K.P., and Mitchell, R.B. (2007). "Internode structure and cell wall composition in maturing tillers of switchgrass (Panicum virgatum L.)." *Bioresour. Technol.*, 98, 2985-2992.

Sarath, G., Mithcell, R.B., Sattler, S.E., Funnell, D., Pedersen, J.F., Graybosch, R.A., and Vogel, K.P. (2008). "Opportunities and roadblocks in utilizing forages and small grains for liquid fuels." *J. Ind. Microbiol. Biotechnol.*, 35, 343-354.

Schmer, M.R., Vogel, K.P., Mitchell, R.B., Moser, L.E., Eskridge, K.M., and Perrin, R.K. (2006). "Establishment stand thresholds for switchgrass grown as an energy crop." *Crop Sci.*, 46, 157-161.

Schmer, M.R., Vogel, K.P., Mitchell, R.B., and Perrin, R.K. (2008). "Net energy of cellulosic ethanol from switchgrass." *Proc. Nat. Acad. Sci.*, 105, 464-469.

Senelwa, K., and Sims, R.E.H. (1999). "Fuel characteristics of short rotation forest biomass." *Biomass and Bioenergy*, 17, 127-140.

Sheehan, J., Aden, A., Paustian, K., Killian, K., Brenner, J., Walsh M., and Nelson, R.
(2004). "Energy and environmental aspects of using corn stover for fuel ethan ol." *J. Ind. Ecol.*, 7(3-4), 117 – 146.

Shelton, H. M., and Brewbaker, J.L., (1994). "Leucaena leucocephala - the most widely used forage tree legume." In: Gutteridge, R.C., and Shelton, H.M. (Ed.) *Forage tree legumes in tropical agriculture*. CAB International, London, pp. 15-29.

Shleser, R. (1994). *Ethanol production in Hawaii*. Prepared for the State of Hawaii, Department of Business, Economic Development and Tourism, Honolulu, Hawaii.

Skerman, P.J., and Riveros, F. (1990). "Tropical grasses. Plant production and protection series 23." *Food and Agriculture Organization*, Rome.

Soil Management Collaborative Research Support Program (SM CRSP). (2008). *A soil carbon accounting and management system for emission trading*. Special publication. SM CRSP 2002-4. University of Hawaii, Honolulu, Hawaii.

Sollenberger, L. E., Williams, M. J., and Jones, C.S. (1990). "Vegetative establishment of dwarf elephant grass: effects of planting date, density, and location." *Proc. Soil Crop Sci. Soc. Fla.*, 50, 47-51.

Sorensson, C.T., and Brewbaker, J. L. (1994). "Interspecific compatibility among 15 Leucaena species (Leguminosae:Mimosoidead) via artificial hybridizations." *Amer. J. Bot*, 81, 240-247.

Strezov, V., Evans, T.J., and Hayman, C. (2008). "Thermal conversion of elephant grass (Pennisetum purpureum Schum) to bio-gas, bio-oil and charcoal." *Bioresource Technology*, 99, 8394-8399.

Tcacenco, F.A., and Botrel, M., de A. (1994). "Identificação e avaliação de acessos e cultivares de capim-elefante". In: Carvalho, M.M., Alvim, M.J., Xavier, D.F., Carvalho, L. de A. (Eds.) *Capim-elefante: Produção e utilização. CoronelPacheco*: EMBRAPA, CNPGL, pp.1-30.

Tilman, D., Socolow, R., Foley, J. A., Hill, J., Larson, E., Lynd, L., Pacala, S., Reilly, J., Searchinger, T., Somerville, C., and Williams, D. (2009). "Beneficial biofuels—The food, energy, and environment trilemma." *Science*, 325, 270-271.

Turn, S., Kinoshita, C., Ishimura, D., Jenkins, B., and Zhou, J. (2003). *Leaching of alkalis in biomass using Banagrass as a prototype herbaceous species*. National Renewable Energy Laboratory Subcontractor Report, NREL/SR-510-35433. Available at http://www.osti.gov/bridge (accessed August 2009).

Tyson, K. S., Riley, C. J., and Humphreys, K. K. (1993). *Fuel cycle evaluations of Biomass ethanol and reformulated gasoline*. NREL/TP-4634950 National Renewable Energy Laboratory: Golden, CO.

Van Den Beldt, R., and Brewbaker, J.L. (1985). "Leucaena Wood Production and Use." NFTA, Waimanalo, HI. 50 pp.

Velez-Santiago, J., and Arroyo-Aguilu, J.A. (1981). "Effect of three harvest intervals on yield and nutritive value of seven napiergrass cultivars." *J. Agric. Univ. P.R.*, 65, 129-137.

Vogel, K.P. (2004). "Switchgrass." In: Moser, L.E., Sollenberger, L. and Burson, B. (Ed.) *Warm –season (C4) grasses*. Madison, Wisconsin: ASA-CSSA-SSSA Monograph.

Vogel, K.P., Brehda, J.J., Walters, D.T., and Buxton, D.R. (2002). "Switchgrass biomass production in the Midwest: harvest and nitrogen management." *Agron. J.*, 94, 413-420.

Vogel, K.P., Dewald, C.I., Gorz, H.J., and Haskins, F.A. (1985). "Development of switchgrass, Indiangrass, and eastern gamagrass: Current status and future. Range improvements in Western North America." *Proc. Mtg. Soc. Ranfe Mgt.,* Salt Lake City, Utah, Feb 14, pp. 51-62.

Wright, L. (2007). *Historical perspective on how and why switchgrass was selected as a "model" high-potential crop.* ORNL/TM-2007/109. Oak Ridge National Laboratory, Oak Ridge, TN.

Appendix A

Table A.1 Switchgrass Production Costs and Returns Per Planted Acre.

Items	Unit	Quantity	Production costs ($/acre/year)		Annual equivalent cost and revenue
			2001	2007	
1. Land preparation	acre	1	$11.48	$13.16	$1.47
Machinery					
Labor					
2. Planting					
Seeds	acre	1	$24.36	$27.93	$3.12
Machinery					
Labor					
Chemicals	acre		$36.59	$41.95	$4.68
Other					
3. Field operations					
Machinery	acre		$16.36	$18.76	$18.76
Labor					
Fertilizer	acre	1	$36.41	$41.75	$41.75
Herbicides	acre	1	$26.01	$29.83	$29.83
Pesticides					
Irrigation					
Other					
4. Harvesting	acre	1	$105.37	$120.82	$120.82
Machinery					
Labor					
Other	acre				
5. Other operations	acre	1	$51	$58.25	$58.25
6. Operating overhead	acre	1	$30.74	$35.24	$27.87
Total costs	acre	1	$338.14	$387.69	$306.54
Fixed cost for machinery	acre	4.44%		$17.20	$13.60
Total variable cost	acre	1		$404.89	$320.14
A. Production of switchgrass					
1. Primary production	tons/ac/year	6.01			
2. Secondary production					
3. Gross revenue per acre	acre	6.01	$75.69	$454.77	$454.77
B. Net revenue from switchgrass					
Net revenue	acre/crop			$49.88	$134.64
C. Production of ethanol					
1. Processing cost	$/gallon	$1.31			

2. Total processing cost	$/acre	402.55		$527.34	$527.34
3. Total production cost	$/acre			$932.23	$847.48
Gross revenue	$/acre	$2.33		$935.94	$935.94
C. Net revenue from ethanol					
Net revenue	$/acre			$3.71	$88.47
Feedstock cost of ethanol	$/1000Btu				$0.003
Feedstock cost of ethanol	$/gallon				$0.80
Break even price of switchgrass	$/ton				$53.28
Break even price of ethanol	$/gallon				$2.11

Notes: Switchgrass

1. Cost of production data are based on Duffy and Nanhou (2001). Note: Prices were inflated to 2007 using an appropriate CPI for the midwest.
2. Cost of other operations include land rent.
3. Irrigation costs were excluded from the analysis as production under rainfed conditions were considered.
4. Operating overhead is assumed to be 10% of total operational cost.
5. Average price of a ton of banagrass dry matter ($80) was based on Kauai Island Utility cooperative. Renewable energy technology assessment report, March 21, 2006 (URL: http://www.kiuc.coop/pdf/KIUC%20RE%20Final%20Report%207 %20-%20Biomass%20&%20MSW.pdf). The price was adjusted to mid-west situation using CPI differences between Honolulu and midwest.
6. Ethanol yield for a ton of switchgrass dry matter is assumed to be similar to banagrass (67 gallons) and was based on the work by Gieskes and Hackett (2003).
7. It is assumed that the harvesting cycle for swithcgrass is 12 months.
8. Processing cost for switchgrass is assumed to be the same as ethanol produced from corn stover. Total processing cost of ethanol from lignocellulose sources like corn stover = $1.50/gallon less $0.49 stover cost/gallon (McAloon et al. 2000) = $1.01/gallon (in 1999$). The cost for switchgrass ethanol processing is inflated to 2007$ by a factor of 1.24 = $ 1.255 (US CPI average 1999-2007) and adjusted to the price in midwest by multiplying a factor of 1.046 (CPI adjusted factor for midwest as compared to the US average) = $ 1.31.
9. Estimated price of ethanol in the midwest for 2007 is $2.33/gallon.
10. Cost and revenue figures may differ slightly due to rounding.
11. Energy per gallon of ethanol = 76,300 Btu (Jaeger et al. 2007).

Table A.2 Banagrass Production Costs and Returns per Planted Acre.

Items	Unit	Quantity	Production costs ($/acre/year)		Annual equivalent cost and revenue
			1999	2007	
1. Land preparation	$/acre	1	$126.00	$159.50	$34.94
Machinery					
Labor					
2. Planting					
Seeds	$/acre	1	$6.00	$7.60	$1.66
Machinery					
Labor					
Chemicals					
Other					
3. Field operations					
Machinery					
Labor					
Fertilizer	$/acre	1	$148.50	$187.98	$187.98
Herbicides	$/acre	1	$111.00	$140.51	$140.51
Pesticides					
Other					
4. Harvesting	$/acre	1	$219.00	$277.23	$277.23
Machinery					
Labor					
Other	$/acre	1	$177.00	$224.06	$224.06
5. Other operations	$/acre	1	$224	$282.92	$282.92
6. Operating overhead	$/acre	1	101.10	$127.98	$114.93
Total costs	$/acre	1	$1,112.10	$1,407.79	$1,264.24
Fixed cost for machinery	$/acre	4.44%		$62.45	$56.08
Variable cost	$/acre	1		$1,345.34	$1,208.16
A. Production of banagrass					
Primary production	tons/ac/year	21.50			
Gross revenue per acre	Acre	21.50		$1,802.99	$1,802.99
Net revenue	$/acre			$395.21	$538.75
B. Production of ethanol					
Processing cost	$/gallon	$1.36			
Total processing cost	$/acre	1440.50		$1,959.08	$1,959.08
Total production cost	$/acre			$3,366.87	$3,223.32
Gross revenue	$/acre	$2.41		$3,465.16	$3,465.16
Net Revenue (ethanol)	$/acre			$98.29	$241.83

Feedstock cost of ethanol	$/1000Btu				$0.012
Feedstock cost of ethanol	$/gallon				$0.88
Break even price of banagrass	$/ton				$58.80
Break even price of ethanol	$/gallon				$2.24

Notes: Banagrass

1. Cost of production data are based on Kinoshita and Zhou (1999). Note: machinery and labor costs are included in the respective operation category (e.g., land preparation, harvesting, etc.).
2. Prices were inflated to 2007 using appropriate CPI for Honolulu.
3. Cost of land preparation includes both soil preparation and planting banagrass.
4. Cost for other operations included road maintenance, crop control research, equipment, landholding, etc.
5. Irrigation costs were excluded from the analysis as banagrass was considered a rainfed crop.
6. Operating overhead is assumed to be 10% of total operational cost.
7. Average price of a ton of banagrass dry matter ($80) was based on Kauai Island Utility cooperative. Renewable energy technology assessment report, March 21, 2006 (URL:http://www.kiuc.coop/pdf/KIUC%20RE%20Final%20Report%207%20-%20Biomass%20&%20MSW.pdf).
8. Ethanol yield for a ton of banagrass dry matter (67 gallons) was based on Gieskes and Hackett (2003).
9. It is assumed that the harvesting cycle for banagrass is 8 months.
10. Processing cost for banagrass is assumed to be the same as ethanol produced from corn stover. Total processing cost of ethanol from a lignocellulose source like corn stover = $1.50/gallon less $0.49 stover cost/gallon (McAloon et al. 2000) = $1.01/gallon (in 1999$). The cost for banagrass ethanol processing is inflated to 2007$ by a factor of 1.24 = $ 1.255 (US CPI average 1999-2007) and adjusted to the price in Honolulu by multiplying by factor of 1.083 (CPI adjustment factor for Honolulu as compared to the US average) = $ 1.36.
11. Price of ethanol in 2007 is rounded to $2.41/gallon.
12. Cost and revenue figures may differ slightly due to rounding.
13. Energy per gallon of ethanol = 76,300 Btu (Jaeger et al., 2007).

Table A.3 Leucaena Production Costs and Returns per Planted Acre.

Items	Unit	Quantity	Production costs ($/acre/year)		Annual equivalent cost and revenue
			1999	2007	
1. Land preparation	$/acre	1	$284.00	$363.19	$8.30
Machinery					
Labor					
2. Planting	$/acre	1	$146.00	$186.71	$4.27
Seeds					
Machinery					
Labor					
Chemicals					
Other					
3. Field operations					
Machinery					
Labor					
Fertilizer	$/acre	1	$33.00	$42.20	$42.20
Herbicides	$/acre	1	$38.00	$48.60	$48.60
Pesticides					
other					
4. Harvesting	$/acre	1	$210.00	$268.55	$268.55
Machinery					
Labor					
Other			$32.00	$40.92	$40.92
5. Other operations	$/acre	1	$43.00	$54.99	$54.99
6. Operating overhead	$/acre	1	$78.60	$95.02	$41.28
Total costs	$/acre	1	$864.60	$1,100.17	$509.12
Fixed cost for machinery	$/acre	10%		$110.02	$50.91
Variable cost	$/acre	1		$990.16	$458.20
A. Production of Leucaena					
Primary production (dry matter)	tons/acre	10.00			
Secondary production					
Gross revenue per acre					
B. Production of Ethanol					
Processing cost	$/gallon		$1.62		
Total processing cost	$/acre	650.00		$1,053.00	$1,053.00
Total production cost	$/acre			$2,153.17	$1,562.12
Gross revenue	$/acre	650.00	$2.41	$1,563.59	$1,563.59
Net revenue (ethanol)	$/acre			-$589.58	$1.47

Feedstock cost of ethanol	$/1000Btu				$0.013
Feedstock cost of ethanol	$/gallon				$0.78
Break even price of ethanol	$/gallon				$2.40

Notes: Leucaena

1. Cost of production data are based on Kinoshita and Zhou (1999). Note: machinery and labor costs are included in their respective operation category (e.g., land preparation, harvesting, etc.).
2. Prices were inflated to 2007 using appropriate CPI differences between 1999 and 2007.
3. Cost for other operations include road maintenance, crop control research, equipment, landholding, etc.
4. Irrigation costs were excluded from the analysis as Leucaena was considered under rainfed conditions.
5. Operating overhead is assumed to be 10% of total operational cost.
6. A 25 year rotation period is assumed for Leucaena.
7. Gross revenue in terms of Leucaena chips were not estimated due to lack of market price.
8. In calculating ethanol yield, it is assumed that on average 1 ton of dry matter yields 65 gallons of ethanol.
9. Processing costs for Leucaena (wood based ethanol) = $98.7/ dry ton (in 2006$) (Jaeger et al., 2007). This cost is converted to per gallon of ethanol in 2007$ as follows: cost/gallon = $98.7/65 = $ 1.518/gallon x 1.032 (average CPI west urban for Oregon 2006-2007) x 1.034 (adjusted CPI for Honolulu as compared to west, urban CPI in 2007) =$1.62/gallon.
10. Price of ethanol in 2007 is rounded to $2.41/gallon.
11. Cost and revenue figures may differ slightly due to rounding.
12. Energy per gallon of ethanol = 76,300 Btu (Jaeger et al., 2007).

CHAPTER 8

Preprocessing of Lignocellulosic Biomass for Biofuel Production

Prachand Shrestha, Buddhi P. Lamsal, and Samir Kumar Khanal

8.1 Introduction

Biomass preprocessing is one of the primary operations in the feedstock supply system. Preprocessing essentially prepares biomass into a form suitable for transportation, storage and handling prior to its conversion to biofuel and other bio-based products (Wright et al., 2006). The importance of biomass preprocessing to the overall bio-based industry lies on its influence on critical cost and quality barriers associated with bulk biomass handling, transportation, and variability in quality.

The harvesting window of biomass depends on geographical locations and prevailing weather conditions like rain, snow, extreme temperatures, and relative humidity among others. In addition, the physical and compositional characteristics of biomass vary significantly. The harvested biomass has low bulk density, high moisture content, and varying composition in addition to different shape and size (USDOE, 2003). Conversion of non-preprocessed biomass to biofuels or other bio-based chemicals is uneconomical due to high delivery costs and greater risk of variability in feedstock quality.

The orderly flow of biomass from field to conversion facilities has the following major requirements:

- Low moisture content
- High bulk density
- Less variability in physical and chemical characteristics
- Less geographical and seasonal variability
- Uniform demands for labor and machine
- Acceptable regulations on storage and transport
- Less variability in price structure

Preprocessing is an essential step in biomass supply and conversion system. Earlier conceptual designs showed that this operation is located at the front-end of the biomass refinery plant. However, research has indicated that the distributed preprocessing at the field-side or in a fixed preprocessing facility can provide significant cost benefits by producing a higher value feedstock with improved handling, transporting, and merchandising potential (Wright et al., 2006). Regardless of its operational location, mechanical preprocessing of biomass is necessary for particle size and density alterations before conversion at the biorefineries. Such steps produce improved materials with bulk and flow characteristics suitable for handling and conveying to the biorefinery. The preprocessing operations typically include the following steps (USDOE, 2003):

(a) ***Cleaning, separation and sorting*** operations remove dirt or foreign materials from the harvested biomass. Depending on requirement of biorefineries, the feedstock can be sorted according to dimensions and density.

(b) ***Mixing/blending*** of different feedstock results in improved biomass quality in terms of the composition that best suits the conversion processes.

(c) ***Moisture*** control is an important preprocessing step. Low moisture improves handling, stability, and transportation of biomass.

(d) ***Densification*** processes like baling, pelletization, briquetting, compaction etc. increase the bulk density of biomass. The handling, storage and transportation of feedstock are greatly enhanced by densification. Some of the densification methods may not be applicable especially for biochemical conversion methods.

(e) ***Chemical/biochemical treatment*** also reduces overall conversion cost of feedstock to biofuels and biochemicals. Ensiling is one of the examples of biochemical treatment.

This chapter focuses on preprocessing of lignocellulosic biomass in improving its post-harvest handling and supply system. The preprocessing steps, discussed in the chapter, are primarily targeted towards biochemical conversion processes; nevertheless, they can also be applied in thermochemical conversion processes. The merits of biomass preprocessing and methods are also elaborated. We provide overview of biomass handling and supply systems, preprocessing technologies. Further research needs in this area are also outlined.

8.2 Merits of Biomass Preprocessing

8.2.1 Stability

An effective preprocessing warrants uniformity in biomass composition, appropriate size and low moisture content. The latter is crucial to minimize or restrict microbial activity on the preprocessed biomass. Moisture less than 20% (w/w) confers aerobic stability of the biomass (Hess et al., 2009). It also minimizes dry weight loss during handling (due to densification) and storage (due to microbial activity). The harvested biomass has a higher moisture content: for example, corn stover 40% w/w, switchgrass 34% w/w. Therefore, biological or chemical amendment would be necessary to stabilize the biomass before processing to biofuels. Solid-substrate fermentation, ensiling or chemical pre-treatment are alternative options.

8.2.2 Storage

Storage is an important part of biomass supply system. There is a small and definite period for biomass harvest. A biomass inventory and supply system needs to be operationally robust. Distributed (and/or intermediate) feedstock storage locations must ensure an uninterrupted supply of homogenous biomass to the biorefineries. Therefore, quality alterations like microbial activity leading to biomass weight loss, dry matter loss due to moisture loss, are not favorable during bulk storage in these facilities. Preprocessing assures minimal or no loss during the long-term storage of biomass (Hess et al., 2007). A smaller space requirement is also attributed to densified biomass (DOE, 2003). Shinners et al. (2006) compared the moisture content and dry matter loss profiles during field and indoor storage of canary grass and switchgrass bales. These biomass were initially field-dried (prior to baling) to reduce the moisture content below 20%. Bales stored (average bulk density: 163 kg m^{-3}) outside had higher dry matter loss compared to less than 3% biomass loss for indoor storage.

8.2.3 Handling

Preprocessing also eases the bulk handling of biomass. When harvested and collected in or nearby the field, the biomass has low bulk density and lesser flowability. This creates problems with the handling of biomass due to capital and operational costs. Size reductions and the densification (chopping, pelleting or grinding) and drying of biomass to a certain size and moisture content help to increase the bulk density of the biomass and enhance the flowability. Such added or formatted attributes to biomass help optimize biomass handling and could be similar to the existing infrastructure of grain handling systems (Sokhansanj and Turhollow, 2004). The overall capital and operational costs are reduced in terms of collection (e.g. conveyor), storage (e.g. ensiling) and transportation either from field to biorefinery or a preprocessing depot to the biorefinery (Sokhansanj and Fenton, 2006).

8.2.4 Transportation

Transportation costs could exceed 40% of the total cost for biomass delivered at the biorefinery. Preprocessing of biomass ensures effective transportation of processed biomass by adequately increasing transportation cycle capacity (Hess et al., 2007), maximizing the quantity of dry biomass hauling. The payload is effective due to reduced a moisture content. The bulk density of processed biomass is higher than raw biomass. Therefore, the hauling cost is greatly optimized for processed biomass (low moisture and high bulk density) as reported by Sokhansanj et al. (2009) in Table 8.1.

Table 8.1 Bulk density of biomass (e.g. switchgrass) after various densification process and cost of transporting the biomass to a biorefinery (capacity: 2000 metric ton per day) (*Adapted from Sokhansanj et al., 2009).*

Form of biomass (densified)	Biomass dimensional characteristics	Bulk density (kg m^{-3})	Transportation cost ($ metric ton^{-1})
Bale	Round or rectangular	140 – 180	15.83
Ground	1.5 mm pack fill with tapping	200	14.42
Pellets	6.24 mm diameter	500 – 700	11.67

8.3 IBSAL Model in Biomass Preprocessing

Integrated biomass supply analysis and logistics (IBSAL) is a computational and mathematical simulation model developed at the Oak Ridge National Laboratory, for biomass supply chains from field to gate (biorefinery) as shown in Figure 8.1. The model embraces the concept that the biomass supply is an integral part of holistic biomass-biorefinery system (Sokhansanj and Fenton, 2006). It is a dynamic model that simulates the biomass supply chain components - harvesting, storage, preprocessing, transportation, etc. (Sokhansanj et al., 2006, 2008). The model considers reported parameters like spatial and yield information of biomass source (location, size and yield), physical characteristics (biomass moisture content), meteorological, machinery and operational data to calculate biomass delivery cost ($ metric ton^{-1}), energy expense (MJ metric ton^{-1}) and environmental impact (carbon emission, kg carbon metric ton^{-1}). Corn stover and switchgrass are two good examples for biomass supply system simulations. For example, IBSAL simulations reported transportation, grinding, stacking and baling (Table 8.2) as the major cost contributors, ca. $53.16 metric ton^{-1} of corn stover, for a biomass supply system. This biomass supply model excludes the payment cost (estimated $10 metric ton^{-1}) to the farmer (Sokhansanj and Fenton, 2006). This cost is commonly known as grower payment (or stumpage fee for forest resources) to cover the cost of the biomass.

Table 8.2 Cost of corn stover collection and supply systems (*Sokhansanj et al., 2006*).

Operation	Cost ($ metric ton^{-1})	Energy input (MJ metric ton^{-1})	Carbon emission (kg C metric ton^{-1})
Combine	1.47	48	1.035
Shredding	1.23	17.6	0.375
Baling	9.01	124.9	2.645
Stacking	9.16	169.1	3.59
Load	3.59	154.5	12.05

Truck travel	13.76	640.6	49.875
Unload	3.58	199.6	15.585
Stack	0.44	7.5	0.585
Grind	10.92	185.7	14.45
Overall	**53.16**	**1547.5**	**100.19**

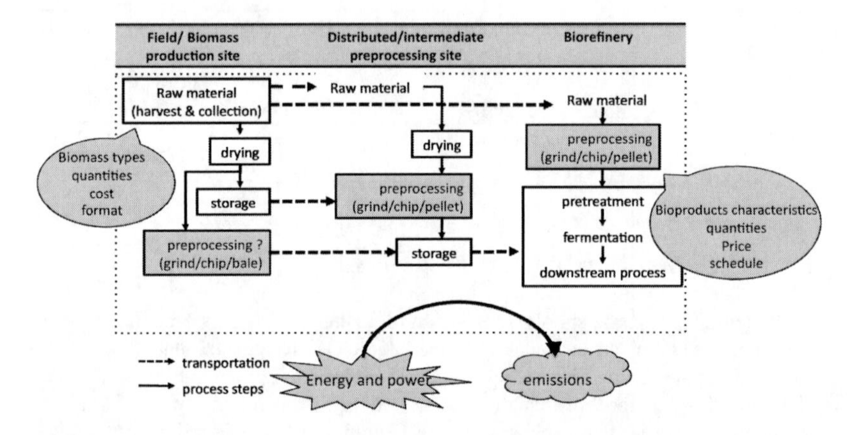

Figure 8.1 Overview of integrated biomass supply systems (*Adapted from Sokhansanj and Fenton, 2006; BRDI, 2007; Hess, 2008*).

The IBSAL model was also applied to the delivery of switchgrass to the biorefinery at varying costs depending on various forms of biomass (Kumar and Sokhansanj, 2007). The cost of switchgrass farming was reported to be \$ 41.5 metric ton^{-1} at yield of 10 metric tons acre^{-1} (Sokhansanj et al., 2009). In addition, the biomass collection, preprocessing and transportation to a 2,000 metric tons day^{-1} biorefinery would roughly cost: (a) \$44-47 metric ton^{-1} for round and square bales, (b) \$37 metric ton^{-1} for loafs (2.4 m x 3.6 m x 6 m), (c) \$40 metric ton^{-1} for chopped biomass, and (d) \$48 metric ton^{-1} for ensiled chops.

8.4 Biomass Preprocessing Methods

8.4.1 Baling

The harvested biomass, primarily dedicated energy crops, and agricultural residues like straw, stover, woody biomass, etc. are baled for easy handling and transport and storage. However, as the bales are bundled up without any densification, their use would largely be limited to short distance hauling to either an on-field preprocessing facility or a biorefinery nearby.

Forage hays have been made into bales after drying on the field to about 15% moisture (w/w), and picking up from windrows with automatic balers. Hay bales can either be round or rectangular, 1.5 m wide by 1.8 m diameter, round bales being popular on most US farms. Similar concepts and equipment have been used to bale agricultural residues and dedicated energy crops. Limited research with switchgrass indicates that round bales may not be the suitable format for large-scale biomass handling (Sokhansanj et al., 2009). Round bales tend to deform under static loads in a stack, and misshaped bales are difficult to secure onto trucks to form a transportable load over open roads. Variations in the density of round bales were shown to be the cause of uneven size reduction and erratic machine operation during de-baling. Sokhansanj et al. (2009) reported large, rectangular bales 0.9 m x 1.2 m x 2.4 m as the size of choice for switchgrass. Dooley et al. (2008) also reported effective baling of urban woody biomass into large square bales and their transportation on conventional flatbed trucks.

8.4.2 Chopping

Chopping is a size reduction operation meant to help transport in bulk. This is carried out for non-bale harvest/ transport of biomass. Once the harvested biomass is dried to less than 15% moisture, the harvester picks it up from windrows and chops in into pieces ranging from 25 to 50 mm in size (Sokhansanj, 2009). The chopped biomass is blown into a forage wagon traveling with the harvester, which is then unloaded into a pile or into a conveyor belt to load into container. Wet chopping is done for the hay going into ensilage.

8.4.3 Grinding

While the grinding of biomass is done as part of pretreatment to make it more amenable to downstream processing, it is increasingly being considered for preprocessing operations also. Biomass size reduction process changes the particle size and shape, increases bulk density, improves flow properties, increases porosity, and generates new surface areas. Grinding preprocessing could be used earlier in the feedstock assembly systems to reduce the size of the biomass prior to delivery to the biorefinery. This will allow the feedstock to be handled with conventional bulk solids handling and transport equipment, enabling the implementation of nonbale harvesting and collection systems (Wright et al., 2006). However, grinding is a major prerequisite for biomass densification operations, especially pelletization, and briquetting, both of which are mass and energy densification processes. Grinders reduce woody biomass particles in size by repeatedly pounding them into smaller pieces through a combination of tensile, shear and compressive forces. Most of the grinders are derivatives of hammer mills and include horizontal feed grinders (Jackson et al., 2007). Most of the herbaceous grass, however, are ground using attrition type of grinders/ mill.

8.4.4 Biomass Densification

In the biological conversion of lignocellulosic biomass, the importance of densification lies more in the handling, storage and transportation, not necessarily during downstream processing, where the biomass is processed as a slurry. However, the thermochemical conversion of biomass requires dry pellets or briquettes for efficient pyrolysis/ gasification of the biomass.

Loose cut biomass has bulk density of less than 150 kg m^{-3} depending on particle size (Gilbert et al., 2009), for example, 60-120 kg m^{-3} for switchgrass (Sokhansanj, 2009). The bulk density of chopped and ground biomass can be increased by up to 25% by vibrating the biomass holder. To further increase density, the biomass must be compacted mechanically into various forms. Various densification operations like compaction, pelletization, briquetting, etc. increase the density of herbaceous plant materials to up to 700 kg m^{-3}. For example, Sokhansanj et al. (2009) reported a bulk density of switchgrass at ~80 kg m^{-3} when chopped at ~50 mm size. When ground, the bulk density was 120 kg m^{-3}, ~180 kg m^{-3} when baled, 350 kg m^{-3} when briquetted at 32 m diameter × 25 mm thick, 400 kg m^{-3} when cubed at 33 mm, and ~700 kg m^{-3} when pelletted at 6.2 mm diameter. The bulk density of wood pellet was reported to be about 591 kg m^{-3} (Obernberger and Thek, 2004). Raw feedstock is typically comminuted, and conditioned to an appropriate moisture content (either by dehydration or moisture addition) prior to the densification process. Table 8.3 compares the bulk density of biomass densified under different conditions.

Table 8.3 Bulk densities of densified biomass.

Types of preprocessed biomass	Dimensions (mm)	Bulk density (kgm^{-3})
Loose cut biomass		>150
Chopped and ground	(through vibration of holder)	188
Baled		128-192
Cubes-small size	16 (W1) x 16 (W2) x 32 (L)	561
Cubes – regular size	32 (W1) x 32 (W2) x 63 (L)	482
Briquette	5(D) x 25 (L)	642

The term 'pellet' is usually referred to materials less than 15 mm in diameter, while 'briquette' is generally the term used for larger units of densified material. The pellet properties and quality depends on the properties of the feedstock, for example, biomass type, moisture content and particle size, pelletizer type and operating conditions, and binding agent. Gilbert et al. (2009) studied the effects of pelletization conditions on product properties for switchgrass and wheat straw using a simple compression pelletizer. Temperature (up to 125°C) had a greater effect on pellet quality than the applied pressure. It is believed that at elevated temperatures, the lignin present in the biomass softens and acts as a binding agent. The cut switchgrass produced more desirable pellets over shredded switchgrass due to an additional binding effect of intertwined fibers. At elevated temperatures, with a grass to tar ratio of 2:1, the pellets were twice as strong as pellets made by cut switchgrass. The increased durability was a result of lignin present in the biomass and the heavy oil dispersing inside the pelletizer, which filled in the gaps between the switchgrass fibbers upon heating.

Briquetting of biomass can be done by mixing it with a binding agent and forcing it through a die or opening (roll and char briquetting, pelletizing) or by direct compacting (piston press, and screw press). Different types of binders e.g., bentonite, hydrated lime, starch, agro-colloids, synthetic binders can be used. The adhesive properties of thermally softened lignin are thought to contribute considerably to the strength characteristics of briquettes made of lignocellulosic materials (Granada et al., 2002). The binderless press-agglomeration of lignocellulosic biomass has been the focus of research recently showing that besides material-specific factors, compaction pressure, compaction temperature, compaction velocity, moisture content and particle size distribution determine the properties of resulting briquettes. Some of these properties include bending and compressive strengths, durability, stability, ignitability, and smoothness (Demirbas and Sahin-Demirbas, 2009). The power requirement varies from 10 to 50 hp/dry metric ton.

Kaliyan et al. (2009) produced almond-shaped briquettes, 29 to 31 mm long, from corn stover and switchgrass in a roll press and showed that high-durability corn stover and switchgrass briquettes with bulk densities of 480 to 530 kg m^{-3} could be produced. This corresponded to about a three- to five-fold increase in bulk densities compared to those of the bales. Comparing the roll press briquettes with pellets from a conventional ring-die machine, the researchers ound that the briquettes had bulk densities (351 to 527 kg m^{-3}), durabilities (39% to 90%), and crushing strengths (28 to 277 N) comparable to that of 10 mm diameter pelletes. The role and necessity of binders in briquetting/pelletizing is an important consideration. With various microstructural studies, including chemical composition analyses, scanning electron microscopy imaging, and UV auto-fluorescence imaging on grinds, briquettes, and pellets, Kaliyan et al. (2009) confirmed that highly dense, strong, and durable briquettes and pellets from corn stover and switchgrass could be produced without adding chemical binders (i.e. additives) by activating (softening) the natural binders such as water-soluble carbohydrates, lignin, protein, starch, and fat in the biomass materials by providing moisture and temperature in the range of glass transition of the biomass materials. As mentioned earlier, pelletizing and briquetting might be more suitable for thermochemical routes of downstream conversion. As the biochemical route of conversion requires that the biomass be slurried for enzyme hydrolysis and fermentation, the rewetting properties of such pellets/ briquettes would be important. Also, how the different densification processes affect downstream processings such as pretreatment, enzyme hydrolysis and fermentation need to be examined. Research in this regard, however, is not available.

8.5 Important Considerations in Biomass Preprocessing

The biomass preprocessing unit(s) can be located at the field, intermediate depot (or satellite station) or biorefinery (in case of big refineries) to optimize the cost of overall handling, storage and transportation (Wright et al., 2006). At the front end, the preprocessing unit(s)/system(s) receive(s) biomass in varying quality (moisture, density, etc.) and quantity (tonnage). The processed biomass should ultimately meet the specification of the biorefineries, which are going to use the feedstock. Therefore, the preprocessing facilities should deliver feedstocks to the biorefineries with acceptable qualities, at competitive prices (US-DOE, 2003).

The major considerations that must be addressed by the preprocessing facilities are: moisture and particle size (Schroeder and Jackson, 2008). A low moisture content helps to improve the shelf life of biomass (during storage) by minimizing the microbial activity (aerobic stability) and dry weight shrinkage. The biomass is densified through appropriate sizing (chopping, grinding or pelleting). Though lower particle size would increase the surface area and hence improve biochemical conversion process, it is not desired to convert biomass into finer particle size at preprocessing stage due to higher energy cost for grinding. Very often, the preprocessing of biomass would also need blending of biomass to provide the mixture with improved characteristics. Such parameters are always required for handling and transporting biomass from source (field or depot) to destination (biorefinery).

8.6 Field (Decentralized) or Plant (Centralized) Preprocessing?

The location of biomass preprocessing and storage facilities depends greatly on the size of the biorefinery, biomass availability and quality, mechanism of biomass supply chain, and transportation costs (Jenkins and Sutherland, 2009). Preprocessing at a central location may be limited to usage of specific feedstock with assured supply and ample biomass storage inventory for the downstream processes. A comparable example would be the corn dry-grind and wet-milling facilities where the corn is hauled (by truck or train), stored (in silos), preprocessed (grinding/milling) and is then converted to products (starch, high fructose corn syrup (HFCS), ethanol and others).

Corn based bioprocessing is a mature industry. So there is a good feedstock supply network for this industry. Supply of forage in animal industries is another example of stable infrastructure. The transportation of cellulosic feedstock from the fields to biorefineries will certainly be influenced by such existing systems in terms of designing new biomass processing facilities and extending the service avenues of the existing systems. Location and the existing infrastructure (storage, transportation network, equipments and labor) have impacts on biomass preprocessing and supply strategies (Plieninger et al., 2006). The biomass feedstock is diverse. The availability and quantity are also spatially discrete. Each of these resources is unique in physical and compositional characteristics. Therefore, the level of preprocessing varies depending on biomass characteristics and demand for the quality and quantity of biomass from the bioenergy industries. The logistics of feedstock collection and supply to biorefineries can be strengthened by densifying the biomass closest to their collection site: at fields or depot, where transportation cost has a minor effect.

Wright et al. (2006) reported that distributed preprocessing at the source or intermediate location (depot) is a promising alternative because higher quality feedstock is expected along with efficient handling and transportation benefits. Selection on the optimal screen size of grinders resulted in improved bulk density of barley straw and its hauling cost (\$12.87 dry metric ton^{-1}). The transportation of biomass (raw or preprocessed) from sources to facilities has to rely on existing infrastructures like truck or rail (and sometimes barges). The hauling cost greatly depends on the tonnage distance and may sometime exceed over 40% of the total of biomass collection, preprocessing, handling and transportation costs.

Modular and decentralized preprocessing of biomass is anticipated to greatly offset the total cost of biomass (dry mass) and transportation (Figure 8.1). The generation of a market potential between the hauling distances (field to gate) is necessary to reduce the transportation cost that ranges between \$0.2 to \$0.6 dry metric ton^{-1} $mile^{-1}$ (USDA-USDOE, 2005). The rural economy and feedstock supply chain can be boosted via the creation of biomass collection, onsite/intermediate processing and supply network schemes (BRDI, 2007).

Regional biomass preprocessing centers (RBPC) is an advanced concept of distributed biomass preprocessing (Carolan et al., 2007). Such decentralized facilities are expected to reduce the life cycle cost of biomass via improved on-site preprocessing of biomass and reducing the hauling cost. Irrespective of preprocessing (e.g. communition) and pre-treatment techniques, a network of RBPCs sustains biomass supply from field to biorefinery gate at a competitive quality but reduced costs. The utilization, preprocessing and blending of local feedstock improves feedstock supply and quality aspects. The local economy is expected to flourish due to dynamic influences in other agro-business and industries by providing local governance for resource utilization and management.

8.7 Advances and Future Research in Biomass Preprocessing

Preprocessing is critical in the success of emerging biorefinery. Presently, this is one of the least researched areas. Some of the advances in biomass preprocessing are discussed below:

8.7.1 Ensiling Wet Biomass

Ensiling is a biomass preservation technique (under anaerobic fermentation condition) to minimize biomass loss during storage. In the meantime, organic acids (like lactic and acetic acid) produced during the fermentation restrict the deterioration of carbohydrate components (Philipp et al., 2007) and enhance the bioconversion of ensiled biomass (Thomsen et al., 2008). Harvested biomass can be ensiled to improve the saccharification and fermentation of cellulosic biomass. Chen et al. (2007) reported ensiling as cost effective biomass treatment. The study concluded that ensiling of wet (60% moisture) agricultural residues (barley, triticale and wheat straws; cottons stalk and triticale hay) with additional cellulase and xylanase enzymes improved the downstream bioconversion of hemicellulose and cellulose to sugars, and therefore, increased the biomass to bioethanol conversion ratio.

8.7.2 Distributed Preprocessing

Distributed preprocessing of biomass is an economically favorable option that certainly needs more research in the future. There is an increased trend to recognize indigenous non-food plants as potential feedstocks for biofuel production because of the adaptability of these plants to local climate, resistance to (a) biotic factors, and a possibility of their additional usage in other areas. Whether it is high-yield indigenous plants or locally grown energy crops, processing on-site or at intermediate facilities may help to minimize the overall processing, handling and hauling costs of feedstock. The diversity of feedstocks is still maintained while the quality and consistency of feedstock supply issues would be eliminated if local biorefineries utilize a blend of locally grown feedstocks. The hauling distance is optimized. Local constraints (labor, climate and crop cycle) are better utilized to develop overall feedstock-biorefinery models.

8.7.3 Green/Wet Preprocessing of Biomass

Encompassing a biorefinery concept, front-end protein recovery from green/wet lignocellulosic biomass is a new concept in the lignocellulose conversion to ethanol. The current biomass-to-biofuel processing model is based on dry biomass. Thus the front-end protein and other high-value biobased product recovery and utilization from wet biomass have not been realized. This concept of wet or green biomass processing could be a suitable option for all crops of the sugarcane family (e.g. miscanthus, energy cane, banagrass, sweet sorghum, etc.) and dedicated energy crops (switchgrass, ryegrass, etc.) that can be harvested during any growth stage and can be processed wet or green. Mature biomass yields more fiber per growing area, but at the same time, it is often more resistant to biochemical conversion, thereby requiring more severe pretreatments. Such severe pretreatments also generate soluble products that are inhibitory to enzyme hydrolysis and subsequent fermentation. The green/wet-pressing generates an organic-rich liquid stream that can be converted into high-value biobased products. The authors have been working currently to preprocess green/wet lignocellulosic biomass, especially switchgrass, banagrass, and alfalfa to examine the downstream processing performance (pretreatment, enzymatic hydrolysis, fermentation). However, many challenges need to be addressed, especially the storage and handling issues that come with processing wet/green biomass. On-farm operations or satellite preprocessing centers could be envisioned for protein recovery, and pressed (and/or dried) fibers could be transported to a biorefinery facility for further processing. Research is currently underway in this direction by the authors at University of Hawaii and Iowa State University.

8.8 Summary

There have been few field trials for biomass collection and densification to optimize the overall cost of delivering feedstock from the field to gate. The generation and utilization of data to simulate real scenarios of feedstock supply chain is necessary via the application of integrated logistic models like IBSAL. Such models should also compare the efficacy of distributed preprocessing for the overall biomass supply logistics. The biomass preprocessing has been considered to be a very important step to provide an uninterrupted supply of biomass, in quality and quantity, from fields to the biorefineries. Tremendous research efforts have been put forward on economic-to-scale conversion of biobased feedstock to biofuels and bioproducts. At the very upstream end, biomass densification and supply infrastructure (transportation and supply network) also need greater research motivation to develop mature systems that successfully channel over a billion tons of various feedstocks to distributed biorefineries.

8.9 References

Biomass Research and Development Initiative – BRDI (2007). *Roadmap for bioenergy and biobased products in the United States*. Available at http://www1.eere.energy.gov/biomass/pdfs/obp_roadmapv2_web.pdf (accessed August, 2009).

Carolan, J.E., Joshi, S.V., and Dale, B.E. (2007). "Technical and financial feasibility analysis of distributed bioprocessing using regional biomass pre-processing centers." *Journal of Agricultural and Food Industrial Organization,* 5, 10.

Chen, Y., Sharma-Shivappa, R.R., and Chen, C. (2007). "Ensiling agricultural residues for bioethanol production." *Appl. Biochem. Biotechnol., 143, 80 – 92.*

Demirbas, K., and Sahin-Demirbas,A. (2009). "Compacting of biomass for energy densification." *Energy Sources*, Part A, 31,1063–1068.

Dooley, J.H., Lanning, D., Lanning, C., and Fridley, J. (2008). "Biomass baling into large square bales for efficient transport, storage, and handling." Paper from *Council on Forest Engineering*, held at Charleston, S.C., June 22-25, 2008, 31st Annual Meeting.

Gilbert, P., Ryu,C., Sharifi, V., and Swithenbank, J. (2009). "Effect of process parameters on pelletisation of herbaceous crops." *Fuel*, 88, 1491-1497

Granada, E., Lopez-Gonzalez, L.M., Miguez, J.L., and Moran, J. (2002). "Fuel lignocellulosic briquettes, die design and products study." *Renewable Energy*, 27, 561–573.

Hess, R.J., Wright, C.T., and Kenney, K.L. (2007). "Cellulosic biomass feedstocks and logistics for ethanol production." *Biofuels Bioproducts & Biorefining,* 1, 181 – 190.

Hess, R. (2008). "Biomass material property challenges in feedstock supply logistics." *Growing the Bioeconomy Conference,* September 9, 2008, Iowa State University.

Hess, R.J., Kenney, K.L., Ovard, L.P., Searcy, E.M., and Wright, C.T. (2009). *Uniform-format solid feedstock supply system: A commodity-scale design to produce an infrastructure-compatible bulk solid from lignocellulosic biomass.* Idaho National Laboratory. INL/EXT -08-14752. Available at https://inlportal.inl.gov/portal/server.pt/gateway/PTARGS_0_1829_37187_0_0_18/Design_Report_Sec_4_Draft%204-06.pdf (accessed July 7, 2009).

Jackson, B., Schroeder, R., and Ashton, S. (2007). "Pre-processing and Drying Woody Biomass." In: Hubbard, W., Biles, L., Mayfield, C., Ashton, S. (Ed.) *Sustainable Forestry for Bioenergy and Bio-based Products: Trainers Curriculum Notebook.* Athens, GA: Southern Forest Research Partnership, Inc., pp. 141–144.

Kaliyan, N., Morey, R. V., White, M. D., and Doering, A. (2009). "Roll press briquetting and pelleting of corn stover and switchgrass." *Transactions of the ASABE*, 52(2), 543 – 555.

Kumar, A., and Sokhansanj, S. (2007). "Switchgrass (*Pinacum vigratum L.*) delivery to a biorefinery using integrated biomass supply analysis and logistics (IBSAL) model." *Bioresource Technology,* 98(5), 1033 – 1044.

Jenkins, T.L., and Sutherland, J.W. (2009). "An integrated supply system for forest biomass." In: Solomon, B.D., and Luzadis, A. (Ed.) *Renewable energy from forest resources in the United States.* New York, NY: Routledge, pp. 92–115.

Obernberger, I., and Thek, G. (2004). "Physical characterisation and chemical composition of densified biomass fuels with regard to their combustion behavior." *Biomass and Bioenergy*, 27, 653 – 669.

Philipp, D., Moore, K.J., Pedersen, J.F., Grant, R.J., Redfearn, D.D., and Mitchell, R.B. (2007). "Ensilage performance of sorghum hybrids varying in extractable sugars." *Biomass and Bioenergy,* 31(7), 492 – 496.

Plieninger, T., Bens, O., and Huttl, R.F. (2006). "Perspectives of bioenergy for agriculture and rural areas." *Outlook on Agriculture,* 35(2), 123 – 127.

Schroeder, R., and Jackson, B. (2008). "Preprocessing and drying." *Forest Encyclopedia Network.* Available at http://www.forestencyclopedia.net/p/p1295/view (accessed July, 2009).

Shinners, K.J., Boettcher, G.C., Muck, R.E., Wiemer, P.J., and Casler, M.D. (2006). "Drying, harvesting and storage characteristics of perennial grasses as biomass feedstocks." *American Society of Agricultural and Biological Engineers Meeting Presentation. ASABE- AIM paper number:* 061012.

Sokhansanj, S., and Turhollow, A.F. (2004). "Biomass desnification – cubing operations and costs for corn stover." *Applied Engineering in Agriculture,* 20(4), 495 – 499.

Sokhansanj, S., and Fenton, J. (2006). "Cost benefit of biomass supply and pre-processing." *BIOCAP (Canada), Research Integration Program Synthesis Paper.* Available at http://www.biocap.ca/rif/report/Sokhansanj_S.pdf (accessed August 2009)

Sokhansanj, S., Kumar, A., and Turhollow, A.F. (2006). "Development and implementation of integrated biomass supply analysis and logistics model (IBSAL)." *Biomass and Bioenergy.* Available at http://www.extendsim.com/downloads/papers/sols_papers_biomass.pdf (accessed August 2009).

Sokhansanj, S., Turhollow, A., and Wikerson, E. (2008). *Development of the integrated biomass supply analysis and logistics model (IBSAL).* Oakridge National Laboratory (ORNL). *ORNL/TM – 2006/57.*

Sokhansanj, S., Mani, S., Turhollow, A., Kumar, A., Bransby, D., Lynd, L., and Laser, M. (2009). "Large-scale production, harvest and logistics of switchgrass (*Panicum virgatum L.*) – current technology and envisioning a mature technology." *Biofuels Bioproducts and Biorefining,* 3, 124 – 141.

Thomsen, M.H., Holm-Nielsen, J.B., Oleskowicz-Popiel, P., and Thomsen, A.B. (2008). "Pretreatment of whole-crop harvested, ensiled maize for ethanol production." *Appl. Biochem. Biotechnol.,* 148, 23 – 33.

Wright, C.T., Pryfogle, P.A., Stevens, N.A., Hess, J.R., and Radtke, C.W. (2006). "Value of distributed preprocessing of biomass feedstocks to a biorefinery industry." Annual International Meeting (AIM), *American Society of Agricultural and Biological Engineers Meeting Presentation.* ASABE- AIM paper number: 066151.

United States Department of Agriculture/Department of Energy. (2005). *Biomass as feedstock for a bioenergy and bioproducts industry: the technical feasibility of a billion-ton annual supply.* Available at http://www1.eere.energy.gov/biomass/pdfs/final_billionton_vision_report2.pdf (accessed August 2009).

United States Department of Energy (US-DOE). (2003). *Roadmap for Agriculture Biomass Feedstock Supply in the United States.* Available at http://devafdc.nrel.gov/pdfs/8245.pdf (accessed August 2009).

CHAPTER 9

Lignocellulosic Biomass Pretreatment

Devin Takara, Prachand Shrestha, and Samir Kumar Khanal

9.1 Introduction

With the ever rising demand for more energy and the limited availability of depleting world resources, many are beginning to look for alternatives to fossil fuels. Liquid fuel, in particular, is of key interest to reduce our dependency on imported petroleum. A recent study by the Energy Information Administration (EIA) projects an increase in transportation-related energy consumption by 5 quadrillion British thermal units (Btu) from 2006 to 2030 (USEIA, 2008). Concerns with climate change and national security have also led many to advocate research on biofuels from renewable resources.

Although monomeric sugars and starch are the ideal feedstocks for biofuel production, their use as raw materials is prohibitively expensive. In addition, the use of staple crops for biofuel production generates warranted concern over fuel versus food. Lignocellulosic biomass, agricultural and forest residues, have gained attention as being a potential source of low-cost feedstock for biofuel production. This non-food renewable energy feedstock is abundantly available. The United States has the potential of producing 1.3 billion dry tons of plant biomass annually, and over 900 million dry tons of this biomass can be farmed on agricultural lands (Perlack et al., 2005). Ideally, one billion tons of biomass can substitute more than 30% of the nation's petroleum consumption, thereby cutting down as much as one-third of our oil imports. Thus, biomass can play an important role in the domestic bio-based economy by producing a variety of biofuels and bio-based products.

Two primary means of utilizing lignocellulose for conversion into liquid fuel, ethanol, are based on thermochemical and biochemical pathways. Thermochemical pathways are a relatively new concept that converts solid feedstocks into an intermediary gaseous state consisting of carbon monoxide (CO), hydrogen (H_2), carbon dioxide (CO_2), methane (CH_4), and other gases, before microbial conversion into ethanol. This method of biofuel production requires a complex ensemble of unit operations that is energetically intensive. Biochemical pathways focus on employing chemical pretreatments to lignocellulosic biomass, followed by enzymatic hydrolysis and fermentation by microbes to produce ethanol. Although the technologies employed with this method are relatively less complex, the main advantage that biochemical pathways have over thermochemical pathways is its maturity. A techno-economic study comparing these two types of processing noted that aside from generating more energy (per measure of energy consumed), the biochemical production of ethanol, because of its maturity, provides more faith for investors, and ultimately lowers the selling price of ethanol (per gallon) on the consumer market (Piccolo and Bezzo, 2009). Both factors have a large impact on the viability of biofuels as a substitute for gasoline and as such, an emphasis of biochemical pathways will be considered here.

The heterogeneity and crystallinity of lignocellulose not only provide plants with structural rigidity and protection against occasional cuts and infections, but also severely impede the direct utilization of biomass by microbes. It is therefore necessary that the biomass first be pretreated to disrupt the lignin-hemicellulose-cellulose interactions. Several techniques have been developed to promote this structural destabilization and increase the overall efficiency of the sugar-to-ethanol conversion.

Biomass pretreatments encompass a multitude of approaches that can be broadly classified into three major categories: *1) physical, 2) chemical, and 3) biological*. In some cases a fourth category, *thermal*, is included (Yang and Wyman, 2007). Combinations of these approaches have also been adopted in several studies.

In general, physical pretreatment attempts to reduce the size of the substrate, increasing the pore volume of the biomass structure and the accessibility of degradative enzymes to cellulose. Chemical pretreatments on the other hand, focus primarily on breaking down the hemicellulose or lignin components by dissolving them in acid or base, respectively. Biological pretreatments enhance saccharification by exploiting the natural degradative properties of certain fungal species, especially wood-rot fungi (Shrestha et al., 2008). The goal of pretreatment processes is to breakdown obstinate structural components and the increase in the vulnerability of the polysaccharides to enzymatic hydrolysis.

This chapter seeks to summarize the state-of-the-art technologies for lignocellulosic biomass pretreatment. The optimal conditions for the leading pretreatment methods, generation of inhibitory chemicals, strategies for lignin recovery, and scale-up are discussed. In addition, mass transport at a high solids level, and water usage are also considered with some insight into future research and development.

9.2 Lignocellulosic Biomass Structure

Lignocellulosic biomass, as the name implies, include agricultural and forest residues, which consists of cellulose (35-50%), hemicellulose (20-35%), and lignin (10-25%) (Figure 9.1). Other compounds like ash, protein, and oils occur in relatively minuscule amounts (Liu et al., 2008).

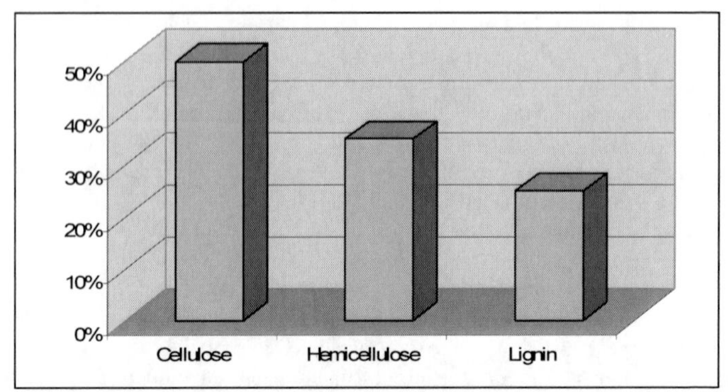

Figure 9.1 Primary compositions of typical lignocellulosic biomass.

Although the composition of lignocellulose can be generalized into three major parts, the way in which these components are arranged is highly complex. Cellulose is a homopolymer that has the unique ability to form highly ordered, crystalline structures that are stabilized by hydrogen bonding (Figure 9.2). Linear strands of cellulose aggregate into structures known as protofibers, which then organize to form microfibrils (Lynd et al., 2002). Current models suggest that hemicellulose, a chain of five and six carbon sugars with xylose as the sugar backbone, forms a protective barrier (often referred to as sheaths) around the microfibrils of cellulose (Figure 9.3) and are held in place by a glue-like material known as lignin (Mousdale, 2008).

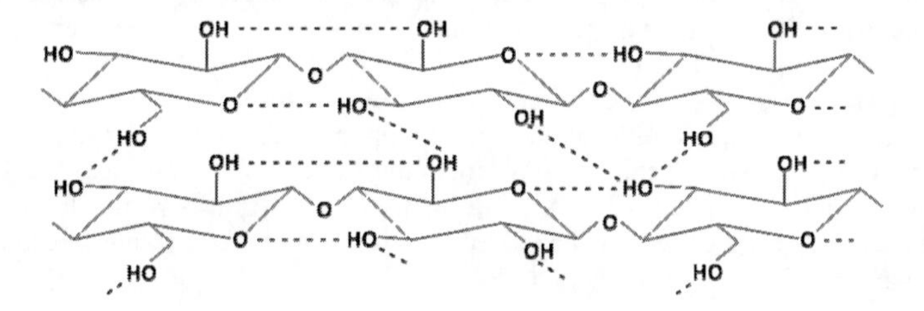

Figure 9.2 Crystalline structure of cellulose stabilized by hydrogen bonding.

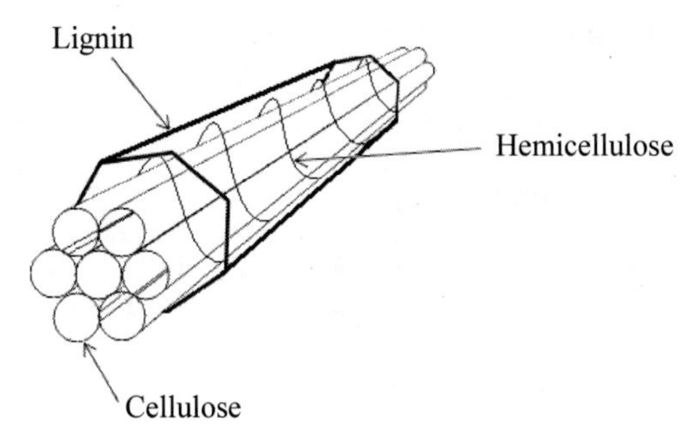

Figure 9.3 Strands of cellulose encased by hemicellulose and lignin.

9.3 Recalcitrance of Lignocellulosic Biomass

The lignin-(hemi)cellulose protective sheath acts as barrier in the microbial degradation of lignocellulosic feedstock, and is one of the main factors preventing the efficient release of fermentable sugars from structural polysaccharides. Lignin is phenolic compound that is formed from the free radical polymerization of an alcohol and varying methoxyl constituents (Ramos, 2003). It is a highly complicated macromolecule that originates from the free-radical polymerization of three precursor alcohols: p-hydroxycinnamyl, 4-hydroxy-3-methoxycinnamyl, and 3,5-dimethoxy-4-hydroxycinnamyl (Ramos, 2003; Lee, 1997). Typically, lignin is characterized by its aromatic groups (p-hydroxybenzyl, guaiacyl, or syringyl) which differ in their degree of methoxylation (Ramos, 2003). Lignin's inherent resistance to chemicals and enzymes, coupled with its ability to crosslink with hemicellulose, obstinately limits access to cellulosic structures. Similarly, hydrogen bonding between cellulose chains makes it more rigid, inhibiting the flexing of the molecules that must occur in the hydrolytic breaking of the glycosidic linkages.

The pretreatment of biomass is necessary to increase the vulnerability of cellulose to enzymatic attack. Although numerous methods have been developed and employed, the underlying concept remains the same. All pretreatments, in some way, act to disrupt organized structures of biomass to increase the exposure of cellulose (Liu, 2008). Often the target compounds of pretreatment are lignin and hemicellulose, which can be solubilized using acid, base and/or heat, to promote structural destabilization and pore opening. By weakening the protective layer surrounding cellulosic structures a group of enzymes called cellulases are able to hydrolyze β-glycosidic bonds, allowing sugar release.

9.4 Biomass Pretreatment

9.4.1 Goals of Pretreatment

The main objective of pretreatment is to prepare the biomass for efficient downstream processing. Currently, there is no particular preferred biomass pretreatment technology. Therefore, it is important to select the technology that fulfills the most, if not all, of the performance goals outlined below (IEA, 2008; Johnson and Elander, 2008):

- maximize the yields of both C-5 and C-6 sugars for downstream processing with minimal degradation of produced sugars;
- facilitate the recovery of lignin as a high-value product;
- minimize the formation of inhibitory soluble chemicals including furfural, hydroxymethylfurfural, organic acids and phenolic compounds in the hydrolyzate for downstream enzyme hydrolysis and fermentation;
- be capable of effectively accommodating diverse feedstocks without significant change in process or the product quality;
- require little or no preprocessing steps, e.g. size reduction by grinding, milling, pulverizing, etc.;
- have a high degree of simplicity;
- utilize low-cost chemicals with minimal generation of waste streams;
- utilize less energy and have low capital and operating costs

9.4.2 Pretreatment Technologies

Different pretreatment options have been developed, each attempting to exploit a specific property of lignocellulosic material. It is often hard to compare one technology with another due to the non-uniformity in the treatment conditions. A USDOE funded Biomass Refining *Consortium for Applied Fundamentals and Innovation (CAFI)* recently evaluated different pretreatment methods undertaken by researchers in North America. The aim of the study was to obtain a comparative process performance and economic data for leading pretreatment technologies. The details can be found in Wyman et al. (2005).

It is important to recall that all biochemical pretreatments can be broadly classified into the following four categories: 1) physical, 2) chemical, 3) biological, and 4) a combination of methods (Yang and Wyman, 2007).

9.4.3 Physical Pretreatment

For the purposes of this discussion, physical pretreatment will refer only to techniques that cause structural destabilization of biomass without chemical or biological assistance. Examples of this type of pretreatment are comminution, steam explosion and hot water washing.

Comminution. Intensive size reduction commonly known as comminution involves various physical means of size reduction such as ball milling (wet, dry and vibratory processes), attrition milling, compression milling, and wet or dry disk refining. High energy costs associated with the mechanical equipment is the major issue with comminution method (Johnson and Elander, 2008). Communition and other forms of biomass size-reduction results in lower sugar yields and they are often combined with other technologies for enhanced sugar release.

Steam Explosion. Steam explosion exists in two forms that are conceptually simple to understand. In the first technique of steam explosion, biomass is heated with a high temperature and pressure (140-240°C and 0.32-3.4 MPa) before it is released instantly (and violently) into a collection chamber (Ramos, 2003). In the second technique, the high temperature and pressure of the substrate are allowed to slowly reach equilibrium with the surrounding conditions, followed by mild blending (Ramos, 2003). Steam explosion without the addition of chemical or biological supplements is commonly known as autohydrolysis. Under such conditions, the acetylated groups found on hemicellulose generate acids (acetic, formic, levulinic) that are believed to increase the overall efficiency of the pretreatment. Unfortunately, these same compounds are inhibitors for microbes used during the fermentation process. Thus, attempts are being made to detoxify hydrolyzates.

The main mechanism behind the effectiveness of this type of pretreatment is the sudden decompression and subsequent breaking of chemical bonds between the macromolecular structures that maintain the integrity of the biomass (Zimbardi et al., 2007). The β-O-4 ether bonds and acidic linkages of lignin are particularly susceptible to certain temperatures and pressures, and are highly dependent on the type of biomass (Ramos, 2003). Steam explosion without the use of an external catalyst is uncommon due to the low sugar yields obtained from the recovery of hemicellulose, which is often less than 65% (Balat et al., 2008). Higher sugar yields can be obtained when steam explosion is used in combination with other types of pretreatments, and will be discussed in a later section.

Hot Water Washing. Hot water washing is attractive because it does not require the input of expensive chemical catalysts like acid or base that often need additional processing steps for chemical recovery and/or neutralization prior to disposal. In a study conducted by Mosier et al. (2005), the optimal conditions for pH-controlled hot water washing of 16% corn stover slurry was determined to be 190°C for 15 minutes with a cellulase loading of 15 filter paper units (FPU) of enzyme per gram of glucan. Hemicellulose is partly solubilized in this pretreatment, facilitating enzymatic digestion. Oligomeric, rather than monomeric, subunits are released from the substrate thereby decreasing the likelihood of complete sugar degradation into inhibitory compounds. A cellulose yield of 90% was obtained with an 88% total xylose and glucose conversion to ethanol (Mosier et al., 2005). It is important to note that the enzyme used in the study was not optimized for xylan degradation; the primary constituent of hemicellulose.

Although hot water washing reduces the pretreatment cost by eliminating the need for chemical additives, like acid or base, it may not be feasible for scale-up due to the large amount of heat that would be required for treatment at high biomass loadings. Water in particular would require a fair amount of energy due to its high heat capacity, 4.186 kJ/kg·K. Maintaining water in the liquid phase at 190°C (the optimal treatment temperature) might also prove to be difficult as it would require a custom-designed, pressurized system.

9.4.4 Chemical Pretreatment

Lignin and hemicellulose are the primary targets in the chemical pretreatment of lignocellulosic biomass. Hemicellulose is the most susceptible to solubilization by acid and to some extent to base, due to some of its inherent properties: a noncrystalline structure and branched functional groups. The following section discusses various chemical pretreatment methods.

Dilute Acid Pretreatment. Dilute acid has long served as a primary candidate for the chemical pretreatment of biomass due to its low cost and high conversion of hemicellulose into fermentable sugars (Yang and Wyman, 2007). Its effect on improving the breakdown of polysaccharides has been investigated over a range of varying types of lignocellulosic material including corn stover, poplar, bagasse, and switchgrass.

Dilute acid pretreatment functions by hydrolyzing hemicellulose components of biomass into soluble mono- and oligosaccharides consisting primarily of xylose and glucose with comparatively minute quantities of other sugars (Torget et al., 1992). It is believed that the hydronium ions (H_3O^+) donated by the acid play a key role in interrupting the bonds that maintain the structure of the hemicellulose chains (Lloyd and Wyman, 2005). This transfer of hemicellulose into the liquid phase generates pores within the remaining cellulose-lignin structure, leaving it vulnerable to enzymatic attack by cellulases (Torget et al., 1992). The released sugars can be fermented into ethanol.

Among the different acids, dilute-sulfuric acid is the most extensively studied chemical based upon its low cost and known ability to break apart hemicellulose into monomeric sugars. Unfortunately, the effect of dilute-sulfuric acid on the saccharification of lignocellulosic biomass is highly variable with respect to the types of biomass, and the optimal conditions for a particular substrate must be examined individually. For example, it was demonstrated that rye straw is harder to degrade into smaller sugar units than bermudagrass. With a sulfuric acid concentration of 1.2% w/w and at a treatment time of 60 min, about 33% of the glucan from bermudagrass was converted to glucose compared to only 10% of the glucan in rye straw (Sun and Cheng, 2005). The reproducibility of the values obtained in this study was proposed to also be dependent on the growing and harvesting conditions of the respective biomass, adding to the complexity of applying the same type of treatment parameters over multiple species of biomass.

For comparative studies, corn stover is often used as the substrate to evaluate the relative effectiveness of varying pretreatment strategies. Lloyd and Wyman (2005) demonstrated that a maximum of 93% of the total sugars in corn stover can be recovered at 140°C for 40 min, with a sulfuric acid concentration of 0.98% w/w prior to hydrolysis at an enzyme loading of 60 FPU/g glucan. For scale-up purposes, however, such high enzyme loadings are not economically feasible. Thus, a 7 FPU/g glucan dosing was applied and resulted in an 88% yield of the total sugars.

Several disadvantages are associated with the use of dilute acid pretreatment that prevent this method from gaining popularity in an industrial setting, most notably the production of inhibitors. Although over a hundred different inhibitors have been identified, furfural and 5-hydroxymethylfurfural (HMF) are often used as representative compounds since their presence in solution has detrimental effects on the microbes responsible for ethanol fermentation (Liu et al., 2005). Furfural and HMF are essentially the degraded products of xylose and glucose, respectively. While they have economic value outside of the biofuel-production industry (Mamman et al., 2008), these are undesirable to ethanol producers, and indicate inefficiency in the pretreatment system. Severe conditions (e.g. extremely low pH and/or high temperature) enable the complete hydrolysis of hemicellulose into its monomeric constituents, but it also promotes the degradation of glucose and xylose. Thus, a careful balance between the pretreatment conditions and hydrolysis of polysaccharides is necessary to optimize the sugar yields.

Eggeman and Elander (2005) proposed a cost analysis on the use of a dilute-acid system as a pretreatment option for varying types of biomass, particularly corn stover. Although dilute-acid was determined to be one of the cheapest treatment methods (compared to hot water, ammonia fiber expansion, ammonia recycle percolation, and lime), a significant amount of the cost was associated with the equipment required to resist the corrosive sulfuric acid (Yang and Wyman 2007; Eggeman and Elander, 2005).

Alkaline Pretreatment. Alkaline pretreatments include the use of chemicals such as sodium hydroxide, urea, potassium hydroxide, lime, and ammonia; some of the most well-studied of which being the latter two. These types of pretreatments act to solubilize lignin and/or hemicellulose into the liquid phase and subsequently destabilize the structural components.

Ammonia has been utilized in a number of applications which include ammonia recycle percolation (ARP), soaking in aqueous ammonia (SAA), and ammonia fiber expansion (AFEX).

Unlike the previous methods of alkaline pretreatment mentioned earlier, soaking with aqueous ammonia attempts to retain hemicellulose in the solid fraction by keeping temperatures relatively low (Sun and Cheng, 2005). In general, hemicellulose is more reactive in acidic and basic conditions because of the functional groups that branch off of the sugar backbone of the polysaccharide. Kim et al. (2008) found that at optimal conditions (15% ammonia w/w, at 75°C for 48 hours), destarched barley hull retained 67% of its xylan while liberating 61% of its lignin.

Ammonia percolation, like its alkaline brethren, tends to dissolve both lignin and hemicellulose into the liquid phase. A two-stage system was proposed by Kim and Lee (2005) to solubilize hemicellulose (from corn stover) with hot water in the first stage, followed by delignification of the biomass in the second stage. A single stage system would not be feasible due to the production of a liquid stream containing a mixture of hemicellulose and lignin that is difficult to separate for subsequent fermentation. At optimal conditions (190°C, 30 min of water treatment and 170°C, 60 min of ARP at 2.3 MPa, 5mL/min of flow rate), up to 86% of the xylan was recovered, and up to 85% of the cellulose remained in the solid fraction with 75-81% lignin removal.

Due to the chemical and physical nature of AFEX, it will be discussed in further detail in the hybrid pretreatment section.

9.4.5 Biological Pretreatment

The ability of microorganisms to degrade lignocellulosic material has been known for over centuries, as evidenced by the array of fungal species that can propagate on woody surfaces found in forests. Biological pretreatment of biomass is appealing because it avoids the use of toxic chemical, requires less energy, and the reactor designs are generally less complex (Kuhar et al., 2008). Interest in applying microbial techniques for ethanol production however, has not been widely popular until recently due to the slow rate and possible loss of sugars associated with microbial consumption. Advances in technology and an improved understanding of the molecular mechanisms that govern degradation processes have led to renewed interest in the field.

Bacteria. Different species of bacteria are capable of producing cellulases in both aerobic and anaerobic environments. Some of the investigated strains include: *Rhodospirillum rubrum*, *Pyrococcus furiosus*, *Saccharophagus degradans*, *Cellulomonas fimi* and *Clostridium* spp (Kumar et al., 2008). Unlike fungi, anaerobic bacteria have the unique ability to produce a multi-enzyme complex known as a cellulosome (Kumar et al., 2008; Ding et al., 2008). Cellulosomes are composed of two subunits (the non-catalytic scaffolding region and the catalytically-active region) and are thought to enhance the efficiency of hydrolysis by maintaining key enzymatic components in close proximity (Ding et al., 2008). While attempts have been made to engineer cellulosomes, success has been limited due to a lack of understanding of molecular interactions.

Fungi. The most common organisms for the bioconversion of lignocellulosic biomass are fungal species, most notably white-rot fungi. White-rot fungi are of particular interest in the ethanol-production industry because of their ability to secrete extracellular enzymes that are capable of degrading plant structures (Shrestha et al., 2008). Unfortunately, each species differs in the quantity of enzyme produced which ultimately affects the efficiency and extent of hydrolysis of biomass. For example, fungi from the *Apergillus* family have high β-glucosidase activities and low endoglucanase activities, while fungi from the *Trichoderma* family have high endo/exoglucanase activities but low β-glucosidase activities (Kuhar et al., 2008). Recall that β-glucosidase is required to hydrolyze cellobiose which limits endo/exoglucanase activity via feedback inhibition (Kumar et al., 2008). Co-culturing of two species of fungi has been conducted with relative success, improving the cellulase:β-glucosidase ratio with the use of *Trichoderma reesei* and *Aspergillus* spp (Tengerdy and Szakacs, 2003).

Other species of white-rot fungi have been applied to a number of substrates. Balan et al. (2008) investigated the effects of *Pleurotus ostreatus* on mushroom spent straw, a byproduct of the food industry. Partial fungal degradation in combination with AFEX pretreatment at mild conditions (1:1 ammonia loading, 80% moisture, 100°C) released 92% of the glucan (Balan et al., 2008). The team concluded that after pretreatment and hydrolysis, 98% glucan and 75% xylan yields could be achieved.

Solid substrate fermentation is favored over submerged fermentation because it is simpler to operate, lowers raw material and input costs, and reduces environmental issues associated with wastewater streams (Shi et al., 2008). Solid substrate conditions also better resemble the natural habitat of most white-rot species, making the environment more conducive to enzyme production and lignocellulose degradation. Shrestha et al. (2008) reported the use of white-rot fungus, *Phanerochaete chrysosporium*, for the saccharification of corn fiber to produce sugar-rich hydrolysate for subsequent ethanol fermentation. The team concluded that the fungus was able to breakdown the corn fiber under solid state conditions, but with limited success. The yield for the conversion of cellulose alone was around 9 g of ethanol per 100 g of corn fiber which was much higher than the observed yield of 3 g of ethanol per 100 g of corn fiber (Shrestha et al., 2009). Further research is required to optimize ethanol production.

As mentioned previously, two of the major disadvantages associated with microbial pretreatment of lignocellulosic biomass are the slow rate of saccharification and the loss of sugars due to microbial consumption (Kuhar et al., 2008; Kumar et al., 2008; Balan et al., 2008; Shi et al., 2008). Fungi utilize the sugar that they breakdown from the biomass and incorporate the constituents into their metabolic processes. Other likely obstacles include the startup and maintenance of the pretreatment environment. The germination of fungal propagules (spores) is typically is slow, as the spore must first be acclimatized to its surroundings, and keeping the fungi alive by providing it with ample substrate and nutrients, as well as consistent temperatures, is arduous at best.

9.4.6 Hybrid (Combination) Pretreatment

Hybrid pretreatments seek to use and combine the advantages associated with the various types of biomass pretreatment options. Most common is perhaps the pairing of physical pretreatments with chemical or microbial assistance. The following section will seek to summarize some of the major studies.

Ball Milling. As the name implies, ball milling employs the use of spherical objects to crush (and grind) biomass into finer particles. The resulting increase in surface area allows for more locations for catalytic breakdown of constituent sugars by enzymes.

Mais et al. (2002) conducted a study to evaluate the effectiveness of the simultaneous ball milling and hydrolysis of Douglas-fir wood chips in a ball milling reactor. They found that there was a positive correlation between the amount saccharification of the substrate and the number of balls added to the reactor. At an enzyme loading of 10 FPU/g cellulose, 45°C, and 10% (w/v) solid concentration, an 85% conversion of the original substrate to monomeric glucose was observed (Mais et al., 2002). Higher yields were possible by increasing the enzyme loading or decreasing the substrate loading. It is important to note that in this ball milling study, the biomass was first steam exploded with sulfur dioxide and delignified with an alkaline-peroxide solution before ball milling and enzymatic saccharification (Mais et al., 2002). Cost is most often the deterring factor in this type of pretreatment.

Hammer Milling. Hammer milling like ball milling, acts to reduce the size of lignocellulosic biomass and improve enzymatic digestion. The primary difference between the two methods is the use of metal hammers (instead of balls) to induce shear and impact stress on the incoming substrate (Mani et al., 2004). Electrical inputs generally drive the cost for this type of pretreatment. A study conducted by Mani et al. (2004) used a mill with 22 swinging hammers attached to a 1.5 kW electric motor. The power consumption of the mill was determined for a number of different substrates: wheat, barley straws, corn stover, and switchgrass. This study confirmed that the power consumption of hammer milling is primarily a function of the biomass type and moisture content. Corn stover was found to require the least amount of energy while switchgrass required the most. For a screen mesh size of 0.8mm, the specific energy consumption for corn stover was 22.07 $kWht^{-1}$ compared to the 62.55 $kWht^{-1}$ used by switchgrass (Mani et al., 2004). Because hammer milling (and similar types of size reduction) are common as a pre-processing step for research on biomass pretreatment, this study did not report obtainable sugar yields from hammer milling alone or in conjunction with other forms of pretreatment.

Wet-oxidative Alkaline. A process called wet-oxidative alkaline pretreatment seeks to combine wet-oxidative and alkaline conditions to the pretreatment of lignocellulosic biomass. Unlike in dilute acid pretreatment, wet-oxidative alkaline conditions do not allow the formation of microbial inhibitors (furfural and HMF) that limit ethanol fermentation (Klinke et al., 2002). This can be explained by the high reactivity of aliphatic aldehydes and saturated carbon bonds found in oxidative conditions (Bjerre et al., 2002). Carboxylic acids are typically the main degradation products observed from the breakdown of hemicellulose and lignin, and have the potential to be neutralized (Klinke et al., 2002). When exposed to wet-oxidative alkaline conditions, hemicellulose and cellulose undergo a β-elimination and begin to degrade from the reducing ends of the polysaccharide chains (Kim and Holtzapple, 2005). Klinke et al. (2002) found that the overall glucose yield of 65%

from wet-oxidative alkaline pretreatment of wheat straw, which was higher than that of alkaline hydrolysis, steam, and ammonia pretreatments.

One of the main deterring factors with this pretreatment is the cost associated with start-up and operation. High temperatures and pressures are typically necessary to obtain the yields cited in literature, and pure oxygen is required as the oxidative component.

To combat costs, some researchers have turned to lime in their wet-oxidative alkaline pretreatments. Relative to other alkaline chemicals, lime is inexpensive and has the ability to be recycled. Calcium in the liquid stream from lime pretreatments can be recovered by neutralization with carbon dioxide to form insoluble calcium carbonate, and the lime can be regenerated by the use of a lime kiln (Kaar and Holtzapple, 2000).

Kim and Holtzapple (2005) conducted a study on the enzymatic hydrolysis of alkaline-treated corn stover using low temperatures and pressures, air (instead of pure oxygen), and lime to reduce the costs. They found that under oxidative conditions as much as 87.5% of the initial lignin content and over 96% of the acetyl groups of the biomass can be removed at a temperature of 55°C. Under optimal conditions (55°C for 4 weeks) the overall yields of glucose and xylose (from corn stover) were 91.3 and 51.8 respectively, at an enzyme loading of 15 FPU/g cellulose (Kim and Holtzapple, 2005).

Although the yields and operating conditions are impressive in the laboratory setting, it becomes difficult to imagine the challenges that will be associated with scale up. Some suggestions were proposed as a plausible design for an industrial size system, however the main issue of mass transfer limitations was not addressed; namely establishing consistent and uniform concentrations of air, lime, and heat to mirror the optimal conditions reported in literature.

Ammonia Fiber Expansion (AFEX). Ammonia fiber expansion (AFEX) was first developed and patented by Dr. Bruce Dale at Michigan State University (Dale, 1986). It employs the use of ammonia in combination with explosive decompression that is analogous to that of steam explosion. Similar to physical pretreatments, AFEX utilizes forceful means to create openings within the biomass to facilitate enzymatic digestion.

Although the set-up of an AFEX system can be slightly complex than other treatment strategies, the basic concept behind the success of this method is simple to understand. Reactors are loaded with the biomass of interest and liquid ammonia before being heated for a specified time period. The liquid ammonia works its way in between the structural sugars of the substrate, and as the temperature rises, pressure begins to build up within the reactor. After a set residence time, the pressure is suddenly released causing the instant volatilization of a fraction of the liquid ammonia. The sudden explosion creates physical openings within the biomass structure (Teymouri et al., 2005; Alizadeh et al., 2005). The advantages of AFEX pretreatment of corn stover were summarized by Teymouri et al. (2005) as the following: 1) Virtually all of the ammonia is recoverable, 2) A wash stream is not necessary, 3) Cellulose and hemicellulose are retained with little degradation, and 4) Neutralization of treated biomass is not necessary.

A recent study has focused its attention on a less studied species of biomass, *Miscanthus x giganteus*, for its biological robustness and high concentration of cellulose compared to the commonly investigated switchgrass (Murnen et al., 2008). Murnen et al. (2008) found that 90-95% of glucan and 80-85% of xylan conversions were obtainable with an enzyme loading of 15 FPU/g glucan along with additives like xylanase, pectinase, and Tween-80 (to inhibit irreversible binding of lignin to enzymes) at 160°C and a 5-minute residence time. Research to reduce enzyme loadings and improve overall yields is ongoing.

Spent ammonia is harmful to release into the environment so recovery and or recycle streams are often implemented into processes that use this chemical. To date, a rigorous cost analysis on the recovery of ammonia has not been reported.

Acid Impregnation and Steam Explosion. The use of acid (as a catalyst) during steam explosion pretreatments has been evaluated over a number of feedstocks. Some of the promising features of this type of pretreatment are: 1) improved sugar release from hemicellulose in the liquid fraction, 2) enhanced enzymatic hydrolysis, and 3) lower energy inputs by a decrease in the temperature and time required (Ramos, 2003). Two common acids that are used in this type of pretreatment are H_2SO_4 and SO_2. Gaseous SO_2 is favored since it can be evenly distributed throughout the biomass, and it does not consume as much steam as H_2SO_4 (Ramos, 2003), but storage and usage problems associated with sulfur dioxide increase the appeal of sulfuric acid (Zimbardi et al., 2007).

Zimbardi et al. (2007) examined the effects of dilute-sulfuric acid on steam exploded corn stover by soaking the biomass in an aqueous acid solution for 10 minutes. The acid concentration throughout the biomass was estimated using a model that was developed by Kim and Lee in 2002, which considers mass transfer within the biomass (Zimbardi et al., 2007; Kim and Lee, 2002). The model had to be adjusted however, to account for acid desorption. It was determined that the best glucose yields (85%) were obtained with a 3% H_2SO_4 loading at 190°C for 5 minutes.

Realistically, the type of feedstock and acid used in this type of pretreatment, as well as in others, will rely upon the region of the world in question, and the climate, availability and cost of production of ethanol within those respective areas. In 2008, Viola et al. evaluated the acid catalyzed steam explosion of seaweed, *Zostera marina* (Eel grass), found on the beaches of Britain and France. The steam explosion was conducted with 0-2% oxalic acid which is known for its ability to bind to metal ions found within the plant, assisting in structural destabilization. The optimal conditions reported for this study were an oxalic acid loading of 2% at 180°C and 5 min, producing 243 g ethanol per kg of water insoluble fiber; a 1.5 % (v/v) ethanol concentration in the fermentation broth. Higher biomass loadings were also tested, and it was found that up to 4.7 % (v/v) ethanol concentrations were possible with 200 grams of biomass per liter, allowing for distillation (Viola et al., 2008).

Ultrasound. The use of ultrasonic techniques as a means of deconstructing lignocellulosic material for ethanol production is a relatively new concept. This method aims to break apart recalcitrant structures by promoting cavitation in the biomass by ultrasound-induced vibration of liquid media. Khanal et al. (2007)

investigated the release of glucose from sonicated corn. They determined that a number of parameters (e.g. power input, sonication time were directly related to sugar release during enzyme hydrolysis. About 32% more glucose was released when enzymes were loaded prior to sonication for 40 seconds of sonication. Ultrasound alone may not be enough to obtain viable yields of sugars from lignocellulose, but the possibility of combining such technology with chemical or other physical pretreatments may have potential.

Microwave. Microwave technology is an innovative method for pretreating biomass since it exploits two key features of radio waves. First, microwaves allow for the volumetric heating of biomass. This allows for quicker heating times since conduction and convection are not the main methods of heat transfer. Secondly, microwaves induce rapid vibrations in biomass molecules that may aid in the disruption of obstinate structural components.

In 2008, a study conducted by Hu and Wen combined the use of a domestic microwave with alkali pretreatment (sodium hydroxide) of switchgrass. The microwave, a Sharp/R-21 HT, was modified such that the temperature of a custom made vessel housing the biomass could be monitored. Observable differences between the control and microwave-treated samples were witnessed, and it was speculated that inhomogeneous heating between polar and nonpolar molecules promoted structural destabilization (Hu and Wen, 2008). Note that polar molecules (like water) absorb microwave energy, while nonpolar molecules do not.

The optimal conditions for microwave-assisted alkali pretreatment of switchgrass were found to be: 0.1 g alkali/g biomass, 190°C, 50g/L solid content, and 30 minutes. About 99% of the theoretical total sugar yield was realized (53.4 g/100 g biomass) (Hu and Wen, 2008).

Despite having impressive yields, microwaves consume a lot of energy and may not be feasible in large-scale operations. The reactor design would be particularly interesting and costly since microwaves cannot be used with traditional industrial metals. More research is required to determine the actual viability of this type of pretreatment approach.

9.4.7 Comparison of Different Pretreatment Technologies

As discussed above, a wide range of pretreatment technologies are at different stages of development. There is no ideal or preferred pretreatment method currently available for diverse feedstocks. Table 9.1 summarizes different pretreatment methods. The merits and demerits of different pretreatments are presented in Table 9.2.

Table 9.1 Summary of prospective pretreatment methods.

Method	Process (Temp)	Feedstock	Enzyme Loading	Yields	Inhibitors	Limitations	Reference
Physical	Hot water washing (190°C)	Corn stover	15 FPU/g glucan	90% glucose	Yes	Large water requirement, high temperature	(Mosier et al., 2005)
Chemical	Dilute sulfuric acid (140°C)	Corn stover	60 FPU/g glucan	93% total sugars	Yes	Cost of equipment, need for neutralization	(Lloyd and Wyman, 2005)
	Ammonia Recycle Percolation (190°C- 170°C)	Corn stover	10 FPU/g glucan	92.5% glucose	Negligible	Requires two stages – increased cost/space	(Kim and Lee, 2005)
Biological	*Pleurotus ostreatus* preceded by AFEX (100°C during AFEX)	Rice straw	15 FPU/g glucan	>98% glucose	Negligible	Extremely slow and require hybrid pretreatments	(Jones et al., 2008)
Hybrid	Ammonia fiber expansion (90°C)	Corn stover	60 FPU/g glucan	100% glucose and 80% xylose	Negligible	Ammonia cost and recovery	(Teymouri et al., 2005)
	Wet-oxidative lime (55°C)	Corn stover	15 FPU/g cellulose	91.3% glucose and 51.8% xylose	Negligible	Slow, possible mass transfer limitations with oxygen	(Kim and Holtzapple, 2005)
	Acid-steam explosion (190°C)	Corn stover	15 FPU/g glucan	85% glucose	Yes	Inhibitor generation, equipment cost	(Zimbardi et al., 2007)
	Microwave-assisted alkali (190°C)	Switch-grass	15-20 FPU/g glucan	99% releasable sugars	Not reported	Energy costs of microwave, possible scale-up setbacks	(Hu and Wen, 2008)

Table 9.2 Merits and demerits of important pretreatment technologies.

Pretreatment Technologies	Merits	Demerits
Physical	• Capable of handling coarse or larger size biomass • Eliminates the need of corrosive chemicals • Simplicity in operation • Generates little or none inhibitory chemicals	• Lower yield of sugars • Extremely high energy requirement for mechanical equipments • Requires additional pretreatment step
Dilute acid	• Abundant published data available for techno-economic analysis • Effective for diverse feedstocks • Effective in achieving high sugar yields via both hemicellulose hydrolysis and enzyme hydrolysis of cellulose fraction	• Requires costly acid-resistant bioreactor. • Requires downstream acid neutralization • Degrades sugars due to extreme condition • Generates inhibitory degradation products that could be toxic to fermentation microbes
Steam explosion	• Eliminates the need of extremely corrosion resistant equipment and reactor • Capable of handling high solids loadings • Suitable for hard woods	• Sugar degradation due to steam explosion reaction • Generation of soluble inhibitory products • Less effective on soft woods • Requires washing of pretreated biomass or conditioning of hydrolyzate to remove inhibitory products
AFEX	• Highly effective on agri-residues and herbaceous crops • Sugars are not degraded • High sugar yield • No generation of inhibitory chemicals	• Hemicellulose is not hydrolysed. Thus, both cellulose and hemicellulose fraction need to be hydrolyzed • High capital cost associated with capturing and recycling of ammonia

9.4.8 Inhibitory Compounds

Many compounds either present or generated during the pretreatment and conversion of lignocellulosic biomass into ethanol can adversely affect the overall product yield. For example, naturally occurring cellulases often have a product inhibition mechanism that systematically decreases the efficiency of hydrolysis when sugar concentrations (like glucose) become too high. Simultaneous saccharification and fermentation is one way around this barrier, which couples the consumption of sugars by microbes such as yeast, with the generation of ethanol. However, many other types of impeding factors must be considered before the industrial scale-up of cellulosic ethanol production. A brief discussion of some of the main inhibitory compounds for fermentation will be given in this section.

Hydroxymethylfurfural. 5-hydroxylmethyfufural (HMF) is a fermentation inhibitory compound that is generated by the degradation of six-carbon sugars, like glucose, under severe pretreatment conditions such as high temperatures, pressures, holding times, and concentration of chemicals. Such pretreatment conditions promote three dehydration reactions of the hexose forming the phenolic HMF compound (Lewkowski, 2003).

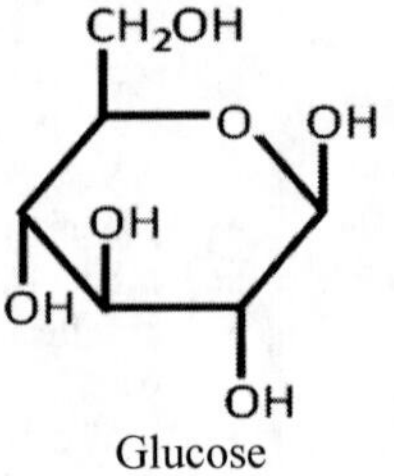

CH_2OH O OH OH OH OH

Glucose

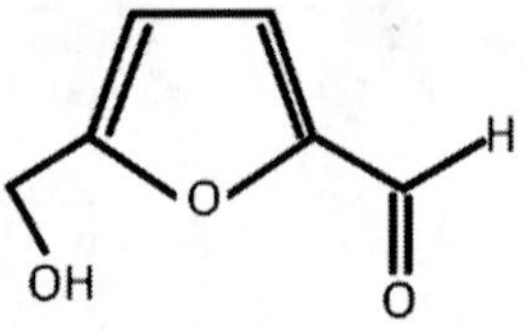

H O OH O

5-hydroxymethylfurfural

An increase in the severity of pretreatment conditions can further degrade HMF into levulinic and formic acid. It has been proposed that these weak acids have the ability to penetrate into microorganisms in their undissociated states and inhibit cellular growth; weak acids are often used as food preservatives (Palmqvist and Hahn-Hagerdahl, 2000). The overall effect is a decrease in ethanol production from a less than optimum amount of sugar-fermenting microbes. The HMF compound itself, can compete with normal, anaerobic metabolic processes and result in a lower yield of ethanol.

Furfural. Furfural, like HMF, is a degradation product formed by the dehydration of pentoses, and is presently produced exclusively from lignocellulosic biomass (Mamman et al., 2008). Under severe pretreatment conditions, furfural can be further degraded into formic acid.

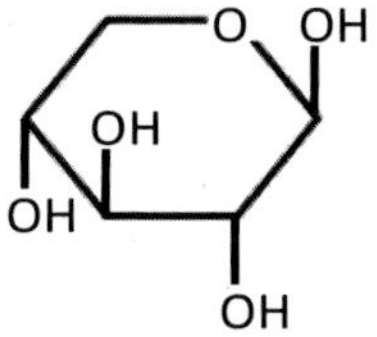

Xylose 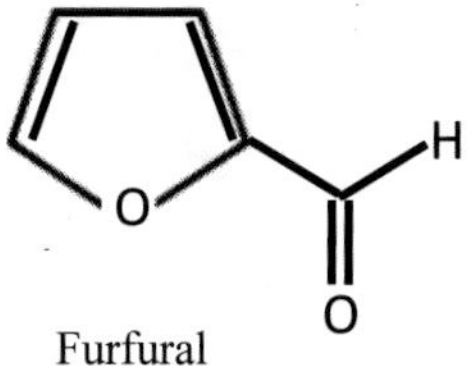 Furfural

Furfural is reduced by *Saccharomyces cerevisiae* into furfuryl alcohols under aerobic, anaerobic, and oxygen-limited conditions (Palmqvist and Hahn-Hagerdahl, 2000). It was proposed that furfural inhibits cell growth by outcompeting normal compounds during metabolism: namely that the reduction of furfural to furfuryl alcohols is prioritized over the reduction of dihydroxyacetone phosphate to glycerol, and that furfural inactivates cellular replication (Palmqvist and Hahn-Hagerdahl, 2000).

A severity factor is often taken into consideration for chemical pretreatments of lignocellulosic material. It was defined by Overend and Chornet (1987) to incorporate the key parameters of temperature and residence time:

$$R_o = t \bullet e^{\frac{T_1 - 100}{14.75}} \tag{9.1}$$

where R_o, t, and T_1 correspond to the severity factor, time, and reaction temperature (°C), respectively. The equation was further modified to account for pH effects by Chum et al. (1990):

$$CS = \log R_o - pH \tag{9.2}$$

where CS represents the combined severity. Liu et al. (2008) reports tolerable concentrations of furfural (and HMF) to be up to 120 mmol/L. In their study, they found that *S. cerevisiae* was able to recover from a growth phase delay after 10, 30, and 60 mmol/L of furfural (and HMF) were added to the yeast media (Liu et al., 2008).

Others. Weak acids are also regarded as inhibitors affecting lignocellulosic conversion into ethanol. Three acids in particular (acetic, formic, and levulinic) are most commonly characterized in the liquid fraction leaving biomass pretreatment. Acetic acid typically originates from hemicellulose, while formic and levulinic acids are products from the degradation of furfural and HMF, as mentioned previously. Although small amounts of weak acid (less than 100 mmol/L) increase ethanol yields, concentrations over 200 mmol/L exhibits adverse effects on microbes (Palmqvist and Hahn-Hagerdahl, 2000).

Phenolic compounds originating from the degradation of lignin can be toxic to *S. cerevisiae* at concentrations near 1 g/L (Palmqvist and Hahn-Hagerdahl, 2000). It has been suggested that these compounds (at high levels) effectively disrupt biological membranes, but further research is required to elucidate the exact mechanism of inhibition with normally occurring concentrations of these molecules (Palmqvist and Hahn-Hagerdahl, 2000).

9.5 Lignin Recovery and Use

9.5.1 Recovery

Lignin recovery occurs in the downstream processing of lignocellulosic biomass and is important for environmental and economic considerations. Dilute-alkaline solutions or organic solvents are typically proposed as the agent of choice for solubilizing the otherwise water-insoluble lignin. Once released into solution, the dark black liquid fraction can be isolated, and its pH can be lowered to precipitate the lignin. Ideally alkalis (like sodium hydroxide or lime), at lower loadings, should be used for this process since recycle/regeneration streams are not possible after lignin precipitation. Peroxides have also been known to separate lignin from biomass, and have the added benefit of sterilization (Gregg and Saddler, 1995). Lignin precipitation is optimal at temperatures between 80-90°C and a pH range of 3-3.5 (Gregg and Saddler, 1995). The solids can then be filtered and sent for further processing. Currently, there are a number of lignin-based products in the development stage from cellulosic-ethanol processes (Lora and Glasser, 2002).

The National Renewable Energy Laboratory (NREL) Clean Fractionation (CF) process as illustrated below is an example of a solvent (Organosolv) pulping process allowing for the recovery of cellulose, hemicellulose and lignin separately (Figure 9.4). The solvent mixture selectively dissolves the lignin and hemicellulose components leaving the cellulose as an undissolved material that can be washed, fiberized, and further purified. The soluble fraction containing the lignin and hemicellulose is treated with water, promoting a phase separation to give an organic phase containing the lignin and an aqueous phase containing the hemicellulose (Bozell et al., 2007).

9.5.2 Lignin Use

The pulping industry is well acquainted with the large production of lignin as a byproduct, and has sought to exploit its chemical properties by using it to supplement (and eventually replace) petroleum-derived compounds in wood adhesives. Lignin from bagasse has already been implemented in particle-board production (Kadam et al., 2008). The generation of lignin however, might exceed the need of the wood adhesive industry if the large scale production of ethanol is implemented worldwide. It would be more ideal to convert the phenolic compound into another source of energy to meet global demands. Burning lignin is one practice that has been implemented, but it is relatively inefficient at exploiting the energy contained within the chemistry of the compound.

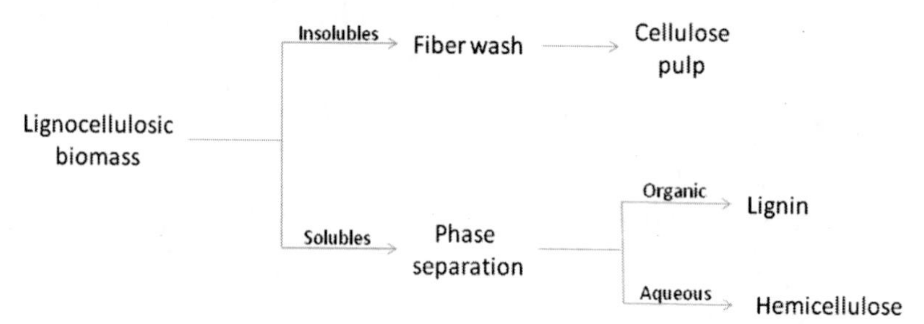

Figure 9.4 Schematic diagram of lignin recovery (adapted from Bozell et al., 2007).

Recent efforts have been made to convert lignin by flash pyrolysis into liquid oil. Unfortunately, successes through this approach have been limited due to a high oxygen: carbon (O/C) and low hydrogen:carbon (H/C) ratios (Kleinhardt and Barth, 2008), which does not compare favorably with crude oil. Kleinhardt and Barth (2008) however, developed a new method to lower the O/C value and increase H/C by employing a novel liquid medium (containing formic acid) during the pyrolysis of lignin. Although the resulting oil does not yet compare equally with crude oil, it is miscible in petroleum-based fuels unlike the products of flash pyrolysis.

Other uses include lignin as an admixture in the concrete industry. The most feasible utilization for sulfur-free lignin is in thermosetting applications in conjuction with phenolic, epoxy or isocyanate resins (Lora and Glasser, 2002). Due to its molecular structure, lignin should be an obvious candidate for applications like a phenol substitute in phenol powder resins, used as binder in the manufacture of friction products.

9.6 High Solids Loading

For cost effective downstream processing, pretreatment should be conducted at high solids levels. For example, Hahn-Hagerdal et al. (2006) reported a minimum of 12 % WIS (water-insoluble solids) to reduce energy cost in distillation. Varga et al. (2004) achieved an ethanol concentration of 52 g/L from acid-pretreated corn stover at 12% dry matter (DM) of the substrate, which met the technical and economic limit of industrial-scale alcohol distillation.

The general consensus is that the ethanol concentration of the broth entering the distillation process must be at least 4% w/w. This translates to a solid concentration of 15% w/w dry matter if we assume a 60% fermentable sugar content, with a 90% sugar monomer release, and an ethanol yield of 0.5 g/g (Jorgensen et al., 2006).

High solids loadings are conceptually simple to understand, and refer to the biomass pretreatment, hydrolysis, and fermentation at a high solid to liquid ratio. The main focus of high solid loadings is aimed at lowering the production costs and increasing the ethanol yield by reducing the size of the necessary equipment and the energy used in processes (e.g. distillation) (Jorgensen et al., 2006). High solid loadings however, are not without key issues, and many, if not all, need to be addressed before being successfully implemented.

9.6.1 Mass Transfer Limitation

A majority of the mass transfer limitations encountered during the bioprocessing of lignocellulosic material occur during the saccharification and fermentation processes, which are downstream from biomass pretreatment. Recent literatures elaborate on some of the accomplishments made on this topic (Jorgensen et al., 2006; Hodge et al., 2008).

Mass transfer limitations however, also exist within the pretreatment of lignocellulosic material and is dependent on the type of pretreatment being conducted. In the case of chemical (and most hybrid) pretreatments, high concentrations of solids limit the diffusion of soluble protons or hydroxyl groups that facilitate the release of sugars or lignin. It has been reported that the viscosity of biomass slurries undergoing pretreatment increases by 4000 centipoises for every 1% increase in solids concentration between loadings of 10-17% (Berson et al., 2006).

9.6.2 Soluble Inhibitors

Many of the studies that involve high solid loadings examine the saccharification of the water insoluble solid (WIS) portions of the pretreated biomass by separating and or washing away the water soluble (WS) portions. The WS fraction often contains inhibitors, released during pretreatment, that have a negative effect on cellulases or microbial growth (during fermentation). In simultaneous saccharification and fermentation process (SSF) situations, however, it is imperative that both WS and WIS components be taken into consideration because a fair amount of fermentable sugar resides in the WS fraction, and it is infeasible to separate the WIS components from the WS.

Hodge et al. (2008) compared washed WIS biomass pretreated with dilute sulfuric acid with unwashed pretreated biomass containing both WS and WIS components. For washed WIS corn stover, it was reported that a WIS fraction of up to 20% was possible before lower yields of cellulose conversion were realized. In contrast, the unwashed slurry exhibited a decrease in cellulose conversion at a WIS loading of less than 10%. The authors concluded that inhibitors like sugars (100-200 g/L), acetic acid (at 15 g/L), furans, and phenolic compounds in the WS fraction are responsible for the significant decrease in the percentage of WIS biomass that can be loaded into contemporary systems and saccharified by enzymes.

It is important to note lignin partially soluble in the chemical pretreatment of biomass can redeposit on the substrate and act as a physical barrier that inhibits enzymatic hydrolysis. This is usually occurrence is usually observed with washed WIS fractions of biomass, and may play a role in capping the conversion yields of cellulose under all solid loading conditions (Hodge et al., 2008).

9.6.3 Effect on Downstream Processing

High solids loadings has the potential to significantly improve the downstream processing of biomass in ethanol production by reducing costs and simplifying the equipment necessary for unit operations. For example, smaller distillation towers and less energy for heating would be required if the liquid content of the processed slurry was reduced drastically. Waste streams would also be reduced, lowering the cost of disposal (if necessary) and the impact of the industry on the ecosystem. In the event that high value products are discovered to be in one of the effluent streams, the isolation and purification of that product are also drastically simplified under high solids loading conditions in upstream processing.

9.7 Water

Water usage in any biorefinery will be one of many concerns. It has been reported (Aden, 2007) that about 6 gallons of water is used to produce one gallon of ethanol in cellulosic ethanol plants utilizing chemical pretreatments. In comparison, corn-based ethanol requires 4 gallons of water per gallon of ethanol, and gasoline requires 2-2.5 gallons of water per gallon of gas (Aden, 2007). Efforts to reduce and optimize water consumption have been the subject of much debate with the construction of pilot plants and industrial scale biorefineries.

Although this paper focuses primarily on the pretreatment of lignocellulosic feedstocks, it is difficult to limit the topic of water usage to a single process within a biorefinery since some operations consume water (pretreatment and hydrolysis) while others generate water (combustion). An in-depth water balance can be better elucidated in other literature (Aden et al., 2002). A report from the National Renewable Energy Laboratory in 2002 proposed a process design that discharged no water to a municipal treatment plant when operated in steady state mode (Aden et al., 2002).

9.8 Economics

Few literatures regarding the current economic feasibility of biomass pretreatment and the overall production of bioethanol exist due to the complexity of the analysis and the large number of variables that change from year to year. For ethanol to be viable, it must be cheaper than petroleum which now averages at roughly $2.01 per gallon (for March 2009); about one dollar less than the previous year (USEIA, 2009).

Eggeman and Elander (2005) defined the minimum ethanol selling price (MESP) as the retail price of ethanol that generates a net zero value for the present worth of a project when cash flows are discounted 10% after tax. The authors used the MESP to compare the performances of the following types of pretreatments: dilute acid, hot water, AFEX, ARP and lime, and compared it with an ideal case and a control. Using the fourth year of operation for comparison, it was suggested that dilute acid pretreatment of biomass may be the most viable due to its lower plant level cash cost and subsequently lower MESP (about \$1.35). The comparison however, is highly speculative since a number of factors must be considered for a rigorous economic analysis of biomass processing, like biomass-specific structural recalcitrance and fermentable sugar content.

Recently, Sendich et al. (2008) revaluated the feasibility of AFEX pretreatment of corn stover in response to a paper published in 2005 (Eggeman and Elander, 2005), which predicted a minimum ethanol selling price (MESP) of about \$1.41 per gallon of ethanol. Sendich et al. (2008) proposed that the utilization of an improved ammonia recovery system and reduced ammonia loadings could reduce the MESP to \$1.03 per gallon for SSFs. It was further suggested that the replacement of the SSF by consolidated bioprocessing (CBP), which eliminates the need to purchase enzymes, would reduce the MESP to as low as \$0.80 per gallon.

The feedstock, enzymes and biomass pretreatment are three of the biggest contributors to ethanol production costs and have subsequently been a subject of research for decades. The feedstock, in particular, contributes up to 35.5% of the final cost of ethanol (Rendleman and Hohmann, 1993). Improvements in technology however, have the potential to reduce expenses significantly and increase the viability of biorefineries. In recent years, enzymes went from \$0.50 per gallon of ethanol to \$0.10-\$0.30 per gallon of ethanol (Hettenhaus et al., 2002), but further cost reductions are still required. Biomass pretreatment-associated costs reflect the strategies employed, but estimates for current technologies are around \$0.11 per gallon of ethanol. The National Resources Defense Council (NRDC) suggests that cost reductions of 22%, 65%, and 89% can be made to feedstock, pretreatment, and enzyme related costs as the bioethanol technology matures; the result being a final cost of \$0.63 per gallon of ethanol with mature ethanol processing compared to the presently achievable \$1.26 per gallon of ethanol (Green and Mugica, 2005).

9.9 Scale-Up

The goal of any lab-scale experiment is ultimately to scale up for industrial applications. Unfortunately, the task is anything but trivial. Many considerations have to be made to simply replicate laboratory conditions. Furthermore, the complexity of the chemistry and biology involved increases with size and may become a safety issue depending on the system set-up. Attempts have been made to model the various pretreatment schemes to offer insight for ultimate scale-up (Kim and Lee, 2002). However, due to the complexity of modeling a biochemical system, work in this field is speculative at best.

As of January 2009, there are 21 companies that are heading projects on the construction and development of cellulosic ethanol plants in the United States (RFA, 2006). The technologies to be employed for lignocellulosic biomass pretreatment are as diverse as the parent companies and include: concentrated acid, enzymes, thermo-chemical conversion, and proprietary processing, among others. Although an economic analysis conducted in 2005 suggests that dilute acid pretreatment may be the most viable at a large scale, this has not yet been confirmed in practice (Eggeman and Elander, 2005). More information on upcoming cellulosic biorefineries can be found at the renewable fuels association's website: http://www.ethanolrfa.org/resource/cellulosic/.

9.10 Future Research

Although some technologies are able to bring sugar and ethanol yields close to theoretical values, improvement in this field is necessary to ensure that ethanol is produced cheaply and consistently year round. High solids loadings, as mentioned earlier, is an engineering design approach to increasing the revenue to the production cost ratio, but it is important to mention that other advances are being made even at the microscopic level.

The development of transgenic plants marks a cornerstone in human achievement, demonstrating our endless capability to modify and improve the genetic information of feedstock to meet our current needs. Perhaps the most innovative approach is the impregnation of cellulases into the subcellular compartments of lignocellulosic material to promote the proper folding and activation of the protein (Sticklen, 2008). The advantages of this type of transgenic plants are multifold. Current pretreatment tactics may be rendered obsolete and replaced with the simple crushing and grinding of biomass to release cellulases already present within the organelles within each cell. However, there are certain obstacles that still need to be addressed. The ability to transform wild type crops to a transgenic species is highly dependent on the type of feedstock being used. Also, the long term effects and population control remain to be determined with outdoor, in-situ observation. Perhaps the biggest obstacle that impedes the progress of transgenic plants is the lack of public support. Genetically modified organisms, in virtually every form, carry a negative connotation with the general public, and are often met with strong oppositions in the political arena.

Another development worth mentioning is the wet processing of biomass to ethanol. Aside from the obvious cost reductions associated with a lower water and energy requirement (for drying and grinding), wet processing may also have the potential to generate more revenue by separating the naturally occurring solid and liquid fractions of the biomass, and using each stream for producing a specific product.

9.11 Summary

Depleting resources and the pressing effects of human activity on the global climate are becoming increasingly prevalent as the populations begin to rise, and as countries begin to modernize. The concept of ethanol as a replacement for petroleum based products is not a new concept, but improvements, particularly in the domain of lignocellulosic pretreatment technologies, must be made before production levels and costs meet consumer expectations.

For decades, pretreatment has served as a bottleneck for the scale up of laboratory testing and pilot plants into an industry. Chemical costs, capital investments, and an overall uncertainty of the viability of ethanol as a replacement for gasoline have served as a deterrent for investors and researchers alike. Fortunately, as our knowledge of the molecular structure and genetic composition of feedstock increases, there exists a plethora of opportunities for us to greatly improve the technologies that were handed down from previous generations and solve the growing problems of today.

9.12 References

Aden, A. (2007). *Water Usage for Current and Future Ethanol Production.* Available at http://www.swhydro.arizona.edu/archive/V6_N5/feature4.pdf (accessed May, 2009).

Aden, A., Ruth, M., Ibsen, K., Jechura, J., Neeves, K., Sheehan, J., Wallace, B., Montaque, L., Slayton, A, and NREL. (2002). *Lignocellulosic Biomass to Ethanol Process Design and Economics Utilizing Co-Current Dilute Acid Prehydrolysis and Enzymatic Hydrolysis for Corn Stover.* NREL/TP-510-32438, Golden, CO.

Alizadeh, H., Teymouri, F., Gilbert, T., and Dale, B. (2005). "Pretreatment of switchgrass by ammonia fiber explosion (AFEX)." *Applied Biochemistry and Biotechnology*, 124, 1133–1141.

Balan, V., da Costa Sousa, L., Chundawat, S.P.S., Vismeh, R., Jones, A.D., and Dale, B. (2008). "Mushroom spent straw: a potential substrate for an ethanol-based refinery." *Journal of Industrial Microbiology and Biotechnology*, 35, 293–301.

Balat, M., Balat, H., and Öz, C. (2008). "Progress in bioethanol processing." *Progress in Energy and Combustion Science*, 34, 551–573.

Berson, R. E., Dasari, R.K., and Hanley, T.R. (2006). "Modeling of a Continuous Pretreatment Reactor Using Computational Fluid Dynamics." *Applied Biochemistry and Biotechnology*, 129-132, 621–630.

Bjerre, A.B., Olesen, A.B., Fernqvist, T., Ploger, A., and Schmidt A.S. (2002). "Pretreatment of wheat straw using combined wet oxidation and alkaline hydrolysis resulting in convertible cellulose and hemicelluloses." *Biotechnology and Bioengineering*, 49, 568–577.

Bozell, J. J., Holladay, J. E., Johnson, D., and White, J. F. (2007). *Top Value Added Chemicals from Biomass- Volume II: Results of Screening for Potential Candidates from Biorefinery Lignin*, PNNL-16983, Pacific Northwest National Laboratory (PNNL) and the National Renewable Energy Laboratory (NREL), Richland, WA.

Chum, H. L., Johnson, D. K., Black, S. K., and Overend, R. P. (1990). "Pretreatment-catalyst effects of the combined severity parameter." *Applied Biochemistry and Biotechnology*, 24/25, 1–14.

Dale, B.E. "Method for increasing the reactivity and digestibility of cellulose with ammonia." U.S. Patent 4 600 590. July 1986.

Ding, S.Y., Xu, Q., Crowley, M., Zheng, Y., Nimlos, M., Lamed, R., Bayer, E.A., Himmel, M.E. (2008). "A biophysical perspective on the cellulosome: new opportunities for biomass conversion." *Current Opinion in Biotechnology*, 19, 218–227.

Eggeman, T., and Elander, R.T. (2005). "Process and economic analysis of pretreatment technologies." *Bioresource Technology*, 96, 2019–2025.

Greene, N., and Mugica, Y. (2005). *Bringing Biofuels to the Pump*. Natural Resources Defense Council. Available at http://www.bio.org/ind/background/NRDC.pdf (accessed May, 2009).

Gregg, D., and Saddler, J.N. (1995). "Bioconversion of lignocellulosic residue to ethanol: Process flowsheet development." *Biomass and Bioenergy*, 9, 287–302.

Hahn-Hagerdal, B., Galbe, M., Gorwa-Grauslund, M.F., Liden, G., and Zacchi, G. (2006). "Bio-ethanol- The fuel of tomorrow from the residues of today." *Trends in Biotechnology*, 24, 549–556.

Hettenhaus, J., Wooley, R., and Ashworth, J. (2002). *Sugar Platform Colloquies*. NREL/SR-510-31970. Available at http://permanent.access.gpo.gov/lps31319/www.ott.doe.gov/biofuels/pdfs/sugar_platform.pdf (accessed May, 2009).

Hodge, D., Karim, M.N., Schell, D.J., and McMillan, J.D. (2008). "Soluble and insoluble solids contributions to high-solids enzymatic hydrolysis of lignocellulose." *Bioresource Technology*, 99, 8940–8948.

Hu, Z., and Wen, Z. (2008). "Enhancing enzymatic digestibility of switchgrass by microwave-assisted alkali pretreatment." *Biochemical Engineering Journal*, 38, 369–378.

International Energy Agency (IEA). (2008). *From 1st– to 2nd - generation biofuel technologies*. Available at http://www.iea.org/textbase/papers/2008/2nd_Biofuel_Gen_Exec_Sum.pdf (accessed May, 2009).

Johnson, D.K., and Elander, R.T. (2008). "Pretreatments for enhanced digestibility of feedstocks." In: Himmel, M. (Ed.) *Biomass recalcitrance: Deconstructing the plant cell wall for bioenergy*. London: Blackwell Publishing, pp. 436-453.

Jones, A.D., Dale, B., da Costa Sousa, Vismeh, R., Chundawat, S.P.S., and Balan, V. (2008). "Mushroom spent straw: a potential substrate for an ethanol-based refinery." *Journal of Industrial Microbiology & Biotechnology*, 35, 293–301.

Jorgensen, H., Vibe-Pederson, J., Larsen, J., and Felby, C. (2006). "Liquefaction of Lignocellulose at High-Solids Concentrations." *Biotechnology and Bioengineering*, 96, 862–870.

Kaar, W.E., and Holtzapple, M.T. (2000). "Using lime pretreatment to ffacilitate the enzymatic hydrolysis of corn stover." *Biomass and Bioenergy*, 18, 189–199.

Kadam, K.L., Chin, C.Y., and Brown, L.W. (2008). "Flexible biorefinery for producing fermentation sugars, lignin and pulp from corn stover." *Journal of Industrial Microbiology and Biotechnology*, 35, 331–341.

Khanal, S.K., Montalbo, M., Van Leeuwen, J. (Hans), Srinivasan, G., and Grewell, D. (2007). "Ultrasound enhanced glucose release from corn in ethanol plants." *Biotechnology and Bioengineering*, 98, 978–985.

Kim, S., and Holtzapple, M.T. (2005). "Lime pretreatment and enzymatic hydrolysis of corn stover." *Bioresource Technology*, 96, 1994–2006.

Kim, S.B., and Lee, Y.Y. (2002). "Diffusion of sulfuric acid within lignocellulosic biomass particles and its impact on dilute-acid pretreatment." *Bioresource Technology*, 83, 165–171.

Kim, T.H., and Lee, Y.Y. (2005). "Pretreatment and fractionation of corn stover by ammonia recycle percolation process." *Bioresource Technology*, 96, 2007–2013.

Kim, T.H., Taylor, F., and Hicks, K.B. (2008). "Bioethanol production from barley hull using SAA (soaking in aqueous ammonia) pretreatment." *Bioresource Technology*, 99, 5694–5702.

Kleinhardt, M., and Barth, T. (2008). "Towards a Lignincellulosic Biorefinery: Direct One-Step Conversion of Lignin to Hydrogen-Enriched Biofuel." *Energy & Fuels,* 22, 1371–1379.

Klinke, H.B., Ahring, B.K., Schmidt, A.S., and Thomsen, A.B. (2002). "Characterization of degradation products from alkaline wet oxidation of wheat straw." *Bioresource Technology*, 82, 15–26.

Kuhar, S., Nair, L.M., and Kuhad, R.C. (2008). "Pretreatment of lignocellulosic material with fungi capable of higher lignin degradation and lower carbohydrate degradation improves substrate acid hydrolysis and the eventual conversion to ethanol." *Canadian Journal of Microbiology*, 54, 305–313.

Kumar, R., Singh, S., and Singh, O.V. (2008). "Bioconversion of lignocellulosic biomass: biochemical and molecular perspectives." *Journal of Industrial Microbiology and Biotechnology*, 35, 377–391.

Lee, J. (1997). "Biological conversion of lignocellulosic biomass to ethanol." *Journal of Biotechnology*, 56, 1–24.

Lewkowski, J. (2003). "Synthesis, chemistry and applications of 5-hydroxymethylfurfural and its derivatives." *ChemInform*, 34.

Liu, Z. L., Slininger, P.J., and Gorsich, S.W. (2005). "Enhanced biotransformation of furfural and hydroxymethylfurfural by newly developed ethanologenic yeast strains." *Applied Biochemistry and Biotechnology*, 121, 451–460.

Liu, Z., Saha, B.C., and Slininger, P.J. (2008). "Lignocellulosic biomass conversion to ethanol by Saccharomyces." In: Wall, J., Harwood, C., Demain, A., (Ed.) *Bioenergy*. Chapter 4. Washington, DC: ASM Press, pp. 17–36.

Lloyd, T.A., and Wyman, C.E. (2005). "Combined sugar yields for dilute sulfuric acid pretreatment of corn stover followed by enzymatic hydrolysis of the remaining solids." *Bioresource Technology*, 96, 1967–1977

Lora, J. H., and Glasser, W. G. (2002). "Recent industrial applications of lignin: A sustainable alternative to nonrenewable materials." *Journal of Polymers and the Environment*, 10, 39–48.

Lynd, L.R., Weimer, P. J., van Zyl, W. H., and Pretorius, I.S. (2002). "Microbial Cellulose Utilization: Fundamentals and Biotechnology." *Microbiology and Molecular Biology Reviews*, 66, 506–577.

Mamman, A.S., Lee, J.M., Kim, Y.C., Hwang, I.T., Park, N.J., Hwang, Y.K., Chang, J.S., and Hwang, J.S. (2008). "Furfural: Hemicellulose/xylose-derived biochemical." *Biofuels, Bioproducts and Biorefining*, 2, 438–454.

Mais, U., Esteghlalian, A.R., Saddler, J.N., and Mansfield, S.D. (2002). "Enhancing the enzymatic hydrolysis of cellulosic materials using simultaneous ball milling." *Applied Biochemistry and Biotechnology*, 98-100, 815–832.

Mani, S., Tabil, L.G., and Sokhansanj, S. (2004). "Grinding performance and physical properties of wheat and barley straws, corn stover and switchgrass." *Biomass and Bioenergy*, 27, 339–352.

Mosier, N., Hendrickson, R., Ho, N., Sedlak, M., and Ladisch, M. (2005). "Optimization of pH controlled liquid hot water pretreatment of corn stover." *Bioresource Technology*, 96, 1986–1993.

Mousdale, D. (2008). *Biofuels*. CRC Press, USA.

Murnen, H.K., Balan, V., Chundawat, S.P.S., Bals, B., da Costa Sousa, L., and Dale, B. (2008). "Optimization of Ammonia Fiber Expansion (AFEX) Pretreatment and Enzymatic Hydrolysis of *Miscanthus x giganteus* to Fermentable Sugars." *Biotechnology Progress*, 23, 846–850.

Nitayavardhana, S., and Khanal, S.K. (2009). "Biofuel residues/wastes: Boon or Ban?" *Critical Reviews in Environmental Science and Technolopgy* (in-review).

Overend, R. P., and Chornet, E. (1987). "Fractionation of lignocellulosics by steam-aqueous pretreatments." *Philosophical Transactions of the Royal Society London*, 321 (A), 523–536.

Palmqvist, E., and Hahn-Hagerdahl, B. (2000). "Fermentation of lignocellulosic hydrolysates. II: inhibitors and mechanisms of inhibition." *Bioresource Technology*, 74, 25–33.

Perlack, R. D., Wright, L.L., Turhollow, A. F., Graham, R. L., Stokes, B.J., and Erbach, D. C. (2005) *A Billion-Ton Feed Stock Supply for Bioenergy and Bioproducts Industry: Technical Feasibility of Annually Supplying 1 Billion Dry Tons of Biomass*. DOE/GO 102005-2135, ORNL/TM-2005/66, http://www1.eere.energy.gov/biomass/pdfs/final_billionton_vision_report2.pdf (accessed May, 2009).

Piccolo, C., and Bezzo, F. (2009). "A techno-economic comparison between two technologies for bioethanol production from lignocellulose." *Biomass and Bioenergy*, 33, 478–491.

Ramos, L. (2003). "The Chemistry Involved in the Steam Treatment of Lignocellulosic Materials." *Quim. Nova*, 26, 863–871.

Rendleman, C.M., and Hohmann, N. (1993). "The impact of production innovations in the fuel ethanol industry." *Agribusiness*, 9, 217–231.

Renewable Fuel Association (RFA). (2009). *Cellulosic Ethanol*. Available at http://www.ethanolrfa.org/resource/cellulosic/ (accessed May, 2009).

Sendich, E., Laser, M., Kim, S., Alizadeh, H., Laureano-Perez, L., Dale, B., and Lynd, L. (2008). "Recent process improvements for the ammonia fiber expansion (AFEX) process and resulting reductions in minimum ethanol selling price." *Bioresource Technology*, 99, 8429–8435.

Shi, J., Chinn, M.S., and Sharma-Shivappa, R.R. (2008). "Microbial pretreatment of cotton stalks by solid state cultivation of *Phanerochaete chrysosporium*." *Bioresource Technology*, 99, 6556–6564.

Shrestha, P., Rasmussen, M., Khanal, S., Pometto, A. L., and van Leeuwen, J. (Hans). (2008). "Solid-Substrate Fermentation of Corn Fiber by Phanerochaete chrysosporium and Subsequent Fermentation of Hydrolysate into Ethanol." *Journal of Agricultural and Food Chemistry*, 56, 3918–3924.

Sticklen, M.B. (2008). "Plant genetic engineering for biofuel production: towards affordable cellulosic ethanol." *Nature Reviews Genetics*, 9, 433–443.

Sun, Y., and Cheng, J.J. (2005). "Dilute acid pretreatment of rye straw and bermudagrass for ethanol production." *Bioresource Technology*, 96, 1599–1606.

Tengerdy, R.P., and Szakacs, G. (2003). "Bioconversion of lignocellulose in solid state fermentation." *Biochemical Engineering Journal*, 13, 169–179.

Teymouri, F., Laureano-Perez, L., Alizadeh, H., and Dale, B.E. (2005). "Optimization of the ammonia fiber explosion (AFEX) treatment parameters for enzymatic hydrolysis of corn stover." *Bioresource Technology*, 96, 2014–2018.

Torget, R., Himmel, M., and Grohmann, K. (1992). "Dilute-acid pretreatment of two short-rotation herbaceous crops." *Applied Biochemistry and Biotechnology*, 34–35, 115–123.

United States Energy Information Administration (USEIA). (2008). *Annual Energy Outlook 2008 with Projections to 2030*. DOE/EIA-0383(2008), http://www.eia.doe.gov/oiaf/aeo/pdf/0383(2008).pdf (accessed October 2008).

United States Energy Information Administration (USEIA). (2009). *Weekly Retail Gasoline and Diesel Prices*. Available at http://tonto.eia.doe.gov/dnav/pet/pet_pri_gnd_dcus_nus_m.htm (accessed April 2009).

Varga, E., Klinke, H.B., Reczey, K., and Thomsen, A.B. (2004). "High Solid Simultaneous Saccharification and Fermentation of Wet Oxidized Corn Stover to Ethanol." *Biotechnology and Bioengineering*, 88, 567–574.

Viola, E., Cardinale, M., Santarcangelo, R., Villone, A., and Zimbardi, F. (2008). "Ethanol from eel grass via steam explosion and enzymatic hydrolysis." *Biomass and Bioenergy*, 32, 613–618.

Wyman, C.E., Dale, B.E., Elander, R.T., Holtzapple, M., Ladisch, M.R., and Lee, Y.Y. (2005). "Coordinated development of leading biomass pretreatment technologies." *Bioresource Technology*, 96, 1959–1966.

Yang, B., and Wyman, C. (2007). "Pretreatment: the key to unlocking low-cost cellulosic ethanol." *Biofuels, Bioproducts and Biorefining*, 2, 26–40.

Zimbardi, F., Viola, E., Nanna, F., Larocca, E., Cardinale, M., and Barisano, D. (2007). "Acid Impregnation and steam explosion of corn stover in batch processes." *Industrial Crops and Products*, 26, 195–206.

CHAPTER 10

Enzymatic Hydrolysis of Lignocellulosic Biomass

Buddhi P. Lamsal, Prachand Shrestha, and Samir Kumar Khanal

10.1 Introduction

Lignocellulosic biomasses are non-food energy resources, the worldwide terrestrial availability of which is estimated to be around 200 x 10^{12} kg (220 billion ton) annually (Foust et al., 2008). A USDA and USDOE report estimates that the United States has the potential of producing 1.3 billion dry tons of biomass annually after meeting food, feed and fiber demands, and exports (USDA and USDOE joint report, 2005), which could theoretically substitute more than 30% of the nation's petroleum consumption. Thus, biomass may play an important role in the domestic bio-based economy through production of a variety of biofuel and biomolecules. Major lignocellulosic biomass sources include forest and woody products along with agricultural residues, agricultural processing byproducts, and energy crops.

Broadly, there are two major pathways for producing biofuels from lignocellulosic feedstocks: biochemical conversion and thermochemical conversion. The biological route to obtain ethanol from lignocellulosic biomass is based on microbial fermentation of sugars derived from saccharification of cellulose biomass. A simplified process overview of ethanol production from lignocellulosic biomass via the biochemical route is shown in Figure 10.1. As cellulose is protected by lignin and intertwining hemicellulose and pectin, it is not easily accessible to the enzymes for saccharification, thus necessitating pretreatment. Pretreatment is the first operation in lignocellulosic ethanol production, which essentially prepares biomass to enzyme hydrolysis. Various forms of pretreatment are: physical, e.g. mechanical comminution, extrusion; physical-chemical, e.g. steam explosion, ammonia fiber explosion, CO_2 explosion; chemical, e.g. ozonolysis, high temperature acid treatment, alkali hydrolysis, organosolv; and biological pretreatment, e.g. fungal treatments of biomass. They are well reviewed by several researchers (See Chapter 9 on Biomass Pretreatment). Depending on the type of pretreatment, the following can be accomplished: lignin breakup and crystallinity reduction in cellulose (mechanical), hemicellulose removal (acid treatment), delignification (alkaline, oxidative delignification, biological), breakup of internal lignin and hemicelluloses bonds (organosolv), size reduction (comminution), or the generation of high shear, high pressure (extrusion) conditions. In all cases, pretreatment prepares the biomass for the hydrolysis of cellulose into fermentable sugars, either enzymatically or chemically, by acids. Hydrolysis of pretreated biomass is a key step in biomass processing to ethanol, and enzymatic hydrolysis is the preferred one.

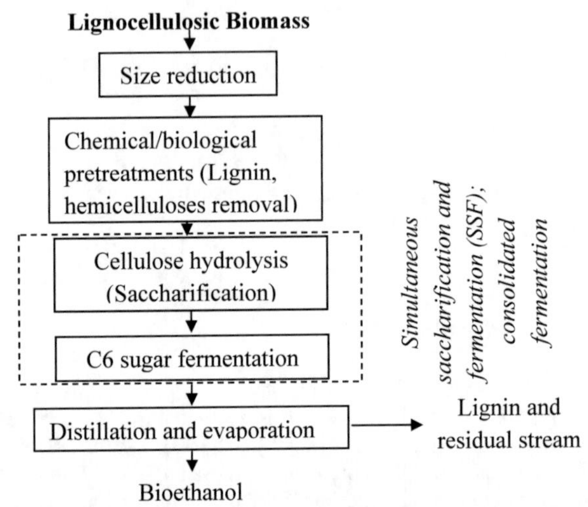

Figure 10.1 Schematic of ethanol production from lignocellulosic biomass via biochemical processing route (*adapted from Lin and Tanaka, 2006; Balat et al., 2008*).

A combination of enzyme cellulases (endo-β-1,4-glucanases, exo-β-1,4-glucanases, and β-1,4-glucosidases) is utilized for systematic degradation of pretreated biomass from long-chain glucose polymers to monomeric units. Enzymes like hemicellulases or even acids hydrolyze the hemicellulose to release its component sugars. The sugars thus released are fermented by microbial catalyst. Six-carbon sugars like galactose and mannose are also readily fermented to ethanol by many naturally occurring organisms, but the five-carbon sugars such as xylose and arabinose (pentoses) are fermented to ethanol by few native strains with relatively low yields (Mosier et al., 2005).

Many proposed full-scale lignocellulose ethanol plants plan to employ a biochemical route. This is primarily due to its potential for perpetual cost reductions as is apparent from historical evidence. Biochemical production of biofuels is possible at prices cheaper than that of thermochemical routes (IEA, 2008). Although enzymes or acid can be used to facilitate the release of monomeric sugars from pretreated biomass, the former is the preferred method due to the low cost of enzymes, better sugar yield and less input of chemicals. Thus enzymatic hydrolysis is considered to be an integral part of biochemical methods for ethanol production.

This chapter covers enzyme systems for cellulose degradation, the mechanism and kinetics of enzyme hydrolysis, factors affecting enzyme hydrolysis of cellulose, enzyme systems for hemicellulose degradation, and the cost of enzymes. In addition, major research advances in enzyme hydrolysis of lignocellulosic biomass, especially enzyme immobilization, enzyme recycling, in-situ enzyme production/biological pretreatment, and progress towards single-step consolidated bioprocessing are discussed. Various challenges and future research direction in enzyme hydrolysis are proposed.

10.2 Enzymatic Hydrolysis of Lignocellulosic Biomass

The hydrolysis of cellulose requires a cocktail of several enzymes, collectively referred to as cellulases, for the effective conversion of cellulose into fermentable sugars. Cellulase enzymes are currently being produced at pilot- and industrial-scales, using fungi *Trichoderma reesei*, *Penicillium funiculosum,* and *Aspergillus niger.* Several other cellulase-producing fungal species are *Trichoderma sp., Penicillium sp., Fusarium sp., Aspergillus sp., Chrysosporium pannorum,* and *Sclerotium rolfsii* (Philippidis, 1994). Several bacteria such as *Acidothermus cellulolyticus, Micromonospora bispora, Bacillus sp., Cytophaga sp., Streptomyces flavogriseus, Thermomonospora fusca, Thermomonospora curvata, Clostridium stercorarium and Clostridium thermocellum,* and *Ruminococcus albus* also produce enzymes that are capable of hydrolyzing cellulose (Philippidis, 1994). Termites and other wood-feeding insects (e.g. bark beetles) are also being examined as potential candidates for cellulase production due to their ability to digest lignocellulosic biomass (Watanabe et al., 1998). Rumen microbes (fungi, bacteria and protozoa) have been studied for their cellulolytic activities (Wood et al., 1986; Weimer, 1992).

10.2.1 Enzyme Systems for Cellulose Degradation

Lignocellulosic degradation involves stepwise removal of hemicelluloses or lignin prior to cellulose degradation. Even though enzyme systems are available for the degradation of lignin and hemicelluloses, they are usually removed via chemical pretreatments like acid pretreatment (Wyman, 1999; Saha, 2003). Two categories of dilute acid pretreatments are used: high temperature (> 160 °C) continuous flow for low-solids loading (0–5% w/w), and low temperature (< 160 °C) batch processes for high solids loadings (10–40% w/w). Dilute acid pretreatment usually hydrolyzes hemicelluloses to its soluble monomeric sugars, with the residues containing cellulose and lignin. The lignin can be extracted with organic solvents like ethanol, butanol or formic acid prior to enzyme hydrolysis. Alternatively, hydrolysis of cellulose with the lignin presence is carried out. The enzymatic hydrolysis of cellulose is a key operation in biochemical conversion, which determines the quality and quantity of the end-product obtained, thus impacting the downstream processing.

Cellulase. Cellulase typically consists of a mixture of three different enzyme systems: i) endo-β-1,4-glucanases, ii) exo-β-1,4-glucanases (cellobiohydrolase), and iii) and β-1,4-glucosidases. Table 10.1 explicitly lists major cellulases used in enzyme systems, their specificity, and end products. Even though all cellulases act on the chemically identical bonds, i.e., the β-1,4-linkage between two glucose units, they differ in terms of their site of attack on the cellulose chain. They are accordingly classified into two distinct groups: endoglucanases, the enzymes that act in the middle of the cellulose chain and exoglucanases, enzymes that act at either end of the cellulose chain.

Table 10.1 Major cellulase enzyme systems, their specificity, and end products (*Adapted from Maija et al., 2003*).

Action	Endo	Exo	Exo
Trivial names	Cellulase, endoglucanase	Cellobiohydrolase	Cellobiase
Systematic names	1,4-β-D-glucan-4-glucanohydrolase	1,4-β-D-glucancellobio-hydrolase	β-Glucosidase
Substrate	Cellulose, 1,3-1,4-β -glucans	Cellulose, 1,3-1,4-β -glucans	β -Glucosidase
Bonds hydrolyzed	1,4-β	1,4-β	1,4-β, 1,3-β, 1,6-β
Reaction products	1,4-β-dextrins, mixed 1,3-1,4-β - dextrins	Cellobiose	Glucose

(i) Endo-β-1,4-glucanases: The endoglucanases (EG) randomly cut in cellulose chains, thereby producing shorter and shorter cello-oligomers, also called cellodextrins, which can be further degraded by exoglucanases. EG account for 15–20% of the extracellular protein when *T. reesei* is grown on cellulose, with up to 6 EG enzymes reported. They differ in the degree of randomness in their attack on cellulose polymers, the degree of synergism with cellobiohydrolase (CBH), and in the range of final products (Marsden et al., 1985). Endoglucanases are mainly active on amorphous (or less-ordered) part of the cellulose, producing soluble oligomers. The degradation of cello-oligomers is catalyzed by endoglucanase and cellobiohydrolase, resulting in glucose production.

(ii) Exo-β-1,4-glucanases (cellobiohydrolase) (CBH): Cellobiohydrolase comprises between 35 and 85% of the extracellular protein produced by *T. reesei.* Cellobiohydrolase cleavage predominantly produces cellobiose units from either the reducing or non-reducing ends of cellulose chains and cello-oligosaccharides (Maija et al., 2003). While CBH mainly acts on crystalline regions of cellulose, there are conflicting accounts of specificity of the enzyme with regards to the source and substrate to act on (Wood, 1989). Cellulase systems of *T. reesei* and *Penicillium pinophilum* are reported to exist in two forms of CBH: I and II. The degradation of crystalline celluloses, cotton, and Avicel was reported to be extensive by a purified CBH-I of *T. reesei* acting alone, but was negligible by a CBH-I from *T. koningii.* Also, carboxymethylcellulose, which was reported not to be a substrate for some purified CBH, has been reported to be extensively hydrolyzed by similar enzymes from other sources (Wood, 1989). While these differences clearly indicate differences in the substrate specificities of cellulase enzymes from different sources, they may also be a consequence of contamination of CBH by trace endoglucanases, in an enzyme complex that is stable under conditions normally studied.

(iii) β-glucosidases: β-glucosidases make up less than 1% of the extracellular protein produced by *T. reesei*, a majority being intracellular. The fungal and bacterial β-glucosidases hydrolyze cellobiose and a range of other β-glucosides producing glucose. While the term β-glucosidase is taken almost synonymously with cellobiase, these enzymes act more readily on higher oligomers than on cellobiose (Marsden et al., 1985). The enzyme systems discussed thus far are produced either by fungal or bacterial microorganisms. However, only a small number of these microorganisms produce extracellular enzymes that can degrade cellulose. Noted fungal species are *T. reesei, T. viride, T. koningii, P. funiculosum*, and bacterial species of *Clostridium* and *Cellulomonas* genus. The enzyme β-glucosidases generally governs the overall cellulolytic process and is considered to be a rate-limiting step in the enzyme hydrolysis of cellulose because the accumulation of cellobiose inhibits both endoglucanase and cellobiohydrolase activities. β-glucosidase activity is also subjected to end-product (glucose) inhibition.

10.2.2 Mechanism of Enzyme Hydrolysis of Cellulose and Kinetics

Mechanism of Enzyme Hydrolysis of Cellulose. While this section is not intended to be a complete review of cellulose hydrolysis mechanisms and kinetics, some important mechanisms reported are discussed. As cellulose is comprised of amorphous and crystalline regions the types of enzymes that hydrolyze these regions also differ. Figure 10.2 shows a schematic of degradation of these regions in cellulose, in which endoglucanases attack amorphous regions, and cellobiohydrolase acts on crystalline regions. Early works (prior to 1980) proposed the mechanism of cellulase action as seen from an enzymological point of view in which the amorphous component was attacked first, followed by hydrolysis of the crystalline component at a slower rate; a so-called 'serial' mechanism. However, since then, a 'parallel' mechanism of cellulase action has been widely accepted, in which amorphous and crystalline regions are attacked simultaneously (Marsden, 1985). In a parallel mechanism (Figure 10.3), amorphous cellulose is acted upon by endoglucanases producing cello-oligomers. These cello-oligomers are then acted upon by β-glucosidases that produce glucose and cellobiose. Simultaneously, the crystalline cellulose is acted upon by a combination of cellobiohydrolase and endoglucanase to produce cellobiose. The cellobioses thus produced are hydrolyzed by cellobiase to produce glucose. It is also accepted that glucose is obtained by a process other than the cleavage of cellobiose, especially by the action of the exoglucosidase on cello-oligomers. A similar model of enzyme degradation of highly ordered cellulose (crystalline regions) and amorphous region is discussed by Rabinovich et al. (2002) in their comprehensive review of the structure and mechanism of cellulase action on cellulose.

A rather similar mechanistic hypothesis of enzyme hydrolysis of cellulose was put forward by Zhang and Lynd (2004), but incorporating the physical and chemical changes occurring in the residual (not yet solubilized) cellulose. The three processes that were proposed to occur simultaneously upon enzymatic action were: i) physical and chemical changes in the residue, ii) primary hydrolysis involving the release of soluble intermediates from cellulose surfaces, and iii) secondary hydrolysis of soluble intermediates to lower molecular intermediates, ultimately to glucose. The physical changes include the increased area accessible to enzymes due to 'amorphogenesis' (swelling, segmentation, of destratification of cellulose), and also the decreased area accessible by heterogeneous hydrolysis. The chemical changes include increased polymer ends due to cleavage by endo-enzymes, and at the same time, decreased polymer ends due to hydrolysis by exoglucanase. The rest of the model proposed primary and secondary hydrolyses in the liquid phase due to endo-, exo-, and β-glucanase. This mechanism of action by endo-glucanase in the solid phase to produce oligosaccharides of various degrees of polymerization is also supported by the similar model put forward by Gupta and Lee (2009) based on their study on various pure cellulosic substrates–microcrystalline cellulose (Avicel), commercial cellulose, filter paper, cotton, and non-crystalline cellulose (NCC). Endo-glucanase reacted rapidly on amorphous NCC producing cello-oligosaccharides with a degree of polymerization ranging from 30 to 60, after which the endo-glucanase reaction with NCC ceased. Cello-oligosaccharides with lower degrees of polymerization are produced by exo-glucanase. Production of monomeric sugars from these cello-oligosaccharides is due to β-glucosidase and cellobiase.

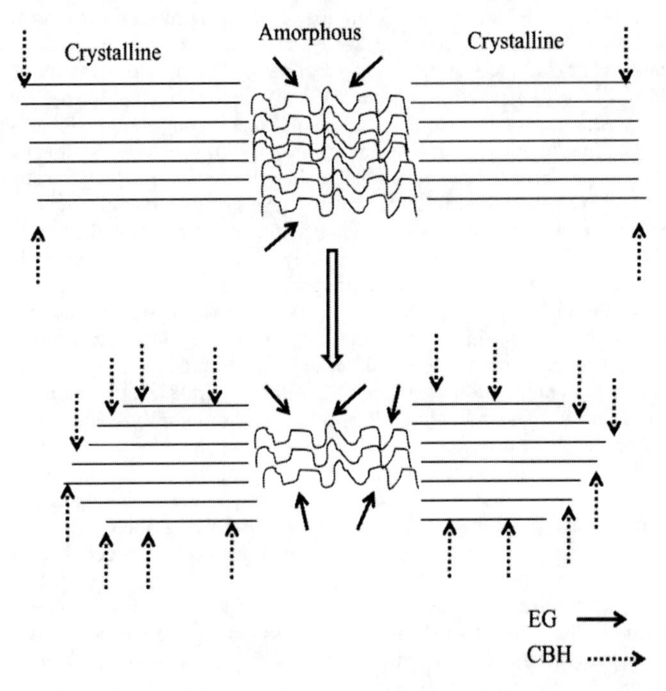

Figure 10.2 Schematic of cellulose structure and its degradation by cellulase enzyme systems. CBH: cellobiohydrolase (--->), and EG: endoglucanase (→) (*adapted from Maija et al., 2003*).

Kinetics of Enzyme Hydrolysis of Cellulose. The cellulase kinetics refers to the rate and extent of cellulose hydrolysis by an enzyme or a combination of enzymes which are affected by many factors like enzyme sources, substrate source and variability in preparation, cellulose crystallinity, etc. Various kinetic models and studies on cellulose from different sources and treatments have revealed useful information on enzyme hydrolysis and the behavior of crystalline and amorphous celluloses. The general model of enzymatic hydrolysis of cellulose was developed after the Michaelis-Menten kinetics. However recently there has been use of other models, like adsorption. The reaction kinetics of enzymatic cellulose hydrolysis is affected by a number of factors such as enzyme adsorption to the substrate, the enzyme-substrate intermediate complex formation and degradation rates, and product inhibitions, among others. Synergism between the individual components of a cellulase system acting upon insoluble cellulose further complicates the study of the mechanisms of action of cellulases. In fact, the action of a mixture of two or more individual cellulolytic components is greater than the sum of the action of each component. Even so, the extent of synergism obtained is not easy to predict as it varies depending on which of the multiple forms of cellulases and on the cellulose samples (amorphous or crystalline, to mention the two extreme variants) used for the experiment.

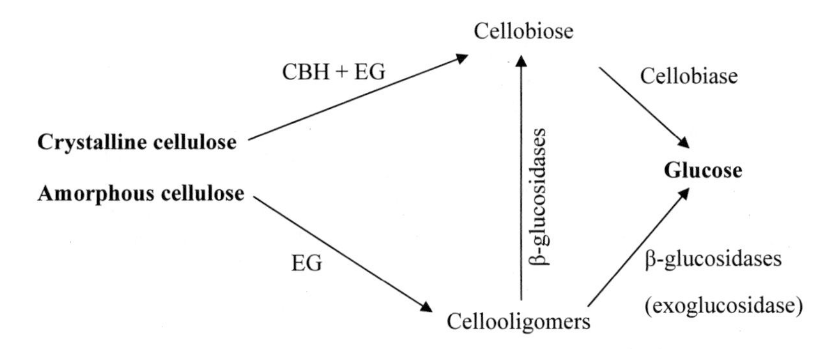

Figure 10.3 Parallel mechanism of cellulase action. EG: endoglucanase, CBH: cellobiohydrolase (*adapted from Marsden et al., 1985*).

In a review of early work on the kinetics of cellulose hydrolysis, the reaction was reported as pseudo first-order (with milled rice hulls) to first-order (as with Solka Floc) (Marsden et al., 1985) with deviations at higher substrate concentrations. The value of the activation energy with rice hulls suggested that cellulase hydrolysis was either diffusion-limited or chemical-reaction limited with the likelihood that both were contributing factors. Marsden et al. (1985) cited van Dyke (1972) on the kinetic study of the hydrolysis of powdered cellulose (Solka Floc) generating some important conclusions. There exists an optimal cellulose concentration that is proportional to the absorbable protein, for the maximum rate of hydrolysis at a given enzyme concentration. The specific hydrolysis rate of the amorphous region is independent of the initial cellulose concentration at cellulose concentrations above the optimum, whereas in the crystalline regions, the hydrolysis rate is inversely proportional. At below optimum concentrations, the enzyme concentration involved in hydrolysis was proportional to the initial substrate concentration. A product inhibition model was developed by Howell and Stuck (1975) to describe the hydrolysis of cellulose (Solka Floc) by the *Trichoderma viride* enzyme system, which assumed that noncompetitive inhibition by cellobiose dominated the reaction kinetics. This was shown to be a reasonable assumption for initial cellulose concentrations of up to 15 g/liter and hydrolysis extents up to 65%. Caminal et al. (1985), however, found that a competitive product inhibition model best describes the hydrolysis reactions involving a two-step process that included the conversion of the intermediate cellobiose to glucose by β-glucosidase.

The presence of both exo- and endo-glucanase enzyme systems is a key requirement for enhanced hydrolysis (synergism) to glucose monomers from cellulose polymers (Suga et al., 1975) even though the degradation rate for the endo-enzyme was much greater. Medve et al. (1994) studied the adsorption and synergism during the hydrolysis of microcrystalline cellulose (Avicel) by CBH I and CBH II from *Trichoderma reesei*. When Avicel was hydrolyzed by increasing amounts of CBH I and/or CBH II, either alone or in reconstituted equimolar mixtures, a linear correlation was found between substrate conversion and the amount of adsorbed enzyme. The enzyme adsorption process however, was slow, taking 30-90 min to reach 95% of the equilibrium binding. Although more CBH I was adsorbed than CBH II, CBH I had a lower specific activity. Synergism between these cellobiohydrolases during the degradation of amorphous Avicel was at a maximum as a function of the total enzyme concentration. The two enzymes were shown to compete for the adsorption sites with stronger competition shown by CBH I (Medve et al., 1994).

Fujii and Shimizu (1986) proposed a kinetic model for the hydrolysis of water-soluble cellulose derivatives by mixed endo- and exoenzyme systems. The authors assumed that at an early stage of the reaction, endo-enzymes split the substrate molecule in order to supply the newly formed non-reducing ends to exo-enzymes until the molecular weight of the substrates reaches a low value. After that point, the reaction kinetics obeyed only the rate equation of the exo-enzymes reaction in which the reaction parameters change linearly with a decrease in the molecular weight of the substrates. In experiments with soluble cellulose derivatives (carboxymethyl cellulose and hydroxyethyl cellulose) this critical molecular weight of the substrate, after which the endo-enzyme action can be neglected, was determined to be 4000 Da. Synergism by these enzymes enhanced the early rate of the reaction. Visual evidence of the synergism between endo- and exo-glucanase was offered by White and Brown (1981), who used cellulose from the gram-negative bacterium *Acetobacter xylinum* as a model substrate for visualizing the action of cellulase enzymes from the fungus *Trichoderma reesei*. Scanning electron microscopy revealed that the enzymes initially bound to the cellulose ribbon produced by bacteria. Within 10 min, the ribbon was split along its long axis into bundles of microfibrils which were subsequently thinned until they were completely dissolved (after about 30 min). Purified 1,4-β-D-glucan CBH I produced no visible change in the cellulose structure. Purified endo-1,4-β-D-glucanase IV produced some splaying of cellulose ribbons into microfibril bundles. In both cases, whole ribbons were present even after 60 min of incubation, visually confirming the synergistic mode of action of these enzymes. In another study, the action of an exoglucanase, CBH I, and an endoglucanase, EG II- alone and in combination on cotton fibers were visualized using atomic force microscopy (Lee et al., 2000). The action of CBH I resulted in the appearance of distinct physical pathways or tracks along the length of the macrofibril, whereas the action of EG II appeared to cause peeling and smoothing of the fiber surface. In combination, their effect was observed to be greatest when both enzymes were present simultaneously, again affirming their synergistic action on cellulose.

Studies on the adsorption of cellulase by cellulose fibers have shown that the enzymatic hydrolysis rate of textile fibers is determined by the concentration of the enzyme-substrate (ES) complex formed. The structural differences and cellulose crystallinity of fibers of different sources lead to differences in cellulase adsorption and activity. Based on the adsorption and ES formed, Shen and Wang (2004) derived a kinetic equation to predict the hydrolysis rate on which they were able to relatively rank the enzyme digestibility of textile fibers. Peri et al. (2007) studied the individual contributions of cello-oligosaccharides, cellobiose, and glucose on the inhibition of cellulase during the enzymatic hydrolysis of noncrystalline cellulose. At higher initial oligomer concentrations, they observed that the initial hydrolysis rate lasted for a relatively shorter time resulting in a reduced extent of hydrolysis after 96 h, thus indicating inhibition by oligomers. At a 5% (w/w) addition of cellobiose, the higher initial hydrolysis rate lasted longer with a higher degree of hydrolysis. However, 10% (w/w) cellobiose addition resulted in a reduction in the final hydrolysis extent compared to the pure non-crystalline cellulose. A similar trend was reported for glucose addition for up to 10% (w/w).

Kinetics of Enzyme Hydrolysis of Lignocellulosic Biomass. The kinetics of cellulase hydrolysis of cellulose is mostly investigated on pure substrates. Enzyme hydrolysis behavior in pretreated lignocellusic biomass (in presence of other components) has not been investigated in great lengths. Recently, Kumar and Wyman (2009) conducted a study on the adsorption of cellulose on variously pretreated lignocellulosic biomass to discern its role in furthering hydrolysis. The adsorption of cellulase on Avicel and of cellulase and xylanase on corn stover solids resulting from pretreatments (such as ammonia fiber expansion (AFEX), ammonia recycled percolation (ARP), controlled pH, dilute acid, lime, and sulfur dioxide (SO_2)) were measured at 4°C. The data was then fitted into a Langmuir adsorption model, and the cellulase accessibility to cellulose was estimated based on the adsorption data for pretreated solids and lignin left after carbohydrate digestion. In all cases, the adsorption equilibrium for cellulase in their experiments was reached in less than two hours. The substrate pretreated by the controlled pH technology showed the greatest desorption upon dilution followed by solids from dilute acid and SO_2 pretreatments. The lowest desorption was reported for Avicel cellulose followed by AFEX treated corn stover solids. The authors relate that the higher cellulose accessibility for AFEX and lime pretreated solids could account for the good digestion reported in the literature for these pretreatments. Lime pretreated solids had the greatest xylanase capacity and AFEX solids the least, hinting that pretreatment pH possibly did not play a role. The 24-hour glucan hydrolysis rate had a strong relationship to cellulase adsorption capacities, but xylan hydrolysis showed no relationship to xylanase adsorption capacities. Delignification greatly enhanced the enzyme effectiveness but had a limited effect on cellulose accessibility. Reducing the acetyl content in corn stover solids significantly improved both cellulose accessibility and enzyme effectiveness.

Meunier-Goddik and Penner (1999) studied the relevance of enzyme adsorption and partitioning between the cellulose and noncellulose components (lignin) of pretreated biomasses to cellulose saccharification rates. The noncellulosic component of most interest in this context was lignin. The experimental system included three cellulose preparations differing in physical-chemical properties: a lignin-rich noncellulosic residue prepared from dilute acid-pretreated switchgrass, an acid-extracted lignin preparation, and a complete *T. reesei* cellulase preparation. The enzyme hydrolysis of cellulose was found to be dependent on both the lignin and cellulose preparations used. Cellulase adsorption to cellulose did not fulfill the strict requirements for the Langmuir model (equilibrium process and absence of a saturation effect), but the model provided a good fit to the adsorption isotherm data. Lignin-rich residue was found to bind relatively little to protein, whereas the adsorption capacity of acid-extracted lignin was greater than that of microcrystalline cellulose. The authors attribute the contrasting effects that the two lignin preparations had on cellulose hydrolysis to this difference in adsorption capacity. Relatively little protein is associated with lignin-rich residue, and thus does not appreciably diminish the amount of catalytically active cellulose-cellulase complexes. Much more protein associates with acid-extracted lignin, thus its addition to a reaction mixture is likely to significantly decrease the amount of cellulase available for hydrolysis. The noncellulosic lignacious residue, when supplemented at up to 40% (w/w) in cellulose-cellulase reaction mixtures, had little effect on rates and the extent of cellulose saccharification. The authors believed that the enzyme partitioning between cellulose and the non-cellulosic components of a pretreated feedstock is not likely to have a major impact on cellulose saccharification in typical biomass-to-ethanol processes.

Zheng et al. (2009) reported a predictive model for enzymatic hydrolysis of a creeping wild ryegrass based on assumptions like the Langmuir adsorption behavior by an enzyme system with endoglucanase/cellobiohydrolase. The authors proposed one homogeneous reaction of cellobiose-to-glucose and two heterogeneous reactions of cellulose-to-cellobiose and cellulose-to-glucose in their model development. The study also incorporated the negative role of lignin (nonproductive adsorption) using the Langmuir-type isotherm adsorption of cellulase onto lignin. Over a range of experiments (solid loading 4–12%, w/w, dry basis, enzyme concentration 15–150 FPU/g-cellulose, glucose inhibition of 30 and 60 mg/mL, and cellobiose inhibition of 10 mg/mL), their model predicted results very close to experimental results. A similar study was reported by Liao et al. (2008) with dairy manure on the kinetic study incorporating dynamic adsorption based on the Langmuir-type isotherm, enzymatic hydrolysis, and product inhibition. The effect of manure proteins on the cellulase enzyme adsorption during hydrolysis was also discussed. A first-order reaction model predicted the behaviors of enzyme adsorption, hydrolysis, and product inhibition for pretreated manure fibers in the range of substrate concentrations 10–50 g/L and enzyme loadings 7–150 FPU/g at the reaction temperature of 50 °C. An earlier study on enzyme degradation of cellulose-rich agro-forestry products (unbarked piece of pine timber and corn stalks) at lower temperatures (22 to 40 °C) had also reported the reaction rate to be first-order (Cendrowska, 1997).

10.2.3 Factors Affecting Enzyme Hydrolysis of Lignocellulosic Biomass

The factors affecting the enzymatic hydrolysis of lignocellulosic biomass could be categorized as physical properties of the substrate (composition, crystallinity, degree of polymerization, etc.), enzyme synergy (origin, composition, etc.), mass transfer (substrate adsorption, bulk and pore diffusion, etc.), and intrinsic kinetics. There is a general consensus that the most important biomass property affecting its enzymatic hydrolysis are the degree of crystallinity and the nature and amount of associated constituents (lignin, hemicelluloses etc.); both of which are structural and compositional properties of the biomass. Other structural and chemical factors include surface area and particle size, degree of polymerization, the nature, concentration and distribution of substituent groups, pore volume, and conformation and steric rigidity of the anhydroglucose units (Chang and Holtzapple, 2000; Laureano-Perez et al., 2005; Marsden et al., 1985). Yet, enzyme-related factors and processing parameters also affect the rate of cellulosic hydrolysis, and they include end-product inhibition, enzyme to substrate ratio, incubation time, temperature, pH, pretreatment types and conditions, and pretreatment byproducts, especially if acid was used during the pretreatment. Some important factors affecting enzyme hydrolysis of lignocellulosic biomass are discussed below.

Cellulose Crystallinity. Cellulose in biomass is composed of amorphous and crystalline regions, the relative ratio of which is expressed in terms of crystalline index (CrI). In an x-ray diffractogram, CrI is determined by the ratio of the crystalline peak to the valley area. The crystalline cellulose exists in the form of microfibrils, which are assemblies of several dozen 1-4-β-D-glucan chains hydrogen-bonded tightly to one another along their length (Laureano-Perez et al., 2005). These crystalline celluloses are water insoluble and resistant to depolymerization compared to the amorphous regions. The initial rapid rate of enzyme hydrolysis followed by slower and sometimes incomplete hydrolysis was proposed. This was due to the rapid hydrolysis of the amorphous constituents of the cellulosic substrates. The recalcitrance of the residual material was thought to be due to a higher inherent degree of crystallinity (Mansfield et al., 1999). Early data presented on kinetic parameters of cellulose hydrolysis by cellulase also supports this observation, even though the enzyme affinity constant for amorphous cellulose was one-tenth that for crystalline cellulose, the maximum reaction rate was 40 times for amorphous cellulose compared to crystalline cellulose (Ryu et al., 1981). These early studies, based on preferential hydrolysis rates of different cellulose regions, reported that the crystallinity index increased during cellulose hydrolysis indicating that the amorphous region was being hydrolyzed more readily. Several studies, however, found that cellulose crystallinity does not increase during enzymatic hydrolysis (Zhang and Lynd, 2004). Nonetheless, there is a common consensus about the inverse linear relationship between glucose production from various cellulosic substrates and the crystallinity index (Fan et al., 1980; Marsden et al., 1985). At the same time, crystallinity alone is insufficient to prevent significant hydrolysis if sufficient amounts of enzyme and reaction time is allowed. Mosier et al. (2005) cited a report in which the hydrolysis of microcrystalline Avicel cellulose proceeded to 80% in 6 days when incubated with 72 units of a cellulase enzyme per gram of cellulose.

Biomass Composition and Associated Constituents. Enzyme digestibility of cellulose in native biomass is very low, < 20%, unless an excess amount of enzyme is used. Lignin and hemicelluloses are the other two major cell wall components that hinder the enzyme accessibility to cellulose. Hemicellulose hydrogen bonds to cellulose microfibrils forming a network that provides the structural backbone for plant cells (Mosier et al., 2005). The degree of branching and heterogeneity in substituent groups in hemicelluloses warrant a large range of specific enzymes for the complete breakdown of hemicellulose. The presence of lignin in outer parts of the cell wall imparts further strength, but at the same time it also serves as an impediment to enzymatic accessibility to cellulose and hemicelluloses. The aims of biomass pretreatment have been to break the lignin seal of cell wall, and to reduce cellulose crystallinity via physical and chemical means. Chang and Holtzapple (2000) identified an empirical model that describes the role of biomass lignin content, acetyl contents, and the cellulose crystalline index. They found that the lignin content and crystalline index had the greatest impact on enzymatic digestibility of biomass. Lignin has been found to interfere with enzyme hydrolysis by blocking the access of cellulase enzyme to cellulose and irreversibly binding the enzymes (McMillan, 1994).

Yoshida et al. (2008) reported the effects of cellulose crystallinity, hemicellulose and lignin content on the enzymatic hydrolysis of ball mill ground *Miscanthus sinensis*, ranging from 355 microns to less than 63 microns. When hydrolyzed with commercially available cellulase and β-glucosidase, the yield of monosaccharides increased as the crystallinity of the substrate decreased. Delignification of miscanthus by sodium chlorite (incubation of 2.5 g miscanthus powder with 1 g sodium chlorite, 0.2 ml acetic acid and 150 ml deionized water at 70 ^{o}C for 1 h) improved the initial rate of hydrolysis by cellulolytic enzymes significantly resulting in a higher yield of monosaccharides. When delignified biomass was hydrolyzed with a combination of cellulase, β-glucosidase, and xylanase, the glucan conversion rate to glucose was 90.6%, and hemicellulose was hydrolyzed completely. Sewalt et al. (1997) utilized a lignin genetic engineering approach to down-regulating the enzymes involved in lignin monomer synthesis in the cell wall of tobacco stems. They reported that the enzymatic digestibility of cell walls from stem internodes was improved in the transgenic lines and was negatively correlated with the lignin concentration. Although the lignin composition was also affected, lignin concentration was the overriding factor influencing cell wall digestibility by enzymes, providing a basis for new strategies for lignin modification to improve digestibility of forages.

Biomass Pretreatment. The main objective of biomass pretreatment is to prepare the biomass for efficient enzyme hydrolysis. The biomass pretreatment could be classified into two broad categories, physical and chemical, and accordingly cause physical and chemical changes in the substrate. Physical methods, which include comminution, steam explosion and hydrothermolysis (Mosier et al., 2005), aim at altering the physical attributes such as particle size, surface area, and cellulose crystallinity, whereas chemical methods make use of chemicals (acid/alkali) to alter the chemical structure of the substrate. A combination of pretreatment methods is usually employed. A coarse or moderately fine grinding is usually followed by acid or base pretreatment that promotes cellulose hydrolysis by removing hemicelluloses or lignin. Liquid hot water pretreatment has also received considerable research attention, which utilizes water under elevated pressure (up to 400 psig) and temperatures (up to 250 °C) to dissolve hemicelluloses. This process is both helped and hindered by the cleavage of hemicellulosic *O*-acetyl and uronic acid substitutions to generate acetic and other organic acids (Mosier et al., 2005). The acids generated help catalyze formation and removal of cello-oligosaccharides. The further hydrolysis of cello-oligosaccharides to monomeric sugars, however, is possible which could degrade into aldehydes that are inhibitory to saccharifying enzymes and microbial fermentation.

The common chemical pretreatment of biomass consists of acid and alkali treatments. Acid pretreatment makes use of mineral acids, most commonly sulfuric and hydrochloric acids, which hydrolyze hemicelluloses to xylose and other sugars, and continue to break xylose down to form furfural. Degradation of sugar products is highly time dependent so most of the acid or auto hydrolysis procedures use shorter times and higher temperatures. There are mainly two types of dilute acid pretreatment processes that affect the downstream saccharification process: low solids (<10% w/w) high temperature (> 160°C) and high solids (10-40%) low temperature (<160°C) processes (Balat et al., 2008). Dilute acid (up to 2% w/w) pretreatment is done at elevated temperatures of 160-220°C for periods ranging from minutes to seconds (Mosier et al., 2005). In general, higher temperature and shorter residence time favor higher hemicelluloses (xylose) hydrolysis and better cellulose digestibility. Another chemical pretreatment is alkali treatment (like with lime and ammonia), which dissolve lignin and cause the depolymerization and cleavage of lignin-carbohydrate linkages. Since lignin is one of the key barriers to enzyme hydrolysis of biomass substrates, the removal of lignin lowers the enzyme requirement. Hemicellulose and/or lignin removal makes way for easier and more cost effective hydrolysis of cellulose into glucose. Chemical pretreatments, while necessary to enhance enzyme accessibility to cellulose fibrils, can also be detrimental to the very enzyme system due to their degradation products like furfurals. A detailed discussion of mechanism of alkali treatment on lignin removal, and of acid treatment on hemicelluloses hydrolysis could be found in Marsden et al. (1985).

End-product Inhibition. In many enzymatic reactions, the accumulation of end products inhibits the rate of the forward reaction. Cellulase activity is inhibited by cellobiose and to a lesser extent by glucose (; Lee et al., 1980; Sun and Cheng, 2002). Glucose inhibition is much milder - 40% inhibition by 30% glucose concentration in comparison to cellobiose inhibition which is quite severe, up to 50% inhibition even at very low concentrations (Ghose and Das, 1971). It is thus understandable that product inhibition does not significantly affect the initial rate of hydrolysis, but the final reaction rates. The cellobiose inhibition of cellulase was reported to be competitive, and the degree of inhibition depended on the concentrations of the inhibitor and substrate, and their relative affinities to active sites. Lee et al. (1980) further cited other researchers who reported that the product inhibition in the cellulose-cellulase system was large when the large soluble oligomer (cellopentose) was present and decreased rapidly with a decrease in the size of product molecules. Several strategies can be adapted to minimize product inhibition, including the use of high enzyme concentrations, the supplementation of β-glucosidase during hydrolysis, the removal of sugars during hydrolysis by ultrafiltration, simultaneous saccharification and fermentation (SSF) which is meant to convert glucose into ethanol, and consolidated bioprocessing, where enzyme production, hydrolysis and fermentation take place in a single step.

10.2.4 Enzyme Systems for Hemicellulose Degradation

Hemicelluloses are heterogeneous polymers of pentoses (xylose, arabinose), hexoses (mannose, glucose, galactose), and sugar acids. Various agricultural biomass, e.g. wheat and rice straws, corn stover, corn fiber, and sugarcane bagasse, contain about 20-40% (w/w) hemicelluloses. With the current focus being on ethanol production from cellulose in lignocellulosic biomass, hemicelluloses and lignins are treated as hindrances to the overall process. In an integrated biorefinery, however, producing fuel, chemicals, and biomolecules from lignocellulosic biomass, hemicellulose hydrolysis products are recovered to produce value-added products, such as xylitol, 2,3-butanediol, lactic acid, gum, etc. (Saha, 2003). Also, if microorganisms capable of fermenting mixed sugar streams (hexoses and pentoses) were to be developed, the importance of enzyme systems that hydrolyze hemicelluloses would be immense.

Depending on the source, woody biomass hemicelluloses contain mostly xylans (e.g. hard wood) or glucomannans (e.g. softwood). Xylans of many plant materials are heteropolysaccharides with homopolymeric backbone chains of 1-4 linked β-D-xylopyranose units. Beside xylose, xylans may contain side chains of arabinose, glucuronic acid or 4-*O*-methyl ether, acetic, ferulic, and *p*-coumaric acids. The total biodegradation of xylan would require a cocktail of endo-β-1,4-xylanase, β-xylosidase, and accessory enzymes to hydrolyze side chains like α-L-arabinofuranosidase, α-glucoronidase, ferulic acid esterase, acetylxylan esterase, and *p*-coumaric acid esterase. Endo-xylanase hydrolyzes mainly interior β-1,4-xylose linkages of the xylan backbone that works to reduce the viscosity and degree of polymerization. The mode of action of different xylanase and the hydrolysis products vary according to the source of the enzyme. Xylanase from *Aspergillus niger* hydrolyzed soluble xylan more rapidly than insoluble branched xylan (Puls and Poutanen, 1989). Exo-xylanase hydrolyzes β-1,4-xylose linkages releasing xylobiose. β-xylosidase releases xylose from xylobiose and short chain xylooligosaccharides. The side chains and ester links are hydrolyzed by respective esterase. A detailed discussion on enzyme systems for hydrolysis of xylans can be found in Puls and Poutanen (1989), Wang et al. (1998), and Saha (2003).

Many organisms, such as *Penicillium capsulatum* and *Talaronyces emersonii* possess complete xylan degrading enzyme systems (Saha, 2003). There have been reports of significant synergistic interaction among endo-xylanase, β-xylosidase, α-arabinofuranosidase, and acetylxylan esterase of the thermophilic actinomycete *Thermomonosopora fusca*. The side chains must be cleaved before the xylan backbone can be completely hydrolyzed, so an interaction between depolymerizing and side-group cleaving enzymes is beneficial and has been verified with acylated xylan as the substrate (Poutanen and Puls, 1989). Complete degradation of xylan takes higher importance and would be essential for the efficient and cost effective conversion of lignocellulosic material to fuel ethanol. Depending on the source and the extent of hemicellulose degradation, the sugar mixture may contain any combination of xylose, arabinose, glucose, galactose, mannose, fucose and rhamnose. Although traditional *Saccharomyces cerevisiae* ferments glucose to ethanol rapidly and efficiently, it cannot ferment other sugars such as xylose and arabinose to ethanol (Saha, 2003). The yeasts *Pachysolen tannophilus, Pichia stipitis*, and *Candida shehate* have the capability to ferment xylose to ethanol however, their commercial usage is limited by their low ethanol tolerance, slow fermentation rates, and sensitivity to other pretreatments and the generation of certain hydrolysis end-products. In fact, improving the performance of C5- and C6- degrading microorganisms (e.g. increased fermentation rate), improved tolerance and sensitivity is one of the key areas of research identified in the U.S. DOE's research agenda (US DOE, 2006).

10.2.5 Cost of Enzymes

The cost of enzyme accounts for a significant part of cellulosic ethanol production costs. β-glucosidase supplementation is essential since commercial cellulases often lack this enzyme. β-glucosidase reduces cellobiose to glucose thus minimizing the accumulation of cellobiose which in aqueous phase, inhibits endogluconase and cellobiohydrolase enzymes. This subsequently increases the costs of enzyme. Nevertheless, the cost of enzyme has gone down by nearly 20-fold in recent years. Recent US DOE grants to two major enzyme producers–Genencor and Novozyme- was a major push toward reducing the cost of enzymes. In recent years, enzyme costs have gone down from $0.50 per gallon of ethanol to $0.10-$0.30 per gallon of ethanol (Hettenhaus and Ashworth, 2002). Another study reported enzyme costs of around $0.25/gallon for lignocellulosic feedstock in California (Williams et al., 2007). Genencor, a major enzyme developer and producer, projects that the cost of enzyme would go down to $0.30 per gallon from current cost of $1.00/gallon (Steele, 2009). For the economic viability of cellulosic ethanol, the enzyme cost should be $0.045-0.09/gal. For corn-based ethanol plants, the enzyme cost is just around $0.03-0.04/gal. Thus, further cost reductions are still needed for the economic viability of lignocellulose-ethanol plants. Enzyme recycling may be adopted to hydrolyze multiple batches of substrate and may reduce the overall enzyme cost. One way to achieve this is by immobilizing enzymes in or onto an inert carrier. Consolidated bioprocessing (CBP) is another approach of lowering the enzyme cost. One study projects an enzyme cost of $0.042/gal for CBP compared to $0.189/gal for simultaneous saccharification and co-fermentation (Zhang and Lynd, 2008).

10.3 Major Advances in Enzymatic Hydrolysis of Lignocellulosic Biomass

Significant advances have been made in the enzyme hydrolysis of lignoccellulosic biomass with aims of cutting down the overall costs of hydrolysis, improving the enzymes' activity and selectivity, and lowering inhibitory effects of the substrate and product. Some of the major advances are outlined here.

10.3.1 Enzyme Immobilization

Immobilization makes it possible to reuse the enzymes many times, thus, bringing the cost of enzymes down over a longer period of operation (Sheldon, 2007). Immobilization also improves enzyme activity, selectivity, and stability (Mateo et al., 2007), including thermal stability (Elnashar et al., 2008; Paljevac et al., 2007). Basically there are three methods of enzyme immobilization: binding to a carrier support, entrapment or encapsulation, and cross linking. The support binding can be physical (hydrophobic and van der Waal interactions), ionic or covalent in nature; the latter being the strongest. While relatively simple in manufacturing, immobilization has the disadvantage of both carrier and enzyme being unusable if the enzyme is inactivated during the process. Entrapment requires the synthesis of a polymeric network (gel lattice of synthetic polymer, silica sol-gel, or a membrane device) in the presence of the enzyme. Although physical restraints like pore size could keep the enzyme entrapped, additional covalent attachment to the network is required to prevent enzyme leaching. A relatively newer concept, cross-linking of enzyme involves the preparation of enzyme aggregates or crystals using a bifunctional reagent to prepare micro/nanoparticles (Sheldon, 2007) without the need of carriers. The addition of salts or water-miscible organic solvents or non-ionic polymers to aqueous a solution of protein (enzyme) leads to their precipitation as physical aggregates of protein molecules, held together by non-covalent bonding, without the perturbation of their tertiary structure; that is, without denaturation. Subsequent cross-linking of these enzyme precipitates would render them permanently insoluble, while maintaining their native structure and subsequently, their catalytic activity.

10.3.2 Enzyme Recycling

This involves recovery and recycling of cellulase enzymes after a batch of enzyme hydrolysis. During the enzymatic hydrolysis of lignocellulosic substrates, enzymes are adsorbed onto the substrate. As the hydrolysis proceeds, free enzymes are released into solution which can be recovered and reused for the subsequent hydrolysis. Thus with the treatment of multiple batches of substrates with the same batch of enzymes, there is possibility of significant cost reduction. Various factors affect the efficiency of the recycled enzyme, e.g. degree of pretreatment, fractionation and hydrolysis (Gregg and Saddler, 1996). The recovery and recycling of cellulase from the supernatant or substrate greatly depend on the time of recovery process and cost of chemicals used. Cellulase recovery from the substrate may be impeded by the structural composition of the substrate. For example Lu et al. (2002) conducted digestion studies with Douglas fir substrates and reported limited enzyme recovery from high lignin substrates in comparison to substrates with low lignin.

The extent of savings in enzyme costs depends on the fraction and activity of enzymes recovered from the system. At a higher recovery rate, the enzymes can be recycled multiple times, and hence, the cost associated with enzymes is greatly reduced. Steele et al. (2005) reported a 15 to 50% cost reduction as the enzyme recovery increased from 60 to 90% at enzyme loadings of 15 FPU/g glucan. However, it is also important to consider additional capital and operational costs associated with recycling that will impact the overall processing cost. The enzyme recovery and recycling process needs to be optimized for appropriate enzyme activity levels and recycling frequencies.

10.3.3 In-situ Enzyme Production: Biological Pretreatment

Alternative strategies to improve lignocellulosic biomass hydrolysis, and subsequently reduce the process cost, needs to be explored. One such potential process is *in situ* enzyme production. By growing mono- or mixed cultures of different fungi, yeasts, or bacteria on cellulosic substrates, it is possible to induce the secretion of extracellular hydrolytic enzymes to be used in the hydrolysis of lignocellulosic biomass. A consortia of cellulose- and hemicellulose- degrading enzymes act synergistically to biodegrade a complex polysaccharide matrix of plant cell wall, releasing monomeric sugars required for ethanol fermentation and other bio-based products. *In situ* enzyme production using good strains of fungi could minimizes the overall cost of biomass hydrolysis due to higher enzyme yields. A higher yield of cellulase, from 250 to 430 IU/g cellulose, was reported by solid substrate fermentation of mutant *Trichoderma reesei* QMY-1 in wheat straw (Chahal, 1985). The celluase titer (IU/ml) was greater (~ 72 %) in the case of solid substrate fermentation compared to liquid fermentation. Tengerdy (1996) compared the production cost of cellulase in submerged and solid-substrate fermentation and concluded that the crude enzyme from solid substrate fermentation can be directly used in lignocellulose hydrolysis. Recent reports of *in situ* cellulolytic enzymes production include using industrial co-product like corn fiber to grow wood rot fungi and *T. reesei* for bio-conversion and subsequent fermentation of sugar to ethanol (Shrestha et al., 2008 & 2009).

10.3.4 Improvements in Saccharification and Fermentation Processes

Simultaneous saccharification and fermentation (SSF), and consolidated bioprocessing (CBP) are two different strategies that aim to reduce the steps required to produce ethanol from lignocellulosic biomasses. In SSF, glucose produced by hydrolyzing enzymes is consumed immediately by the fermenting microorganism present in the culture (Taherzadeh and Kairimi, 2007). Apart from reducing the cost, the inhibitory effects of products such as cellobiose and glucose on enzyme activity are minimized in this scheme. An important consideration in SSF is to have the optimum conditions, especially pH and temperature for the enzymatic hydrolysis and fermentation as close as possible. Nevertheless, the difference in optimum temperatures of hydrolyzing enzymes and fermenting microorganisms is still a drawback of SSF. The use of thermo-tolerant yeast and bacteria e.g. *Candida acidothermophilum* and *Kluyveromyces marxianus* have been proposed for the use in SSF since they operate in conditions close to hydrolysis temperatures (Hong et al, 2007; Krishna et al., 2001). Yet another improvement in the SSF scheme is the cofermentation of C5 sugars into ethanol, instead of just C6 sugars known as simultaneous saccharification and co-fermentation (SSCF).

Consolidated bioprocessing takes the integration of hydrolysis and fermentation one step closer by consolidating the production of the enzyme being used in hydrolysis to the fermentation process. Ethanol, together with all of the required enzymes is produced in a single bioreactor by a single microorganism's community (Taherzadeh and Kairimi, 2007). The four steps in lignocellulosic ethanol production, that is, the production of saccharolytic enzymes (cellulases and hemicellulases); the hydrolysis of carbohydrate components present in pretreated biomass to sugars; the fermentation of hexose and the fermentation of pentose sugars occur in CBP configuration (Lynd et al, 2005). The limited tolerance of enzymes toward ethanol and temperature tolerance of fermentation microorganisms, low yields of fermentation and significant generation of by-products like lactic and acetic acids production, are some of the key challenges to this concept. Mascoma Corporation based in Lebanon, New Hampshireis currently pursuing the commercialization of CBP technology.

10.4 Challenges and Future Research Direction in Enzyme Hydrolysis

As have been discussed in the previous sections, the effectiveness of enzyme hydrolysis is governed by the degree of disruption of lignin-cellulose-hemicellulose interactions during pretreatment with minimal generation of inhibitory soluble products. Thermal and chemical (acid/alkaline) pretreatments make the cellulose more readily accessible to enzymes by opening up cleavage sites. Such pretreatments also cause the redeposition of thermally labile lignin that hinders the access of cellulase enzymes to cellulose. One recent study revealed up to a 20% loss of enzymatic saccharification efficacy due to the surface deposition of lignin droplets on thermal/acid-pretreated biomass (Selig et al., 2007). The presence of lignin interferes with the enzyme hydrolysis by either blocking the enzymes' access to cellulose or irreversibly binding the enzymes. Extreme chemical pretreatment is often needed to break lignin-cellulose-hemicellulose interactions for efficient enzyme digestion. However, such pretreatment also generate soluble inhibitory products (such as acetic acid, furfural, hydroxymethylfurfural, formic acid, phenolic compounds, etc.) (Hodge et al., 2008). Cellulose hydrolysis is also subjected to product inhibition. For example, the product cellobiose inhibits the enzymes endogluconase and cellobiohydrolase while an accumulation of glucose inhibits the enzyme β-glucosidase. End-product inhibition can be alleviated by combining enzyme hydrolysis and fermentation known as simultaneous saccharification and fermentation (SSF). Commercial grain-to-ethanol processes adopt SSF to minimize product inhibition. Some of the major barriers at present to the commercialization of enzyme hydrolysis of lignocellulosic biomass are:

- Substrate and product inhibitions
- Thermal inactivation
- Low product yield
- High cost of enzymes

10.5 Summary

The grain-based ethanol industry produced 7.8 billion gallon in year 2007, however, this is not sustainable and is estimated to plateau at 15 billion gallons. To meet future energy challenges, including a congressional mandate of 36 billion gallons of annual renewable fuel use by 2022, lignocellulosic resources are expected to play a key role in producing more bioenergy and other chemicals. For this, enzymatic hydrolysis of lignocellulosic into simple sugars has to be realized efficiently and cost-effectively. This chapter reviewed important aspects of enzyme hydrolysis of lignocellulosic biomass, including their composition with emphasis to agricultural residues, enzyme systems to degrade cellulose and hemicelluloses and mechanisms of doing so, and various factors that affect the enzyme hydrolysis. Importantfuture research areas in lignocellulosic hydrolysis have also been identified and discussed.

10.6 References

Balat, M., Balat, H., and Oz, C. (2008). "Progress in bioethanol processing." *Progress in Energy and Combustion Science*, 34, 551–573.

Caminal, G., López-Santin, J., and Sola, C. (1985). "Kinetic modeling of the enzymatic hydrolysis of pretreated cellulose." *Biotechnology and Bioengineering*, 27(9),1282–1290.

Cendrowska, A. (1997). "Hydrolysis kinetics of cellulose of forest and agricultural biomass." *Holz als Rob- und Werkstoff*, 55, 195 – 196.

Chahal, D.S. (1985). "Solid-state fermentation with *Trichoderma reesei* for cellulase production." *Applied and Environmental Microbiology*, 49(1), 205 – 210.

Chang, V. S., and Holtzapple, M.T. (2000). "Fundamental factors affecting biomass enzyme reactivity." *Applied Biochemistry and Biotechnology*, 84-86, 5 – 37.

Elnashar, M. M., Yassin, M. A., and Kahil, T. (2008). "Novel thermally and mechanically stable hydrogel for enzyme immobilization of penicillin G acylase via covalent technique." *Journal of Applied Polymer Science*, 109, 4105 – 4111.

Fan L.T., Lee Y.H., and Beardmore D.H. (1980). "Mechanism of the enzymatic-hydrolysis of cellulose - effects of major structural features of cellulose on enzymatic-hydrolysis." *Biotechnology and Bioengineering*, 22(1), 177 – 199.

Foust, T. D., Ibsen, K. N., Dyton, D.C., Hess, J. R., and Kenney, K.E. (2008). "The biorefinery." In: Himmel, M. (Ed.) *Biomass Recalcitrance: Deconstructing the Plant Cell Wall for Bioenergy.* United Kingdom: Blackwell Publishing, pp. 7 – 37.

Fujii, M., and Shimizu, M. (1986). "Synergism of endoenzyme and exoenzyme on hydrolysis of soluble cellulose derivatives." *Biotechnology and Bioengineering*, 28(6), 878 – 882.

Ghose, T.K., and Das, K. (1971). "A simplified kinetic approach to cellulose-cellulase system." In: Ghose, T.K. and Fiechter, A. (Ed.) *Advances in Biochemical Engineering*. Berlin: Springer-Verlag, pp. 55 – 76.

Gregg, D.J., and Saddler, J.N. (1996). "Factors affecting cellulose hydrolysis and the potential of enzyme recycle to enhance the efficiency of an integrated wood to ethanol process." *Biotechnology and Bioengineering*, 51, 375 – 383.

Gupta, R, and Lee, Y.Y. (2009). "Mechanism of cellulase reaction on pure cellulosic substrates." *Biotechnology and Bioengineering*, 102 (6), 1570 – 1581.

Hettenhaus, J., Wooley, R., and Ashworth, J. (2002). *Sugar Platform Colloquies*. NREL/SR-510-31970. Available at http://permanent.access.gpo.gov/lps31319/www.ott.doe.gov/biofuels/pdfs/sugar_platform.pdf (accessed May, 2009).

Hodge, D.B., Karim, M.N., Schell, D.J., and McMillan, J.D. (2008). "Soluble and insoluble solids contributions to high-solids enzymatic hydrolysis of lignocellulose." *Bioresource Technology*, 99(18), 8940 – 8948.

Hong, J., Wang, Y., Kumagai, H., and Tamaki, H. (2007). "Construction of thermotolerant yeast expressing thermostable cellulase genes." *Journal of Biotechnology*, 130(2): 114 – 123.

Howell, J.A., and Stuck, J.D. (1975). "Kinetics of solka floc cellulose hydrolysis by *Trichoderma viride* cellulase." *Biotechnology and Bioengineering*, 17(6), 873 – 893.

International Energy Agency (IEA). (2008). *From 1st– to 2nd - generation biofuel technologies*. Available at http://www.iea.org/textbase/papers/2008/2nd_Biofuel_Gen_Exec_Sum.pdf (accessed May, 2009).

Krishna S.H., Reddy T.J., and Chowdary G.V. (2001). "Simultaneous saccharification and fermentation of lignocellulosic wastes to ethanol using a thermotolerant yeast." *Bioresource Technology*, 77(2), 193 – 196.

Kumar, R., and Wyman, C.E. (2009). "Cellulase adsorption and relationship to features of corn stover solids produced by leading pretreatments." *Biotechnology and Bioengineering*, 103(2), 252 – 267.

Laureano-Perez, L., Teymouri, F., Alizadeh, H., and Dale, B.E. (2005). "Understanding factors that limit enzyme hydrolysis of biomass." *Applied Biochemistry and Biotechnology*, 121-124, 1081 – 1099.

Lee, Y. H., Fan, L. T., and Fan, L.S. (1980). "Properties and mode of action of cellulase." *Advances in Biochemical Engineering*, 17, 132 – 168.

Lee, I., Evans, B. R., and Woodward, J. (2000). "The mechanism of cellulase action on cotton fibers: evidence from atomic force microscopy." *Ultramicroscopy*, 82, 213 – 221.

Liao, W., Liu, Y., Wen, Z. Frear, C., and Chen, S. (2008). "Kinetic modeling of enzymatic hydrolysis of cellulose in differently pretreated fibers from dairy manure." *Biotechnology and Bioengineering*, 101(3), 441 – 451.

Lin, Y., and Tanaka, S. (2006). "Ethanol fermentation from biomass resources: current state and prospects." *Applied Microbiology and Biotechnology*, 69(6), 627 – 642.

Lu, Y., Yang, B., Gregg, D., Saddler, J.N., and Mansfield, S.D. (2002). "Cellulase adsorption and an evaluation of enzyme recycle during hydrolysis of steam-exploded softwood residues." *Applied Biochemistry and Biotechnology*, 98 – 100, 641 – 654.

Lynd, L.R., van Zyl W.H., McBride J.E., and Laser, M. (2005). "Consolidated bioprocessing of cellulosic biomass: an update." *Current Opinion in Biotechnology*, 16(5), 577 – 583.

Maija, T., Marja-Leena, N., Markus, L., and Lisa, V. (2003). "Cellulases in food processing." In: Whitaker, J.R., Voragen, A.G.J., and Wong, D.W.S. (Ed.) *Handbook of Food Enzymology*. New York: Marcel Dekker.

Mansfield, S. D., Mooney, C., and Saddler, J. N. (1999). "Substrate and enzyme characteristics that limit cellulose hydrolysis." *Biotechnology Progress,* 15, 804 – 816.

Marsden, W. L., Gray, P. P., and Mandels, M. (1985). "Enzymatic hydrolysis of cellulose in lignocellulosic materials." *Critical Reviews in Biotechnology*, 3(3), 235 – 276.

Mateo, C., Palomo, J. M., Fernandez-Lorente, G., Guisan, J. M., and Fernandez-Lafuente, R. (2007). "Improvement of enzyme activity, stability and selectivity via immobilization techniques." *Enzyme and Microbial Technology,* 40,1451 – 1463.

McMillan, J. D.(1994). "Pretreatment of lignocellulosic biomass." In: Himmel, M. E., Baker, J.O. and Overend, R. P. (Ed.) *Enzymatic Conversion of Biomass for Fuels Production*. Washington, DC: American Symposium Series 566, pp 292 – 324.

Medve, J., Stahlberg, J., and Tjerneld, F. (1994). "Adsorption and synergism of cellobiohydrolase I and II of *Trichoderma reesei* during hydrolysis of microcrystalline cellulose." *Biotechnology and Bioengineering*, 44(9), 1064 – 1073.

Meunier-Goddik, L., and Penner, M.H. (1999). "Enzyme-catalyzed saccharification of model celluloses in the presence of lignacious residues." *Journal of Agricultural and Food Chemistry*, 47(1), 346 – 351.

Mosier, N., Wyman, C., Dale, B., Elander, R., Lee, Y.Y., Holtzapple, M., and Ladisch, M. (2005). "Features of promising technologies for pretreatment of lignocellulosic biomass." *Bioresource Technology*, 96(6), 673 – 686.

Paljevac, M., Primozic, M., Habulin, M., Novak, Z., and Knez, Z. (2007). "Hydrolysis of carboxymethyl cellulose catalyzed by cellulase immobilized on silica gels at low and high pressures." *Journal of Supercritical Fluids*, 43, 74 – 80.

Peri, S., Karra, S., Lee, Y.Y., and Karim, M. N. (2007). "Modeling intrinsic kinetics of enzymatic cellulose hydrolysis." *Biotechnology Progress*, 23, 626 – 637.

Philippidis, G. P. (1994). "Cellulase production technology." In: Himmel, M. E., Baker, J.O. and Overend, R. P. (Ed.) *Enzymatic Conversion of Biomass for Fuels Production.* Washington, DC.: American Symposium Series 566, pp. 188 – 217.

Poutanen, K., and Puls, J. (1989). "The xylanolytic enzyme system of *Trichoderma reesei*." In: Lewis, G. and Paice, M. (Ed.) *Biogenesis and biodegradation of plant cell wall polymers.* Washington, D.C.: American Chemical Society, pp. 630 – 640.

Puls, J., and Poutanen, K. (1989). "Mechanisms of enzymic hydrolysisi of hemicelluloses (xylans) and procedures for determination of the enzyme activities involved" In: Coughlan, M.P. (Ed.) *Enzyme Systems for Lignocellulose Degradation.* Elsevier Applied Science, pp. 151 – 165.

Rabinovich, M. L., Melnick, M. S., and Bolobova, A.V. (2002). "The structure and mechanism of action of cellulolytic enzymes." *Biochemistry (Moscow),* 67(8), 850 – 871.

Ryu, D., Lee, S. B., and Tassinari, T. (1981). "Effect of crystallinity of cellulose on enzymatic hydrolysis kinetics." Abstracts of papers of the American Chemical Society, 182 (August): 58-MICR.

Saha, B.C. (2003). "Hemicellulose bioconversion." *Journal of Industrial Microbiology and Biotechnology*, 30, 279 – 291.

Selig, M.J., Viamajala, S., Decker, S.R., Tucker, M.P., Himmel, M.E., and Vinzant, T.B. (2007). "Deposition of lignin droplets produced during dilute acid pretreatment of maize stems retards enzymatic hydrolysis of cellulose." *Biotechnology Progress*, 23, 1333 – 1339.

Sewalt, V.J. H., Ni, W., Jung, H.G., and Dixon, R.A. (1997). "Lignin impact on fiber degradation: increased enzymatic digestibility of genetically engineered tobacco (*Nicotiana tabacum*) stems reduced in lignin content." *Journal of Agricultural and Food Chemistry*, 45, 1977 – 1983.

Sheldon, R. A. (2007). "Enzyme Immobilization: The quest for optimum performance." *Advances in Synthetic Catalysts*, 349,1289 – 1307.

Shen, Y., and Wang, L. M. (2004). "Kinetics of the cellulase catalyzed hydrolysis of cellulose fibers." *Textile Research Journal*, 74(6), 539 – 545.

Shrestha, P., Rasmussen, M., Khanal, S.K., Pometto III, A.L., and Van Leeuwen, J. (2008). "Solid-substrate fermentation of corn fiber by *Phanerochaete chrysosporium* and subsequent fermentation of hydrolyzate into ethanol." *Journal of Agricultural and Food Chemistry*, 56(11), 3918 – 3924.

Shrestha, P., Khanal, S.K., Pometto III, A.L., and Van Leeuwen, J. (2009). "Enzyme production by wood-rot and soft-rot fungi cultivated on corn fiber followed by simultaneous saccharification and fermentation." *Journal of Agricultural and Food Chemistry*, 57(10), 4156 – 4161.

Steele, L. (2009). "Advancing enzymatic hydrolysis of lignocellulosic biomass." Extracted from technical presentation by Genencor in the *Sixth World Congress on Industrial Biotechnology and Bioprocessing*. Jul 19 -22, 2009, Montreal Canada.

Steele, B., R.S., Nghiem, J., Stowers, M. (2005). "Enzyme recovery and recycling following hydrolysis of ammonia fiber explosion-treated corn stover." *Applied Biochemistry and Biotechnology,* 121-124, 901 – 910.

Suga, K., Van Dedem, G., and Moo-Young, M. (1975). "Degradation of polysaccharides by endo and exo enzymes: A theoretical analysis." *Biotechnology and Bioengineering*, 17(3), 433 – 439.

Sun, Y., and Cheng, J. (2002). "Hydrolysis of lignocellulosic materials for ethanol production: a review." *Bioresource Technology*, 83, 1 – 11.

Taherzadeh M.J., and Karimi, K. (2007). "Enzyme-based hydrolysis processes for ethanol from lignocellulosic materials: a review." *Bio-Resources,* 2(4), 707 – 738.

Tengerdy, R.P. (1996). "Cellulase production by solid substrate fermentation." *Journal of scientific and industrial research,* 55 (5-6), 313 – 316.

US DOE. (2006). *Breaking the biological barriers to cellulosic ethanol: a joint research Agenda*. A research roadmap resulting from the U.S. Department of Energy's Biomass to Biofuels Workshop, Dec 7-9, 2005, Rockville, Maryland.

USDA and USDOE Joint Report. (2005). *A billion-ton feed stock supply for bioenergy and bioproducts industry: technical feasibility of annually supplying 1 billion dry tons of biomass*. Joint Report-U.S. Department of Agriculture and U.S. Department of Energy.

Van Dyke, B. H. (1972). "Enzymatic hydrolysis of cellulose - a kinetic study." (Ph.D. thesis, Massachusetts Institute of Technology).

Wang, C.H., Ye, H.L., Hong, H.W., and Gao, H.H. (1998). "Xylanase production and its application in degradation of hemicellulose materials." In proceedings of *The 7th International Conference on Biotechnology in the Pulp and Paper Industry* - Poster Presentations Vol. C, pp. C65 – C67, Canadian Pulp & Paper Assoc.

Watanabe, H., Noda, H. , Tokuda, and Lo, N. (1998). "A cellulase gene of termite origin." *Nature,* 394, 330 – 331.

Weimer, P.J. (1992). "Cellulose degradation by ruminal microorganisms." *Critical Reviews in Microbiology,* 12(3), 189 – 223.

White, A.R., and Brown, R. M. Jr. (1981). "Enzymatic hydrolysis of cellulose: Visual characterization of the process." *Proceedings of National Academy of Science*, 78(2),1047 – 1051.

Williams, R.B., Jenkins, B. M., and Gildart, M.C. (2007). "Ethanol production potential and costs from lignocellolosic resources in California." 15th European Biomass Conference and Exhibition, Berlin, Germany.

Wood, T.M., Wilson, C.A., McCrae, S.I., and Joblin, K.N. (1986). "A highly active extracellular cellulase from the anaerobic rumen fungus *Neocallismastix frontalis." FEMS Microbiology Letters,* 34(1), 37 – 40.

Wood, T.M. (1989). "Mechanism of cellulose degradation by enzymes from aerobic and anaerobic fungi." In: Coughlan, M.P. (Ed.) *Enzyme Systems for Lignocellulose Degradation.* Elsevier Applied Science, pp. 17 – 35.

Wyman, C.E. (1999). "Biomass ethanol: technical progress, opportunities, and commercial challenges." *Annual Review of Energy and the Environment*, 24, 189 – 226.

Yoshida, M., Liu, Y., Uchida, S., Kawarada, K., Ukagami, Y., Chinose, H., Kaneko, S., and Fukuda, K. (2008). "Effects of cellulose crystallinity, hemicellulose, and lignin on the enzymatic hydrolysis of *Miscanthus sinensis* to monosaccharides." *Bioscience Biotechnology and Biochemistry*, 72 (3), 805 – 810.

Zhang, Y-H. P., and Lynd, L.R. (2008). "New generation biomass conversion: consolidated bioprocessing." In: Himmel, M. (Ed.) *Biomass Recalcitrance: Deconstructing the Plant Cell Wall for Bioenergy*, United Kingdom: Blackwell Publishing, pp. 480 – 494.

Zhang, Y-H. P., and Lynd, L.R. (2004). "Towards and aggregated understanding of enzymatic hydrolysis of cellulose: non-complexed cellulose systems." *Biotechnology and Bioengineering*, 88, 797 – 824.

Zheng, Y., Zhongli P., Ruihong, Z., and Jenkins, BM. (2009). "Kinetic modeling for enzymatic hydrolysis of pretreated creeping wild ryegrass." *Biotechnology and Bioengineering*, 102(6), 1558 – 1569.

CHAPTER 11

Syngas Fermentation to Ethanol: Challenges and Opportunities

Randy S. Lewis, Douglas R. Tree, Peng Hu, and Allyson Frankman

11.1 Introduction

Alternative fuels seek to improve national security, decrease the trade deficit, build the economy, supply a renewable fuel source, and provide a smaller environmental footprint relative to conventional fossil fuels. The US Energy Information Administration (EIA) reports that despite being the world's third largest producer of crude oil, the United States imports more than half of their yearly consumption with approximately one-third of the imports coming from the Persian Gulf (U.S. Energy Information Administration, 2008a). Fossil fuels consist of 98% of all transportation fuels which are primarily gasoline, diesel fuel, and natural gas. Petroleum products alone amounted to a total national consumption of 20.7 million barrels per day in 2007 (U.S. Energy Information Administration, 2008b).

The above facts have led many to suggest that the United States' national security is jeopardized by the large dependence on oil imports. This line of reasoning has also suggested that protecting foreign oil markets has created large hidden costs to petroleum usage by the necessity of a military presence in unstable areas (Sneller, 2007). Many also contend that the current rate of consumption is not sustainable for the long term future, and that an alternative source for transportation fuels needs to be found. Finally, many proponents of alternative fuels suggest that fossil fuels cause unacceptably large amounts of emissions in the form of greenhouse gases and aromatic hydrocarbons (Sneller, 2007). One proposed solution to these concerns is fuel ethanol from renewable resources. As of the year 2000, fuel ethanol comprised 1.2% of the transportation fuels consumed in the United States, mainly as an oxygenate additive to gasoline (Yacobucci and Womach, 2000). Ethanol production cost is reported to be around $1.00 per gallon (before subsidy) but has a government subsidy $0.51 per gallon (Curtis, 2008). In comparison, gasoline has an approximate production cost of $0.50 per gallon (Yacobucci and Womach, 2000).

From an environmental standpoint, fuel ethanol is predicted to reduce the environmental impact compared to gasoline by reducing greenhouse emissions (CO_2, CH_4, N_2O), carbon monoxide (CO), and volatile organic compounds (VOCs). A

2007 study on the greenhouse gas emissions for several types of corn ethanol plants concluded that fuel ethanol from corn could range from a 3% increase in greenhouse gas production to a 28% decrease depending on the type of process. If fuel ethanol were derived from other sources than corn grain, greenhouse gas emissions could be reduced by as much as 86% (Wang et al., 2007). The reasons for greenhouse gas reductions are attributed to the uptake of carbon in biomass, agricultural soil carbon sequestration, and co-generated electricity in ethanol plants (Brinkman et al., 2005). While providing a reduction in greenhouse gases, reports are varying to the other potential environmental benefits of fuel ethanol. Brinkman predicts that use of fuel ethanol would result in a net increase in NO_x, VOCs and PM_{10} emissions in an attempt at a complete "well-to-wheel" study (Brinkman et al., 2005). Other sources indicate a decrease in CO, VOCs and PM_{10} emissions, but do not dispute the NO_x emissions claims (Sneller, 2007).

Ethanol's ability to reduce dependence on fossil fuels is contingent on whether ethanol can reasonably assume a significant portion of the nation's transportation fuel, and whether ethanol is net energy positive. With US oil imports above 57% of oil consumption, projections show that oil imports will grow to 68% in 2025 (Renewable Fuels Association, 2003). Ethanol production from biomass has a potential production of 130 billion gallons per year, which is 16 times higher than the current ethanol production in the world and could replace 32% of the global gasoline consumption (Kim and Dale, 2004). Thus, ethanol cannot completely reduce the dependence on foreign oil although it can substantially reduce imports. One of the primary reasons for the slow acceptance of biomass to ethanol conversion is the lack of a reliable and sustainable lignocellulosic supply infrastructure and these issues must be resolved in order to provide a consistent supply of high quality, low cost biomass (US Department of Energy, 2003).

General concerns about ethanol include the energy used for production, the energy content and transportation of ethanol, and the impact on the food supply. The majority of recent studies conclude that ethanol production from corn has a ratio of fossil energy in to total energy out of approximately 1.3, despite several studies indicating the opposite (Wang, 2007). The low, and sometimes controversial, energy ratio is likely not high enough to develop a commercial platform on corn ethanol alone. However, it is estimated that ethanol from lignocellulosic materials, such as switchgrass, can have an energy ratio around 4.5 or higher (McLaughlin and Walsh, 1998). With regards to energy content, the heating value of ethanol is 76,000 Btu/gallon as compared to 115,000 Btu/gallon for gasoline. Thus, more ethanol is required to travel the same distance although some argue that there is statistically no mileage difference due to a higher combustion efficiency of ethanol as compared to gasoline (US Department of Energy, 1999). From an infrastructure standpoint, ethanol cannot be transported in current gasoline pipelines due to corrosion concerns, so it likely will have to be manufactured locally (Yacobucci and Womach, 2000).

Finally, the sustainability of ethanol has been questioned based on concerns over rising food prices attributed to corn ethanol and questions about sustainable land use. United States corn consumption is divided between food (12%), fuel (19%), animal feed (50%), and exports (19%) (Sneller and Durante, 2007). Thus, ethanol is a major source of demand for corn and can displace both food production and animal feed demand. However two facts make food prices not likely to be strongly dependent on ethanol demand. Firstly, corn production is steadily rising due to both increased yields and increasing acreage dedicated to corn production. Secondly, corn costs are a relatively small part of most food costs, with predictions that a doubling of feed grain prices will lead to an average increase in food prices of less than 4% (Sneller and Durante, 2007).

While industry and government decide whether fuel ethanol is a legitimate alternative fuel, United States ethanol production is currently on the rise (Yacobucci and Womach, 2000; Curtis, 2008). There are three main categories of processes for producing ethanol from various feedstocks: ethylene hydration, sugar/starch crop fermentation, and lignocellulosic processes (Yacobucci and Womach, 2000; Curtis, 2008; Wade, 2008). Figure 11.1 shows a general overview of the latter two processes. Feedstocks for these latter processes may include sugar-based crops such as beets and cane, starch-based crops such as corn, barley, wheat, and potatoes, and lignocellulosic-based materials such as wood, corn stover, and grasses. Most of the fuel ethanol in the United States is produced by fermentation of corn (Yacobucci and Womach, 2000). Wet milling and dry milling are the two main processes for corn fermentation and lead to the formation of sugars that can be used for fermentation. The wet milling process involves liquid extraction and yields secondary products of corn gluten feed, corn gluten meal and corn oil. The dry milling process is generally smaller and simpler and gives distiller grains that can be used for animal feed besides the ethanol product (Yacobucci and Womach, 2000; Wang et al., 2008).

The fermentation process using sugars increases in difficulty from sugar crops to starchy crops to lignocellulosic materials as a result of the increasing complexity and heterogeneity of sugar components of these biomass resources (Lewis et al., 2005). Sugar crops contain the simplest sources of sugars in the form of sucrose, a disaccharide of glucose and fructose. Starch is a polysaccharide of glucose residues that serves as the nutritional reservoir in plants. Lignocellulosic materials are characterized by varying amounts of cellulose, hemi-cellulose, lignin and small quantities of other extractives. Typically, the composition by weight is 40-50% cellulose, 20-40% hemi-cellulose and 10-30% lignin (McKendry, 2002a). Similar to starch, cellulose is also a polysaccharide of glucose residues. However, the glucose bonds are stronger in cellulose to provide structure to the plant. Thus, cellulose is more difficult to convert into sugar residues as compared to starch. Hemi-cellulose is primarily composed of five-carbon sugars such as xylose and arabinose in addition to six-carbon sugars such as glucose and mannose. Lignin is a group of amorphous, high molecular weight compounds that cannot be fermented.

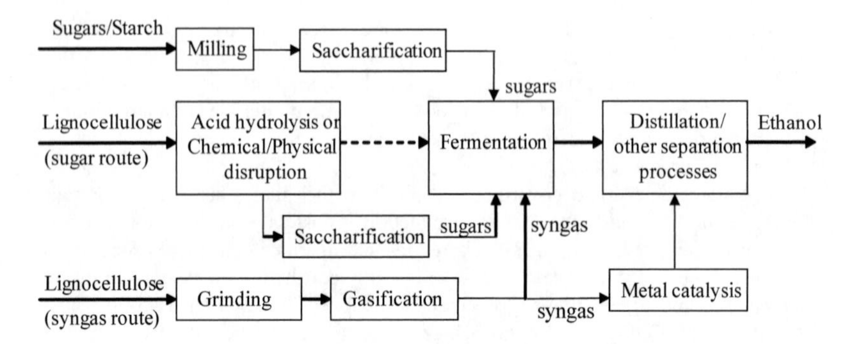

Figure 11.1 Ethanol production processes (Adapted from Lewis et al., 2005).

11.2 Lignocellulosic Processes for Ethanol Production

New technology is being developed to implement lignocellulosic ethanol production. The use of lignocellulosic materials for ethanol production has several advantages including a cheaper feedstock, better carbon conversion, and is non-competitive with food (Kim and Dale, 2004). The downside is that lignocellulosic materials are more complex for fermentation applications.

The most prevalent lignocellulosic processes for ethanol production include: fermentation of sugars, metal catalysis of syngas, and fermentation of syngas (see Figure 11.1). For the process involving fermentation of sugars, cellulose and hemi-cellulose from the lignocellulosic feedstock is converted to simple sugars (saccharification) using enzymes and/or other treatment processes and then the sugars are fermented to ethanol. For the syngas processes, lignocellulosic feedstock is first gasified to produce syngas, primarily consisting of carbon monoxide (CO), carbon dioxide (CO_2), and hydrogen (H_2), along with other hydrocarbons and residual species. Although raw gasification products are generally referred to as producer gas and refined producer gas is generally referred to as syngas, syngas is often used synonymously with producer gas (as is the case for this article). Following syngas generation, syngas is converted to ethanol and other products using either metal catalysts (such as Fischer-Tropsch Synthesis (FTS)) or fermentation.

Fermentation of Sugars. For this process, three main steps are involved in ethanol production: biomass pretreatment, cellulose saccharification, and fermentation. In order for saccharification to occur, pretreatment of the lignocellulosic material is applied due to the crystalline structure of the lignocellulosic material. Pretreatment processes to expose the cellulosic and hemi-cellulosic components to enzymes include acid, ammonia, steam, and alkaline chemicals (Shleser, 1994; Weil et al., 2002; Kadar et al., 2004). The cellulose and hemi-cellulose are then treated with enzymes called cellulases and xylanases to break down the cellulose and hemi-cellulose fractions into fermentable six- and five-carbon sugars, respectively. The sugars are then fermented to ethanol using organisms such as genetically-engineered *Escherichia coli* (Ingram et al., 1997), *Saccharomyces cerevisiae* (Ho et al., 2000) and *Zymomonas mobilis* (Picataggio et al., 1997). In some cases, the saccharification and fermentation occur simultaneously (Kadar et al., 2004). Disadvantages of this process include the high cost of enzymes, the formation of waste streams (including lignin), and the need for economic subsidies, although major efforts are being extended to reduce or eliminate these disadvantages (Weil et al., 2002; Wyman, 2003; Novozymes, 2004).

11.3 Metal Catalysis of Syngas

Catalytic conversion of syngas using metal catalysts, such as Fe, Co, Ru and Ni, is currently a primary industrial method to produce fuels from syngas. The most common process is FTS involving the production of liquid hydrocarbons from CO and H_2 mixtures over a transition metal catalyst. However, large scale commercial processes do not currently employ the utilization of syngas from cellulosic materials, although the potential exists for using cellulosic-generated syngas. The first FTS plant began operation in Germany in 1938, playing an important role in supplying the fuel needs of Germany during World War II. From 1955 to the early 1990s, Sasol (South Africa) was the only large-scale FTS operation in the world. Currently, many oil companies such as Shell Oil, Chevron, ExxonMobil and ConocoPhillips have been conducting research and have built pilot plants (U.S. Department of Energy, 2008). With regards to FTS using lignocellulosic-generated syngas, there are still some barriers. The severe reaction conditions with high temperature and high pressure limit the applications and require high energy inputs. The heat removal resulting from the strong exothermal reactions during the process is also a serious challenge. In addition, the selectivity of catalytic conversion is low and some catalysts are very sensitive to biomass-generated syngas contaminants and are easily poisoned (Subramani and Gangwal, 2008).

Fermentation of Syngas. An alternative route to metal-catalytic conversion is the bio-catalytic conversion of syngas to fuels. Several microbial catalysts, primarily acetogenic bacteria, are capable of consuming syngas and producing useful end-products including alcohols and acids. *Clostridium ljungdahlii* (Phillips et al., 1994), *Butyribacterium methylotrophicum* (Bredwell et al., 1999) and *Clostridium autoethanogenum* (Abrini et al., 1994) are a few examples of the bacteria that have been utilized. Fermentation of syngas has advantages over fermentation of sugars in that softwoods that are normally difficult to handle can be utilized (Dayton and Spath, 2003) and gasification can convert all lignocellulosic material to syngas, providing a greater carbon conversion potential towards ethanol (McKendry, 2002b). Although there are still some challenges, fermentation of syngas circumvents problems associated with high pressure, high temperature, or catalyst poisoning that are more problematic for metal catalysis of syngas (Ragauskas et al., 2006). The remainder of this review focuses on the process of syngas fermentation.

11.4 Syngas Generation

Gasification is defined as the "conversion of a carbonaceous fuel to a gaseous product with a useable heating value" (Higman and van der Burgt, 2003). Gasification processes are broadly categorized as pyrolysis, partial oxidation, and hydrogenation with partial oxidation being the most common. In partial oxidation, a fuel is heated with an oxidant at less than a stoichiometric concentration. Common oxidants include oxygen, air, and steam (Higman and van der Burgt, 2003). Syngas from partial oxidation is composed of mainly CO, CO_2, and H_2 but will include a variety of additional major and minor species depending on the feedstock and type of gasifier. Syngas is an intermediate product in the formation of a number of useful products including additional H_2, Fischer-Tropsch liquids, methane, electricity, and ethanol. The potential flexibility offered by variable gasification processes, differing feedstocks, and end products has made gasification a recent subject of interest and study.

11.4.1 Types of Gasifiers

There are three main types of gasifiers used for gasification: moving-bed (a.k.a. fixed-bed), fluidized-bed, and entrained-flow gasifiers (Nowacki, 1981; Higman and van der Burgt, 2003). In moving-bed gasifiers, particles of feedstock are fed to the top of the reactor and slowly flow downward driven by its consumption and gravity. Ash exits at the bottom of the reactor. Oxidant is fed either counter-current (updraft), co-current (downdraft), or cross-current to the feedstock flow (McKendry, 2002c). Fluidized-bed gasifiers use smaller particles of feed that become fluidized by an upflow of oxidant. Fluidized-bed gasifiers also implement solid heat carriers (sand, ash, or char) to increase the heat transfer in the bed. Finally, in entrained-flow gasifiers, oxidant and feed are fed co-current into the reactor and the feed is carried in the flow. In order to ensure good conversion and high mass transfer, the reactor is run at high temperatures and the feedstock particles are greatly reduced in size (Nowacki, 1981; Higman and van der Burgt, 2003).

The unique characteristics of these processes affect the qualities of the exiting syngas. Table 11.1 lists the unique characteristics, advantages, and disadvantages of each of the three classes of gasifiers. Moving-bed gasifiers are the traditional and most simple class of gasifiers. One drawback is that they have a non-uniform temperature distribution and can produce a large amount of tars. Fluidized-bed gasifiers facilitate a higher feed rate, are easy to control the temperature, and are able to be pressurized. However, fluidized-bed gasifiers only convert 95-97% of the carbon compared to 99% for other gasifiers (Higman and van der Burgt, 2003). Finally, entrained-flow gasifiers produce low amounts of CO_2, CH_4, and tars, as well as provide a good carbon conversion (Higman and van der Burgt, 2003; McKendry, 2002a).

Table 11.1 Characteristics of gasifiers (McKendry, 2002c; Higman and van der Burgt, 2003).

Gasifier	Process Characteristics	Advantages	Disadvantages
Moving Bed	• Large Feed Size (6 – 50 mm) • Low Exit Temp (125 – 650 °C)	Simple and inexpensive	Large tar production
Fluidized Bed	• Intermediate Feed Size (6 – 10mm) • Intermediate Exit Temp (900 –1050 °C)	Uniform temperature	Lower carbon conversion and large tar production
Entrained Flow	• Small Feed Size (<100μm) • High Exit Temp (1250 – 1600°C)	Pure gas; High carbon conversion and very low tar	

Feedstocks. One of the principal advantages of gasification is the flexible nature of its feed. Feedstocks for gasification include coal, natural gas, asphalt, petcoke, bitumen, tar sands residues, liquid organic residues, refinery gas, vegetable biomass, animal biomass, and black liquor (Higman and van der Burgt, 2003). By far the most common feedstock for commercial gasification is coal, which is used for about 55% of the syngas produced worldwide (Gasification Technology Council, 2008). However, due to the renewable nature of biomass, interest has increased in using biomass more widely in gasification processes.

As the traditional gasification feed, coal has many desirable properties. Coal has a high bulk density, high heating value, low moisture content, and is well studied. Biomass fuels vary widely in their properties. Compared to pulverized coal fuels, biomass is usually larger in size, exhibits wider ranges in moisture content, has higher H_2 and O_2 content, exhibits wide ranges in amount and composition of inorganic components, and has key impurities in different amounts and form (Baxter et al., 2006). Table 11.2 shows a side by side comparison of the heating value, moisture content, bulk density and fixed carbon content between coal and biomass. As can be seen in the table, biomass generally is more volatile (i.e. has less fixed carbon), has more moisture, and has a lower heating value than coal.

Table 11.2 Coal and biomass properties (McKendry, 2002a; Higman and van der Burgt, 2003).

Property	Coal	Biomass
Fixed Carbon Content	Lignite: 65-73 wt% Bituminous: 78-92 wt%	Wood: 17 wt% Straw: 18-21 wt%
Moisture Content	Lignite: 35-60 wt% Bituminous: 5-15%	Wood: 10-60 wt% Straw: 10 wt% Maize: 10-20 wt%
Heating Value of Syngas	Lignite: 26-28 MJ/kg Bituminous: 32-36 MJ/kg	Wood: 10-20 MJ/kg Straw: 14-16 MJ/kg Maize Stalk: 13-15 MJ/kg
Bulk Density	600 - 900 kg/m^3	450 kg/m^3

Different properties of the feedstock can lead to a variety of effects in the gasification process and associated economics. Bulk density, for instance, directly impacts transportation costs. Biomass is generally more expensive to ship due to the low bulk density and because the biomass must be shipped as solids (Higman and van der Burgt, 2003). Also, biomass has a high alkaline content that reacts with ash to form a mobile, sticky liquid which can block lines upon cooling. Higman and van der Burgt report that biomass ash has a melting point near 800°C (McKendry, 2002a; Higman and van der Burgt, 2003). Another important property of the feedstock is the ratio of volatiles to fixed carbon content. In coals, this ratio is low, always less than one. In biomass, however, this ratio is as high as four. This quantity describes how easily a fuel is volatilized and affects the gasification mechanism and subsequent products of the system (Tillman, 2000). The high volatility is considered an advantage for biomass fuels and allows the fuel to burn at a high power output (Demirbas, 2004).

In order to mitigate the costs of a pure biomass feed, some have suggested co-firing coal and biomass together. Co-firing has been proposed as a first step to biomass utilization and many have hopes that co-firing will lead to reduced CO_2, NO_x, and trace metals emissions (Tillman, 2000). Co-fired biomass and coal are typically blended to a mass percentage around 90-95 wt% coal and 5-10 wt% biomass. Even at low percentages, mixed fuels present problems to the solids handling and feed stages (Tillman, 2000). However various reports show that co-firing may provide advantages in combustion, including reduced CO_2 and NO_x emissions (Tillman, 2000; Demirbas, 2004).

Products of Gasification. One of the most important considerations in gasification is the chemical composition of the syngas. There are a few solids (mostly ash), condensable volatiles, and gases in the raw syngas stream. The major species in syngas are CO, CO_2, H_2, H_2O, and CH_4 (Higman and van der Burgt, 2003). The rest of the species can be classified by their elemental makeup into compounds containing carbon, nitrogen, sulfur, chlorine, oxygen and heavy metals. Table 11.3 shows the most common raw syngas vapor phase components and typical concentrations.

Table 11.3 Typical gasification gas composition, on a dry and N_2-free basis (Adapted from Higman and van der Burgt, 2003; Bain et al., 1998).

Product	Coal (Bituminous)	Biomass
CO	5-10 %	~35 - 40%
CO_2	~50 %	~15%
H_2	~35 %	~25-30%
H_2O	Variable	Variable
CH_4	2-3 %	~15-20%
C_2^+	< 1 %	~5%
NH_3	~1000 ppmv	2200 mg/Nm^3
HCN	22 ppmv	< 25 mg/Nm^3
NO_x	< 0.02 ppmv	~ 0 ppmv
H2S	15300 ppmv	~100 – 400 ppmv
COS	180 ppmv	
Mercaptan, S	600 ppmv	
SO_x	~ 0 ppmv	~ 0 ppmv
CS_2	100 ppmv	

Thiophenes	5 ppmv	
Metal Carbonyls	~1 ppmv	
O_2	Variable	Variable
As	1-10 ppmw	

There are many carbon containing compounds that can exit a gasifier in trace amounts, particularly from the volatilized but ungasified hydrocarbons from the fuel. These compounds include tars, unsaturated hydrocarbons, and aromatics (Higman and van der Burgt, 2003). Nitrogen compounds include molecular nitrogen, ammonia, hydrogen cyanide, and NO_x. Most of the non-molecular nitrogen comes from fuel nitrogen. The fuel nitrogen is decomposed to ammonia or hydrogen cyanide and, if conditions permit, it is oxidized to NO_x (Higman and van der Burgt, 2003). Major sulfur compounds include H_2S, COS, and SO_x. Chlorine compounds include metal chlorides and ammonium chloride.

11.5 Syngas Fermentation: Metabolic Pathway

Clostridium ljungdahlii (Phillips et al., 1994), *Butyribacterium methylotrophicum* (Bredwell et al., 1999) and *Clostridium autoethanogenum* (Abrini et al., 1994) are known examples of bacteria that have been shown to produce ethanol from syngas. *Clostridium carboxidivorans* (carbon dioxide devouring), a bacteria isolated in an agricultural settling lagoon, has been found to produce higher than average rates of ethanol and butanol from CO-rich mixtures (Liou et al., 2005). *Clostridium carboxidivorans* can grow through the fermentation of a myriad of substrates including: CO, H_2/CO_2, glucose, galactose, fructose, xylose, mannose, cellobiose, trehalose, cellulose, starch pectin, citrate, glycerol, ethanol, propanol, 2-propanol, butanol, glutamate, aspartate, alanine, histidine, asparagines, serine, betaine, choline and syringate (Liou et al., 2005).

Ethanol (C_2H_5OH) and acetic acid (C_2H_3COOH) are produced from syngas via the following four overall reactions (Vega et al., 1989):

$$6CO + 3H_2O \rightarrow C_2H_5OH + 4CO_2$$

(11.1)

$$6H_2 + 2CO_2 \rightarrow C_2H_5OH + 3H_2O$$

(11.2)

$$4CO + 2H_2O \rightarrow CH_3COOH + 2CO_2$$

(11.3)

$$2CO_2 + 4H_2 \rightarrow CH_3COOH + 2H_2O$$

(11.4)

Additional products, such as butanol, have also been identified as products of syngas fermentation (Rajagopalan et al., 2002). It is important to note that in the absences of H_2, only one-third of the available carbon in CO can be converted to ethanol (Equation 11.1) and one-half of the available carbon in CO can be converted to acetic acid (Equation 11.3). This is due to CO being used as both a carbon source and a source for electrons.

During acetogenesis (acetic acid production coupled with growth) and solventogenesis (ethanol production—primarily occurring during non-growth), acetyl-CoA is formed as an intermediate of the metabolic pathway. The acetyl-CoA pathway differs from other CO_2 fixation pathways as it is linear as opposed to cyclic like the Calvin cycle and the reductive tri-carboxylic acid cycle. There are a variety of pathways through which acetyl-CoA can be produced. However, syngas fermentation usually occurs via the "Wood-Ljungdahl" pathway shown in Figure 11.2. In the metabolic pathway, two molecules of CO_2 are reduced to acetyl-CoA (Drake and Daniel, 2004), following which acetyl-CoA is converted to cell mass and/or products such as ethanol and acetic acid.

In the methyl branch, one CO_2 molecule is converted to a methyl moiety. During the conversion, one ATP and six electrons are utilized. Within the methyl branch there are six key enzymes: formate dehydrogenase, formate-THF ligase, methenyl-THF cyclohydrolase, methylene-THF dehydrogenase, methylene-THF reductase, and methyl transferase. In the carbonyl branch, two electrons are utilized to reduce one CO_2 molecule to CO via the carbon monoxide dehydrogenase (CODH) enzyme. CO is then incorporated with the methyl moiety to form acetyl CoA via the acetyl-CoA synthase enzyme. CO can also be utilized directly, rather than being produced from CO_2 reduction. Research has shown that CODH and acetyl-CoA synthase form a complex (Drennan et al., 2004). Following formation of acetyl CoA, acetic acid formation occurs during cell growth. ATP is produced during acetic acid formation. Ethanol is formed via the incorporation of four electrons, either directly from acetyl CoA or from acetic acid.

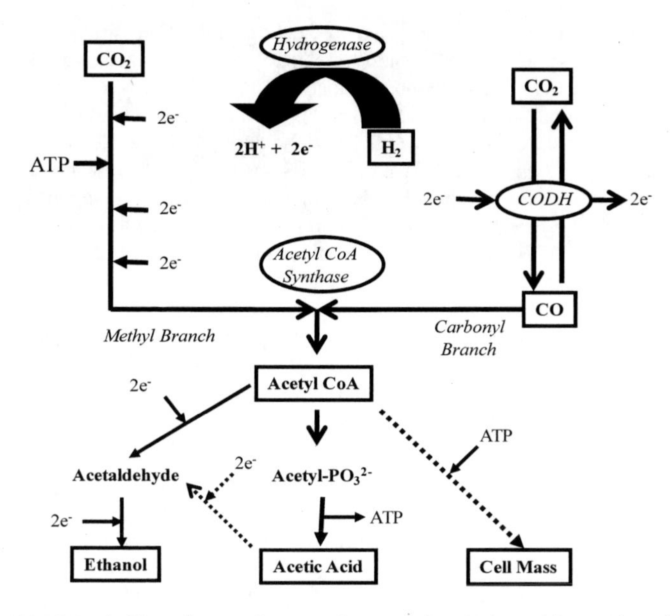

Figure 11.2 Metabolic pathway of syngas fermentation (Adapted from Ahmed and Lewis, 2007).

Electrons required for the metabolic process are generated from H_2 via the hydrogenase enzyme and/or from CO via the CODH enzyme. If H_2 is present in the gas, the most efficient carbon conversion process involving CO would involve the production of electrons from H_2 such that the carbon in CO could then be used as a carbon source. This is evident as shown in the following reactions:

$$6CO + 6H_2 \rightarrow 2C_2H_5OH + 2CO_2$$
(11.5)

$$4CO + 4H_2 \rightarrow 2CH_3COOH$$
(11.6)

where Equation 11.5 is a combination of Equations 11.1 and 11.2 and Equation 11.6 is a combination of Equations 11.3 and 11.4. In Equation 11.5, two-thirds of the carbon in CO is converted to ethanol whereas all of the carbon is converted to acetic acid. If H_2 is not present or hydrogenase is inhibited, such as was observed in the presence of nitric oxide in syngas produced during gasification (Ahmed and Lewis, 2007), electrons can still be produced through the oxidation of CO to CO_2 via CODH at the expense of reducing the carbon conversion efficiency. It is known that the redox potential for the CO_2/CO couple (electrons from CODH) is much more negative than for the $2H^+/H_2$ couple (electrons from hydrogenase) (Ragsdale, 2007). This suggests that when both H_2 and CO are present and hydrogenase and CODH enzymes are active, preference is given to electron production from CO. This has major implications since CO utilization for electron production comes at the expense of using the carbon of CO for ethanol production.

It is important to note, as seen in Figure 11.2, that one ATP molecule is consumed in the methyl branch in addition to the one ATP molecule produced in the acetic acid branch. Since ATP is required for growth, ATP must additionally be produced from outside of the Woods-Ljungdahl pathway. Two chemiosmotic mechanisms have been proposed, either a proton gradient is generated to synthesize proton-dependent ATP production or a sodium gradient is generated to synthesize sodium-dependent ATP. The proton-gradient mechanism has been shown to involve an electron transport system and has often been associated with the presence of enzymes which generate electrons (Drake and Kusel, 2005). Thus, electron production from hydrogenase, or possibly CODH, can contribute to ATP production. It should be noted that electrons utilized in the production of ATP are not consumed and can therefore still be utilized in the metabolic pathway.

11.6 Syngas Fermentation: Current Progress and Challenges

Syngas fermentation is often hindered by low cell density, lack of regulation of metabolic pathways to yield only the desired product, inhibition of the biological catalysts by products and substrates, and low gas-liquid mass transfer rates (Bredwell et al., 1999). Currently, ethanol from syngas fermentation is not at commercial scale although a 40,000 gallon per year pilot plant is in construction, with a 40 to 100 million gallon per year facility planned for 2011 (Keegan, 2008). Most syngas fermentation is conducted at laboratory scale, often using syngas obtained from mixed commercial gases. The disadvantage of using commercial gases is that they do not truly mimic biomass-generated syngas which has additional components besides CO, CO_2, and H_2 that may affect the fermentation process.

Since the main carbon source for syngas fermentation is a gas, the transport of gas is a key issue that needs to be addressed for commercialization. Mass transfer rates are affected by fluid dynamics (which includes diffusional aspects), the transport area, and the gas concentration gradient (which includes gas solubility and partial pressure aspects). From a diffusional aspect, diffusivities of CO, CO_2, and H_2 in water at 37 °C are 3.26×10^{-5} cm^2/s, 2.74×10^{-5} cm^2/s, and 6.48×10^{-5} cm^2/s, respectively (Wise and Houghton, 1968; Verhallen et al., 1984; Tamimi and Rinker, 1994). For comparison, the diffusivity of O_2 in water at 37 °C is 3.05×10^{-5} cm^2/s. Thus, H_2 would have the best mass transfer in solution from a diffusional viewpoint. Sparingly soluble gases, such as CO and H_2, lead to a low concentration gradient, and hence a low mass transfer rate (Bredwell et al., 1999). CO, H_2, and CO_2 have aqueous solubilities of approximately 75%, 69%, and 2400%, respectively, as compared to O_2 on a molar basis (Incropera and DeWitt, 1985; Cooney, 1976). Thus, from a solubility standpoint, CO_2 would provide a far greater driving force for mass transfer at equivalent partial pressures of the three gases. If membranes are used in the reactor design, such as hollow fiber units, additional aspects of mass transfer through the membranes must be considered for assessing mass transfer effects.

Reactor Design. Several types of reactors can be utilized for syngas fermentation although particular attention on how the reactor affects gas utilization and product formation should be an important consideration. Trickle-bed reactors (TBR), consisting of cells attached to a solid support material retained within the reactor, have been used for the production of H_2 from CO (Amos, 2004; Wolfrum and Watt, 2002). The direction of fluid-flow is normally counter current, with the liquid trickling downwards and the gas flowing upwards. A TBR can help minimize the liquid resistance to mass transfer since a very thin liquid film flows across the cells. Mass transfer can also be improved in membrane biofilm reactors (Ebrahimi et al., 2005) where cells are attached to a membrane surface and exposed to both the gas and liquid. One application of a membrane biofilm reactor is a hollow fiber module in which cells are attached to the hollow fiber with gas flowing through the hollow fibers and liquid flowing around the hollow fibers (Rittmann et al., 2004).

Reactors involving gas dispersion in the form of bubbles include continuous stirred-tank reactors (CSTR) and bubble-column reactors. In a CSTR, a continuous flow of liquid provides a continual replenishment of nutrients while the gas is continuously bubbled through the reactor (Klasson et al., 1992). To enhance mass transfer, micro-bubbles can be employed (Bredwell et al., 1999). Cell-recycle systems can also be used in conjunction with the CSTR to increase cell-density within the reactor (Klasson et al., 1993). Bubble-column reactors have a large height-to-diameter ratio in which high mass transfer can be obtained even without the use of additional agitation. Smaller bubble sizes and improved gas dispersion can be obtained by using porous fritted discs to disperse the syngas (Vega et al., 1990). High pressure drops at large capacities is a disadvantage of bubble columns. In the CSTR or bubble-column reactor, cells can be immobilized on biocatalyst particles to improve the cell density and cell retention (Bailey and Ollis, 1986).

Media formulations typically contain minerals, trace metals, and vitamins and have been reported in literature (Rajagopalan et al., 2002; Younesi et al., 2005). Reducing agents, such as cysteine or cysteine-sulfide, can also been used to maintain anaerobic conditions or affect the redox potential of the media (Sim and Kamaruddin, 2008; Balusu and Paduru 2005).

Cell Growth and Product Formation. Results from published literature show that temperature, pH, partial pressure, gas compositions, and reducing agents have significant effects on cell growth and ethanol production (Abrini et al. 1994; Gaddy et al., 2007). For example, a study shows that the concentration of reducing agents is correlated with cell growth and butanol formation (Kim and Bajapai, 1988). Another study regarding syngas fermentation via *Clostridium carboxidivorans* shows that a higher CO partial pressure causes faster cell growth and higher ethanol production (Hurst, 2005). However, the mechanisms and optimal conditions of syngas fermentation have not yet been determined.

Table 11.4 shows reported or calculated cell growth rates for several species. As noted, the growth rates are typically much less than the 1 hr^{-1} maximum growth rates observed for bacteria (Blanch and Clark, 1997). It is possible that mass transfer limitations can mask the true growth rate so care should be given when interpreting data. Nevertheless, the reported values show that cells grow much slower than typical bacteria, even in small experimental systems.

Table 11.4 Cell growth rates on syngas fermentation.

Microbial catalyst	Growth rate (hr^{-1})	Reference
C. ljungdahlii	0.18	(Tanner et al., 1993)
C. ljungdahlii	0.07	(Younesi et al., 2005)
Clostridium P7	0.04	(Datar et al., 2004)
C. carboxidivorans	0.11	(Liou et al., 2005)
Butyribacterium methylotrophicum	0.03 to 0.06	(Grethlein et al., 1991; Shen et al., 1999)

With regards to product formation, acetic acid is produced simultaneously with growth and ethanol is generally produced once the cells reach the stationary phase. Yields for the conversion of CO to ethanol, as calculated from laboratory-scale experiments, are shown in Table 11.5. Yields on H_2 and/or CO_2 utilization are not as readily available. It should be noted that yields can vary depending upon the experimental conditions, including gas partial pressures and enzyme activities. In general, most studies showed yields of 15-25%. According to the stoichiometry in Equation 1 where CO alone is used as the carbon and electron source, the maximum theoretical yield is 17%. If H_2 was also present and used as the electron source, the maximum yield would increase to 33% (see Equation 5). Although all studies did include H_2, the studies in which yields approached 17% suggest that the cells preferred growth on CO alone (i.e. CO was used as both an electron source and a carbon source). For the study where the yield was 25%, H_2 must have been used (at least partially) for the electron source.

Table 11.5 Yield of ethanol from CO.

Microorganism	Yield (mol ethanol/mol CO)	Reference
C. ljungdahlii	0.008	Vega et al., 1989
C. ljungdahlii	0.25	Younesi et al., 2005
C. carboxidivorans	0.15	Rajagopalan et al., 2002
C. carboxidivorans	0.16	Liou et al., 2005

As shown in Figure 11.2, CO, CO_2, and H_2 can all theoretically be used simultaneously for syngas fermentation although it is not clear as to which gases are preferred. For instance, both CO and H_2 can provide electrons. As previously stated, the redox potential for the CO_2/CO couple (electrons from CODH) is much more negative than for the $2H^+/H_2$ couple (electrons from hydrogenase) (Ragsdale, 2007) suggesting that when both H_2 and CO are present, preference is given to electron production from CO. Additionally, different enzymes in the metabolic pathway may be affected differently by the partial pressures of the gases since enzyme activities are greatly affected by substrate concentrations (Shuler and Kargi, 2002). These aspects will have a great impact as to which gas and how much is utilized in a reactor. Depending upon gas utilization, multiple reactors may need to be incorporated to utilize a majority of syngas components.

Syngas Effects. As noted in Table 11.3, syngas contains additional species besides CO, CO_2, and H_2. It is feasible that biological systems can be positively or negatively affected by the additional species. Most reported studies involving syngas fermentation have utilized "clean" compressed gas mixtures to obtain desired CO, CO_2, and H_2 concentrations. However, studies have been reported using biomass-generated syngas. The first reported results using biomass-generated syngas showed that the syngas induced cell dormancy, inhibited H_2 consumption, and affected the acetic acid/ethanol product distribution (Datar et al., 2004). Further work showed that a 0.025 μm filter negated the cell dormancy, tars were the likely cause of dormancy and product distribution, and that cells showed an ability to adapt to the tars upon prolonged exposure (Ahmed et al., 2006). With regards to H_2 inhibition, it was shown that nitric oxide (at gas concentrations above 40 ppm) was a non-competitive inhibitor of H_2 consumption, although the loss of hydrogenase activity was reversible (Ahmed and Lewis, 2007). As is evident, biomass-generated syngas can have adverse effects on syngas fermentation and the effects will likely depend upon the syngas source.

Ethanol Recovery. For commercial development, the purification of ethanol from the syngas fermentation process is an important aspect. Depending upon the reactor design, the product stream can contain ethanol, acetic acid, other by-products, cell mass and/or water. Since ethanol is more volatile than water, recovery by distillation is often the best choice. Distillation is a standard unit operation to allow the recovery of dilute volatile products from streams containing a variety of impurities (Madson and Lococo, 2000). The only difficulty with distillation is that ethanol forms an azeotrope with water such that distillation alone will yield only a 95% ethanol concentration (Hamelinck et al., 2005). Ethanol can be further purified via distillation using an entrainer, for example benzene (Lynd, 1996). In addition to distillation, ethanol can be removed with the use of membranes. The utilization of ceramic membranes has been proposed for the filtration of cell biomass and the removal of ethanol during fermentation (Ohashi et al., 1998). Other ethanol recovery methods include vacuum removal, gas stripping removal, and liquid extraction (Cysewski and Wilke, 1977; Taylor et al., 1998; Minier and Goma, 1982).

Regardless of recent progress, the ethanol concentration obtained from syngas fermentation is still relatively low (< 30 g/L; see Gaddy et al., 2007). The low ethanol concentration leads to higher energy costs for ethanol recovery. Thus, for large-scale operation, higher ethanol concentration should be a goal in order to reduce the cost and energy impacts associated with recovery.

11.7 Summary

The development of new processes for the conversion of low-cost renewable biomass to ethanol is important as a possible alternative to the fast depleting petroleum resources. With regards to ethanol from lignocellulosic processes, syngas fermentation is still in the early stages of development compared to fermentation of sugars from lignocellulose. Key challenges that must be addressed include the following:

- Bioreactor designs must minimize mass transfer limitations.
- The effects of biomass-generated syngas constituents (other than CO, CO_2, and H_2) on the fermentation process must be assessed since the syngas composition can vary depending upon the gasification process.
- High cell concentrations must be obtained to minimize reactor size.
- Gas utilization must be optimized through reactor design and metabolic engineering to maximize the use of gas substrates.
- Continued discovery and development of new bacteria strains should be pursued to improve the fermentation.
- Ethanol recovery technology must be optimized.
- By-products from syngas fermentation should be explored to increase the economic viability of the process.

Even while lignocellulosic processes are in development, many are looking beyond these newly labeled 2nd generation biofuels to 3rd generation biofuels. Due to the inconvenience and controversy of ethanol, many would like to produce hydrocarbons or butanol as biofuels (Cascone, 2008). These fuels are considered more advantageous because of their ability to utilize the existing infrastructure and their similarities to current fuels. For instance, in contrast to ethanol, butanol is more easily separated from water, can be shipped in pipelines, and has a similar energy density to gasoline (Cascone, 2008). However, metabolic engineering to develop strains of bacteria that produce these fuels and to produce strains that are resistant to fuel toxicity are still current challenges (Liao and Higashide, 2008). Thus, 3rd generation biofuels are still in the infancy stages.

In summary, there is a growing interest in biofuels and ethanol production from syngas fermentation is one possible source of biofuels. However, syngas fermentation is a relatively new technology that still requires research in order to overcome the current challenges. The future of biofuels development will likely not rely on one process. A variety of processes and biofuels will likely be required to make the global shift towards the use of renewable resources for biofuels.

11.8 References

Abrini, J., Naveau, H., and Nyns, E.J. (1994). "*Clostridium autoethanogenum*, sp-nov, an anaerobic bacterium that produces ethanol from carbon-monoxide." *Arch. Microbiol.*, 161(4), 345–351.

Ahmed, A., and Lewis, R.S. (2007). "Fermentation of biomass-generated synthesis gas: Effects of nitric oxide." *Biotechnol. Bioeng.*, 97(5), 1080–1086.

Ahmed, A., Cateni, B.G., Huhnke, R.L., and Lewis, R.S. (2006). "Effects of biomass-generated syngas constituents on cell growth, product distribution and hydrogenase activity of *Clostridium carboxidivorans* $P7^T$." *Biomass Bioenergy*, 30 (7), 665–672.

Amos, W.M. (2004). "Biological water-gas shift conversion of carbon monoxide to hydrogen." No. NREL/MP-560-35592. Golden, Colorado: National Renewable Energy Laboratory.

Bailey, J.E., and Ollis, D.F. (1986). *Biochemical Engineering Fundamentals (2nd ed.).* San Francisco, California: McGraw-Hill.

Bain, R.L., Ratcliff, M., Deutch, S., Feik, C., French, R., Graham, J., Meglen, R., Overly, C., Patrick, J., Phillips, S., and Prentice, B. (1998). "Biosyngas characterization test results." Milestone Completion Report—Center for Renewable Chemicals Technology and Materials. Biomass Power Fuel Cell Testing (BP712130).

Balusu, R., and Paduru, R.R. (2005). "Optimization of critical medium components using response surface methodology for ethanol production from cellulosic biomass by Clostridium thermocellum SS19." *Process Biochem.*, 40(9), 3025–3030.

Baxter, L., Ip, L., Lu, H., Tree, D. (2006). "Distinguishing Biomass Combustion Characteristics and Their Implications for Sustainable Energy." 20th Annual ACERC Conference Abstracts. Provo, UT.

Blanch, H.W., and Clark, D.S. (1997). *Biochemical Engineering.* New York : Marcel Dekker, pp. 187.

Bredwell, M.D., Srivastava, P., and Worden, R.M. (1999). "Reactor design issues for synthesis-gas fermentations." *Biotechnol. Progr.*, 15 (5), 834–844.

Brinkman, N., Wang, M., Weber,T., and Darlington, T. (2005). Well-to-Wheels Analysis of Advanced Fuel/Vehicle Systems -- A North American Study of Energy Use, Greenhouse Gas Emissions, and Criteria Pollutant Emissions.

Cascone, R. (2008). "Biobutanol – A replacement for bioethanol?" *Chem. Eng. Prog.*, 104 (8), S4–S9.

Cooney, D.O. (1976) *Biomedical Engineering Principles*. New York : Marcel Dekker, pp. 352.

Curtis, B. (2008). *2007 Year in Review U.S. Ethanol Industry: The Next Inflection Point*. Washington, DC: B. Curtis Energies and Resource Group, pp. 52.

Cysewski, G.R., and Wilke, C.R. (1977). "Rapid ethanol fermentations using vacuum and cell recycle." *Biotechnol. Bioeng.*, 19(8), 1125–43.

Datar, R.P., Shenkman, R.M., Cateni, B.G., Huhnke, R.L., and Lewis, R.S. (2004). "Fermentation of biomass-generated producer gas to ethanol." *Biotechnol. Bioeng.*, 86, 587–594.

Dayton, D.C and Spath, P.L (2003). "Preliminary screening —Technical and economic assessment of synthesis gas to fuels and chemicals with emphasis on the potential for biomass-derived syngas." 80401-3393. Golden, Colorado:National Renewable Energy Laboratory.

Demirbas, A. (2004). "Combustion characteristics of different biomass fuels." *Prog. Energy Combust. Sci.*, 30, 219–230.

Drake, H. L., and Daniel, S.L. (2004). "Physiology of the thermophilic acetogen *Moorella thermoacetica*." *Res. Microbiol.*, 155, 422–436.

Drake,H.L., and Kusel,K. (2005). In: Durre, P. (Ed.) *Acetogenic Clostridia.* Boca Raton, Florida: CRC Press, pp.710–746.

Drennan, C.L., Doukov, T.I, and Ragsdale, S.W. (2004). "The metalloclusters of carbon monoxide dehydrogenase/acetyl-CoA synthase: a story in pictures." J. *Biol. Inorg. Chem.*, 9, 511–515.

Ebrahimi, S., Picioreanu, C., Xavier, J.B., Kleerebezem, R., Kreutzer, M., Kapteijn, F., Moulijn, J.A., and van Loosdrecht, M.C.M. (2005). "Biofilm growth pattern in honeycomb monolith packings: Effect of shear rate and substrate transport limitations." *Catal. Today*, 105(3-4), 448–454.

Gaddy, J. L., Arora, D.K., Ko, C.W., Phillips, J.R., Basu, R., Wikstrom, C., and Clausen, E.C. "Methods for increasing the production of ethanol from microbial fermentation." U.S. Patent 7 285 402 B2. 2007.

Gasification Technology Council. (2008). *Gasification: Redefining clean energy.* Available at http://www.gasification.org/Docs/Final_whitepaper.pdf (accessed Jan 6, 2008).

Grethlein, A.J., Worden, R.M., Jain, M.K., and Datta, R. (1991). "Evidence for production of n-butanol from carbon monoxide by Butyribacterium methylotrophicum." *J Ferment Bioeng.*, 72(1), 58–60.

Hamelinck, C.N., van Hooijdonk, G., and Faaij, A.P.C. (2005). "Ethanol from lignocellulosic biomass: techno-economic performance in short-, middle- and long-term." *Biomass Bioenergy*, 28(4), 384–410.

Higman, C., and van der Burgt, M. (2003). "Gasification." Elsevier, 41–219.

Ho, N.W.Y., Chen, Z., Brainard, A.P., and Sedlak, M. (2000). "Genetically engineered Saccharomyces yeasts for conversion of cellulosic biomass to environmentally friendly transportation fuel ethanol." In: *ACS Symposium Series.* ACS, pp.143–159.

Hurst, K. M. (2005). "Effects of carbon monoxide and yeast extract on growth, hydrogenase activity and product formation of *Clostridium carboxidivorans* $P7^T$." (Master thesis, Oklahoma State University).

Incropera, F.P., and DeWitt, D.P. (1985). *Fundamentals of Heat and Mass Transfer.* New York: Wiley, pp. 778.

Ingram, L.O., Lai, X., Moniruzzaman, M., Wood, B.E., and York, S.W. (1997). "Fuel ethanol production from lignocellulose using genetically engineered bacteria." In: *American Chemical Society Symposium Series 666*. Washington, D.C.: American Chemical Society Press, pp. 57–73.

Kadar, Z., Szengyel, Z., and Reczey, K. (2004). "Simultaneous saccharification and fermentation (SSF) of industrial wastes for the production of ethanol." *Ind. Crops Prod.*, 20 (1), 103–110.

Keegan, M. (2008). *Making ethanol the Coskata way.* Available at http://thearticlewriter.com/autowriter (accessed May 2, 2008).

Kim, J., Bajapai, R., and Iannotti, E.L. (1988). "Redox potential in acetone-butanol fermentations." *Appl. Biochem. Biotechnol.*, 18, 175–186.

Kim, S., and Dale, B.E. (2004). "Global potential bioethanol production from wasted crops and crop residues." *Biomass Bioenergy*, 26 (4), 361–375.

Klasson, K. T., Ackerson, M. D., Clausen, E. C., and Gaddy, J. L. (1992). "Bioconversion of synthesis gas into liquid or gaseous fuels." *Enzyme Microb. Technol.*, 14(8), 602–608.

Klasson, K. T., Lundback, K. M. O., Clausen, E. C., and Gaddy, J. L. (1993). "Kinetics of light limited growth and biological hydrogen-production from carbon monoxide and water by *Rhodospirillum-Rubrum*." *J. Biotechnol.*, 29(1-2), 177–188.

Lewis, R.S., Datar, R.P, and Huhnke, R.L. (2005). "Biomass to Ethanol." In: Lee, S. (Ed.) *Encyclopedia of Chemical Processing* New York, NY: Marcel Dekker.

Liao, J.C., Higashide, W. (2008). "Metabolic engineering of next-generation biofuels." *Chem. Eng. Prog.*, 104 (8). S19–S23.

Liou, J.S.C., Balkwill, D.L., Drake, G.R., and Tanner, R.S. (2005). "*Clostridium carboxidivorans* sp. nov., a solvent-producing clostridium isolated from an agricultural settling lagoon, and reclassification of the acetogen *Clostridium scatologenes* strain SL1 as Clostridium drakei sp. nov." Int. *J. Syst. Evol. Microbiol.*, 55(5), 2085–2091.

Lynd, L. R. (1996). "Overview and evaluation of fuel ethanol from cellulosic biomass: Technology, economics, the environment, and policy." *Annu. Rev. Energy Env.*, 21, 403–465.

Madson, P.W., and Lococo, D.B. (2000). "Recovery of volatile products from dilute high-fouling process streams." *Appl. Biochem. Biotechnol.*, 84(6), 1049–1061.

McKendry, P. (2002a). "Energy production from biomass (Part 1): Overview of biomass." *Bioresour. Technol.*, 83 (1), 37–46.

McKendry, P. (2002b). "Energy production from biomass (Part 2): Conversion Technologies." *Bioresour. Technol.*, 83 (1), 47–54.

McKendry, P. (2002c). "Energy production from biomass (Part 3): gasification technologies." *Bioresour. Technol.*, *83* (1), 55–63.

McLaughlin, S.B., and Walsh, M.E. (1998). "Evaluating environmental consequences of producing herbaceous crops for bioenergy." *Biomass Bioenergy*, 14 (4), 317–324.

Minier, M., and Goma, G., (1982). "Ethanol production by extractive fermentation." *Biotechnol. Bioeng*, 24(7), 1565–1579.

Novozymes. (2004). "Novozymes cuts biomass ethanol enzyme cost." *Ind. Bioproc.*, *26* (3), 4.

Nowacki, P. (1981). "Coal Gasification Processes." Park Ridge, New Jersey, USA: Noyes Data Corporation.

Ohashi, R., Kamoshita, Y., Kishimoto, M., and Suzuki, T., (1998). "Continuous production and separation of ethanol without effluence of wastewater using a distiller integrated SCM-reactor system." *J. Ferment. Bioeng.*, 86(2), 220–225.

Phillips, J. R., Clausen, E.C., and Gaddy, J.L. (1994). "Synthesis gas as substrate for the biological production of fuels and chemicals." *Appl. Biochem. Biotechnol.*, 45-6, 145–157.

Picataggio, S.K., Zhang, M., Eddy, C., Deanda, K.A., and Finkelstein, M. (1997). "Recombinant zymomonas for pentose fermentation." U.S. Patent 5 514 583.

Ragauskas, A. J., Williams, C. K., Davison, B. H., Britovsek, G., Cairney, J., and Eckert, C. A. (2006). "The path forward for biofuels and biomaterials." *Science*, 311(5760), 484–489.

Ragsdale, S.W. (2007). "Nickel and the carbon cycle." *J. Inorg. Biochem.*, 101, 1657–1666.

Rajagopalan, S., Datar, R.P., and Lewis, R. S. (2002). "Formation of ethanol from carbon monoxide via a new microbial catalyst." *Biomass Bioenergy*, 23(6), 487–493.

Renewable Fuels Association (2003). "Ethanol industry outlook 2003: Building a secure energy future." Washington D.C., 2003, 1–14.

Rittmann, B.E., Nerenberg, R., Lee, K.C., Najm, I., Gillogly, T.E., Lehman, G.E., and Adham, S.S. (2004). "Hydrogen-based hollow-fiber membrane biofilm reactor (MBfR) for removing oxidized contaminants." *Water Sci. Technol.: Water Supply*, 4(1), 127–133.

Shen, G.J., Shieh, J.S., Grethlein, A.J., Jain, M.K., and Zeikus, J.G. (1999). "Biochemical basis for carbon monoxide tolerance and butanol production by Butyribacterium methylotrophicum." *Appl. Microbiol. Biotechnol.*, 51(6), 827–832.

Shleser, R. (1994). "Ethanol production in Hawaii: Processes, feedstocks and current economic feasibility of fuel grade ethanol production in Hawaii." HD9502.5.B54.S4, Hawaii State Department of Business, Economic Development & Tourism: Honolulu, HI.

Shuler, M.L., and Kargi, F. (2002). "Bioprocess Engineering: Basic Concepts." New Jersey: Prentice Hall, pp. 61.

Sim, J. H., and Kamaruddin, A. H. (2008). "Optimization of acetic acid production from synthesis gas by *chemolithotrophic bacterium - Clostridium aceticum* using statistical approach." *Bioresour. Technol.*, 99(8), 2724–2735.

Sneller, T. (2007). *Ethanol Fact Book*. Haigwood, B. (Ed.) Clean Fuels Development Coalition.

Sneller, T., and Durante, D. (2007). *The Impact of Ethanol Production on Food, Feed, and Fuel*. Ethanol Across America and U.S. Dept. of Agriculture.

Subramani, V., and Gangwal, S.K. (2008). "A review of recent literature to search for an efficient catalytic process for the conversion of syngas to ethanol." *Energy Fuels.*, 22(2), 814-839.

Tamimi, A., and Rinker, E.B. (1994). "Diffusion-coefficients for hydrogen-sulfide, carbon-dixoide, and nitrous-oxide in water over the temperature-range 293-368 K." *J. Chem. Eng. Data*, 39(2), 330–332.

Tanner, R.S., Miller, L.M., and Yang, D. (1993). "*Clostridium ljungdahlii* sp. nov., an acetogenic species in clostridial ribosomal-RNA homology group-I." *Int J Syst Bacteriol.*, 43(2), 232–236.

Taylor, F., Kurantz, M.J., Goldberg, N., and Craig J.C. (1998). "Kinetics of continuous fermentation and stripping of ethanol." *Biotechnol. Lett*, 20(1), 67–72.

Tillman, D.A. (2000). “Biomass cofiring: the technology, the experience, the combustion consequences.” *Biomass Bioenergy*, 19, 365–384.

U.S. Department of Energy. (1999). *Ethanol: Separating fact from fiction.* DOE/GO-10099-736. Washington D.C.

U.S. Department of Energy. (2003). *Roadmap for agriculture biomass feedstock supply in the United States*. DOE/NE-ID-11129. Office of Energy Efficiency and Renewable Energy, Biomass Program: Washington D.C.

U.S. Department Of Energy. (2008). "ischer-Tropsch Fuels. Available at http://www.netl.doe.gov/publications/factsheets/rd/R&D089.pdf.

U.S. Energy Information Administration. (2008a). *Energy In Brief: Are we dependent on Foreign Oil?* Available at http://tonto.eia.doe.gov/energy_in_brief/print_pages/foreign_oil_dependence.pdf (accessed December 22, 2008).

U.S. Energy Information Administration (2008b). *Petroleum Products: Consumption.* Available at http://www.eia.doe.gov/neic/infosheets/petroleumproductsconsumption.html (accessed December 22, 2008).

Vega, J.L., Prieto, S., Elmore, B.B., Clausen, E.C., and Gaddy, J.L. (1989). “The biological production of ethanol from synthesis gas.” *Appl. Biochem. Biotechnol.*, 20/21, 781–797.

Vega, J.L., Clausen, E.C., and Gaddy, J.L. (1990). "Design of bioreactors for coal synthesis gas fermentations." *Resour. Conserv. Recycl.*, 3(2-3), 149–160.

Verhallen, P.T.H.M, Oomen, L.J.P., Vanderelsen, A.J.J.M., Kruger, A.J., and Fortuin, J.M.H. (1984). "The diffusion-coefficients of helium, hydrogen, oxygen and nitrogen in water determined from the permeability of a stagnant liquid layer in the quasi-steady state." *Chem. Eng. Sci.*, 39(11), 1535–1541.

Wade, L. G. (2008). “Alcohol.” *Encyclopedia Britannica.* Encyclopedia Britannica Online, pp. 24.

Wang, M. (2007). *Ethanol: the complete energy lifecycle picture.* U.S. Department of Energy, Energy Efficiency and Renewable Energy.

Wang, M., Wu, M., and Huo,H. (2007). "Life-cycle energy and greenhouse gas emission impacts of different corn ethanol plant types." *Environ. Res. Lett.*, 2(2): 13.

Weil, J.R., Dien, B., Bothast, R., Hendrickson, R., Mosier, N.S., and Ladisch, M.R. (2002). “Removal of fermentation inhibitors formed during pretreatment of biomass by polymeric adsorbants.” *Ind. Eng. Chem. Res.*, 41 (24), 6132–6138.

Wise, D. L., and Houghton, G. (1968). "Diffusion coefficients of neon, krypton, xenon, carbon monoxide, and nitric oxide in water at 10-60.deg." *Chem. Eng. Sci.*, 23(10), 1211–16.

Wolfrum, E.J., and Watt, A.S. (2002). "Bioreactor design studies for a hydrogen-producing bacterium." *Appl. Biochem. Biotechnol.*, 98, 611–625.

Wyman, C.E. (2003). “Potential synergies and challenges in refining cellulosic biomass to fuels, chemicals, and power.” *Biotechnol. Progr.*, 19 (2), 254–262.

Yacobucci, B. D., and Womach, J. (2000). “Fuel Ethanol: Background and Public Policy Issues.” *CRS Issue Brief for Congress*. Washington, D.C.: NCSE.

Younesi,H., Najafpour,G., and Mohamed, A.R. (2005). "Ethanol and acetate production from synthesis gas via fermentation processes using anaerobic bacterium, *Clostridium ljungdahlii*." *Biochem. Eng. J.*, 27(2), 110–119.

CHAPTER 12

Lignin Recovery and Utilization

Kasthurirangan Gopalakrishnan, Sunghwan Kim, and Halil Ceylan

12.1 Introduction

Lignocellulosic biomass such as agri-residues (e.g., corn stover, and wheat and barley straws), agri-processing by-products (e.g., corn fiber, sugarcane bagasse, seed cake etc.), and energy crops (e.g. switch grass, poplar, banagrass, Miscanthus etc.) are abundantly available in the United States. During biochemical conversion process, only cellulosic and hemicellulosic fractions are converted into fermentable sugar. Lignin, which represents the third largest fraction of lignocellulosic biomass, is not convertible into fermentable sugars. It is therefore extremely important to recover and convert biomass-derived lignin into high-value product to maintain economic competitiveness of cellulosic-ethanol processes. Traditionally, lignin has been used as a boiler fuel in the production of octane boosters, and in bio-based products and chemical productions.

Different types of lignin exist depending upon isolation protocol and biomass sources. This chapter discusses different types of lignin, their properties and several examples of industrial application with focus on civil engineering infrastructure applications. Special attention is given to the recovery and utilization of sulfur-free lignins, an emerging class of lignin products from biomass conversion processes, solvent pulping, and soda pulping.

12.2 Composition of Lignocellulosic Biomass

Lignin is the essential natural glue that holds all plants together. The term lignin defines only the polymeric material located in the woody cell wall. Various processes can be used to remove and isolate lignin from these cells. Each process, however, produces material of different composition and properties. Because of these differences, unique names are applied to each lignin preparation and should be used when describing these materials.

Lignocellulosic or woody biomass is composed of carbohydrate polymers (cellulose and hemicellulose), lignin and a smaller fraction of extractives, acids, salts and minerals (Hamelinck et al., 2005). The cellulose and hemicellulose, which typically comprise two-thirds of the dry mass, are polysaccharides that can be hydrolyzed to simple sugars for subsequent ethanol fermentation. The lignin is highly calcitrant substrate and cannot be used for ethanol production. The cellulose, hemicellulose, and lignin contents in typical lignocellulosic materials are presented in Table 12.1.

Table 12.1 Cellulose, hemicellulose and lignin contents in typical lignocellulosic materials[a].

Lignocellulosic materials	Cellullose (%)	Hemicellulose (%)	Lignin (%)
Hardwoods stems	40 – 55	24 - 40	18 - 25
Black locust	41 – 42	17 - 18	26 - 27
Hybrid Poplar	44 – 45	18 - 19	26 - 27
Eucalyptus	49 – 50	13 - 14	27 - 28
Softwood stems	45–50	25–35	25–35
Pine	44 – 45	21 - 22	27 – 28
Nut shells	25–30	25–30	30–40
Bagasse	48-32	24-19	32-23
Corn stover	38-40	28	21-7
Corn cobs	45	35	15
Grasses	25–40	35–50	10–30
Switchgrass	31- 32	25- 26	18 - 19
Paper	85–99	0	0–15
Wheat straw	30	50	15
Rice straw	36-28	28-23	14-12
Barley straw	45-31	38-27	19-14
Sorghum stalks	27	25	11
Sorted refuse	60	20	20
Leaves	15–20	80–85	0
Cotton seed hairs	80–95	5–20	0
Newspaper	40–55	25–40	18–30
Waste papers from chemical pulps	60–70	10–20	5–10
Primary wastewater solids	8–15	N/A[b]	24–29
Swine waste	6.0	28	N/A[b]
Solid cattle manure	1.6–4.7	1.4–3.3	2.7–5.7
Coastal Bermuda grass	25	35.7	6.4

[a] Source: Reshamwala et al. (1995), Cheung and Anderson (1997), Boopathy (1998), Dewes and Hünsche (1998), Sun and Cheng (2002), Hamelinck et al. (2005), Reddy and Yang (2005). [b] N/A: not available.

Cellulose (40–60% dry weight) is a linear polymer of cellobiose (glucose–glucose dimers); the orientation of the linkages and additional hydrogen bonding make the polymer highly stable and crystalline. During enzyme hydrolysis, the polysaccharide is broken down to free sugar molecules by the addition of water and is known as saccharification. Cellulose essentially produces glucose, a six-carbon sugar or hexose (Hamelinck et al., 2005).

Hemicellulose (20–40% dry weight) consists of short highly branched chains of five- and six-carbon sugars. Xylose and arabinose are the major 5-carbon sugars, whereas, glucose, galactose, and mannose are the six-carbon sugars. It also contains smaller amounts of non-sugar compounds such as acetyl groups (Lynd et al., 1999).

Hemicellulose, because of its branched and amorphous nature, is relatively easy to hydrolyze (Hamelinck et al., 2005).

Lignin (10–25% dry weight) is present in most of the plant-based biomass. Ethanol production through biochemical route will leave behind lignin as a residue. It is a large complex polymer of phenylpropane and methoxy groups, a non-carbohydrate polyphenolic substance that encrusts the cell walls and cements the cells together. The combination of hemicellulose and lignin provides a protective sheath around the cellulose, which must be modified or removed before efficient hydrolysis of cellulose can occur. The crystalline structure of cellulose makes it highly insoluble and resistant to biological attack. Therefore, to economically hydrolyze (hemi) cellulose, pre-treatment is needed.

12.3 Lignin Chemistry

Since the chemical structure of lignin is complex and changes according to the biomass source and the recovery technique (Glasser et al., 1983; Glasser and Kelley, 1987; Glasser, 2001), it is not possible to define the precise structure of lignin as a chemical molecule. Regardless of its complexity, lignin's unifying chemical structural feature is a branched and cross-linked network polymer of phenylpropenyl units (Bozell et al., 2007; ILI, 2008).

The principal building blocks in lignin structure are three hydroxycinnamyl alcohol monomers differing in their degree of methoxylation, *p*-coumaryl alcohol, coniferyl alcohol, and sinapyl alcohol (Figure 12.1) (Freudenberg, 1968). These monolignols produce, respectively, *p*-hydroxyphenyl, guaiacyl, and syringyl phenylpropanoid units when incorporated into the lignin polymer (Boerjan et al., 2003). The relative proportion of these units depends on the lignin source (Jeffries, 1994; Bozell et al., 2007). In softwoods, guaiacyl propanol is the principal unit (90%) with the remainder derived from *p*-coumaryl and sinapyl alcohols. The lignin of hardwoods is composed of guaiacyl and syringyl propanol units. Grass lignins are primary derived from guiacyl and syringyl units with 10 – 20% *p*-coumaryl alcohol (Bozell et al., 2007).

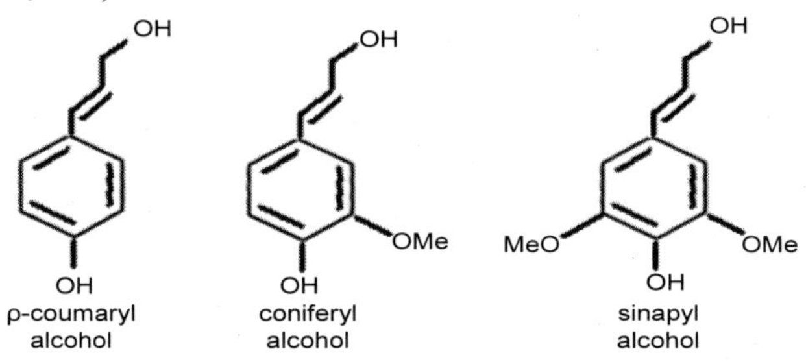

Figure 12.1 Major monomers of lignin (adapted from Boerjan et al., 2003).

More than 20 different inter-unit linkage types have been identified (Whetten et al., 1998). Some of the most important inter-unit linkage types in lignin are β-O-4, α-O-4, 4-O-5, β-5, 5-5, β-1, and β-β as shown in Figure 12.2, along with their common designation (Dence and Lin, 1992; Sjöström, 1993). The β–O–4 (β-aryl ether type) linkage is known to dominate in native lignin and estimated to make up to 58% of the native plant lignin linkages present (Boerjan et al., 2003; Bozell et al., 2007). It is also the one most easily cleaved chemically, providing a basis for industrial processes, such as chemical pulping, and several analytical methods (Boerjan et al., 2003). The other linkages are all more resistant to chemical degradation. The variety of linked units provides numerous branching sites and ring units, and lignin becomes a highly irregular, complex polymer (Dence and Lin, 1992; Sjöström, 1993).

Due to the lignin's complex nature and the difficulties inherent to lignin analysis, only hypothetical structure models have been proposed from the estimation of the relative proportions of each lignin unit and each linkage types (Boerjan et al., 2003).

It is important to recognize that the native lignin undergoes profound structural changes and dramatic modification of molecular weight profiles depending on the recovery technology employed. The lignin modified by various recovery technologies can be divided into two principal categories: the sulfur bearing lignin and the sulfur -free lignin (ILI, 2008).

The sulfur bearing lignin includes the sulfite lignin (lignosulphonates) generated from traditional sulfite process and the kraft lignin generated from kraft pulping in alkaline medium. The sulfite lignin is water soluble and contains sulfonated lignin polymers, sugars, sugar acids, and small amounts of wood extractives and inorganic compounds (Lignin Institute, 2008). The kraft lignin contains a small number of aliphatic thio- groups that give the isolated product a characteristic order (Lora and Glasser, 2002).

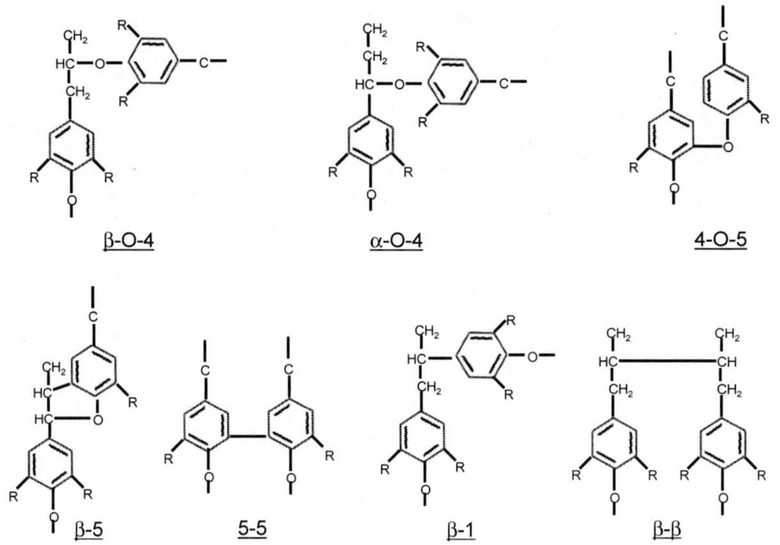

Figure 12.2 Major inter-unit linkage types in lignin (adapted from Boerjan et al., 2003).

The sulfur -free lignins come from mainly three sources, biomass conversion processes for ethanol production, solvent pulping, and soda pulping (Lora and Glasser, 2002). The sulfur bearing lignins have been commercialized while sulfur-free not been yet commercialized due to the lack of suitable industrial processes (ILI, 2008).

12.4 Recovery of Lignins

Even though numerous methods exist to separate/isolate lignin from plant biomass in order to obtain the desirable materials for the paper and the ethanol production, only limited methods are recognized for lignin recovery and utilization. This is because lignin is generally considered as a waste product and the chemical structure of lignin changes during recovery.. In the following sections, the existing state-of-the-art lignin recovery technology is reviewed and presented with a particular focus on recovery from chemical wood pulping and biomass conversion processes for biofuel production.

12.4.1 Recovery from Chemical Wood Pulping

The main objective of chemical wood pulping is to remove enough native lignin to separate cellulosic fibers one from another, producing a pulp suitable for the paper manufacturing and other related products (Pulp and Paper Manufacture, 1987). Figure 12.3 presents a typical chemical wood pulping process diagram. The four processes principally used in chemical wood pulping are kraft, sulfite, soda, and solvent. The kraft process alone accounts for over 80 percent of the chemical pulp produced in the United States (EPA, 1995).

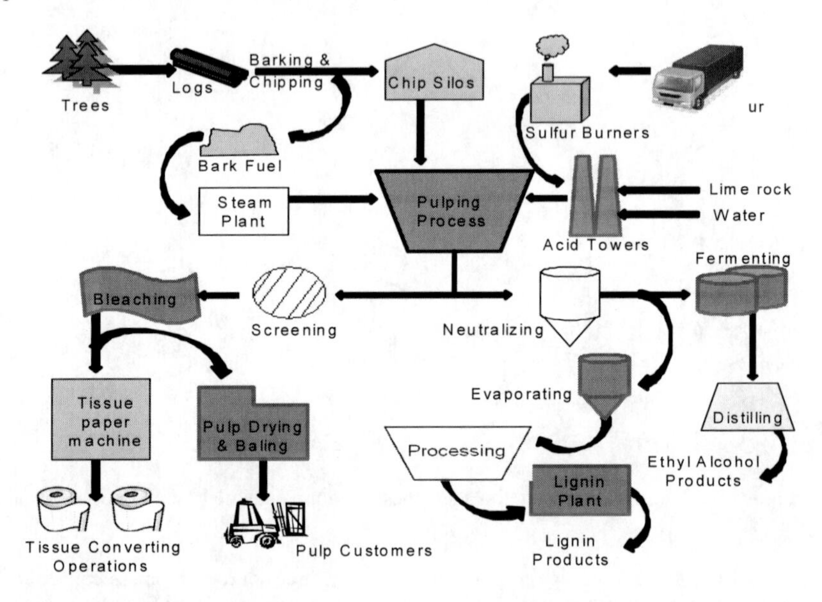

Figure 12.3 Chemical wood pulping process diagram (adapted from Lignin Institute, 2008).

12.4.1.1 Kraft Lignin Recovery

The kraft pulping process of wood involves the digesting of wood chips at an elevated temperature of 150-180 °C and pressure in "white liquor", which is a water solution of sodium sulfide and sodium hydroxide (EPA, 1995). The kraft process as show in Figure 12.4 produces a residue known as "black liquors", and lignin is the major component together with other chemical compounds. Usually, the black liquors are burned after evaporation to recover energy and any remaining chemical reactant. However, the separation of kraft lignin from black liquors could be an alternative to incineration, allowing several commercial possibilities (Caballero et. al, 1997).

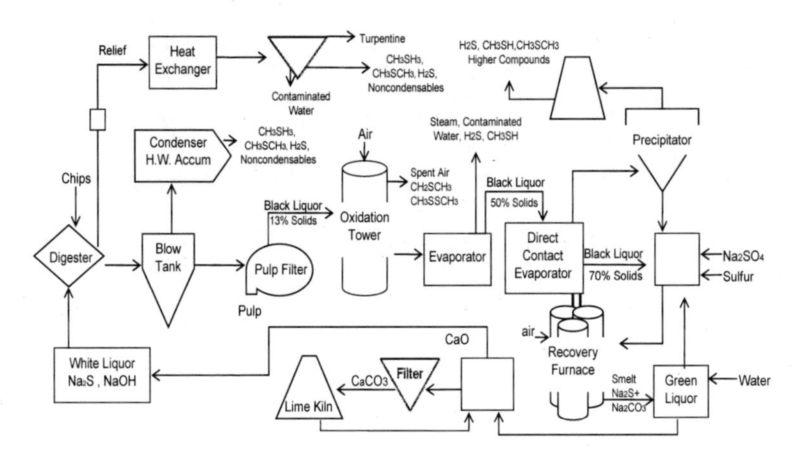

Figure 12.4 Typical kraft sulfate pulping process (adapted from EPA, 1995).

Two approaches are commonly adopted for lignin separation in the kraft pulp process (Wallberg et al., 2005), acid precipitation and ultrafiltration (UF). Generally, acid precipitation is used to extract lignin from kraft black liquor (Alén et al., 1979; Alén et al., 1985; Uloth and Wearing, 1989; Loutfi et al., 1991; Davy et al., 1998). Figure 12.5 shows schematic picture of one of the lignin precipitation processes. The pH of the black liquor is lowered by injecting carbon dioxide (CO_2). On lowering the pH, a substantial portion of the kraft lignin is precipitated and can be recovered by filtering and washing. Given the high-sulfur environment of kraft pulping, the sulfur content of precipitated and washed kraft lignin is surprisingly quite low, typically less than 1-2% which is consistent with the number of –SH linkages created during the kraft pulping process. Moreover, it can be almost free of sugars (Bozell et al., 2007).

UF has been used mainly to purify the lignin fraction in cooking liquor from sulfite and kraft pulp mills in order to use it as a valuable chemical product (Eriksson, 1979; Forss et al., 1979; Kovasin and Nordén, 1984). Process improvements obtained after UF have also been studied (Kirkman et al., 1986; Hill et al., 1988; Cortinas et al., 2002; Liu et al., 2004; Keyoumu et al., 2004). Since the concentration of the liquor to be treated by UF is not that critical unlike precipitation, there is a considerable freedom in the choice of liquor for the treatment with UF (Wallberg et al., 2005).

12.4.1.2 Sulfite Lignin Recovery

The sulfite pulping process is similarly to kraft pulping, except that different chemicals are used in the cooking liquor (EPA, 1995). In place of the caustic solution used to dissolve the lignin in the wood, sulfurous acid is employed. The sulfite pulping can be carried out over the full pH range and any of four different bases, sodium, magnesium, ammonium and calcium might be used to buffer the cooking solution (Grace, 1987). Figure 12.6 shows the simplified magnesium based sulfite lignin pulping process. The sulfite process produces a water-soluble polymeric derivative in admixtures with degraded carbohydrates. An aliphatic sulfonic acid (HSO_3) function becomes part of the lignin backbone ensuring ready water solubility in the presence of a suitable counter ion (Na, Ca, Mg, NH_4, etc.) (Lora and Glasser, 2002). Lignins produced by the sulfite pulping process are known as lignosulfonates.

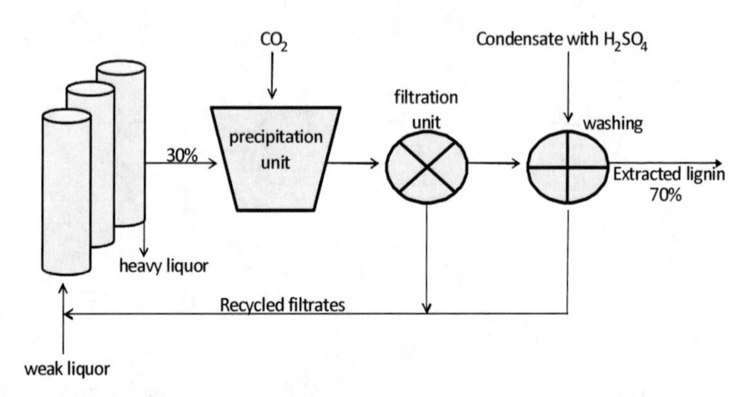

Figure 12.5 Schematic picture of the lignin precipitation process (adapted from Axelsson et al., 2006).

Unlike kraft lignins, it is not possible to precipitate lignosulfonates by changing the pH since sulfite lignin generally is soluble throughout the entire pH range. The recovery of sulfite lignin (lignosulfonate) is commonly done from waste pulping liquor concentrate after stripping and recovery of the sulfur. Precipitation of calcium lignosulfonate with excess lime (Howard process) is the simplest recovery method, and up to 95% of the liquor's lignin may be recovered (Bozell et al., 2007). It appears to be preferred especially when calcium lignosulfonate product is desired.

Amine extraction (AE) and UF are also among the most commonly used lignin recovery methods (Ringena et al., 2005). During isolation with long-chain amines, lignosulfonates are transferred into lignosulfonic acid-amine adducts, which are not soluble in water. The adducts are subsequently extracted by liquid-liquid-extraction and in this way purified lignosulfonates are obtained (Lin, 1992). Typical problems in the amine extraction of lignosulfonates are complete removal of the amine, the formation of NaCl during re-extraction, foam and emulsion problems, as well as the time-consuming separation procedure. In addition, the high lignosulfonate variability as regards to molar mass and the degree of sulfonation, complicates exhaustive extraction (Ringena et al., 2005). Thus, amine extraction can only be used for analytical purposes.

Ultrafiltration is a pressure-driven filtration technique that has long been studied for industrial and analytical purposes. Due to the development of improved membranes, UF became more important over the last decade. Investigations concerning UF in the pulp and paper industry usually focus on the production of fractions rich in sugar or lignin (Boggs, 1973; Collins et al., 1973; Bansal and Wiley,

1974; Tsapiuk et al., 1989). Typical problems discussed are flux, membrane fouling and the formation of a gel layer.

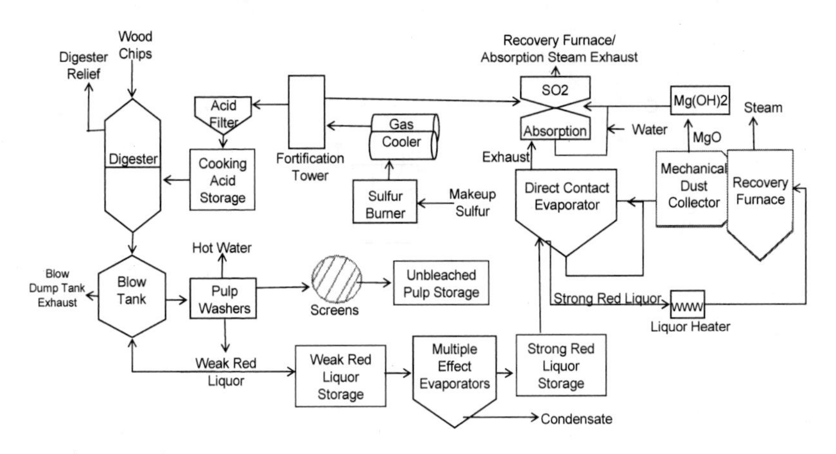

Figure 12.6 Simplified process flow diagram of magnesium – base sulfite pulping process (adapted from EPA, 1995).

12.4.1.3 Solvent (Organosolv) Lignin Recovery

Solvent (Organosolv) pulping is a general term for the separation of wood components through treatment with organic solvents in both acidic and alkaline conditions (Bozell et al., 2007). Due to the absence of sulfur compounds, solvent pulping have been recognized as an alternative of kraft and sulfite lignin, and is considered more environmentally acceptable and less capital intensive (Lora and Glasser, 2002).

Lignin in solvent processes can be recovered from solvent mixture by precipitation, which typically involves heating it in aqueous dioxane at elevated temperatures. Such an approach affords a lignin that retains much of its original structure in the form of β-O-4 inter-unit linkages (Bozell et al., 2007). A wide variety of solvents and combinations have been proposed for organosolv pulping.

The National Renewable Energy Laboratory (NREL) Clean Fractionation (CF) process as illustrated in Figure 12.7 is an example of the solvent (Organosolv) pulping process allowing the recovery of cellulose, hemicellulose and lignin separately. The solvent mixture selectively dissolves the lignin and hemicellulose components leaving the cellulose as an undissolved material that can be washed, fiberized, and further purified. The soluble fraction containing the lignin and hemicellulose is treated with water causing a phase separation to give an organic phase containing the lignin and an aqueous phase containing the hemicellulose (Bozell et al., 2007).

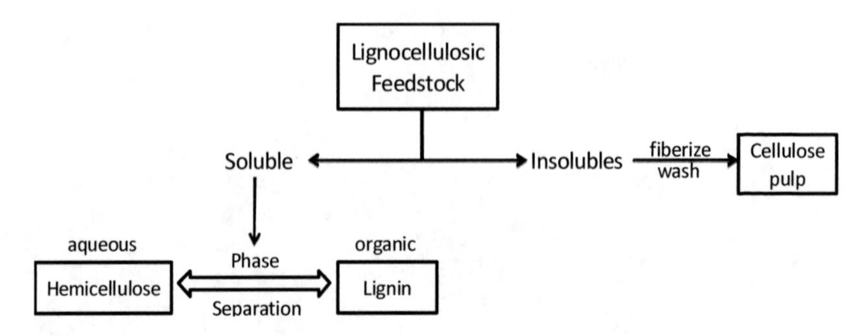

Figure 12.7 The NREL Clean Fractionation (CF) process (adapted from Bozell et al., 2007).

12.4.1.4 Soda Lignin Recovery

The major process elements of soda pulping are similar to the kraft process except that the sodium-base alkaline compound is used to dissolve in pulp preparation. Since sulfide is not used, soda pulping is environmentally acceptable like solvent pulping (Grace, 1987).

Recovery of lignins from soda pulping residual is based on precipitation, followed by liquid-solid separation and drying (Lora and Glasser, 2002). One of main difficulties in this process is that the silica may co-precipitate with lignin when applying to non-wood source materials. The LPS® precipitation process in Europe (Granit S.A. of Lausanne, VD, Switzerland) was recently introduced commercially to solve this problem and LPS® lignins recently became commercially available in powder and solution form.

12.4.2 Recovery from Biomass Conversion Processes during Ethanol Production

Agricultural and forest residues can be converted into ethanol by hydrolysis and subsequent fermentation (Hamelinck et al., 2005). In hydrolysis the cellulosic part of the biomass is converted into fermentable sugars. To increase the yield of hydrolysis, a pre-treatment step is required that softens the biomass and breaks down cell structures to a large extent. The pretreatments not only make the cellulose component susceptible to saccharification but also have the potential of generating sulfur-free lignin with the hemicellulose as a residue (Hamelinck et al., 2005). Figure 12.8 presents the generalized biomass-to-ethanol process.

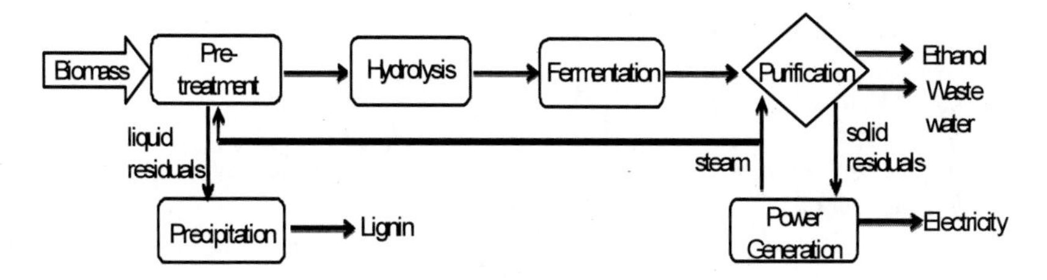

Figure 12.8 Generalized biomass to ethanol process diagram (adapted from Hamelinck et al., 2005).

The pretreatment methods can be classified into physical, chemical, or biological. The physical pretreatment is to clean and size the biomass, and destroy its cell structure to make it more accessible to further chemical or biological treatment. Desired sizes of the biomass after physical treatment vary from a few centimeters (Wooley et al., 1999) to 1–3 mm (Lynd, 1996). Common chemical pre-treatment methods use dilute acid, alkaline, ammonia, organic solvent, sulfur dioxide, carbon dioxide or other chemicals. Biological pre-treatments use fungi to solubilize/degrade the lignin (Graf and Koehler, 2000). Biodelignification is the biological degradation of lignin by microorganisms. Although at that time it was an expensive process, with low yields and long reaction time, and the microorganisms were poisoned by lignin derivatives. Biological pretreatment has the advantages of low energy use and mild environmental conditions. However, the very low hydrolysis rate impedes its implementation (Sun and Cheng, 2002). Biological treatments can be sometimes used in combination with chemical pretreatments (Graf and Koehler, 2000). The steam explosion pretreatment is promising alternative because it allows the recovery of all constitutive wood components without the destructive degradation of any one component in favor of other. However, it will require more development before sufficient conversion yields are guaranteed (Hamelinck et al., 2005).

The lignin from pretreated residues is recovered normally by precipitation (Lora and Glasser, 2002). The lignin recovered is largely insoluble in water under neutral or acidic conditions. It is soluble in organic solvents and in aqueous alkali, and it can be recovered with a low content of contaminants such as sugars and ash. Some lignins originating from the development of various biomass conversion processes have been available. Among these were commercial lignins including Sucrolin and Angiolin, produced on pilot-scale from residuals of the industrial production of furfural from sugarcane bagasse and was the designation used for a steam explosion lignin from hard wood, respectively. Currently, there are a number of lignin-based products in the development stage from cellulosic-ethanol processes (Lora and Glasser, 2002).

12.5 Lignin Utilization

12.5.1 Historical Uses

The usefulness of lignin is derived from the following positive or promoting factors (Lindberg et al., 1989):

- readily available in large amounts.
- if disposed of as waste, black liquor can be a serious environmental hazard.
- high energy content owing to the aromatic nuclei.
- a number of reactive points are present on the carbon skeleton which can be used for a wide range of substitution and addition reactions.
- good compatibility with several important basic chemicals.
- excellent colloidal and rheological properties, especially in the case of lignosulfonic acids.
- good adsorbent and ion exchange and adhesive properties.
- a direct source of various kinds of phenolic and aromatic compounds.

The first utilization of lignin in industry began in the 1880s when lignosulfonates were used in leather tanning and dye baths. Since then, a number of studies have been conducted to expand the use of lignin in many applications including the production of dyes, vanilla, plastics, base-exchange material for water softening, and the cleavage products of lignin from nitration, chlorinate, and caustic fusion (Cooper, 1942). Conventional sulfite lignin (lignosulfonates) is the most mature product among all types of lignin. Lignin Institute (2008) lists the following traditional applications that sulfite lignin (lignosulfonates) recovered from sulfite pulping can serve:

- ***Binder***: lignosulfonates are a very effective and economical adhesive, acting as a binding agent or "glue" in pellets or compressed materials. This binding ability makes it a useful component of coal briquettes, plywood and particle board, ceramics, animal feed pellets, carbon black, fiberglass insulation, fertilizers and herbicides, linoleum paste, dust suppressants, and soil stabilizers.
- ***Dispersant***: lignosulfonate prevents the clumping and settling of undissolved particles in suspensions. By attaching to the particle surface, it keeps the particle from being attracted to other particles and reduces the amount of water needed to use the product effectively. The dispersing property makes lignosulfonate useful in cement mixtures, leather tanning, clay and ceramics, concrete admixtures, dyes and pigments, gypsum board, oil drilling muds, and pesticides and insecticides.
- ***Emulsifier***: lignosulfonate stabilizes emulsions of immiscible liquids, such as oil and water, making them highly resistant to breaking. lignosulfonates are at work as emulsifiers in asphalt emulsions, pesticides, pigments and dyes, and wax emulsions.

- ***Sequestrant***: lignosulfonates can tie up metal ions, preventing them from reacting with other compounds and becoming insoluble. Metal ions sequestered with lignosulfonates stay dissolved in solution, keeping them available to plants and preventing scaly deposits in water systems. As a result, they are used in micronutrient systems, cleaning compounds, water treatments for boilers and cooling systems.

Even though the kraft pulping process is more prevalent compared to the sulfite lignin extraction, the industrial utilization of kraft lignin has been limited since most of the kraft lignin generated during pulping process is burned to generate heat energy for steam and power generation in the pulping plant (Lora and Glasser, 2002). Only very small amount of kraft lignin is isolated from black liquors in the United States and Europe (Gargulak and Lebo, 2000). However, since kraft pulping is the most common source to derive industrial lignin, numerous research studies have been conducted in an effort to find promising uses of kraft lignin.

The industrial utilization of kraft lignin reported in the literature include its use as an additive or extender in thermoset systems (Hatakeyama et al., 1989; Kelley et al., 1989), as phenol formaldehyde adhesives for plywood (Danielson and Simonson, 1998), the incorporation of kraft lignin into epoxy resins (Feldman, 2002), and the production of the thermoplastic materials (Li and Sarkanen, 2002; Gosselink et al., 2004) and carbon fibers (Kadla et al., 2002a; Kadla et al., 2002b) from kraft lignin.

12.5.2 Civil Engineering Infrastructure Applications

12.5.2.1 Pavement Geomaterials

Soil Stabilization. A good road (paved or unpaved) requires a suitable foundation which in turn requires stability. The degree of stability is primarily a function of the road material resistance to lateral movement or flow (USDOT, 1976). Different types of road material employ different mechanism for resisting lateral movement. In general, granular soils count on their particle sizes, angularity, and interlocking ability to develop the internal friction required to resist lateral flow. However, in fine-grained soils such as clay soils, the stability is very much moisture dependent.

There are many varieties of soil available for road construction. Unfortunately, many of the soil deposits do not naturally possess the requisite engineering properties to serve as a good foundation material for roads and highways. As a result, soil-stabilizing additives or admixtures are used to improve the properties of less-desirable road soils (ARBA, 1976). When used these stabilizing agents can improve and maintain soil moisture content, increase soil particle cohesion, and serve as cementing and waterproofing agents (Gow et al., 1961; ARBA, 1976). Unpaved road dust suppressants are considered soil additives because they produce changes in soil characteristics that influence soil stabilization (Gow et al., 1961; Ross and Woods, 1988). Many factors influence soil stabilization. The most notable factors are the physical and chemical properties of the soil and the chemical additive. The stabilization effect of a soil additive is measured in terms of the increase in shear strength of the soil-additive mixture.

Lignin has been implicated as having a positive role in soil stabilization (Kozan, 1955; Nicholls and Davidson, 1958; Johnson et al., 2003). Adding lignin to clay soils increases the soil stability by causing dispersion of the clay fraction (Davidson and Handy, 1960; Gow et al., 1961). According to Gow et al. (1961), the dispersion of the clay fraction benefits stability of the soil-aggregate mix by: a) plugging voids and consequently improving water tightness and reducing frost susceptibility, b) eliminating soft spots caused by local concentrations of binder soil, c) filling voids with fines thus increasing density, and d) increasing the effective surface area of the binder fraction which results in greater contribution to strength.

It has been demonstrated that lignin (a natural polymer) introduces better improvement for ground modification compared to non-organic stabilizers (Palmer et al., 1995). Lignin is also used in combination with other chemicals to achieve soil improvement (Puppala and Hanchanloet, 1999). Lignin as a soil additive causes dispersion of the clay fraction of some soils resulting in the shear strength increase of the soil (due to particle rearrangement) (Addo et al., 2004). Various studies on lignin as a soil additive have concluded that lignin is primarily a cementing agent (Woods, 1960; Ingles and Metcalf, 1973; Landon et al., 1983). In most of these studies, sulfite lignin (lignosulfonates) has been utilized.

Laboratory methods as well as onsite testing have been done to quantify soil stabilization using chemical additives including ligninsulfonate. In one such study, Lane et al. (1984) used laboratory methods to measure soil cohesion increase resulting from the addition of some commercially available chemical additives. The laboratory methods included the unconfined compression test (ASTM test No. C-39) and a modified wet sieve analysis test (ASTM test No. C-117). The testing was performed at three sample-drying conditions, 24-hour air-dried, 24-hour bag cured, and immediate sample testing. Figure 12.9 shows the resulting cohesive strength measured for the 24-hour air-dried test conditions. The results indicate that each additive tested varies in cohesive strength with a range of 4-55 psi. The calcium ligninsulfonate at each of the initial aggregate moisture content (4, 6 and 8%) showed a higher cohesive strength than the petroleum-based additives.

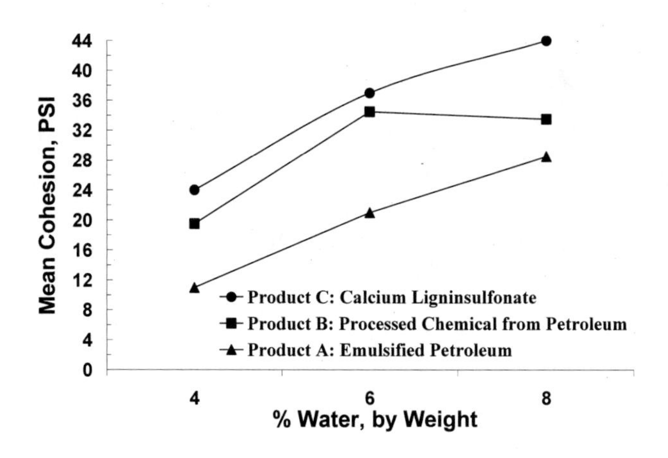

Figure 12.9 Effect of moisture content on the cohesion of treated aggregate for the 24-hour air-dried test condition (adapted from Lane et al., 1984).

In the past, several studies have been conducted at Iowa State University on the use of lignin as stabilizing agents on r Iowa's silty loam and loess soils (Sinha et al., 1957; Nicholls and Davidson, 1958; Hoover et al., 1959; Demirel and Davidson, 1960; Gow et al., 1961). Sinha et al. (1957) found that lignins used alone as admixtures do not show much promise as stabilizing agents for loess or loess-derived soils. However, their investigations indicated that lignins should be much more effective as stabilizing agents for granular soils or soil aggregate mixtures.

The Quebec Department of Roads conducted laboratory tests comparing the engineering properties of lignin-treated aggregate with that of raw aggregate and clay-mixed gravel (Hurtubise, 1953). The bearing capacity of the aggregate treated with 1.2 percent lignin was higher than that of the raw aggregate soil and clay-mixed aggregate. Cohesive strength increased with the addition of 2 percent lignin. The strength increase was also found to be nearly linearly proportional to the amount of lignin used. Water absorption tests indicated that water absorption through capillary action was reduced substantially. Moisture density relationship tests showed that an increase in the amount of lignin added to the soil increased the density and reduced the optimum moisture content.

In a low-volume road study, laboratory methods were used to evaluate the strength and density modification of unpaved road soils because of chemical additives (Palmer et al., 1995). The additives tested included lignin, $CaCl_2$ and $MgCl_2$ at different concentrations. Three different road soil materials with different soil classifications were used. The seven-day air-cured samples exhibited large strength increases for the lignin-treated specimens at all concentration levels. For each of the soils tested, lignin provided the highest increase in strength as determined by the unconfined compression tests. Figure 12.10 illustrates some results for soil 1B – classified as A-1-b by AASHTO designation M-145, and SM by USCS (PI = 0).

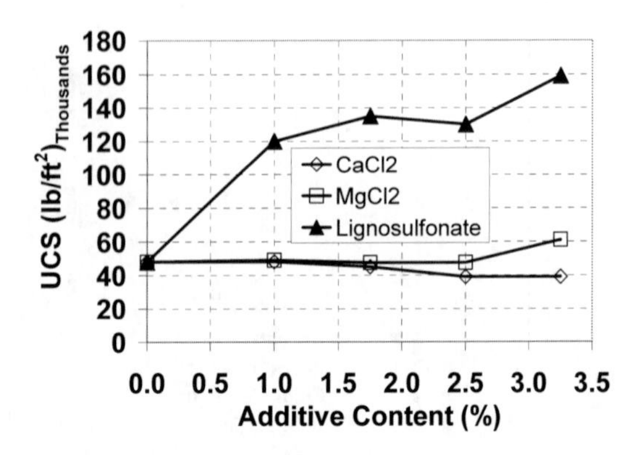

Figure 12.10 Average peak Unconfined Compressive Strength (UCS) for specimens tested dry (adapted from Palmer et al., 1995).

Puppala and Hanchanloet (1999) studied the effect of a new chemical treatment method using a liquid comprising of sulfuric acid and lignosulfonate stabilizer (SA-44/LS-40 or DRP) on the shear strength and plasticity characteristics of soils. Three soil types including a silty clay (raw soil) and two types of lime stabilized raw soils, two chemical dilution rates and curing periods were investigated. The percent increase in unconfined compressive strength (qu) with the SA-44/LS-40 treatment ranged between 30% to more than 130% for the soils evaluated in this study. The increase in strength properties was attributed to the formation of chemical bonds between soil particles. The lignosulfonate-based chemical treatment increased the resilient moduli of soils, which is important from the point of mechanistic design of flexible pavements.

Nicholls and Davidson (1958) confirmed that lignin admixtures indeed do improve some engineering properties related to stability of soils. They also reported that the strength of lignin-treated soil increases rapidly with an increase in the length of air curing. Also, lignin is considered biodegradable, therefore its presence in the environment can be considered less harmful compared to industrial by-products such as fly-ash. An ongoing study at Iowa State University is currently investigating the feasibility of utilizing ethanol co-products and bio-oil as stabilizing agents for pavement geomaterials (Ceylan and Gopalakrishnan, 2007; Ceylan et al, 2009).

Dust control on unpaved roads. A large portion of U.S. road network is made up of unpaved roads that usually carry a very small volume of the nation's vehicular traffic. The use of unpaved roads causes dust emission into the atmosphere, loss of the road surface material over time, and frequent road surface deterioration in the form of ruts, washboarding, and potholes. Influenced by the traffic volume, these problems can lead to high economic cost. For instance, Hoover et al. (1981) reported that in 1978 the secondary road departments of Iowa's 99 counties spent about $32 million for aggregate replacement.

To reduce the loss of road surface fines in the form of dust, chemical additives (dust suppressants) are applied to the unpaved road surface to control dust generation and to improve the road surface stability. Commonly used dust suppressants are chloride compounds, such as magnesium chloride ($MgCl_2$), calcium chloride ($CaCl_2$), and ligninsulfonate (also known as "tree sap"). Hygroscopic salts, such as calcium and magnesium chloride depend on absorbing moisture from the air to reduce dust, while ligninsulfonates work by adding cohesion to the aggregate material for the same purpose.

A major disadvantage with chlorides is that they are water soluble and tend to migrate downwards through the roadway. They are easily washed away by rain, which may imply that more than one treatment per year is required. Another problem regarding chlorides is that they are corrosive to most metals. Ligninsulfonate creates a hard surface crust on the road but the product is highly water soluble and therefore tends to leach from the road during heavy rainfall, which implies that this dust suppressant is much more effective in a dry than in a wet environment (Oscarsson, 2007).

Ligninsulfonate were first utilized to control duct on unpaved roads in Sweden about the 1910s (Arnfelt, 1939). The Institute of Road Research in Sweden in their dust control experiment with ligninsulfonate reported that it reacted well with dust and bound at the particle together if road surface was rich in clay. Several states in the U.S also made much use of ligninsulfonate as a dust suppressants on base with impervious wearing courses in 1930s (Sinha et al., 1957).

Many studies have already confirmed that lignin derivatives act as effective dust suppressants on unpaved roadways. The state of Idaho in 1937 used concentrated ligninsulfonate for surface treatment on several roads and found that the binder gave relief from dust and have conserved surface aggregate during the summer (Sinha et al., 1957). The city of Spokane, Washington, reported in 1943 that diluted sulfite liquor had been used for twelve years as dust layer on secondary city street, and indicated the results obtained were very satisfactory (Sinha et al., 1957).

In a study conducted in Taylor county, Iowa, the effectiveness of lignin was tested on three test sections using 0.75, 1.00 and 1.25 percent by soil weight of Bindtite, a lignosulfonate (Blanc, 1975). The sections ranged in length from 0.25 to 1.25 miles and before the treatments, they were experiencing complaints of dust, potholes, washboarding, frost heaving and general deterioration of the surface. Preconstruction consisted primarily of blade dressing of the road shoulders and ditch line and reclaiming of aggregate pullout. Finally the surface was scarified, mixed with the lignin and compacted into a 6-inch thick wearing course. The treated sections did not receive a seal coat until way over a year after construction. During this period, all the sections remained nearly free of dust, potholes or washboarding and the surfaces remained relatively dry and solid after each rain.

Addo et al. (2004), at Colorado State University, used field-based methods to measure the effect of road soil characteristics on the effectiveness of some commonly used dust suppressants in the context of unpaved road maintenance. The effect of dust suppression on the stabilization of the road soils in general was also examined. Safety and environmental concerns arising from the use of dust suppressants were reviewed as well. Field study results showed that the in-depth application of the lignin produced a road surface that was firm, smooth, dust free, and comfortable to drive for most of the test period of nearly a year. Field observation of the lignin-treated test sections indicated that the lignin acted like cement, binding the soil particles together into a hard surface that show strength gains over time (Addo et al., 2004).

12.5.2.2 Asphalt

Lignin has been used as emulsifiers in asphalt emulsions due to its sequestering and dispersing properties (Lignin Institute, 2008). Several laboratory experiments have been conducted to examine the use of lignin from wood pulping as a substitute or an extender for asphalt in paving mixtures (Terrel and Rimstritong, 1979; Sundstrom et al., 1983; Kandhal, 1992). Sundstrom et al. (1983) found that the effects of increasing lignin content in asphalt and asphalt mixture were to increase the viscosity of asphalt and the stability of asphalt mixture among the performance properties. Terrel and Rimstritong (1979) also reported that the asphalt mixtures containing lignin-asphalt binders can be designed to achieve comparable performance of asphalt mixtures containing conventional materials. The study indicated that the lignins have no adverse effects on asphalt paving materials but none of the lignins investigated were suitable as a complete substitute for asphalt in paving materials.

The oxidation of asphalt contributes to the long term failure in asphalt pavements (Herrington et al., 1994; Liu et al., 1998; Domke et al., 2000). The complex reaction of the various chemicals present in asphalt with atmospheric oxygen causes the age hardening of asphalt during mixing and services, which can eventually lead to brittle failures of asphalt pavement associated with traffic load and moisture (Roberts et al., 1996; Williams and Nicolaus, 2008). The rate of oxidative aging primarily depends on the charter of asphalt and temperature (Roberts et al., 1996). The addition of polymers in asphalt can improve the performance properties in low and high temperature resulting in the increase of failure resistance during oxidative aging (Ruan et al., 2003). However, none of polymers available truly act as an antioxidant by preventing aging from oxygen radicals (Lucena et al., 2004). Recently, attempts have been made to investigate the use of lignin from wood pulping as an antioxidant in asphalt (Bishara et al., 2005; Guffey et al., 2005a; Guffey et al., 2005b). These studies imply that lignin-modified asphalt can decrease the rate of oxidation without adverse effects on the other asphalt performance properties.

12.5.2.3 Cement and Concrete

The workability is the property of fresh concrete that determines the working characteristic in mixing, placing, molding, and finishing. Cement compositions are brought into a workable form by mixing the solid components with an amount of water which is greater than that required to hydrate the cement components therein. The mixed mineral binder composition is poured into a form and allowed to harden at atmospheric temperature. During the hardening, some of the excess water remains, leaving cavities in the formed structural unit and, thus, reduces the mechanical strength of the resultant unit. It is well known that the compressive strength of the resultant structure generally bears an inverse relationship to the water-cement ratio of the starting mix. The need to use smaller quantities of water is limited by the required flow and workability properties of the fresh mixture. High temperature of fresh concrete often reduces setting time of concrete, which has adverse effect on workability.

The use of ligninsulfonate as an admixture in concrete has been known for more than 60 years (Zhor and Bremner, 1999). Ligninsulfonate has been used as a water reducing and a set-retarding admixture to reduce water and offset the effects of high temperature without losing workability (Midness, et al., 2002). The hydrophilic modified aromatic structures of lignin can reduce the amount of water necessary in a concrete to reach a certain fluidity resulting in the improvement of the concrete's final strength (Nadif et al., 2002). However, the dosage of Ligninsulfonate should be controlled to prevent retarding a wanted initial concrete strength. Zhor and Bremner (1999) investigated the effect of ligninsulfonate dosage rate on fresh concrete properties and concluded that the highest dosage rate always cause the high performance in set retardation. Typical dosages of lignosulfonates in concrete are 0.1–0.3% by weight of cement (Plank and Winter, 2003).

The purpose of ligninsulfonate in ready-mix concrete and pre-cast concrete is to produce concrete with an improved pourability at the job-site and to obtain high-strength concrete (Plank, 2004). The effectiveness of lignosulfonate as a water reducer is limited compared to other water-reducing admixtures, but lignosulfonate is mainly used because of its cost-competitiveness.

12.5.3 Utilization of Lignin from Cellulosic-Ethanol Plants

The lignins obtained from cellulosic-ethanol processes are sulfur-free. The other sulfur-free lignins include organosolv lignins and soda lignin. Even though sulfur-free lignins have been known for many years, the use of sulfur-free lignin has recently gained interest as a result of diversification of biomass processing schemes (Lora and Glasser, 2002). Sulfur-free lignin can be used in many thermosetting applications with environmental advantages. Recently attempts are being made to use sulfur-free lignin in civil engineering infrastructure applications. Some recent sulfur-free lignin utilization examples are presented here.

12.5.3.1 Sulfur-Free Lignin in Thermosetting Applications

Several uses of sulfur-free lignin have been proposed and demonstrated. Most feasible utilization of sulfur-free lignin is in thermosetting applications in conjunction with phenolic, epoxy, or isocyanate resins (Lora and Glasser, 2002).

Lignin should be an obvious candidate for application as a phenol substitute in phenol powder resins used as the binder in the manufacture of friction products. However, its chemical heterogeneity is a limiting factor. Several approaches have been introduced and used to overcome this limitation (Stewart, 2008). One of the methods include (bio)chemical modification of the lignin including reaction with enzyme systems to oxidatively crosslink the lignin (Popp et al., 1991) or pre-reaction of the lignin (Doering, 1993). The addition of filler agent is also found to enhance reaction of lignin in phenolic resins (Peng and Riedl, 1994). Lignins derived from advancing processes were shown to be effective phenol diluents in phenol-based resin systems. For example, the organosolv lignins have already been used as a partial replacement for phenolic resins (Lora and Glasser, 2002). A North American producer of Oriented Strand Board (OSB) successfully used organosolv lignins to produce construction panels in the early 1990s. OSB panels pressed with 5 – 25% organosolv lignin loading in the phenol formaldehyde adhesive resin were reported to exhibit no drop in mechanical properties compared to the controls (Lora and Glasser, 2002).

Lignin has also been used in epoxy resins, and many different formulation approaches have been explored (Lora and Glasser, 2002). The lignin to be used in epoxy resins requires freedom from impurities which can be difficult to achieve in wood pulping lignin. Whereas conventional kraft and soda lignins proved to be hampered by high ionic contents, requiring extensive post-isolation purification protocols, organosolv lignin could be used directly in epoxy resins (Lora and Glasser, 2002). The most common utilization of ligin as epoxy resins is the incorporation of lignin into the resin for fabrication of Printed Circuit Boards (PCBs) by IBM (Kosbar et al., 2001). This resin has physical and electrical properties similar to those of common laminate resins (Kosbar et al., 2001).

Polyurethane foams and resins have been made using lignin mixed with poly(ethylene glycol) and isocyanates. From the industrial test results of lignin and non-lignin derived foams, Lora and Glasser (2002) concluded that the use of lignin in foam products has great potential, especially at elevated temperature and humidity, but friability is recognized as a limitation.

12.5.3.2 Sulfur Free-Lignin in Civil Engineering Infrastructure Applications

Attempts have been made to use sulfur-free lignins as water-reducing admixtures in concrete industry (Creamer et al., 1997). Furthermore, sulfur-free lignins were applied as plasticizers of cement (Every and Jacob, 1978) and as concrete admixture (Chang and Chan, 1995). Sulfur-free lignins from a soda pulping mill were also directly used in mortar to increase final concrete strength and to prevent corrosion in armed concrete (Ali and El-Sabbagh, 1998). Nadif et al. (2002) found that all sulfur-free lignins tested in their study improved the flow of the mortar and are comparable to non-sulfur free lignin.

It has been hypothesized that since the by-product of corn stover fermentation is high in lignin, which is thought to play a role in stabilizing soil, soil incorporation of the by-product may help maintain or improve soil structure and stability (Johnson et al., 2004). The authors conducted the first study to evaluate the impact of corn stover fermentation by-product on biological, chemical, and physical properties of soils. The utilization of lignin based by-products from corn (stover or grain) fermentation in pavement base stabilization need to be investigated as it is hypothesized that one may achieve a stronger base stabilization possibly reducing the need for as much base material through this innovative approach. Currently, researchers at Iowa State University are investigating the feasibility of this approach (Ceylan and Gopalakrishnan, 2007; Ceylan et al, 2009). Research studies are also being carried out to develop a sulfur-free lignin-based pavement deicer as well as lignin-based dust control agent for suppressing dust on unpaved roads.

The feasibility of utilizing lignin from ethanol production as an asphalt antioxidant has also been investigated. Williams and McCready (2007) examined the performance properties of asphalt combined with lignin-containing co-products from ethanol production. The authors found that lignin co-products have resulted in improvement of performance in widening the temperature range and decreasing some oxidative aging products.

12.6 Summary

Lignin is considered as nature's most abundant aromatic polymer co-generated during papermaking and biomass fractionation. There are different types of lignins depending on the source (hardwood, softwood, annual crops, etc.) and recovery process. Conventional kraft and sulfite lignins, co-products of papermaking and pulp industry, have been traditionally used as dispersants and binders. Recently, an emerging class of lignin products, namely sulfur-free lignins, extracted during biomass conversion for alcohol production, solvent pulping, and soda pulping, have generated interesting new applications owing to their versatility.

Ethanol derived from biomass is often advocated as a significant contributor to possible solutions to our need for a sustainable transportation fuel. In recent years, research and development efforts are gearing towards the expansion of biofuel production capability worldwide through the use of a range of agricultural and non-agricultural cellulosic biomass (e.g., corn stover, forest products waste, switch grass, municipal solid wastes). Thus, lignin will become available in abundance as a waste product of these biorefineries. Considering the wide range of applications in which cellulosic-ethanol plants-derived lignin could be used, more research is needed to enable broader use of sulfur-free lignins as innovative infrastructure materials which can lead to dramatic breakthroughs for improving the durability, efficiency, economy, environmental impact, and sustainability aspects of civil engineering infrastructure systems.

12.7 References

Addo, J.Q., Sanders, T.G., and Chenard, M. (2004). *Road Dust Suppression: Effect on Maintenance Stability, Safety and the Environment Phases 1-3*. Mountain-Plains Consortium (MPC) Report No. 04-156, Fargo, ND.

Alén, R., Patja, P., and Sjöström, E. (1979). "Carbon dioxide precipitation of lignin from kraft black liquor." *Tappi J.*, 62 (11), 108–110.

Alén, R., Sjöström, E., and Vaskikari, P. (1985). "Carbon dioxide precipitation of lignin from alkaline pulping liquors." *Cell. Chem. Technol.*, 19 (5), 537–541.

Ali, A.H., and El-Sabbagh, B. (1998). "Role of waste products in increasing concrete strength and protecting against reinforcement corrosion." *Corrosion Prevention and Control*, 45 (6), 173–180.

ARBA (1976). *Materials for Stabilization.* American Road Builders Association (ARBA), Washington, DC.

Arnfelt, H. (1939). *Nägra Undersöknigar Av Sulfitlut (Some Investigations on Sulfite Waste Liquor -English translated title)*. Rapport 14, Statens Vaginstitut, Stockholm, Sweden.

Axelsson, E., Olsson, M., and Berntsson, T. (2006). "Increased capacity in kraft pulp mills: lignin separation combined with reduced steam demand compared with recovery boiler upgrade." *Nord. Pulp Paper Res. J.*, (21)4, 485-492.

Bansal, I.K., and Wiley, A.J. (1974). "Fractionation of spent sulfite liquor using ultrafiltration cellulose acetate membranes." *Environ. Sci. Technol.*, 8(13), 1085–1090.

Bishara, S.W., Robertson, R.E., and Mahoney, D. (2005). "Lignin as an antioxidant: a limited study on asphalts frequently used on Kansas roads." *42nd Petersen Asphalt Research Conference*, Cheyenne, WY.

Blanc, T.R. (1975). "Lignosulfonate stabilization." *Presented at the ARTBA-NACE National Conference on Local Transportation*, Des Moines, IA.

Boerjan, W., Ralph, J., and Baucher, M. (2003). "Lignin biosynthesis." *Annual Review of Plant Biology*, 54(1), 519-546.

Boggs, L.A. (1973). "Isolation of reducing sugars from spent sulfite liquor." *Tappi*, 56 (9), 127–129.

Boopathy, R. (1998). "Biological treatment of swine waste using anaerobic baffled reactors." *Bioresour. Technol.*, 64, 1–6.

Bozell, J.J., Holladay, J.E., Johnson, D., and White, J.F. (2007). *Top Value Added Chemicals from Biomass- Volume II: Results of Screening for Potential Candidates from Biorefinery Lignin.* PNNL-16983, Pacific Northwest National Laboratory (PNNL) and the National Renewable Energy Laboratory (NREL), Richland, WA.

Caballero, J.A., Font, R., and Marcilla, A. (1997). "Pyrolysis of kraft lignin: yields and correlations." *J. Anal. Appl. Pyrol.*, 39 (2), 161–183.

Ceylan, H., and Gopalakrishnan, K. (2007). "Improving soil strength under roads with ethanol co-products." *Scitizen: Bringing Science Closer to Society,* A Open Science News Source, Take Part Media Inc., Available at http://scitizen.com/screens/blogPage/viewBlog/sw_viewBlog.php?idTheme=14&idContribution=1322.

Ceylan, H., Gopalakrishnan, K., and Kim, S. (2009). "Use of bio-oil for pavement subgrade soil stabilization" *Proceedings of 2009 Mid-Continent Transportation Research Symposium*, CD ROM, InTrans, Iowa State University, Ames, IA.

Chang, D.Y., and Chan, S.Y.N. (1995). "Straw pulp waste liquor as a water-reducing admixture." *Magazine of Concrete Research* 47 (171), 113–118.

Cheung, S.W., and Anderson, B.C. (1997). "Laboratory investigation of ethanol production from municipal primary wastewater." *Bioresour.Technol.*, 59, 81–96.

Collins, J.W., Boggs, L.A., Webb, A.A., and Wiley, A.A. (1973) "Spent sulfite liquor reducing sugar purification by ultrafiltration with dynamic membranes." *Tappi*, 56 (6), 121–124.

Cooper, H.H. (1942). *Commercial Utilization of Lignin.* Ph.D thesis, Iowa State University, Ames, IA.

Cortinas, S., Luque, S., Álvarez, J.R., Canaval, J., and Romero, J. (2002). "Microfiltration of kraft black liquors for the removal of colloidal suspended matter (pitch)." *Desalination*, 147 (1-3), 49–54.

Creamer, A.W., Blackner, B.A., and Laura, H. (1997). "Properties and potential applications of a low-molecular-weight lignin fraction from organosolv pulping." Presentation in *International Symposium on Wood and Pulping Chemistry, Proceedings*, ISWPC., V2, Canadian Pulp & Paper Assoc., Montreal, Que., Canada, p. 21-1–21-4.

Danielson, B., and Simonson, R. (1998). "Kraft lignin in phenol formaldehyde resin; Part 1, Partial replacement of phenol by kraft lignin in phenol formaldehyde adhesives for plywood." *Journal of Adhesion Science and Technology*, 12(9), 923-939.

Davidson, D.T., and Handy, R.L. (1960). "Soil stabilization." Section 21 in *Highway Engineering Handbook*, Wood, K.B. (Ed.), McGraw-Hill.

Davy, M.F., Uloth, V.C., and Cloutier, J.N. (1998). "Economic evaluation of black liquor treatment for incremental kraft pulp production." *Pulp Paper Can.*, 92 (2), 35–39.

Demirel, T., and Davidson, D.T. (1960). "Stabilization of a calcareous loess with calcium lignosulfonate and aluminum sulfate." Section in *Soil Stabilization with Chemicals*, Joint Publication – Bulletin 193 of the Iowa Engineering Experiment Station and Bulletin 22 of the Iowa Highway Research Board, Davidson, D.T., and Associates (Eds.), Ames, IA, pp. 206-221.

Dence, C.W., and Lin, S.Y. (1992). "Introduction." Section in *Methods in Lignin Chemistry*, Lin, S.Y., and Dence, C.W. (Eds.), Springer-Verlag, Berlin, pp. 3-19.

Dewes, T., and Hünsche, E. (1998). „Composition and microbial degradability in the soil of farmyard manure from ecologically-managed farms." *Biol. Agric. Hortic.*, 16, 251–268.

Doering, G.A. (1993). *Lignin Modified Phenol-formaldehyde Resins*. U.S. Patent 5202403.

Domke, C.H., Davidson, R.R., and Glover, C.J. (2000). "Effect of oxygen pressure on asphalt oxidation kinetics." *Industrial Engineering Chemistry*, 39(3), 592-598.

EPA. (1995). "Chemical wood pulping." Chapter 10.2 in *Compilation of Air Pollutant Emission Factors, Volume 1: Stationary Point and Area Sources*, United States Environmental Protection Agency, Washington, DC.

Eriksson, P. (1979). "Ultrafiltration for recovery of lignosulfonates from spent sulfite liquor." *AlChE Symp. Series*, 76 (197), 316–320.

Every, R.L., and Jacob, J.T. (1978). *Production of Raw Mix Cement Slurries Having Reduced Water Content*. US Patent 4115139.

Feldman, D. (2002). "Lignin and its polyblends – a review." Section in *Chemical, Modification, Properties, and Usage of Lignin*, Hu, T.Q. (Ed.), Kluwer Academic/Plenum Publishers, New York, pp. 81-99.

Forss, K., Kokkonen, R., Sirelius, H., and Sagfors, P.E. (1979). "How to improve spent sulphite liquor use." *Pulp Paper Can.*, 80 (12), 411–415.

Freudenberg, K. (1968). "The constitution and biosynthesis of lignin." Section in *Constitution and Biosynthesis of Lignin*, Neish, A.C., and Freudenberg, K. (Ed.), Springer-Verlag., Berlin, pp. 47-122.

Gargulak, J.D., and Lebo, S.E. (2000). "Commercial use of lignin-based materials." Section in *Lignin: Properties and Materials*, ACS Symp. Ser. No. 742, Scultz, T.P. (Ed.), American Chemical Society, Washington, DC., pp. 304–320.

Glasser, W.G., Barnett, C.A., and Sano, Y. (1983). "Classification of lignins with different genetic and industrial origins." *Applied Polymer Symposium*, 37, 441-460.

Glasser, W.G., and Kelley, S.S. (1987). "Lignin." Section in *Encyclopedia of Polymer Science and Engineering*, Vol. 8, 2nd edition. Mark, H.F., and Kroschwits, J.I.. (Eds.), John Wiley & Sons, Inc., New York, pp. 795-852.

Glasser, W.G. (2001). "Lignin based polymers." Section in *Encyclopedia of Materials: Science and Technology*, Beall, F.C. (Ed.), Elsevier Science Ltd., Oxford, UK, pp. 4504-4508.

Gosselink, R.J.A., Snijder, M.H.B., Kranenbarg, A., Keijsers, E. R. P., Jong, E. D., and Stigsson, L. L. (2004). "Characterisation and application of NovaFiber lignin." *Industrial Crops and Products*, 20(2), 191-203.

Gow, A.J., Davidson, D.T., and Sheeler, J.B. (1961). "Relative effects of chlorides, lignosulfonates and molasses on properties of a soil-aggregate mix." *Highway Research Board Bulletin,* 282, 66-83.

Grace, T.M. (1987). "Chemical recovery technology - a review." *IPC Technical Paper Series*, No. 247, The Institute of Paper Chemistry, Appleton, WI.

Graf, A., and Koehler, T. (2000). *Oregon Cellulose-Ethanol Study*. Oregon Office of Energy, Salem, OR.

Guffey, F.D., Robertson, R.E., and Hettenhaus, J.R. (2005a). "The use of lignin as antioxidant to improve highway pavement performance." *The World Congress on Industrial Biotechnology and Bioprocessing*, Orlando, FL.

Guffey, F.D., Robertson, R.E., Bland, A.E., and Hettenhaus, J. R. (2005b). "Lignin as antioxidant in petroleum asphalt." *27th Symposium on Biotechnology for Fuels and Chemicals*, Denver, CO.

Hamelinck, C.N., Hooijdonk, G.V., and Faaij, A.PC. (2005). "Ethanol from lignocellulosic biomass: techno-economic performance in short-, middle-and long-term." *Biomass and Bioenergy*, 28(4), 384-410.

Hatakeyama, H., Hirose, S., and Hatakeyama, T. (1989). "High-performance polymers from lignin degradation products." Section in *Lignin: Properties and Materials*, ACS Symp. Ser. No. 397, Glasser, W.G., and Sarkanen, S. (Eds.), American Chemical Society, Washington, DC., pp. 402–413.

Herrington, P.R., Patrick, J.E., and Ball, F.A. (1994). "Oxidation of roading asphalts." *Industrial Engineering Chemistry*, 33 (11), 2801-2809.

Hill, M.K., Violette, D.A., and Woerner, D.L. (1988). "Lowering black liquor viscosity by ultrafiltration." *Sep. Sci. Techn.*, 23 (1-2), 1789–1798.

Hoover, J.M., Davidson, D.T., Plunkett, J.J., and Monoriti, E.J. (1959). "Soil-organic cationic chemical-lignin stabilization." *Highway Research Board Bulletin*, 241, 1-13.

Hoover, J.M., Fox, D.E., Lustig, M.T., and Pitt, J.M (1981). *Mission-Oriented Dust Control and Surface Improvement Processes for Unpaved Roads*. Final Report, Iowa Highway Research Board (IHRB) Project H-194, Iowa DOT, Ames, IA.

Hurtubise, J.E. (1953). "Soil stabilization with lignosol." *Canadian Chemical Process*, 37, 58-61.

ILI. (2008). *the International Lignin Institute*. Available at http://www.ili-lignin.com/aboutus.php.

Ingles, O.G., and Metcalf, J.B. (1973). *Soil Stabilization: Principles and Practices*. John Wiley and Sons, New York.

Jeffries, T.W. (1994). "Biodegradation of lignin and hemicelluloses." Section in *Biochmistry of Microbial Degradation*, Ratledge, C. (Ed.), Kluwer Academic Publishers, Dordrecht, Netherlands. pp. 233-277.

Johnson, J.M., Carpenter-boggs, L., and Lindstrom, M.J. (2003). "Humic acid and aggregate stability in amended soils." *Proceedings of the Natural Organic Matter in Soils and Water North Central Region Symposium*, 21.

Johnson, J.M.F., Reicosky, D., Sharratt, B., Lindstrom, M., Voorhees, W., and Carpenter-Boggs, L. (2004). "Characterization of soil amended with the by-product of corn stover fermentation." *Soil Sci. Soc. Am. J.*, 68, 139–147.

Kadla, J.F., Kubo, S., Venditti, R.A, Gilbert, R.D., Compere, A.L., and Griffith, W.(2002a). "Lignin-based carbon fibers for composite fiber applications." *Carbon*, 40(15), 2913-2920.

Kadla, J.F., Satoshi, K., Gilbert, R.D., and Venditti, R.A. (2002b). "Lignin-based carbon fibers." Section in *Chemical, Modification, Properties, and Usage of Lignin*, Hu, T. Q. (Ed.), Kluwer Academic/Plenum Publishers, New York, pp. 121-137.

Kandhal, P.S. (1992). *Waste Materials in Hot-Mix Asphalt – An Overview*. NCAT Report No.92-6, Auburn University, AL.

Kelley, S.S., Glasser, W.G., and Ward, T.C. (1989). "Effect of soft-segment content on the properties of lignin-based polyurethanes." Section in *Lignin: Properties and Materials*, ACS Symp. Ser. No. 397, Glasser, W.G., and Sarkanen, S. (Eds.), American Chemical Society, Washington, DC., pp. 402–413.

Keyoumu, A., Sjödahl, R., Henriksson, G., Ek, M., Gellerstedt, G., and Lindström, M.E. (2004). "Continuous nano- and ultrafiltration of kraft pulping black liquor with ceramic filters: a method for lowering the load on the recovery boiler while generating valuable side-products." *Ind. Crops Products*, 20 (2), 143–150.

Kirkman, A.G., Gratzl, J.S., and Edwards, L.L. (1986). "Kraft lignin recovery by ultrafiltration: economic feasibility and impact on the kraft recovery system." *Tappi J.*, 69 (5), 110–114.

Kosbar, L.L., Gelorme, J., Japp, R.M., and Fotorny, W.T. (2001). "Introducing biobased materials into the electronics industry." *J. Ind. Ecol.*, 4, 93–98.

Kovasin, K., and Nordén, H.V. (1984). "Determination of lignosulfonate rejection from test results in the ultrafiltration of spent sulfite liquor." *Sven. Papperstidn.*, 87 (6), 44–47.

Kozan, G.R. (1955). *Summary Review of Lignin and Chrome-Lignin Processes for Soil Stabilization.* Miscellaneous paper, U.S. Army Engineer Waterways Experiment Station, Vicksburg, MS.

Landon, B., and Williamson, R.K. (1983). "Dust-abatement materials: evaluation and selection." *Transportation Research Record*, 898, 250-257.

Lane, D.D., Baxter, T.E., Cuscino, T., and Coward, C.Jr. (1984). "Use of laboratory methods to quantify dust suppressants effectiveness." *Transactions of the Society of Mining Engineers of the American Institute of Mining, Metallurgical, and Petroleum Engineers*, Vol. 274, 2001–2004.

Li, Y., and Sarkanen, S. (2002). "Alkylated kraft lignin-based thermoplastic blends with aliphatic polyesters." *Macromolecules*, 35(26), 9707-9715.

Lignin Institute (2008). *Lignin Institute.* Available at http://www.lignin.org/01augdialogue.html.

Lin, S.Y. (1992). "Commercial spent pulping liquors." Section in *Methods in Lignin Chemistry*, Lin, S.Y., and Dence, C.W. (Eds.), Springer, Berlin, pp. 75–80.

Lindberg, J.J., Kuusela, T.A., and Levon, K. (1989). "Specialty polymers from lignin." Section in *Lignin: Properties and Materials*, ACS Symposium Series No. 397, Glasser, W.G., and Sarkanen, S (Eds.), American Chemical Society, Washington, DC., pp. 190- 204.

Liu, G., Liu, Y., Ni, J., Shi, H., and Qian, Y. (2004). "Treatability of kraft spent liquor by microfiltration and ultrafiltration." *Desalination*, 160 (2), 131–141.

Liu, M., Ferry, M.A., Davidson, R.R., Glover, C.J., and Bullin, J.A. (1998). "Oxygen uptake as correlated to carbonyl growth in aged asphalts and corbett fractions." *Industrial Engineering Chemistry*, 37 (12), 4669-4674.

Lora, J.H., and Glasser, W.G. (2002). "Recent industrial applications of lignin: A sustainable alternative to nonrenewable materials." *Journal of Polymers and the Environment*, 10(1-2), 39-48.

Loutfi, H., Blackwell, B., and Uloth, V. (1991). "Lignin recovery from kraft black liquor: preliminary process design." *Tappi J.*, 74 (1), 203–210.

Lucena, M.C., Soares, S.A., and Soares, J.B. (2004). "Characterization of thermal behavior of polymer-modified asphalt." *Materials Research*, 7 (4), 529-534.

Lynd, L.R. (1996). "Overview and evaluation of fuel ethanol from cellulosic biomass technology, economics, the environment, and policy." *Annual Review of Energy and the Environment*, 21, 403–465.

Lynd, L.R., Wyman, C.E., and Gerngross, T.U. (1999). "Biocommodity engineering." *Biotechnology Progress*, 15 (5), 777-793.

Mindess, S., Yong, J.F., and Darwin, D. (2002). *Concrete*. Prentice Hall, NJ.

Nadif, A., Hunkeler, D., and Käuper, P. (2002). "Sulfur-free lignins from alkaline pulping tested in mortar for use as mortar additives." *Bioresource Technology*, 84(1), 49-55.

Nicholls, R.L., and Davidson, D.T. (1958). "Polyacids and lignin used with large organic cations for soil stabilization." *Highway Research Board Proceedings*, 37, 517-537.

Oscarsson, K. (2007). *Dust suppressants for Nordic gravel roads.* Department of Civil and Architectural Engineering, Royal Institute of Technology, KTH, Stockholm, Sweden.

Palmer, J.T., Edgar, T.V., and Boresi, A.P. (1995). *Strength and Density Modification of Unpaved Road Soils due to Chemical Additives*. Mountain-Plains Consortium (MPC) Report No. 95–39, Fargo, ND.

Peng, W., and Riedl, B. (1994). "The chemorheology of phenol–formaldehyde thermoset resin and mixtures of the resin with lignin fillers." *Polymer*, 35 (6), 1280–1286.

Plank, J. (2004). "Applications of biopolymers and other biotechnological products in building materials." *Applied Microbiology and Biotechnology*, 66(1), 1-9.

Plank, J., and Winter, C. (2003). "Adsorption von flie-β mitteln an zement in gegenwart von verzögerern. " *GDCh Monogr*, 27, 55–64.

Popp, J.L., Kirk, T.K., and Dordick, J.S. (1991). "Enzymic modification of lignin for use as a phenolic resin." *Enzyme Microb. Technol.*, 13 (12), 964–968.

Pulp and Paper Manufacture (1987). *Aklaline Pulping*. Vol. 5, Tappi Press, Atlanta, GA.

Puppala, A.J., and Hanchanloet, S (1999). "Evaluation of a chemical treatment method (sulphuric acid and lignin mixture) on strength and resilient properties of cohesive soils." *78th Transportation Research Board Annual Meeting*, CD ROM, National Research Council, National Academy of Science, Washington, DC.

Reddy, N., and Yang, Y. (2005). "Biofibers from agricultural byproducts for industrial applications." *Trends in Biotechnology*, 23(1), 22-27.

Reshamwala, S., Shawky, B.T., and Dale, B.E. (1995). "Ethanol production from enzymatic hydrolysates of AFEX-treated coastal Bermuda grass and switchgrass." *Appl. Biochem. Biotechnol.*, 51/52, 43–55.

Ringena, O., Saake, B., and Lehnen, R. (2005). "Isolation and fractionation of lignosulfonates by amine extraction and ultrafiltration: a comparative study." *Holzforschung*, 59 (4), 405-412.

Roberts, F.L., Kandhal, P.S., Brown, E.R., Lee, D.Y., and Kennedy, T.W. (1996). *Hot Mix Asphalt Materials, Mixture, Design, and Construction*. 2nd edition, National Asphalt Pavement Association Research and Education Foundation, Lanham, MD.

Ross, S.S., and Woods, K.B. (1988). *Highway Design Reference Guide*. McGraw-Hill.

Ruan, Y., Davison, R.R., and Glover, C.J. (2003). "Oxidation and viscosity hardening of polymer-modified asphalts." *Energy and Fuels*, 17 (4), 991-998.

Sinha, S.P., Davidson, D.T., and Hoover, J.M. (1957). "Lignins as stabilizing agents for northeastern Iowa loess." *Iowa Academy of Science Proceedings*, 64, 314-347.

Sjöström, E. (1993). *Wood Chemistry Fundamentals and Applications*. 2nd edition, Academic Press, Inc., SanDiego, CA.

Stewart, D. (2008). "Lignin as a base material for materials applications: chemistry, application and economics." *Industrial Crops and Products*, 27(2), 202-207.

Sun, Y., and Cheng, J. (2002). "Hydrolysis of lignocellulosic materials for ethanol production a review." *Bioresource Technology*, 83 (1), 1–11.

Sundstrom, D.W., Herbert E.K., and Daubenspeck, T.H. (1983). "Use of byproduct lignins as extenders in asphalt." *Ind. Eng. Chem. Prod.*, 22(3), 496-500.

Terrel, R.L., and Rimstritong, S. (1979). "Wood lignins used as extenders for asphalt in bituminous pavements." *Journal of Association of Asphalt Paving Technologists*, 48, 111-134.

Tsapiuk, E.A., Bryk, M.T., Medvedev, M.I., and Kochkodan, V.M. (1989). "Fractionation and concentration of lignosulfonates by ultrafiltration." *J. Membr. Sci.*, 47 (1-2), 107–130.

Uloth, V.C., and Wearing, J.T. (1989). "Kraft lignin recovery: acid precipitation versus ultrafiltration: part I. laboratory test results." *Pulp Paper Can.*,90 (9), 67–71.

US DOT. (1976). *America's Highways, 1776-1976*. Federal Highway Administration, U.S. Department of Transportation, Washington, DC.

Wallberg, O., Holmqvist, A., and Jönsson, A.S. (2005). "Ultrafiltration of kraft cooking liquors from a continuous cooking process." *Desalination*, 180 (1-3), 109–118.

Whetten, R.W., MacKay, J.J., and Sederoff, R.R. (1998). "Recent advances in understanding lignin biosynthesis." *Annual Review of Plant Physiology and Plant Molecular Biology*, 49(1), 585-609.

Woods, K.B. (1960). *Highway Engineering Handbook.* 1st Edition, McGraw-Hill.

Wooley, R., Ruth, M., Sheehan, J., Ibsen, K., Majdeski, H., and Galvez, A. (1999). *Lignocellulosic Biomass to Ethanol—Process Design and Economics Utilizing Co-current Dilute Acid Prehydrolysis and Enzymatic Hyrolysis—Current and Futuristic Scenarios*. Report No. TP-580-26157, National Reneawable Energy Laboratory, Golden, CO.

Williams, R.C., and McCready, N.S. (2008). *The Utilization of Agriculturally Derived Lignin as an Antioxidant in Asphalt Binder*. CTRE Project 06-260, Iowa State University, Ames, IA.

Zhor, J., and Bremner, T.W. (1999). "Lignosulfonates in concrete: effect of dosage rate on basic properties of fresh cement-water systems." Section in *Modern Concrete Materials: Binders, Additions and Admixtures*, Dhir, R.K., and Dyer, T.D. (Eds.), Thomas Telford, London, pp. 419-431.

CHAPTER 13

Bioreactor Systems for Biofuel/Bioelectricity Production

Venkataramana Gadhamshetty, Nagamany Nirmalakhandan, Samir Kumar Khanal, and Glenn R. Johnson

13.1 Introduction

Effective reactor systems enable reproducible and controlled changes of environmental conditions to optimize microbial growth and align fermentation pathways toward biofuel production. An assessment of the kinetics, reaction rate, and morphological characteristics of the microbes will help in the selection of appropriate bioreactor configuration and operating parameters for cost effective biofuel production. This chapter summarizes bioreactor design criteria relevant to the production of bioenergy and biofuels (e.g. hydrogen, methane, butanol, ethanol, biodiesel), and the conversion of chemical energy (biomass) into electricity in microbial fuel cells.

13.2 Biofuel/Bioenergy Production

Microbial cells assimilate organic substrates for cell growth and product formation. Transformation of organic substrates into biofuels/bioenergy products is accompanied by a series of enzymatic reactions while the desired products are produced intracellularly (within the cells) and extracellularly (secreted outside the cells). An understanding of fermentation mechanisms assists in the identification of required pathways and necessary environmental conditions to obtain the desired biofuel/bioenergy. Clear criteria for the selection of specific microbial communities to optimize the yield of desired products in a bioreactor should be established (Kleerebezem and Loosdrecht, 2007). Non-oxidative pathways, such as anaerobic fermentation, is often used to stabilize organic wastes and produce methane simultaneously. The fermentation pathways can also be directed towards production of alternate energy carriers such as hydrogen, ethanol, and butanol.

13.2.1 Design Criteria

Bioreactors differ from conventional chemical reactors in that a microbial population (biomass) is used for transformation of organic feedstock into a desired product. As such bioreactors should enable and control the growth of selected microbes that produce the final product. Robust bioreactors are required to sustain operation during upsets and maintain the selectivity of a particular strain and the final product. Bioreactors should be designed to promote aseptic conditions and retain high biomass levels, while operating continuously with minimal mass transfer limitations. Basic bioreactor design considerations are summarized below.

13.2.2 Environmental Conditions

An in-depth understanding of the microbiology and biochemical pathways of the selected bacterial strain and morphological characteristics of biomass facilitates the identification of optimal macro-environmental conditions that affect cellular metabolism in a bioreactor. Small changes in environmental conditions can alter the fermentation pathways, microbial composition and product yield even when axenic cultures are employed in the bioreactor (Zeikus, 1980). Some important operational parameters in biofuel production are temperature, pH, oxygen and hydrogen partial pressure, loading rate, and hydraulic retention time (HRT). Sensors integrated with microprocessors can be used to maintain relevant parameters at set-point values (Margaritis and Wallace, 1984).

13.2.3 Hydrodynamic Characteristics

The hydrodynamic characteristics influence mass transfer and heat transfer capabilities of the bioreactor. An appropriate geometry of the bioreactor coupled with effective mixing is vital to minimize gradients in pH, temperature and concentration within the bioreactor. The viscosity of the culture medium dictates the choice of the impeller and heat transfer mechanisms. Hydrodynamic characteristics must be further tailored based on whether the cells are cultivated in suspension or attached to fixed or mobile media. The choice of an impeller system influences the shearing effects and care has to be taken to not disrupt cell propagation. For example, shearing effects are detrimental to the formation of flocculant cell mass found in methanogenic reactors. However, efficient mass transfer is important when it becomes the rate limiting step. Bioreactor design should mitigate mass transfer limitations between the feedstock and the microbes responsible for its bioconversion by ensuring contact and adequate exposure time by using appropriate mixing mechanisms, immobilization techniques and reactor configurations. For example, attached growth systems may pose mass transfer limitations due to uncontrolled biofilm growth and increased biosolids accumulation in the packing materials, and require periodic maintenance to eliminate short-circuiting.

13.2.4 Biomass Retention

Bioreactors should incorporate solids retention mechanisms to ensure adequate microbial population even at high dilution rates. Solids retention time, SRT, (the residence time of bacteria in the bioreactor) can be improved by either solids recycling or immobilization techniques. The latter includes formation of biofilm onto solid matrices and/or the entrapment of bacteria into a solid matrix. Favorable conditions that sustain immobilized biomass in the bioreactor include moderate turbulence, fixed surfaces and carrier materials for enhanced biofilm formation, the development of high settling granules, and quiescent conditions for the development of biomass aggregates. Desirable characteristics of the packing material that improve the adherence of the biomass include a high volume/surface area ratio, rough texture for bacterial adhesion, biological inertness, mechanical strength, appropriate shape, porosity and particle size. The selection of the immobilized media is largely dependent on the specific application. For example, microbial fuel cell operation requires immobilized media to be conductive, while photofermentation require transparent materials to facilitate light diffusion. Commonly used immobilization media include zeolite, glass beads, baked clay, activated carbon, anthracite and cellulose acetate. Membranes can also be used for solids retention in the reactor.

13.2.5 Bioreactor Configuration

A rigorous assessment of the physiological requirements and metabolic pathways of the selected bacterial strain, physical and chemical characteristics of the feedstock and its loading rate, and the selected microbial strain often paves a clear path towards the selection of a bioreactor configuration. The selected bioreactor design should optimize biochemical processes as well as physico-chemical processes such as ion association/dissociation, gas-liquid transfer and the mass transfer of substrates and products.

Traditionally, completely stirred tank reactors (CSTR), based on the chemostat principle, fitted with pH and temperature controllers, have been used as bioreactors. The application of attached growth and non-attached growth reactors for biofuel production has emerged recently. Attached growth bioreactors include fixed-film reactors, fluidized bed reactors, anaerobic filters, and anaerobic moving-bed reactors. Non-attached growth bioreactors include chemostats, anaerobic contact reactors, and upflow anaerobic sludge blanket reactors (UASB). Therefore, multiple bioreactor configurations are available that offer appropriate cell retention mechanisms for specific applications (Figure 13.1a-i).

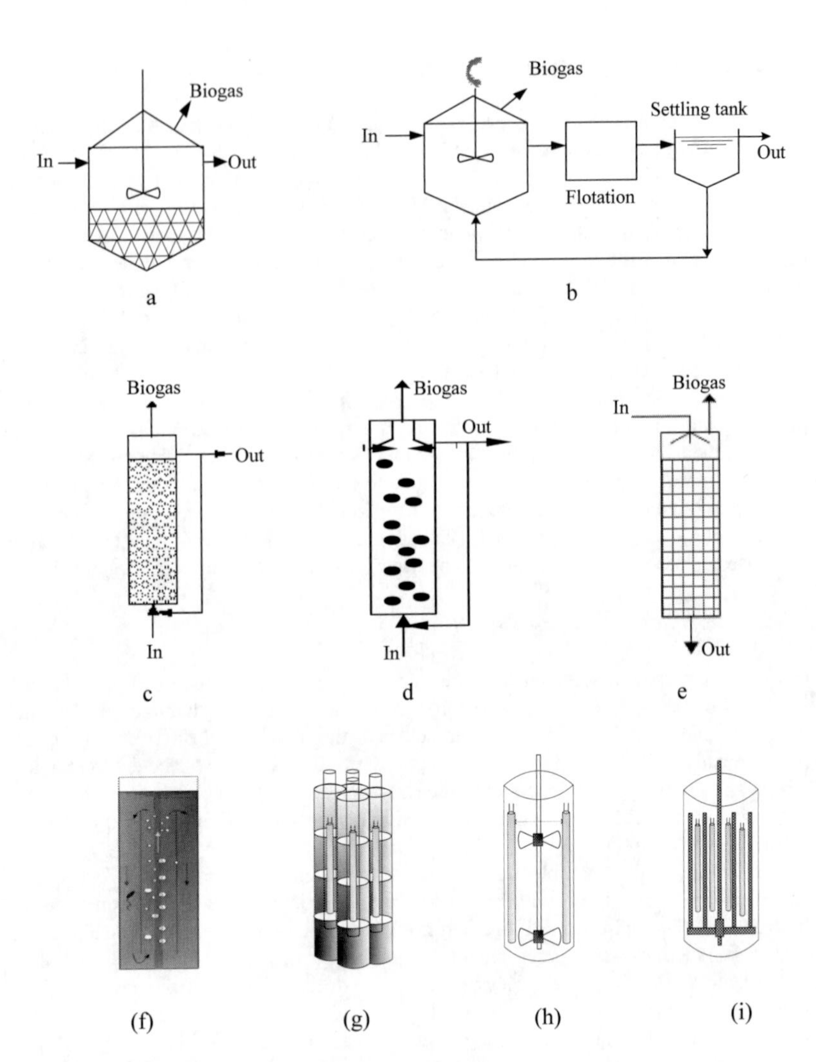

Figure 13.1 Bioreactors for biofuel production: (a) immobilized-cell CSTR; (b) CSTR with recycle; (c) packed bed; (d) fluidized bed; (e) trickling filter; (f) air lift PBR (g) unit concept PBR; (h) internally illuminated PBR with turbine impeller; and (i) internally illuminated PBR with modified paddle impeller.

Modified bioreactors may be required in special cases, such as photobioreactors (PBRs), where an efficient light emitting device has to be incorporated in the bioreactor (Figure 13.1h, 13.1i).

Bioreactor configurations influence the biomass washout rates and hence, the requirements on the hydraulic retention time, HRT. For example, suspended growth bioreactors such as CSTRs and UASBs are more prone to washout compared to attached growth bioreactors. Higher values of SRTs are also required at lower temperature due to the reduced rates of hydrolysis and growth. In completely mixed systems, the SRT is equal to the HRT, while in systems with built-in sludge retention, the SRT is higher than the HRT. Other factors that influence the upper limit on the loading rate include mass transfer limitations between the incoming feed and retained biosolids, the amount of biosolids retained in the bioreactor, temperature and level of toxicity in the wastewater, pH and the bioreactor configuration. A good discussion on different bioreactors available for bioenergy/biofuel production can be found in Khanal et al. (2008).

13.3 Biomethane

The fundamental science underlying fermentative methane production has been well explored in the past four decades or so.

13.3.1 Process Fundamentals

Biomethane production entails the microbial decomposition of organic substrates in non-oxidative environments to sustain syntrophic activities of multiple microbial species responsible for hydrolysis, fermentation, acidogenesis, anaerobic oxidation of volatile fatty acids and methanogenesis (Mata-Alvarez, 1987).

1. Hydrolysis involves enzyme mediated, extracellular biological transformations of complex biopolymers (proteins, carbohydrates and lipids) into simple monomers such as amino acids, sugar and long-chain fatty acids by hydrolytic bacteria

$$A - X + H_2O \Leftrightarrow X - OH + HA \quad \text{(Eq.13.1)}$$

2. Fermentation involves the microbial production of volatile fatty acids from sugars in the absence of external electron acceptors.

Acetate $$C_6H_{12}O_6 + 2H_2O \rightarrow 2CH_3COOH + 2CO_2 + 4H_2 \quad \text{(Eq.13.2)}$$

Propionate $$C_6H_{12}O_6 + 2H_2 \rightarrow 2CH_3CH_2COOH + 2H_2O \quad \text{(Eq.13.3)}$$

Butyrate $$C_6H_{12}O_6 \rightarrow 2CH_3CH_2CH_2COOH + 2CO_2 + 2H_2O \quad \text{(Eq.13.4)}$$

Ethanol $$C_6H_{12}O_6 \rightarrow 2CH_3CH_2OH + 2CO_2 \quad \text{(Eq.13.5)}$$

Lactate $$C_6H_{12}O_6 \rightarrow 2CH_3CHOHCOOH \quad \text{(Eq.13.6)}$$

3. Acidogenesis of long-chain fatty acids into acetate and hydrogen which obligately depend on hydrogen removal by methanogens via interspecies hydrogen transfer.

4. Anaerobic oxidation of volatile fatty acids into acetate and hydrogen.

$$4CH_3CH_2COOH + 3H_2O \rightarrow 4CH_3COOH + CO_2 + H_2O + 3CH_4 \quad \text{(Eq.13.7)}$$

$$2CH_3CH_2CH_2COOH + HCO_3^- + H_2O \rightarrow 4CH_3COOH + CH_4 \quad \text{(Eq.13.8)}$$

$$2CH_3CH_2OH + HCO_3^- \rightarrow 2CH_3COOH + CH_4 + H_2O \quad \text{(Eq.13.9)}$$

5. Aceticlastic methanogenesis of acetate into methane by acid-utilizing methanogens

$$CH_3COOH + H_20 \rightarrow HCO_3^- + CH_4 \quad \text{(Eq.13.10)}$$

$$CO_2 + 4H_2 \rightarrow 2H_2O + CH_4 \quad \text{(Eq.13.11)}$$

Sustained methane production by the above reactions (Eq 13.10-13.11) is possible only when syntrophic relations between acetogenesis (Eq 13.7-13.9) and hydrogen-utilizing methanogenesis is maintained. Such balanced reactions can be attained under optimal conditions to support the growth of methanogenic bacteria and prevent the accumulation of fermentation products responsible for cessation of methane production.

13.3.2 Environmental Conditions for Methane Production

The optimal pH conditions for acid formation and methane production phases are 4-6.5 and 6.5-8.2, respectively. Due to the wide difference in optimal pH ranges, the two phases are often carried out in two separate reactors (Ghosh, 1984). Methane production can be achieved under psychrophilic (10-20 ºC), mesophilic (20-40 ºC), and thermophilic (50-60 ºC) conditions. The optimum values for C/N and N/P are 4.3 and 7.6 respectively. Among the micronutrients, iron is required in higher concentrations while magnesium, sodium, zinc, iron, cobalt, nickel, calcium, copper, manganese, molybdenum are reported to enhance bacterial metabolism, growth and the activity of the anaerobes (McCarty, 1964).

13.3.3 Bioreactor Design for Methane Production

Bioreactor designs for methane production are well established and numerous full-scale units are available for waste remediation and biomethane production from high strength waste streams. The solid content of the raw waste renders two design choices for bioreactors: low-rate systems and high-rate systems (de Mes et al., 2002). Low rate systems are appropriate for the treatment of concentrated wastes such as sewage sludge and manure slurries that require longer HRTs, and are designed to have SRTs equal to HRTs. Low rate systems are further classified according to the proportion of total solids (TS) in the waste: i) wet digestion (15-25% TS) ii) dry digestion (>30% TS). High-rate systems are appropriate for low-solids (<10%)

wastewater with larger flow rates and require cell immobilization techniques to maintain a high biomass concentration and minimize HRT (de Mes et al., 2002).

13.3.3.1 Low Rate Systems–Wet (15-25% TS)

Complete-Mix. Conventional low rate anaerobic systems for the treatment of solid wastes (15-25% total solids, TS) and slurries (2-10% TS) using CSTRs are widely used due to their simplicity and low costs. Typically, in CSTRs, where HRTs (= SRT) range from 30-60 days, providing for an ample safety factor, process stability may vary slightly depending on the nature of the feedstock and process temperature (Parkin and Owen, 1986). CSTR design has been applied for the treatment of animal manure, kitchen waste, sewage sludge, and household waste.

Plug Flow. Plug flow bioreactors (PFBRs) typically treat wastes with a total solid content of 10-15%, and basic designs consist of a long trough often built below the ground level with a gas tight, expandable cover. Mes et al., (2005) modified this design by separating the hydrolysis and methanogenesis steps over the length of the reactor. Typical HRTs (= SRTs) range from 20 to 30 days. The pretreatment of the incoming waste to screen larger particles such as sand and wood-based materials improves the feasibility of PFBR operations. Incorporating vertical mixing inside the PFBR can minimize problems due to the formation of floating layers.

13.3.3.2 Low Rate Systems–Dry: Two Stage Leach-Bed (> 30 % TS)

The two-stage leach-bed configuration has been proposed to separate the acidogenic step from the methanogenic step so that optimal conditions can be maintained for the two steps. In the first stage, which is a leach-bed containing the solid wastes, acidogenesis produces a leachate low in suspended solids and rich in volatile acids. This leachate is fed to the second-stage, which is a suspended growth digester for methane production. The liquid effluent from the second-stage is recycled back to first-stage. This design can be applied to solid wastes of TS >35% (Ghosh, 1984).

13.3.3.3 High Rate Systems (< 10% TS)

In high-rate systems, SRTs are increased with improved solids retention and/or immobilization to minimize the HRT and hence, reduce the reactor volume. This is achieved by fixed bacterial films on solid surfaces as in packed bed reactors and fluidized bed reactors. Schematics of pilot scale and full-scale designs are shown in Figure 13.1. A brief summary of the different designs follows.

Anaerobic Contact. Anaerobic contact processes (McCarty, 1964) incorporate a separation unit that follows the completely mixed reactor to separate and recycle biomass. The biomass separation can be achieved by gravity settling, centrifugation, and flotation processes. The process can be operated with low HRTs (~24 h) and is applicable to high strength waste streams containing fats, oils and grease concentrations higher than 150 ppm. Operational difficulties include floc retention and stability problems (Callander and Barford, 1983).

Anaerobic Filter (AF). Anaerobic filter processes rely upon the entrapment of the bacteria within the packing material and adhesion to the external surface of the packing material. This process has been applied to both, dilute and high strength wastewaters. The key design criterion for AFs is to facilitate the withdrawal of accumulated biomass in order to prevent short-circuiting. The design has to ensure that the packing material plugged with biomass can be removed by draining the liquid contents, refilling and purging the bioreactor with nitrogen gas (Speece, 1996). There are two types of anaerobic filters: upflow and down flow, both of which are characterized by i) a quiescent inlet region with dense biomass aggregates not prone to washout and ii) a suitable surface to facilitate biofilm accumulation. Random packing is used in upflow AFs, while downflow AFs orient packing in the form of vertical channels. Advantages of AFs include good adaptation to different types of wastewater, an insensitivity to load fluctuations and a fast restart after shut down. Major disadvantages of AFs include high expenditure and operational problems associated with the packing materials.

Upflow Anaerobic Sludge Bed (UASB). UASB reactors maintain an anaerobic sludge blanket with a residing bacterial population for the hydrolysis and acidogenesis of raw waste. This design eliminates the need for packing media and carrier materials by using the granulation of biomass as a means of immobilization (Lettinga et al., 1983). UASBs consist of the following four compartments (from the bottom to the top): the sludge bed, the fluidized zone, the gas-liquid separator, and the settling compartment. Feed inlet systems at the bottom of the bioreactor distributes wastewater to the reactor vessel that passes upward through a dense anaerobic sludge bed where hydrolytic and acidogenic bacteria convert wastewater into volatile fatty acids, carbon dioxide and hydrogen, which are then subsequently converted to methane by methanogens. Biogas and liquid pass through the fluidized zone before being separated in the gas-liquid separator section. The key to the successful operation of UASBs is to maintain aggregates of active biomass as granules with good settling properties. Further, excess sludge has to be discharged periodically to maintain the fluidized zone and prevent the wash out of biomass. Settler sections allow for effective degasification so that sludge particles devoid of attached gas bubbles sink to the bottom, establishing a return downward circulation. Upward flow of gas-containing sludge through the blanket, combined with a return downward flow of degassed sludge, creates a continuous convection to ensure the effective contact of sludge and wastewater. This unique feature eliminates the need for energy consuming agitation and facilitates the retention of a highly active biomass in relation to soluble organic solids passing through the sludge bed for handling high loading rates (short hydraulic retention time). High biomass concentrations (30-50 g VSS/L) can be achieved by operating UASBs at a low organic loading during the initial start-up time, with slightly acidified feed and high calcium concentrations. Organic loading rates of 13-39 kg COD/m^3 were reported to be degraded with removal efficiencies (75-90%) in the range of 2-7 hours (Torkian et al., 2003).

Advantages of UASB include a low investment, operational ease, improved sludge disposal, and a reduced space requirement, while some of the disadvantages include poor granular sludge formation, an accumulation and slow hydrolysis of suspended solids, and decreased methane (Kallogo and Verstraete, 1999).

Expanded Granular Sludge Bed. The Expanded Granular Sludge Bed (EGSB) reactor incorporates two or more UASB compartments on top of each other, and was introduced to increase the hydraulic mixing intensity (DeMan et al., 1988). Higher upward flow velocities (5-10 m/h), that are 5 to 10 fold that of UASBs, ensure sludge expansion, CSTR behavior, an improved contact between wastewater and biomass, and a decreased apparent substrate saturation constant. ESGBs are ideal choice for loading conditions <1 kg COD/m^3/day (Lettinga et al., 1983) and for low-strength wastewater, and can serve as a robust polishing reactor (Kallogo and Verstraete, 1999). However, the quality of the effluent can be lower than that of UASBs due to the washout of granular sludge.

Anaerobic Hybrid. Anaerobic filters suffer from dead zones and channeling in the lower part of the filter, while a potential development of loose, fluffy, unsettled particles and sludge washout is a problem when a UASB is used to treat high-solids waste. The hybrid system was developed to overcome the problems in UASB and AF systems by integrating attributes of anaerobic filters (packing material in the upper portion) and UASBs (unpacked in lower portion). The performance can be sustained in absence of granular sludge if sufficient flocculent sludge is retained in the AF section to maintain a high rate of degradation. Full-scale plants have been installed to treat sludge, thermal conditioning liquid, landfill leachate and domestic sewage. A major advantage is the prevention of sludge flotation, while a lengthy start-up time serves as a drawback (Speece, 1996).

13.3.4 Outlook on Methane Production

Anaerobic digestion has been well demonstrated as a waste treatment process and is a mature technology with nearly 2000 full-scale systems in operation treating industrial wastewaters, landfill leachate, sewage sludge, etc. (de Mes et al., 2002). However, there is still a resistance for the commercialization of this technology solely for biofuel (methane) production due to political and economical reasons. Legislative and financial incentives may be key to rapid growth and the adaptation of this technology.

13.4 Biohydrogen through Dark Fermentation

The classical anaerobic fermentation process can be engineered to produce biohydrogen by suppressing methane formation- a process known as dark fermentation (Gadhamshetty et al., 2009c). The first step towards fermentative hydrogen production involves the enrichment of spore-forming hydrogen, producing bacteria and the inhibition of hydrogen consuming bacteria. The second step involves the identification and reinforcement of bioreactor conditions that support the germination of hydrogen producing spores. Optimized environmental conditions are required to both nurture hydrogen yielding reactions (Eq. 13.12-13.13) and inhibit hydrogen consuming bacteria such as propionic acid–formers, ethanol formers, homoacetogens and methanogens (Eq 13.14-13.17).

$$C_6H_{12}O_6 \rightarrow 2CH_3COOH + 2CO_2 + 4H_2 \quad \text{(Eq.13.12)}$$

$$C_6H_{12}O_6 \rightarrow CH_3CH_2CH_2COOH + 2CO_2 + 2H_2 \quad \text{(Eq.13.13)}$$

$$C_6H_{12}O_6 + 2H_2O \rightarrow 2CH_3CH_2COOH + 2H_2O \quad \text{(Eq.13.14)}$$

$$C_6H_{12}O_6 \rightarrow 2C_2H_4OH\,COOH \quad \text{(Eq.13.15)}$$

$$CH_3COOH + H_2 \rightarrow 2CH_3CH_2OH + H_2O \quad \text{(Eq.13.16)}$$

$$CO_2 + 4H_2 \rightarrow CH_4 + 2H_2O \quad \text{(Eq.13.17)}$$

A properly engineered bioreactor can theoretically yield a maximum of four moles of hydrogen per mole of glucose when acetate is the only by-product (Eq.13.12). All other pathways with lower yields due to loss of hydrogen to undesirable by-products should be avoided.

13.4.1 Inocula for Fermentative H_2 Production

Bacterial strains belonging to families of *Clostridia, Enterobacteriacae and Citrobacter* can be readily sporulated, and therefore produce hydrogen. Hydrogen producing strains can be obtained from natural sources such as soil, sludge and compost, and enriched for spore formers by acid and heat treatment. Acid-treatment of anaerobic sludge entails the acidification to a pH of 3-4 for 24 hrs and restoring the pH to 7.0. Heat-treatment entails boiling for 15 min by steam injection or exposure of the inocula source to ~120^0 C for 2 hours (Hawkes et al., 2002).

13.4.2 Environmental Conditions for Fermentative H_2 Production

Since there is lack of experience on full-scale fermentative biohydrogen production, results from laboratory-scale studies that have a potential for full-scale applications are summarized here. Mesophilic and thermophilic ranges are preferable while recommended C/N ratio and C/P ratio are 6.7 and 130 (Lin and Fang, 2007). Iron (10 mg/L) is the most important trace metal due to the presence of hydrogenase enzymes involved in biohydrogen production. Low pH (<6.0) and shorter retention times (3-8 hours versus 1-2 days for methanogens) are key to methane inhibition and hydrogen production (Gadhamshetty et al., 2009b; Hawkes et al., 2002). Other conditions favorable for biohydrogen production by dark fermentation include a headspace partial pressure of hydrogen of 30-60 kPa, and nitrogen sparging from the liquid space to maintain strict anaerobic conditions. Carbohydrates are the preferred substrates and nutrient requirements are similar to that of the biomethane process (Lin and Lay, 2005).

13.4.3 Bioreactor Design for Fermentative H_2 Production

As in anaerobic digesters, biohydrogen reactors should facilitate biosolids retention at higher loading rates. Literature reports include immobilized-cell systems such as *Enterobacter cloacae* supported by lignocellulosic material support, combined systems of *Phormidium valderianum, Halobacterium halobium,* and *Escherichia coli* entrapped by a PVA-alginate film, and sewage sludge on activated carbon, polyurethane and acrylic latex on silicone (Wu et al., 2002). Established designs for methane bioreactors such as fixed-bed, fluidized-bed, and upflow anaerobic sludge blanket and continuous stirred tank reactors have been successfully adapted in biohydrogen studies.

CSTR. This is a commonly used suspended-mass bioreactor in biohydrogen production and has been tested in batch, semi-batch and continuous modes. Liquid to headspace ratios along with gas pressure release techniques are often manipulated to maintain a desired hydrogen partial pressure (<60 kPa). Growth rate differences between hydrogen producing bacteria versus methanogenic bacteria has to be exploited by operating CSTRs at low HRTs and low pH to eliminate methanogenic bacteria (Kotsopoulos et al., 2005). Even though stable H_2 production and CSTR operation for over 1 year have been demonstrated (Lin and Lay, 2005), precise control of operating parameters such as HRTs and pH is a major challenge in CSTR operations. Biomass retention is another major problem, and the incorporation of a recycle line has been reported to be less feasible in continuous modes (Chen et al., 2001).

Upflow Anaerobic Sludge Blanket (UASB). The major obstacle in adapting UASBs for hydrogen production is a longer SRT and start-up time required which can trigger methanogens reducing the hydrogen yield. Strategies suggested for the successful operation of UASBs include the addition of chemical inhibitors and operation under thermophilic operations at low pH (4.5-6), optimum substrate concentrations (<10 g COD/L), and a flow recycle to enrich hydrogen producing organisms and inhibit methanogens (Kotsopoulos et al., 2005). This process shows promise for biohydrogen production and has been tested with synthetic and winery wastewaters in mesophilic, hypothermophilic and thermophilic ranges (Gavala et al., 2006).

Packed Bed. Packed bed bioreactors with porous packing media such as a loofah sponge, expanded clay, and activated carbon can be used to produce biohydrogen (Chang et al., 2002). While packed bed bioreactors are relatively inexpensive and easy to operate, they have been found to be limited by inefficient mass transfer. Gaseous hydrogen as the final fermentation product may further damage the packing media (Wu et al., 2002).

Biofilm. Biofilm based reactors, such as trickling filters, have achieved high H_2 production rates of nearly ~7.4 L H_2/hr/L, but their long-term stability has not been demonstrated. Oh et al. (2004) reported a thermophilic biohydrogen production of 0.98 L/L-h from glucose in a trickling filter with nearly a 25% conversion efficiency. Advantages of trickling filters include high cell density, reduced gas hold-up, and an ease of control of pH and temperature (Oh et al., 2004b).

Fluidized Bed. In a fluidized bed bioreactors (FBBR), fine carrier particles with entrapped biofilm are fluidized by high upflow fluid velocities generated by the recirculation effect of the influent and effluent. Based on the percentage of expansion, the bioreactor is classified as expanded (<30%) and fluidized (>30% expansion) (Saravanane and Murthy, 2000). FBBRs offer the advantages of superior mass and heat transfer characteristics, good mixing and relatively low energy requirements. The low shear rates in FBBRs make them suitable for shear sensitive cells. Examples of carrier particles include activated carbon, sand, and filter sponge. Wu et al. (2002) have developed a fluidized bed with acrylic latex, sodium alginate, and activated carbon. The synthetic waste was fed from the bottom of the fluidized bed reactor into the immobilized media (1/3rd of the reactor comprised with immobilized media) and the effluent was passed through a gas-liquid separator. Reasonable hydrogen yields ~33% were achieved with HRT values as low as 2 hours.

Sequencing Batch Reactor (SBR). Anaerobic sequencing batch reactors operated with self-immobilized biomass are advantageous because they do not require settling tanks and sludge return systems. They also benefit from controlled operating conditions. Dal-Yeol et al. (2007) have obtained prolonged and stable biohydrogen production (4.5-5.5 L/L/h) and biomass retention (13 g VSS/L) with a feed, reaction, settle, and decant time of 20, 190, 30, and 20 minutes. The performance of H_2-producing SBRs is on par with other bioreactors and has been tested on wastewaters from distillery and dairy, showing promise for SBR applications in biohydrogen production.

Membrane-Type Reactors (MBR). Advantages of MBRs include higher solids retention time that translate to smaller reactor volumes, higher organics removal efficiency, smaller excess sludge production and a high quality effluent due to the complete removal of bacteria by the membrane. Disadvantages include expensive membranes, membrane fouling, and energy intensive operations. In one study, MBRs incorporating a cross-flow membrane (alumina of 0.2-0.8μ) achieved improved solids retention time (> 10 h), solids concentration (> 5800 mg/L) and reasonable hydrogen yields (25%) (Oh et al., 2004a) Techniques for reducing membrane fouling include the addition of coagulants, the operation of the system below a critical flux, backwashing, backpulsing, and jet aeration (Oh et al., 2004a).

13.4.4 Outlook on H_2 Production

Maximum biohydrogen yields are realized with effective immobilization techniques, a low pH and high organic loading rates, proper management of spore formers and the inhibition of methanogens. However, an ideal bioreactor can recover only ~15% of the electron equivalents in a high-carbohydrate wastewater as hydrogen, restricting this process to a pre-treatment step in a larger bioenergy train.

Biohydrogen production should be integrated with other processes such as methanogenic anaerobic digestion, photofermentation or bioelectrochemically-assisted microbial system. Some strategies for improved biohydrogen production can be found in Khanal et al. (2008).

13.5 Photofermentative Hydrogen Production

Hydrogen can be produced photosynthetically by purple non-sulfur bacteria in the presence of external light. Optimal combination of photoparameters such as the light source, intensity, duration, and wavelength based on the photo pigments in the bacterial strain is an extra step required in photobioreactor (PBR) design. Available PBR designs limit light-hydrogen conversion efficiencies (PEs) to 10% due to problems of light attenuation, light saturation, and ineffective light utilization (Gadhamshetty et al., 2009a; Gadhamshetty et al., 2008).

13.5.1 Process Fundamentals

Purple non-sulfur bacteria (PNS) are anoxygenic photoheterotrophs that can utilize organic acids for the production of hydrogen (H_2) in the presence of sunlight (Eq. 13.18).

$$C_2O_2H_4 + 2H_2O + \text{Light} \rightarrow 2CO_2 + 4H_2 \qquad \text{(Eq.13.18)}$$

Generation of ATP and reduced ferredoxin are the two major requirements for nitrogenase catalyzed hydrogen production in PNS bacteria (Eq. 13.19).

$$N_2 + 8H^+ + 8e^- + 16ATP \rightarrow 2NH_3 + H_{2gas} + 16ADP + 6P_i \qquad \text{(Eq.13.19)}$$

13.5.2 Environmental Conditions for Photofermentative H_2 Production

Purple non-sulfur (PNS) bacteria generate hydrogen under anaerobic conditions, in the presence of optimal illumination, with suitable nutrients and stressful concentrations of nitrogen sources, and in the absence of oxygen and nitrogen. Optimum conditions of pH (6.5-7.0), temperature (27-35^0 C), C/N=2-6 on molar basis, light wavelength (450-900 nm) and light intensities (150-250 W/m^2) are recommended (Gadhamshetty et al., 2009a).

13.5.3 Photobioreactor Design for H_2 Production

Unique requirements of photobioreactors include a high illuminated surface to volume ratio, homogenous internal illumination and reasonable volumes. Janssen (2002) have summarized advances made in PBR designs for outdoor algal cultivation. Such systems can be adapted for photofermentative biohydrogen production as well. However, since the product here is in gaseous form, the reactors have to closed systems. Commonly used PBRs can be broadly classified as draft tube reactors, bubble column reactors, flat panel reactors or tubular reactors (Janssen, 2002).

Draft Tube. Light enters on one side of the PBR wall and the depth of penetration zone, depending on the biomass concentration and column diameter, creates light and dark regions within the PBR. Internal draft tube systems deliver gas bubbles via compressors which are responsible for turbulent liquid mixing that promote the movement of biomass between light and dark zones. Air injected via draught tubes cause liquid to rise in the draught tube (riser) which comes down again in the space between draught tube and PBR wall (downcomer). The liquid flow in the downcomer resembles plugflow while flow regime in the riser is close to plug flow with small lateral velocity differences. Successful manipulation of the reactor design and biomass concentration, required to overlap the dark zone with the riser and create photic volumes in the downcomer, render the liquid circulation velocity as the determining factor on the frequency of light/dark (L/D) fluctuations. While the liquid circulation velocity is influenced by the PBR design and superficial gas velocity, cell adhesion on the PBR walls create unmixed (still) zones in PBRs. Coiled spargers can be used to cover entire PBR base area to remove such still zones but at an expense of energy intensive pressurized air flow.

Bubble Column. Direct air bubbling is an alternative to draft tube in vertical PBRs. Large circulatory flows often characterize the heterogeneous flow regime of the liquid in bubble columns. Higher gas hold-up values in the center of the PBR are responsible for the upward velocity in the center and a downward velocity close to the PBR wall leading to the liquid circulation. There is limited research on bubble column PBRs and are often limited to gas flow rates of less than 0.01 m/second and no cell circulation was reported to occur under these ranges. The average circulation time in bubble-column PBRs was roughly 0.5-15 s and further design improvements are necessary to operate them at higher superficial velocities to achieve shorter cycle times (<1.5 s) and better PEs. Air bubbling techniques also suffer from the disadvantage of a reduced gas supply capacity when PBRs are tilted at an angle to maximize incident solar radiation.

Flat Panel. Laboratory scale flat panel PBRs are typically rectangular transparent boxes with a depth of 1–5 cm, height and width smaller than 1 m, fitted with gas spargers. In order to create a high degree of turbulence, 2.8–4.2 l of sparging gas per litre of reactor volume per minute is recommended. The panels are illuminated on one side by direct sunlight; the panels are oriented vertically, or inclined to maximize illumination. Light/dark cycles are short in these PBRs and this is likely the key factor leading to a high PE. A disadvantage of these systems is that the power consumption of aeration (or mixing with another gas) is high, although mixing is always necessary in any reactor. Large-scale flat-plate reactors based on a rectangular air-lift photobioreactor with a large number of light re-distributing plates fixed a few centimeters from each other is suggested by (Akkerman et al., 2002).

Tubular. Tubular PBRs consist of long transparent tubes with diameters ranging from 3 to 6 cm, and lengths ranging from 10 to 100 m, manifolded at the ends. The culture is pumped through these tubes by mechanical or air-lift pumps. The tubes can be oriented in different ways: in a horizontally as a plane surface or vertically as a fence coiled around cylindrical or conical framework. Biomass yields in tubular reactors are reported to be comparable to that of vertical PBRs while these PBRs are superior in terms of light regime due to shorter frequency of L/D cycles.

Internally Illuminated. PBRs can be modified in such a way that they can utilize both solar and artificial light system (Ogbonna et al., 1999), where, the artificial light source is switched on whenever the solar light intensity decreases below a set value (during cloudy weather or at night). There are some reports on the use of fiber optics to collect and distribute solar light in cylindrical photobioreactors (Mori, 1985). One of the major advantages of internally-illuminated photobioreactors is that it can be heat-sterilized under pressure to minimize contamination. Furthermore, the supply of light to the photobioreactor can be maintained continuously (both day and night) by integrating artificial and solar light devices.

13.5.4 Outlook on Photofermentative H_2 production

The future of large-scale photofermentative hydrogen production will depend on advances in genetic engineering, solar collection technologies, and efficient PBR engineering to maximize light-to-hydrogen conversions by microorganisms. In addition, fossil fuel costs, energy security, environmental concerns, social acceptance, and the growth of a hydrogen economy can indirectly contribute to the economic viability of PBRs.

13.6 Biological Electricity Production

In microbial fuel cells (MFC), organic compounds are microbially converted directly into electrical energy by exploiting respiratory pathways of the microbes. Maximum power levels in MFCs require i) a stable biofilm formation on the anode surface ii) optimum environmental conditions, and iii) efficient electrode materials and iv) bioreactor configurations that promote the generation and transport of the electrons and protons.

13.6.1 Fundamentals of Bioelectricity Production

Microbial fuel cells are similar to conventional fuel cells where protons from anode compartment travel towards the cathode via electrolyte membrane with electrons going in the same direction via conducting wires and resistor (Figure 13.2). However, in MFCs, bacteria act as an electrochemical catalyst in the anode to liberate electrons from an organic substrate such as glucose and conduct it directly to the anode surface in the absence of soluble electron acceptors (Eq. 13.20). Electrons and protons combine with oxygen on the cathode surface to generate water (Eq. 13.21).

Anaerobic oxidation of glucose for electricity production completing the counter electrode reactions are shown below:

$$\text{Anode}: C_6H_{12}O_6 + 6H_2O \rightarrow 6CO_2 + 24H^+ + 24e^- \quad \text{(Eq.13.20)}$$

$$\text{Cathode}: 6O_2 + 24H^+ + 24e^- + 12H_2O \quad \text{(Eq.13.21)}$$

Microbes transfer electrons to the anode surface via a series of electron carriers in an extracellular matrix either with or without a soluble redox mediator (e.g. flavins, quinones). Since the outer layers of major microbial species are composed of non-conductive membranes, extracellular electron transfer is proposed to use at least three mechanisms: i) set membrane cytochromes near the electrode surface; ii) secrete active soluble redox mediator iii) synthesize protein structures called nanowires. For example, when redox mediators are the only means of electron transport (Figure 13.2), mediators diffuse across the membrane back and forth to consume electrons within the cell and liberate electrons to the anode surface via cyclic redox reactions (Logan et al., 2006).

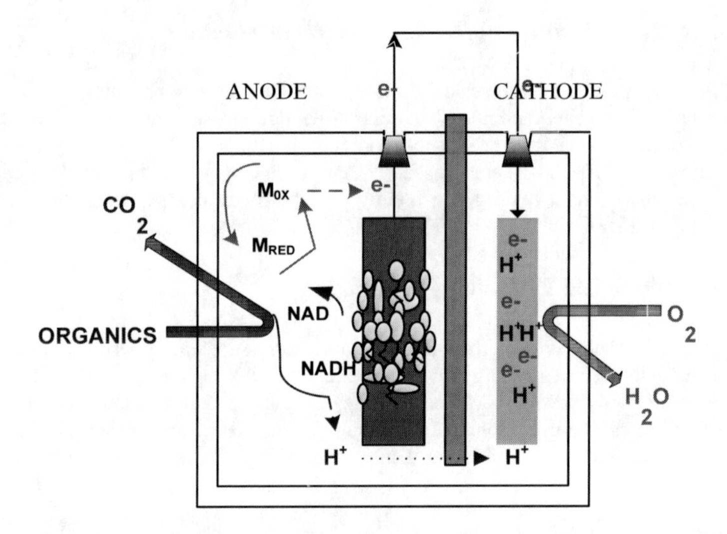

Figure 13.2 Schematic for electricity production from microbial fuel cell.

13.6.2 Power Losses in Microbial Fuel Cells

Total electromotive force (cell potential) that can be developed in a microbial fuel cell is determined by the difference between anodic and cathodic potentials. However, attainable voltages are at least 30% lower than that of the theoretical value, due to overpotentials (transfer losses) at individual electrodes. Such losses can be mitigated by using effective electrode materials and efficient MFC designs (Logan et al., 2006).

Ohmic Losses. The ohmic losses in an MFC include the resistance to the flow of electrons through the electrodes and interconnections, and the resistance to the flow of ions through the cation exchange membrane and the electrolyte. Ohmic losses can be reduced by minimizing the electrode spacing, using a membrane with a low resistivity, and increasing the solution conductivity to the maximum tolerated by the bacteria.

Activation Losses. Activation losses occur during the transfer of electrons from or to a compound reacting at the electrode surface due to the requirements of the activation energy for any redox reaction. The extent of activation losses can be mitigated by increasing the roughness and surface area of the electrode, improving catalysts for fuel oxidation, electron transport at the anode with mediator supplements, and ensuring enriched biofilms on the electrode.

Metabolic Losses. To generate metabolic energy, bacteria transport electron from a substrate at a low potential through the electron transport chain to the final electron acceptor at a higher potential. In an MFC, the anode is the final electron acceptor and its potential determines the energy gain for the bacteria. The higher the difference between the redox potential of the substrate and the anode potential, the higher the possible metabolic energy gain for the bacteria, but the lower the maximum attainable MFC voltage. To maximize the MFC voltage, therefore, the potential of the anode should be kept as low as possible. However, if the anode potential becomes too low, electron transport will be inhibited and the fermentation of the substrate may provide greater energy for the microorganisms (Rabaey et al., 2005b).

Concentration Losses. Concentration losses occur when the rate of mass transport of a species to or from the electrode limits current production. Concentration losses occur mainly at high current densities due to limited mass transfer of chemical species by diffusion to the electrode surface. At the anode, concentration losses are caused by either a limited discharge of oxidized species from the electrode surface or a limited supply of reduced species toward the electrode. This increases the ratio between the oxidized and the reduced species at the electrode surface, which can produce an increase in the electrode potential. At the cathode side the reverse may occur, causing a drop in cathode potential. In poorly mixed systems, diffusional gradients may also arise in the bulk liquid. Mass transport limitations in the bulk fluid can limit the substrate flux to the biofilm, which is a separate type of concentration loss.

13.6.3 Materials for Microbial Fuel Cells

Unique features of microbial fuel cells that differ from that of conventional bioreactors are the electrochemical parameters such as electrode surface area, redox potentials of the anode and cathode along with migration effects.

Electrodes. Electrode used in MFCs should have large surface area and be conductive, non-corroding, and porous. Carbon is a widely used electrode and is available in the form of graphite plates, rods, granules, brush, and fibrous material (felt, cloth, paper, fibers, and foam). The electrode serving as the anode should be robust and should support biofilm formation. The cathode provides a meeting point for electrons, protons and oxygen to react on the catalytic surface to form water. Catalyst (platinum, iron (II) and cobalt--tetramethoxyphenylporphyrin) will help the cathode surface to promote oxygen reduction. In certain cases, poly-tetraflouro-ethylene (PTFE) is used to achieve hydrophobic surfaces on the cathode to prevent water loss from the MFCs (Logan et al., 2006).

Membrane. Membranes used to separate the anode and cathode should be inexpensive and exhibit selectivity for the proton and stability towards the nutrient rich anolyte and the bacteria. Commonly used membranes include cation exchange membranes (CEMs) such as Nafion, Ultrex CMI-7000. Nafion exhibits high selectivity and low stability, while Ultrex shows larger stability and offers larger internal resistance. While the life of these two membranes is limited to three months, membranes can be eliminated in single chambered and sediment MFCs. Problems with membranes involve high expenditure and its permeability to chemicals such as oxygen, ferricyanide, ions, and organic matter used as the substrate.

13.6.4 MFC Designs

Some of the latest MFC designs are shown in Figure 13.3 and briefly explained below.

H-Shaped. This is a traditionally used design based on the two chamber design built in an "H" shape. Two chambers consisting of anode and cathode electrodes are connected by a tube containing membranes such as salt bridge, Nafion or Ultrex. The substrate along with biocatalyst is placed in the anode chamber while dissolved oxygen is made available in the cathode chamber. The purpose of the membrane is to allow proton transfer while separating protons, substrate, products and the terminal electron acceptor. The amount of power generated in these systems is dependent on the surface area of the cathode relative to that of the anode and the surface of the membrane. H-shaped MFCs have low power densities and high internal resistances and are limited to laboratory scale research for the examination of new parameters and microbial communities (Milliken and May, 2007).

Two-Chambered. In the modified version of H-shaped MFCs, two-chambered MFC designs, the chambers are replaced with two rectangular compartments with the membrane in between to obtain a larger surface area for proton transfer. Several versions of two-chambered MFCs have emerged with variations in the electron acceptors, surface to volume ratios of the electrodes, and reactor configurations. Rabaey et al. 2003 have achieved the highest power densities (~250 W/m^3) using acclimated bacterial cultures, closely spaced electrodes, cation exchange membrane, roughened graphite electrodes, and ferricyanide catholyte. Another version of two-chambered MFCs, called the mini MFC, was developed with an anode chamber that holds only 1.2 mL of anolyte. Larger surface area to volume ratios, the direct contact of the electrodes with PEM, and higher cathodic potentials of ferricyanide used in this reactor enabled higher power densities (500 W/m^3) (Ringeisen et al., 2006).

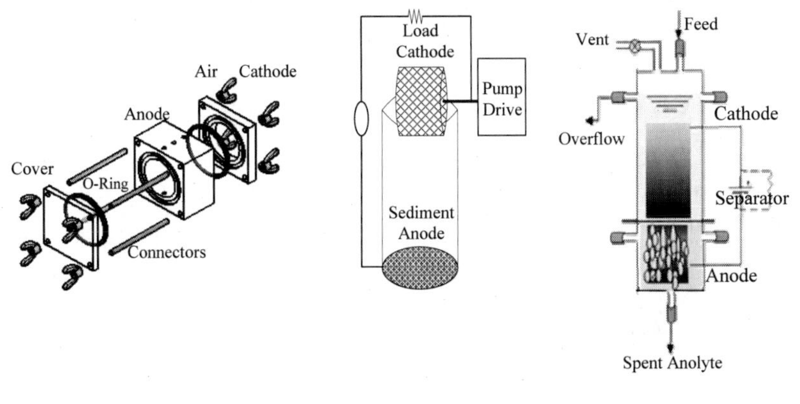

a) Single chambered MFC b) Sediment MFC c) Gravity-fed MFC

Figure 13.3 MFC bioreactors.

Single-Chambered. It is not essential to place the cathode in water or in a separate chamber when using oxygen at the cathode. The cathode is placed in direct contact with air either in the presence or absence of a membrane (Figure 13.3a). Larger power densities have been achieved when aqueous-cathodes are replaced with air-cathodes. In the simplest configuration, the anode and cathode are placed on either side of a tube, with the anode sealed against a flat plate and the cathode exposed to air on one side, and water on the other. When a membrane is used in this air-cathode system, it serves primarily to keep water from leaking through the cathode, although it also reduces oxygen diffusion into the anode chamber. The utilization of oxygen by bacteria in the anode chamber can result in a lower Coulombic efficiency (defined as the fraction of electrons recovered as current versus the maximum possible recovery). Hydrostatic pressure on the cathode will induce water leak, but that can be minimized by applying hydrophobic coatings such as polytetrafloroethylene (PTFE) on the outside of the cathode surface that permit oxygen diffusion but limit bulk water loss.

Sediment MFC. Sediment MFCs can be used to produce electricity from organic-rich aquatic sediments of remote bodies of water and power certain marine devices. Electricity production from sediment MFC designs can be realized by placing the anode electrode in marine sediments rich in organic matter and sulfides, and the cathode electrode in the overlying oxic water. Electricity producing bacteria in the sediments transfer electrons, produced from oxidation of organic matter, to the anode, and subsequently to the cathode for oxygen reduction. A cation exchange membrane may not be required due to natural potential difference obtained from decreasing the oxygen gradient over the depth of water (He et al., 2007). A rotating biological contactor based sediment MFC consists of vertical discs mounted on a horizontal shaft partially immersed in the organic waste while rotating continuously (Figure 13.3b). Biofilm is formed on the disks and the rotation provides oxygen to the cathode. External energy inputs for the rotation are eliminated when the disks are driven by natural river currents or river tides (He et al., 2007).

Tubular. Tubular bioreactor based MFCs inherit the operational principles of upflow granular bed reactors and facilitate higher power density levels and continuous flow operations. Rabaey et al. (2005) developed the design based on graphite granules packed into a cation exchange membrane soldered to form a tube. Inner tubes formed anode chambers where influent flowed upward and out of the reactor, while ferricyanide was recirculated in the outer tube. Variations of tubular MFCs include outer cylindrical reactors with a concentric inner tube that is the cathode (Liu et al., 2004), and with an inner cylindrical reactor (anode consisting of granular media) with the cathode on the outside (Rabaey et al., 2005a). Another variation is to design the system like an upflow fixed-bed biofilm reactor, with the fluid flowing continuously through porous anodes toward a membrane separating the anode from the cathode chamber (He et al., 2005). Jang et al. (2004) developed a novel tubular reactor with an anode and cathode within the same chamber separated by glass-wool and glass beads, where feed was directed upward into the anode chamber followed by the cathode chamber. This design may be less feasible in real-time applications due to possible oxygen demand in the cathode chamber from incomplete organic degradation in the anode chamber. While all the upflow tubular configurations achieved higher organic conversion efficiencies based on efficient conductive biomass granule formation, the energy inputs required for the continuous pumping of anolyte and catholyte exceeds power outputs and reduces the sustainability of the MFC technology.

Gravity Fed (GMFC). This design, developed at Air Force Research laboratory on Tyndall Air Force Base, consists of a tubular glass reactor that contains an anode and cathode separated using a polycarbonate filter (Figure 13.3c) (Gadhamshetty et al., 2009d). A downflow configuration was accomplished by maintaining a definite volume of feedstock in the cathode chamber that diffuses into the anode chamber via PC filter and subsequently trickles over the anode reactor bed before leaving the GMFC. Design allows the dissolved oxygen in the feed to serve as the electron acceptor, and being stripped off in the cathode compartment before passing into the anode compartment. A vent is provided on the cathode side (instead of anode) to provide atmospheric air exchange and allow feed to flow by gravity. An overflow line facilitates constant catholyte levels.

By eliminating proton exchange membranes and energy intensive mechanical pumps for aeration and feed supply, this MFC configuration has the potential to produce electricity and treat wastewater simultaneously.

13.6.5 Outlook on MFC Technology

Microbial fuel cell technology as an alternative to activated sludge systems or trickling filters, offers advantages by eliminating aeration requirements, reducing sludge production, and improving odor control, all at cost of waste treatment for electricity production. To make realistic comparisons of MFC technology with those currently available for wastewater treatment, the power production from methane-based anaerobic digestion is estimated to be ~400 W/m^3 while power density levels with available MFC design ranges from 20-100 W/m^3 (Pham et al., 2006). Therefore, the power densities of MFCs have to be improved by at least by a factor of 3 before they can be considered for full-scale applications

13.7 Biobutanol

Biobutanol is one of the well-known, industrialized, fermentation products that has recently gained renewed interest due to its potential to replace gasoline directly in existing internal combustion engines. Butanol as a biofuel is superior to ethanol in several aspects: higher energy content, less volatile, less hygroscopic, and less corrosive. Butanol producing bacteria often follow mixed culture pathways, and the slightest deviation from optimum bioreactor conditions can have serious implications in terms of a reduced yield and productivity. Further, the economic feasibility of biobutanol production is improved only when inexpensive feedstocks and economic product purification methods are opted.

13.7.1 Microbial Species

While the choice of a bacterial strain employed for large-scale butanol production depends on the nature of feedstock, desirable product yield, and need for additional nutrients, well-known butanol producing bacterial strains belong to *Clostridum* family*: C.acetobutylicum, C. butylicum, and C. tetanomorphum.* Biologists often use genetic manipulations techniques to obtain higher butanol yield and productivity. This is achieved by modifying metabolic pathways using recombinant DNA technology integrated with traditional mutagenesis and selection techniques (Zeikus, 1980).

13.7.2 Process Fundamentals

Anaerobic ABE (acetone-butanol-ethanol) fermentation, typical to butanol producers such as *Clostridia* is described as biphasic: i) Acidogenic phase for production of acetate, butyrate, hydrogen, and carbon dioxide is characterized by an exponential cell growth pattern, acid accumulation and pH decline; ii) Solvent production marked with the onset of a pH increase in the culture medium, due to acid-reassimilation in the stationary growth phase, is associated with the production of acetone/isopropanol, butanol and ethanol. Zeikus (1980) termed butanol producers as "ripe-cultures" available at end of the exponential growth phase. These "ripe-cultures" are reported to be responsive to endangered conditions of low pH due to acid accumulation by changing the gene expression pattern to perform the following: i) slow-down acid production; ii) assimilate acetate and butyrate; iii) continue glucose assimilation for the production of solvents (butanol and acetone). Butanol production is also closely related to sporulation as both these processes are initiated by a common transcription factor (Lee et al., 2008). Readers are directed to the literature to gain an in-depth understanding on the biochemical pathways and physiology of butanol producing strains (Jones and Woods, 1986).

13.7.3 Environmental Conditions for Butanol Production

Microbes such as *Clostridia* are common to both biohydrogen and butanol production, and controlled bioreactor conditions are required to manipulate fermentation pathways towards butanol production. Excess carbon, sufficient iron levels, nitrogen and phosphorus (0.74 mM) limitations integrated with threshold concentrations of acetate and butyrate (5 mM) are reported to be the optimum conditions for continuous fermentative butanol production (Jones and Woods, 1986). Unlike methane and biohydrogen production, low pH conditions (~4.5), high hydrogen partial pressures (100-250 kPa) and high substrate concentrations (~40 mM glucose) are required for butanol production. A buffering capacity that allows sufficient acid accumulation before pH declines to 4.5 is necessary for the on-set of butanol production. Inhibitory levels of butanol (0.1-0.15 M) that retards cell growth and substrate consumption have to be eliminated to sustain butanol production (Durre, 2007). Variations in operating conditions are further influenced by the choice of selected strain for butanol production. The maximum theoretical value for butanol production is not yet defined in the literature due to complexity involved in branched pathways for solvent production and is generally reported in concentration units.

13.7.4 Bioreactor Design for Butanol Production

A major challenge in butanol production involves maintaining a balanced selection of acidogenic and solventogenic bacterial strains, and the recovery of the butanol from the product mixture. Bioreactor designs for butanol production are selected to exhibit non-back mixing and plug flow behavior to accommodate problems of gas production, induced mass-transfer limitations, and butanol toxicity (Qureshi and Blaschek, 2001).

Batch. Batch reactors are preferred due to simplicity in operation and reduced risk of contamination. However, the productivity achievable in batch operations suffers from the disadvantages of a lag time for start-up, product inhibition, as well as down time for cleaning, sterilizing, and filling. While these problems can be minimized by adapting continuous operation, single stage fermentation is reported to be less feasible due to observations on declined solvent production and increased acid accumulation over time. Reduced performance in single stage continuous bioreactors is attributed to the complexity of butanol fermentation pathways in *Clostridia spp.* (Jones and Woods, 1986).

Fed-Batch. Fedbatch bioreactors are generally applied to processes where a high substrate concentration is inhibitory to the microbes and may be less applicable to butanol production due to toxic levels imposed by butanol to strains such as *C.acetobutylicum*. However, the feasibility of fed-batch systems towards butanol fermentation can be improved when the bioreactor is integrated with in-situ butanol recovery techniques as explained in the latter part of the chapter (Lee et al., 2008).

Continuous. The cascade principle has been suggested as a means to handle biphasic nature of ABE fermentation in continuous reactors. This technique separates production of acids and solvents and has been tested in both pilot and large-scale bioreactors (Jones and Woods, 1986) For example, butanol production was demonstrated in a series of five bioreactors to separate acid formation in the first two and butanol production in the last three bioreactors. Cell-recycle has been suggested as an option to overcome the problems of selection of acid-producing cells and cell degeneration that occurs at high solvent concentrations (Afschar et al., 1986). On similar lines, a two stage configuration was recently suggested by (Angenent and Wrenn, 2008) where butyrate is produced in the first stage, and butyrate along with glucose are assimilated in the second stage to produce butanol. Two-stage configurations facilitate different operating conditions of pH and hydrogen partial pressure suitable for acidogens and solventogens. Higher substrate conversions are realized with butyrate (~100%) versus glucose (~80%) and two-stage configuration therefore improves butyrate productivity (Bahl et al., 1982).

Immobilized Cells. Butanol production is initiated only during non-growing (stationary) phases having implications on the superiority of immobilized systems versus CSTRs with the continuous washout of the cells. As mentioned in other sections, immobilized systems such as FBBRS and PBBRs facilitate improved cell retention, a reduced reactor volume, minimized nutrient depletion and product inhibition. Problems associated with immobilized systems include mass transport limitations and reduced productivity due to an accumulation of gaseous products in form of bubbles within the packing materials (Jones and Woods, 1986). Other disadvantages with immobilized system include loss of cell activity in non-growth media and a difficulty in maintaining homogenous population of complex bacterial consortia: acidogens, soventogens and sporulators (Haggstrom and Molin, 1980). For example, *C. beijerinckii* BA101 was immobilized onto clay brick particles and the butanol production was sustained at a dilution rate of 2 h^{-1} to achieve a solvent productivity of 15.8 g/L/h, yield of 0.38 g/g and a concentration of 7.9 g/L (Qureshi and Blaschek, 2000). However, the process suffered from the fact that only a fraction of the biomass was in the solventogenic state and a significant amount of biomass was present as inactive biomass, spores. A possible remedy to achieve higher productivity in immobilized systems is to block the sporulation.

While, the immobilization media such as coke, kaonline, brick, resin, vermiculite, bagasses, polypropylene, and plastic composite are used to retain cells in FBR and PBBRs, membranes can be used to replace immobilized media in bioreactors. For instance, a hollow-fiber ultrafilter was applied to separate and recycle cells in a continuous fermentation to achieve cell mass, solvent concentration, and solvent productivity of 20 g/L, 13 g/L, and 6.5 g/L/h respectively, at a dilution rate of 0.5 1/h (Pierrot et al., 1986). Problems associated with membrane fouling can be overcome by allowing only the fermentation broth to undergo filtration by using the immobilized cell system (Lipnizki et al., 2000).

13.7.5 Downstream Processing

Butanol-producing bacterial strains exhibit mixed-solvent fermentation characteristics and produce by-products such as ethanol, butyric acid and acetone, requiring expensive product recovery step. Some techniques for butanol purification from the product mixture include: i) distillation ii) pervaporation iii) gas-stripping iv) liquid-liquid extraction v) reverse osmosis. The traditional distillation process, working on a separation principle based on differences in volatilities in liquid mixtures, for butanol separation suffers from high operation cost due to the low concentration of butanol in the fermentation broth. Reverse osmosis (RO) is an economical choice and an established technology but suffers from disadvantages of membrane clogging or fouling (Durre, 2007). The liquid–liquid extraction (LLE) technique that takes advantage of the differences in the distribution coefficients of the chemicals has high capacity and selectivity, although it can be expensive to perform. Further, LLE technique has critical problems such as the toxicity of the extractant to the cell and emulsion formation. These problems can be overcome if the fermentation broth and the extractant are separated by a membrane that provides a surface area for butanol exchange between the two immiscible phases; this is called ''Perstraction''. Gas stripping is another efficient method for butanol recovery where fermentation gas is bubbled through the fermentation broth, then passed through a condenser for solvent recovery; the stripped gas is then recycled back to the bioreactor and the process continues until all the sugar in the fermentor is utilized. Pervaporation is a membrane-based process that allows selective removal of volatile compounds from fermentation broth. The membrane is placed in contact with the fermentation broth and the volatile liquids or solvents diffuse through the membrane as a vapor which is recovered by condensation (Lee et al., 2008).

13.7.6 Outlook on Butanol Production

The economics of butanol technology can be improved only when metabolically engineered strains, with an enhanced yield and productivity and an improved tolerance to butanol, are used to produce butanol from lignocellulosic feedstock in a well-designed bioreactor integrated with an efficient butanol recovery step. Ventures on biobutanol production by giant companies such as DuPont and

British Sugars show great promise for commercialization of microbial butanol technology in near-future.

13.8 Bioethanol Production

There are two major pathways for producing ethanol through fermentation in which the bioreactor design becomes critically important. In the first pathway, the monomeric sugars are fermented into ethanol in stirred-tank reactors operated in a batch-mode. The second one involves producing synthesis gas (syngas) through gasification of biomass which is then converted into liquid biofuel using microbial catalysts, known as syngas fermentation. The bioreactor design is of significant importance in syngas fermentation due to poor solubility of syngas (hydrogen and carbon monoxide) in the aqueous phase and is discussed next.

13.8.1 Bioreactor Design for Ethanol Production

Both batch and continuous-flow bioreactors can be used for syngas fermentation. In batch reactors, syngas fermentation is done in a closed system as the gaseous substrate is supplied continuously. Liquid samples are withdrawn at frequent intervals during fermentation. Vega et al. (1990) examined the kinetic parameters through a series of batch experiments. While, a CSTR has been traditionally used in syngas fermentation, bubble column reactors, monolithic biofilm reactors, trickling bed reactors and microbubble dispersion stirred-tank reactors have also been studied by various researchers, under both continuous and batch operations. Different types of reactor configurations employed in syngas fermentation are briefly discussed here.

CSTR. The continuously stirred-tank reactor (Fig. 13.4a) is the most common bioreactor employed in syngas fermentation. In CSTRs, the gaseous substrate is injected continuously and a liquid nutrient flow (culture media) is fed into the bioreactor to supplement the nutrients for microbial metabolism (Klasson et al., 1992; Vega et al., 1990). The fermentation product is drawn from the system at the same flow rate as the feed. A higher level of agitation or mixing is maintained in the reactor by baffled impellers to increase the mass transfer between the substrate and the microbes. Higher rotational speeds of the impellers tend to break the gas bubbles into finer ones thereby making the gaseous substrate more accessible to the microbes. In addition, the slow rising velocity of the finer bubbles leads to longer gas retention in the aqueous medium, which leads to higher mass transfer rates.

Bubble Column. Bubble column reactors (Fig. 13.4b) are designed mainly for industrial applications with large working volume. Higher mass transfer rates, and low operational and maintenance costs are the merits of the system while back mixing and coalescence are considered as the major drawbacks of bubble columns (Datar et al., 2004).

Monolithic Biofilm. In monolithic biofilm reactors the gaseous substrate is allowed to pass through a bed of carrier media. The microbes grow on the media as biofilm. During the operation, attached microorganisms in the biofilm utilize the gaseous substrates to produce ethanol, acetic acid and other end-products. The monolithic biofilm reactors are operated under atmospheric pressure rendering the process more economically viable.

Trickle-Bed. A trickle-bed reactor (Fig. 13.4c) is a packed bed, continuous flow reactor where a liquid culture flows down through a packing media. The syngas flows

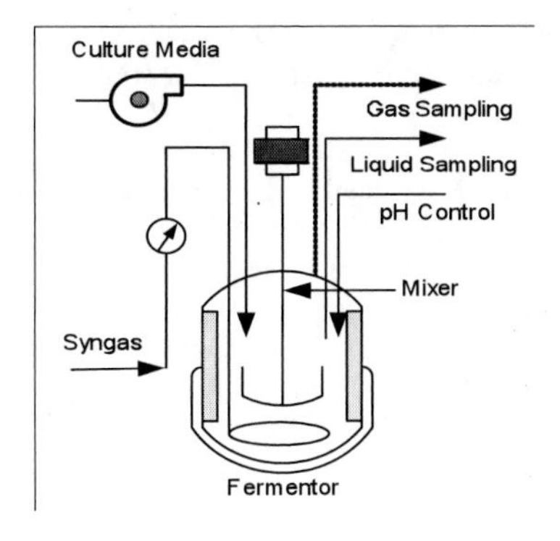

a) Continuous-stirred tank reactor

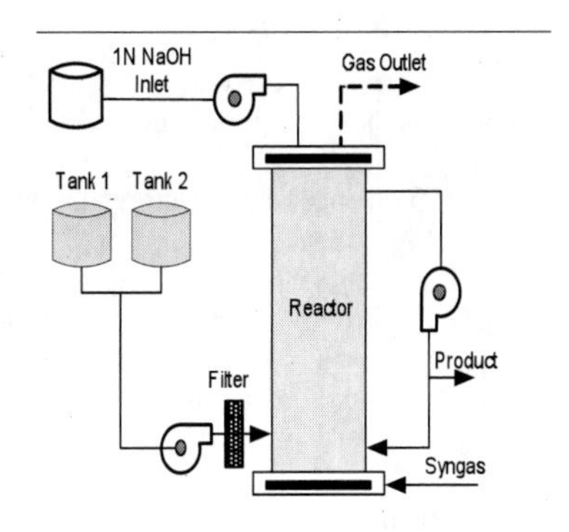

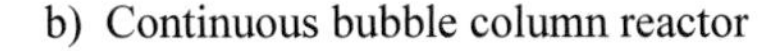

b) Continuous bubble column reactor

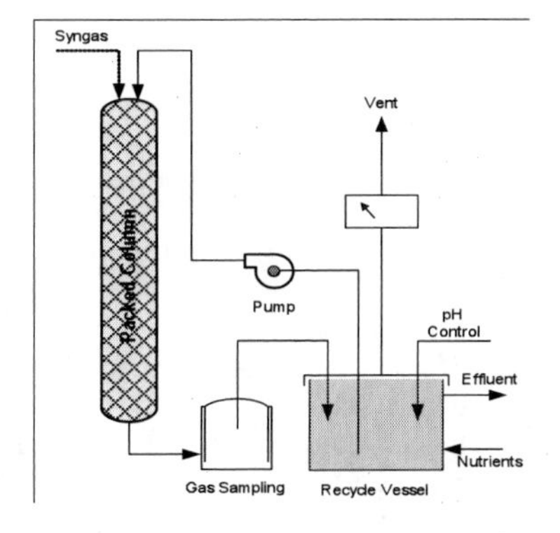

(c) Trickling-bed bioreactor

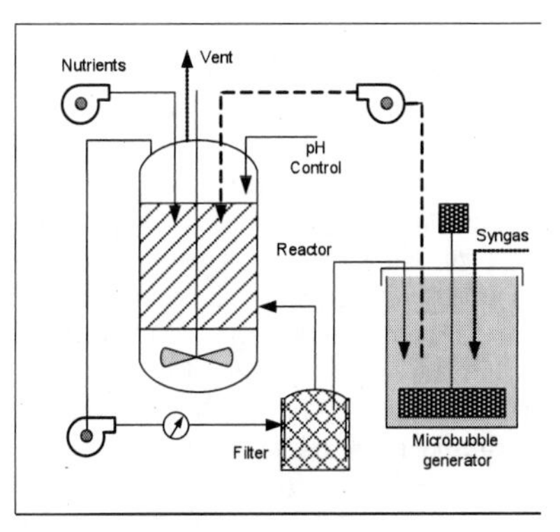

(d) Microbubble-sparged bioreactor with a membrane based cell-recycle system

Figure 13.4 Different types of reactors used in Syngas fermentation.

either downward (co-current) or upward (counter-current). Since these types of reactors do not require mechanical agitation, the power consumption of trickle-bed reactors is lower than that of CSTRs (Bredwell et al., 1999).

13.8.2 Mass Transfer Issues in Syngas Fermentation

One drawback associated with syngas fermentation is the gas-liquid mass transfer limitation, which are primarily due to the low solubility of the major syngas components (CO and H_2) in the aqueous fermentation broth when it contains a high cell concentration. If the cell concentration is too low, the system yield would be low and mass transfer would be kinetically limited.

Microbubble Dispersion Stirred-Tank. Microbubble dispersion stirred-tank reactors (Fig. 13.4d) are a stirred-tank equipped with a microbubble sparger. Bredwell et al. (1999) found that the mass transfer of the system increased in two ways. Firstly, due to the decreasing bubble size the internal pressure increase, which leads to an increase in the driving force. Secondly, the steady state liquid phase concentration gradient at the surface of the bubble is inversely proportional to the diameter. In other words, the flux increases as the diameter of the bubble decreases.

Membrane-Based. Composite hollow fiber membranes (HFM) can be used to facilitate the mass transfer in aqueous culture media. Even though this technique is not adopted exclusively in syngas fermentation, it has been examined for hydrogen and oxygen transfer in water treatment applications (Nerenberg and Rittmann, 2004; Lee and Rittmann, 2001). In the HFM, syngas is diffused through the walls of membranes without forming bubbles. The microbial community grown as a biofilm on the outer wall of the membranes continuously ferment H_2 and CO to ethanol and acetic acid. This innovative approach offers significant advantages in achieving a higher yield and reaction rate, and higher tolerances to toxic compounds present in syngas (tar, acetylene, NO_x, O_2). Moreover, these HFM reactors are operated under high pressure with higher mass transfer rates and reduced reactor volumes. Microorganisms which can tolerate high pressures can be utilized in the fermentation process as microbial catalysts (Madigan et al., 1997).

High gas and liquid flow rates, large specific gas–liquid interfacial areas, and increased gas solubility (through the use of increased pressure or solvents), stimulate gas/liquid mass transfer rates. In order to have higher yields of ethanol, interphase mass transfer has to be maximized to get a high value of volumetric mass transfer coefficient, k_La. For instance, CSTRs offer high gas/liquid mass transfer coefficients (k_La) at high impeller speeds, but at a high power consumption. High impeller speeds effectively break up large bubbles into smaller bubbles with more beneficial surface/volume ratios. Small bubbles additionally have lower rise velocities, and thus a longer liquid contact time. In micro bubble dispersion, extremely small, surfactant-stabilized bubbles are created in a high shear zone, providing a more energy efficient method to increase k_La values. In biotrickling filters the k_La is relatively independent of the gas flow rate for sparingly soluble gasses.

13.9 Biodiesel

Of the current biofuels, biodiesel has the greatest potential for immediate use in the transportation and electric power sectors as it suits the existing infrastructure well. Even though biodiesel can be produced from oil crops, it is not feasible to meet the transportation fuel needs using plant-based biodiesel due to the huge land and other resources required. As such, algal biodiesel is an emerging a more feasible approach. This section focuses on the bioreactor aspects of algal biomass cultivation for biodiesel production. Traditionally, open raceway reactors have been employed for large-scale algal cultivation in batch or semi-continuous mode. While they are simple to operate, engineered photobioreactors have several advantages such as higher biomass densities, smaller footprint, continuous operation, better contamination resistance, efficient harvesting, longer production cycles, and minimal labor input, albeit at a higher initial cost (Carvalho et al., 2006; Chisti, 2007).

13.9.1 Background

Microalgae have the ability to harvest sunlight and transform its radiant energy into valuable end products such as biodiesel, using inexpensive natural resources (viz., CO_2 and H_2O). Microalgae have the same basic photosynthetic mechanism as higher plants. As they are of simple structure, being unicellular, energy is directed into photosynthesis, growth and reproduction rather than maintaining differentiated structures. Thus, the levels of protein in microalgae can be between 30–50% of the dry biomass. Further, since microalgae are microscopic in size and are grown in a liquid culture, nutrients can be maintained near optimal conditions potentially providing the benefits of high levels of controlled, continuous productivity similar to microbial fermentation.

In terms of photosynthetic efficiency, microalgal yields are greater than those of macroalgae and similar to those of higher plants. But, up to 18% of the solar energy can be stored in algal cells in contrast to the 6% by higher plants (Pitt, 1980). Unlike higher plants, microalgal biomass has a uniform cell content and chemistry as there are no leaves, stems, or roots with different chemical compositions. Microalgal biomass usually has little ash content (< 10% dry weight) in contrast to the larger amount of ash (up to 50%) of macroalgae. Unlike other oil crops, microalgae grow extremely rapidly and many contain 20-50% oil. Microalgae commonly double their biomass within 24 h. Biomass doubling times during exponential growth could be as short as 3.5 h. Goldman (1980) has reviewed the outdoor mass culture of microalgae and suggested that yields of 15-25 g dry wt/m^2-day could be attained for reasonably long periods of time. In the context of algal biodiesel production, maximizing oil productivity (the mass of oil produced per unit volume of the microalgal broth per day) is the major goal, which depends on the algal growth rate and the oil content of the biomass.

13.9.2 Light Needs for Algal Cultivation

As with all photosynthetic organisms, the growth rate of microalgae also depends on light levels. A key photoparameter is light saturation characterized by the light saturation constant: the intensity of light at which the specific biomass growth rate is half its maximum value. Light saturation constants for microalgae tend to be much lower than the maximum sunlight level that occurs at midday. For example, the light saturation constants for microalgae *Phaeodactylum tricornutum* and *Porphyridium cruentum* are 185 μE/m^2-s and 200 μE/m^2-s (Molina Grima et al., 2000), respectively. In comparison with these values, the typical midday outdoor light intensity in equatorial regions is about 2000 μE/m^2-s. Above a certain value of light intensity, a further increase in light level reduces the biomass growth rate due to photoinhibition. Light intensities slightly greater than the light level at which the specific growth rate peaks can lead to photoinhibition. Elimination of photoinhibition or its postponement to higher light intensities can greatly increase average daily growth rate of algal biomass.

13.9.3 Bioreactor Design for Biodiesel Production

Bioreactor configurations for algal cultivation fall into two basic categories with varying designs in each category: open raceway-types and the photobioreactor types. The open raceway is the most common and simplest configuration, and has the longest history of application for algal cultivation. Photobioreactors are of closed design that include flat plate and tubular designs. Unlike open raceways, photobioreactors enable single-species cultures of microalgae for prolonged durations, at higher productivity.

Raceway. Raceway ponds for the mass culture of microalgae have been used since the 1950s. A raceway pond is essentially a loop recirculation channel about 0.3 m deep (Figure 13.5a). Mixing and circulation of the broth are produced by a paddlewheels. Raceway channels are built in concrete or compacted earth, and may be lined with white plastic. During daylight, the culture is fed continuously downstream of the paddlewheel where the flow begins, and the broth is harvested behind the paddlewheel on completion of the circulation loop. The paddlewheel operates continuously, maintaining flow/turbulence to prevent sedimentation, and providing for the transfer of carbon dioxide to the broth and stripping of oxygen from the broth.

In raceways, the temperature fluctuates within a diurnal cycle and seasonally, and any cooling is achieved only by evaporation. Evaporative water loss can be significant. Because of significant losses to the atmosphere, raceways use carbon dioxide much less efficiently than photobioreactors. Productivity is affected by the contamination with unwanted algae and predation. Biomass concentrations remain low because raceways are poorly mixed and cannot sustain an optically dark zone. Raceways are perceived to be less expensive than photobioreactors, because they cost less to build and operate. Although raceways are low-cost, they have a low biomass productivity compared with photobioreactors. Raceway ponds and other open culture systems for producing microalgae are further discussed by Terry and Raymond (1985).

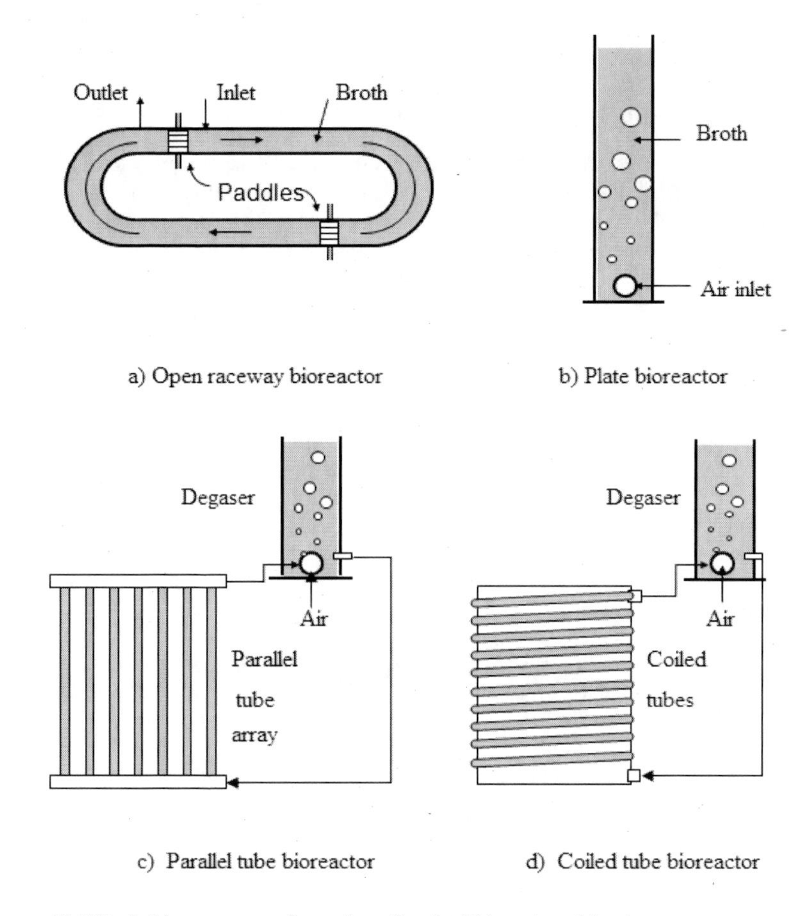

Figure 13.5 Basic bioreactor configurations for algal biomass cultivation.

Flat Panel. Flat plate reactors are built with closely spaced parallel panels of glass or plastic to attain high area-to-volume ratios and to harvest sunlight. The broth is circulated through the space between the panels at a velocity of about 1.2 m/s (Figure 13.5b). This system includes an open gas transfer unit to strip oxygen; however, such an open zone restricts the effectiveness of contamination control, as compared with completely closed bioreactors. A closed system of water spraying was employed to control temperature; the sprayed water was then collected in troughs and recirculated through a ventilated water column for refrigeration.

One such bioreactor, with an overall volume of 1000 L, was tested with various light paths for the cultivation of Nannochloropsis sp.; the maximum volumetric productivity, 0.85 g L-1 d-1, was attained with the minimum light path, i.e., 1.3 cm.

Tubular. Tubular photobioreactors consist of an array of transparent tubes of plastic or glass, generally 0.1 m or less in diameter through which the broth is continuously circulated. This array of tubes is oriented north-south to capture sunlight. The tube diameter is limited to ensure sufficient illumination and maintain high biomass productivity. The array of tubes may be laid parallel to the ground (Figure 13.5c) or coiled around a supporting frame to form helical coil tubular photobioreactors (Figure 13.5d). Other variants of tubular photobioreactors exist (Carvalho et al., 2006; Molina Grima et al., 1999), but are not widely used.

Biomass sedimentation in tubes is prevented by maintaining turbulent flow using either a mechanical pump or a gentler airlift pump. Mechanical pumps can damage the biomass but are easy to design, install and operate. Airlift pumps for use in tubular photobioreactors are designed using the same methods that were originally developed for designing conventional airlift bioreactors. Airlift pumps are less flexible than mechanical pumps and require a supply of air to operate.

Under high irradiance, the maximum rate of oxygen generation in a typical tubular photobioreactor may be as high as 10 g O_2/m^3-min. Dissolved oxygen (DO) levels much greater than the air saturation values inhibit photosynthesis. Furthermore, a high concentration of DO in combination with intense sunlight produces photooxidative damage to algal cells. To prevent inhibition and damage, the maximum tolerable DO level should not generally exceed about 400% of air saturation value. Since oxygen cannot be removed within the photobioreactor tubes, the maximum length of a continuous run of tube is limited. The culture must periodically return to a degassing zone that is bubbled with air to strip out DO.

In addition to removing the accumulated DO, the degassing zone must disengage all the gas bubbles from the broth so that essentially bubble-free broth returns to the solar collector tubes. Because a degassing zone is generally optically deep compared with the solar collector tubes, it is poorly illuminated and, therefore, its volume needs to be kept small relative to the volume of the solar collector. As the broth moves along a photobioreactor tube, the pH increases because of consumption of carbon dioxide. Carbon dioxide is fed in the degassing zone in response to a pH controller. Additional carbon dioxide injection points may be necessary at intervals along the tubes to prevent carbon limitation and an excessive rise in pH (Molina Grima et al., 1999).

Photobioreactors require cooling during daylight hours. Furthermore, temperature control at night is also useful. For example, the nightly loss of biomass due to respiration can be reduced by lowering the temperature at night. Outdoor tubular photobioreactors are effectively and inexpensively cooled using heat exchangers. A heat exchange coil may be located in the degassing column. Alternatively, heat exchangers may be placed in the tubular loop. Evaporative cooling by water sprayed on tubes can also be used and has proven successful in dry climates.

13.9.4 Outlook on Biodiesel Production

Even though biodiesel offers several benefits, current overall production costs of algal-based biodiesel has to be reduced by a factor of 7 for it to be cost-competitive against petroleum-based fuels (Chisti, 2007). The following have been identified as key factors that contribute to the high cost of algal-based biodiesel which are being addressed by current researchers:

i) High cost of algal biomass cultivation due to: sensitivity of biomass growth to temperature, low light-to-biomass conversion rate, variable oil content and high water content, CO_2 availability and delivery strategy, high cost of bioreactors, water loss by evaporation in outdoor ponds, and contamination and predation.

ii) High cost of biomass harvesting and downstream processing due to: highly dilute harvest, excessive water removal requirements, and recycling costs, batch operations, use of large amounts of chemical flocculants for harvesting, excessive water removal from algal biomass in the drying process, selective separation of desirable algal oil components, use of chemical solvents in extraction, purification of lipids, and conversion of lipids and other hydrocarbons to biofuels.

iii) Poor utilization of residual biomass and byproducts for resource/energy recovery.

13.10 Summary

Dwindling fossil fuels, and the exponential rise of global energy needs coupled with fossil-fuel related negative impacts on the environment has already triggered biofuel initiatives in several countries. For example, the US Department of Energy has established the goal of supplementing 30% of gasoline consumption with cellulose ethanol by 2030. Similarly, the European Union directive (2003) intends to substitute 6% of gasoline and diesel fuels with biofuels by 2010. Advances in genomic studies, metabolic engineering, gene regulation and protein engineering coupled with advanced bioreactor designs have a great potential to deliver cost-effective biofuels in near future. While breakthrough developments in reactor engineering and materials are essential to improve feasibility of biofuel production in bioreactors, the use of arable lands and pure organic substrates should be replaced with that of lignocellulosic feedstocks.

13.11 References

Afschar, A. S., K. Schaller., and K.Schugerl. (1986). "Continuous production of acetone and butanol with shear-activated *Clostridum acetobutylicum.*" *Appl microbial biotechnol*, 23, 315-321.

Akkerman, I., Janssen, M., Rocha, J., and Wijffels, R. H. (2002). "Photobiological hydrogen production: photochemical efficiency and bioreactor design." *International Journal of Hydrogen Energy*, 27, 1195-1208.

Angenent, L. T., and Wrenn, B. A. (2008). "Optimizing mixed-culture bioprocesses to convert wastes into biofuels." W. J.D., C. S. Harwood, and A. Demain, eds., Washington.

Bahl, H., Andersch, W., and Gottschalk, G. (1982). "Continuous production of acetone and butanol by *Clostridum acetobuytlicum* in a two-state phosphate limited chemostat." *Eur. J. Appl. Microbial. Biotechnol*, 15, 201-205.

Bredwell, M.D., Srivastava, P., and Worden, R.M. (1999). "Reactor design issues for synthesis-gas fermentations". *Biotechnol. Prog*, 15, 834-844.

Callander, I. J., and Barford, J. P. (1983). "Precipitation, Chelation, and the Availability of metals as nutrients in Anaerobic Digestion. II. Applications." *Biotech and Bioeng*, 25.

Carvalho, A. P., Meireles, L. A., and Malcata, F. X. (2006). "Microalgal Reactors: A Review of Enclosed System Designs and Performances." *Biotechnol. Prog*, 22, 1490-1506.

Chang, J. S., Lee, K. S., and Lin, P. J. (2002). "Biohydrogen production with fixed-bed bioreactors." *Int. J. Hydrogen Energy*, 27, 1167-1174.

Chen, C. C., Lin, C. Y., and Chang, J. S. (2001). "Kinetics of hydrogen production with continuous anaerobic cultures utilizing sucrose as the limiting substrate." *Appl. Microbiol.Biotechnol*, 57(1/2), 56-64.

Chisti, Y. (2007). "Biodiesel from microalgae." *Biotech. Adv*, 25, 294-306.

Dal-Yeol, C., Conly, L. H., and David, K. S. (2007). "Production of Bio-Hydrogen by mesophilic anaerobic fermentation in an Acid-Phase Sequencing batch reactor." *Biotechnol. Bioeng*, 96(3), 421-431.

Datar, R.P., Shenkman, R.M., Cateni, B.G., Huhnke, R.L., and Lewis, R.S. (2004). "Fermentation of biomass-generated producer gas to ethanol". *Biotechnol. Bioeng,* 86, 587–94

de Mes, T. Z. D., Stams, A. J. M., Reith, J. H., and Zeeman, G. (2002). "Methane production by anaerobic digestion of wastewater and solid wastes." Biomethane and Biohydrogen, J. H. Reith, R. H. Wijffels, and H. Barten, eds., Smiet-hague, 58-103.

DeMan, A. W. A., Vander Last, A. R. M., and Letting, G. "The use of EGSB and UASB anaerobic systems for low strength soluble and complex wastewaters at temperatures ranging from 8 to 30 degree centigrade." *Fifth international conference of anaerobic digestion*, Bologna, 197-209.

Durre, P. (2007). "Biobutanol-An attractive biofuel." *Biotechnology Journal*, 2, 1525-1534.

Gadhamshetty, V., Anoop, S., and Nirmalakhandan, N. (2009a). "Review on photoparameters in photofermentative biohydrogen production." *Critical reviews Environmental Science Technology*, Accepted

Gadhamshetty, V., Anoop, S., Nirmalakhandan, N., and Maung, T. M. (2008). "Photofermentation of malate for biohydrogen production-A modeling approach." *Int. J. Hydrogen Energy*, 33, 2138-2146.

Gadhamshetty, V., David, C. J., Nirmalakhandan, N., Geoff, B. S., and Deng, S. (2009b). "Dark and acidic conditions for fermentative hydrogen production." *Int. J. Hydrogen Energy*, 34(2), 821-826.

Gadhamshetty, V., David, C. J., Nirmalakhandan, N., Geoff, B. S., and Deng, S. (2009c). "Feasibility of biohydrogen production at low temperatures in unbuffered reactors." *Int. J. Hydrogen Energy*, 34(3), 1233-1243.

Gadhamshetty, V., Ramaraja, P. R., Susan, S., Lloyd, J. N., Justin, C. B., Brad, R., and Johnson, R. G. (2009d). "*Shewanella* assisted electricity production in gravity fed microbial fuel cells." 238th ACS national meeting, Washington, D.C.

Gavala, H. N., Skiadas, I. V., and K.Ahring, B. (2006). "Biological hydrogen production in suspended and attached growth anaerobic reactor systems." *International Journal of Hydrogen Energy*, 31, 1164-1175.

Ghosh, S. "Solid-phase methane fermentation of solid wastes." *Proceedings of Eleventh American Society of Mecahnical Engineergs national Waste Processing Conference, American society of mechanical engineering*, Orlando, Fl.

Goldman, J. C. (1980). *Water Res*, 13, 1.

Haggstrom, L., and Molin, N. (1980). "Calcium alginate immobilized cells of Clostridium acetobutylicum for solvent production." *Biotechnology Letters*, 2, 241-246.

Hawkes, F. R., R.Dinsdale, Hawkes, D. L., and I.Hussy. (2002). "Sustainable fermentative hydrogen production: challenges for process optimization." *International Journal of Hydrogen Energy*, 27, 1139-1347.

He, Z., Minteer, S. D., and Angenent, L. T. (2005). "Electricity generation from artificial wastewater using an upflow microbial fuel cell." *Environmental Science & Technology*, 39(14), 5262-5267.

He, Z., Shao, H., and T.Angenent, L. (2007). "Increased power production from a sediment microbial fuel cell with a rotating cathode." *Biosensors and Bioelectronics*, 22, 3252-3255.

Jang, J. K., Pham, T. H., Chang, I. S., Kang, K. H., Moon, H., Cho, K. S., and Kim, B. H. (2004). "Construction and operation of a novel mediator- and membrane-less microbial fuel cell." *Process Biochemistry*, 39(8), 1007-1012.

Janssen, M. (2002). "Cultivation of microalgae:effect of light/dark cycles on biomass yield," Wageningen University, Wageningen, The Netherlands.

Jones, D. T., and Woods, D. R. (1986). "Acetone-Butanol Fermentation Revisited." *Microbiological Reviews*, 50(4), 484-524.

Kallogo, Y., and Verstraete, W. (1999). "Development of anaerobic sludge bed (ASB) reactor technologies for domestic wastewater: motives and perspectives." *World Journal of Microbiology & Biotechnology*, 15, 523-534.

Khanal, S.K. (2008). "*Anaerobic Biotechnology for Bioenergy Production: Principles and Application.*" Wiley-Blackwell Publishing, Ames, IA, USA.

Klasson, K.T., Ackerson, C.M.D., Clausen, E.C., and Gaddy, J.L. (1992). "Biological conversion of synthesis gases into fuels". *Int. J. Hydrogen Energy*, 17(4), 281-288.

Kleerebezem, R., and Loosdrecht, M. C. v. (2007). *Current opinion in biotechnology*, 18, 207-212.

Kotsopoulos, T. A., Zeng, R. J., and Angelidaki, I. (2005). "Biohydrogen production in Granular Up-Flow Anaerobic Sludge Blanket Reactors With Mixed

Cultures Under Hyper-Thermophilic Temperature " *Biotechnology & Bioengineering*, 94(2), 296-302.

Lee, K.C., and Rittmann, B.E. (2001). "Applying a novel autohydrogenotrophic hollow-fiber membrane biofilm reactor for denitrification of drinking water". *Water Res,* 36, 2040-2052.

Lee, S. Y., Park, J. H., Jang, S. H., Nielsen, L. K., and Kim, J. (2008). "Fermentative butanol production by Clostridia." *Biotechnology and Bioengineering*, 101(2), 209-228.

Lettinga, G., Roersma, R., and Grin, P. (1983). "Anaerobic treatment of raw domestic sewage at ambient temperatures using a granular bed UASB reactor." *Biotech and Bioeng*, 25(1701-1723).

Lin, C. L., and Fang, H. H. P. (2007). "Fermentative hydrogen production from wastewater and solids wastes by mixed cultures." *Critical reviews in Environmental Science Technology*, 37(1), 1-39.

Lin, C. Y., and Lay, C. H. (2005). "A nutrient formulation for fermentative hydrogen production using anaerobic sewage sludge microflora." *Int. J. Hydrogen Energy*, 30(3), 285-292.

Lipnizki, F., Hausmann, S., Laufenberg, G., Field, R., and Kunz, B. (2000). "Use of pervaporation-bioreactor hybrid processes in biotechnology." *Chemical engineering technology*, 23, 569-577.

Liu, H., Ramnarayan, R., and Logan, B. E. (2004). "Production of electricity during wastewater treatment using a single chamber microbial fuel cell." *Environmental Science & Technology*, 38(7), 2281-2285.

Logan, B. E., Aelterman, P., Hamelers, B., Rozendal, R., Schroder, U., Keller, J., Freguiac, S., Verstraete, W., and Rabaey, K. (2006). "Microbial fuel cells: methodology and technology." *Environmental Science & Technology*, 40(17), 5181-5192.

Madigan, M.T., Martinko, J.M., and Parker J. (1997). "Biology of Microorganisms". Edn 8. Edited by Brock T.D. New Jersey, Prentice-Hall, Inc.

Margaritis, A., and Wallace, J. B. (1984). "Novel Bioreactor Systems and their Applications." *Bitechnology*, 447-452.

Mata-Alvarez, J. (1987). "A dynamic simulation of two-phase anaerobic digestion system for solid wastes." *Biotechnology & Bioengineering*, 30, 844-851.

McCarty. (1964). "Anaerobic Waste Treatment Fundamentals." *Public works*, 95, 107-110.

Milliken, C. E., and May, H. D. (2007). "Sustained generation of electricity by spore-forming gram positive, Desulfitobacterium hafniese strain DCB2." *Applied microbiol biotechno*, 73, 1180-1189.

Molina Grima, E., Acién Fernández, F., García Camacho, F., and Chisti, Y. (1999). "Photobioreactors: light regime, mass transfer, and scaleup." *J Biotechnol* 70, 231-247.

Molina Grima, E., Fernández, J., Acién Fernández, F., and Chisti, Y. (2000). "Scale-up of tubular photobioreactors." *J Appl Phycol*, 12, 355-368.

Mori, K. (1985). "Photoautotrophic bioreactor using visible solar rays condensed by Fresnel lenses and transmitted through optical fibers." 331-345.

Nerenberg, R., and Rittmann, B.E. (2004). "Hydrogen-based, hollow-fiber membrane biofilm reactor for reduction of perchlorate and other oxidized contamitants". *Water Sci. Technol,* 49, 223-230.

Ogbonna, J. C., Soejima, T., and Tanaka, H. (1999). "An integrated solar and artificial light system for internal illumination of photobioreactors." *J Biotechnology*, 70, 189-297.

Oh, S.-E., Iyer, P., Bruns, M. A., and Logan, B. E. (2004a). "Biological Hydrogen Production Using a Membrane Bioreactor." *Biotechnology & Bioengineering.*

Oh, Y. K., Kim, S. H., Kim, M. S., and Park, S. (2004b). "Thermophilic biohydrogen production from glucose with trickling biofilter." *Biotechnol. Bioeng*, 88(6), 690-698.

Parkin, G. F., and Owen, W. F. (1986). "Fundamentals of anaerobic digestion of wastewater sludges." *Journal of Environmental Engineering*, 112, 867-920.

Pham, T. H., Rabaey, K., Aelterman, P., Clauwert, P., De Schamphelaire, L., Boon, N., and Verstraete, W. (2006). "Microbial fuel cells in relation to conventional anaerobic digestion technology." *Engineering Life Science*, 6(3), 285-292.

Pierrot, P., Fick, M., and Engasser, J. (1986). "Continuous acetone-butanol fermentation with high productivity by cell ultrafiltration and recycling." *Biotech Lett*, 8, 253-256.

Pitt, S. J. (1980). *Biochem Soc Trans*, 8, 479.

Qureshi, N., and Blaschek, H. (2000). "Economics of butanol fermentation using hyper-butanol producing Clostridium beijerinckii BA101." *Trans IChemE*, 78(PartC), 139-144.

Qureshi, N., and Blaschek, H. (2001). "Evaluation of recent advances in butanol fermentation, upstream, and downstream processing." *Bioproc Biosys*, 24, 219-226.

Rabaey, K., Clauwaert, P., Aelterman, P., and Verstraete, W. (2005a). "Tubular microbial fuel cells for efficient electricity generation." *Environmental science & technology*, 39, 8077-8082.

Rabaey, K., Geert Lissens, and Verstraete, W. (2005b). "Microbial fuel cells: performance and perspectives." Biofuels for fuel cells: biomass fermentation towards usage in fuel cells, P. N. Lens, P. Westermann, M. Haberbauer, and A. Moreno, eds.

Rabaey, K., Lissens, G., Siciliano, S. D., and Verstraete, W. (2003). "A microbial fuel cell capable of converting glucose to electricity at high rate and efficiency." *Biotechnol. letter*, 25(18), 1531-1535.

Ringeisen, B. R., Henderson, E., Wu, P. K., Pietron, J., Ray, R., Little, B., Biffinger, J. C., and Jones-Meehan, J. M. (2006). "High power density from a miniature microbial fuel cell using *Shewanella oneidensis* DSP10." *Environmental Science & Technology*, 40(8), 2629-2634.

Saravanane, R., and Murthy, D. V. S. (2000). "Application of anaerobic fluidized bed reactors in wastewater treatment: a review." *Environmental science & health*, 11(2), 97-117.

Speece, J. M. R. (1996). "Anaerobic Biotechnology for Industrial Wastewaters." Archae Press, Nashville.

Terry, K. L., and Raymond, L. P. (1985). "System design for the autotrophic production of microalgae." *Enz. Microb. Technol*, 7, 474-487.

Torkian, A., A, E., and Hasheminan, S. J. (2003). "The effect of organic loading rate on the performance of UASB reactor treating slaughterhouse effluent." *Resource conservation and recycling*, 40(1), 1-11.

Vega, J.L., Clausen, E.C., and Gaddy, J.L. (1990). "Design of Bioreactors for Coal Synthesis Gas Fermentation". *Resour. Conserv. Recycling,* 3, 149-160.

Wu, S.-Y., Lin, C.-N., Chang, J.-S., Lee, K.-S., and Lin, P.-J. (2002). "Microbial Hydrogen Production with Immobilized Sewage Sludge." *Biotechnology Program,* 8, 921-926.

Zeikus, J. G. (1980). "Chemical and fuel production by anaerobic bacteria." *Annual Review Microbiology*, 34, 423-464.

CHAPTER 14

Algal Biodiesel Production: Challenges and Opportunities

Puspendu Bhunia, Rojan P. John, S. Yan, R.D. Tyagi, and R.Y. Surampalli

14.1 Introduction

Standing at this twenty first century, it is hard to think of a smoothly running world without depending on fossil fuels. However, the hard reality as predicted by the World Energy Forum is that the fossil-based oil, coal and gas reservoirs will be exhausted in less than a century (Sharma and Singh, 2008). As fossil fuels are limited in supply and take millions of years to form, their availability may be prolonged by reducing overall consumption. Biodiesel fuel has received significant attention in recent years as a biodegradable, renewable, and non-toxic fuel source. It does not produce carbon dioxide (CO_2) or sulfur to the atmosphere and emits less gaseous pollutants than fossil fuels (Krawczyk, 1996; Lang et al., 2001; Antolin et al., 2002; Vicente et al., 2004).

According to US Department of Agriculture report, global biofuel production has tripled from 4.8 billion gallons in 2000 to about 16 billion gallons in 2007, but only accounts for less than 3% of the global transportation fuel supply (Coyle, 2007). Biodiesel provides environmental benefits as its use leads to a decrease in the harmful emissions of carbon monoxide, hydrocarbons and particulate matter and SOx emissions, leading to possible decreased greenhouse effect, in line with the Kyoto Protocol agreement.

Biodiesel is currently produced from oleaginous crops, such as rapeseed, soybean, sunflower, and palm, through a chemical transesterification process of their oils with short chain alcohols, mainly methanol (Siler-Marinkovic and Tomasevic, 1998; Lang et al., 2001; Al-Widyan and Al-Shyoukh, 2002; Antolin et al., 2002). However, concerns about sustainability, energy security, climate change, and soaring oil prices have driven policy makers and scientists to develop alternative energy sources that would allow them to reduce their dependency on foreign oil. To overcome these issues, it is essential to find new sources of biomass for alternative energy. Biodiesel from microalgae is once such alternative.

Microalgae have many desirable attributes as energy producers. The use of microalgae can be a sustainable alternative as algae are the most efficient biological producer of oil on the planet (Campbell, 1997). Algae could be a vital renewal fuel source due to its many positive attributes such as higher photosynthetic efficiency, higher biomass productivity, faster growth rate than higher plants, and highest CO_2 fixation and O_2 production. They can grow in water and be handled easily, can be grown in harsh climatic conditions and non-arable land including the one unsuitable for agricultural purposes, in non-potable water or even as a waste treatment purpose. They use far less water than traditional crops and do not displace food crop cultures, and their production is not seasonal and can be harvested daily (Campbell, 1997; Chisti, 2007; Chisti, 2008; Gouveia and Oliveira, 2009). Considering all these positive attributes of microalgal biodiesel production, average biodiesel production yield from microalgae can be 10 to 20 times higher per unit area than the yield obtained from oleaginous seeds and/or vegetable oils (Tickell, 2000; Chisti, 2007).

This chapter endows with an overview of the technologies in the production of biodiesel from microalgae, including different sources, different modes of microalgal cultivation, downstream processing for biodiesel production. This chapter tries to incorporate advances and prospects of microalgal biofuel technology.

14.2 Properties of Biodiesel

Biodiesel is composed of fatty acid chains that are chemically bonded to one methanol molecule and the glycerol molecules are almost completely removed from the final biodiesel product. Biodiesel is sometimes called fatty acid methyl esters or FAME. When the fatty acid chains break off the triglyceride, they are known as free fatty acids (FFA). The goal of all the technologies is to produce a fuel grade biodiesel whose properties meet ASTM standards. The key quality control issues involve the complete (or nearly complete) removal of alcohol, catalyst, water, soaps, glycerine, and unreacted or partially reacted triglycerides and free fatty acids. Failure to remove or minimize these contaminants causes the methyl ester product to fail one or more fuel standards (Kinast, 2003). Only methyl esters that meet ASTM standards are considered biodiesel. The basic process selected to get required quality is transesterification process and it run at standard atmosphere and temperatures around 60°C. However, some continuous technologies use higher temperatures and elevated pressures, typically in the super critical range of methanol. Distillation is sometimes used for quality control, but not always necessary. The comparison of biodiesel derived from algae, diesel fuel, and ASTM standards have been mentioned in Table 14.1. The location and number of double bonds are important because they influence reactions that can occur to destabilize the fatty acid chain in FAME. The interaction of oxygen molecules with the fatty acid chain, called "oxidation" is the chemical mechanism that destabilizes oil/biodiesel. In biodiesel these degraded chains can cross linked to form polymers like insoluble gums that clog up parts.

Microalgal oils differ from most vegetable oils in being quite rich in polyunsaturated fatty acids with 4 or more double bonds (Belarbi et al., 2000). Fatty acids and FAME with 4 and more double bonds are susceptible to oxidation during storage and this reduces their acceptability for use as biodiesel. This is common for some vegetable oils too. Although these fatty acids have much higher oxidative stability compared to docosahexaenoic acid and eicosapentaenoic acid, the European standard limits linolenic acid methyl ester content in biodiesel for vehicle use. No such limitation exists for biodiesel intended for use as heating oil, but acceptable biodiesel must meet criteria related to the extent of total unsaturation of oil. However, the extent of unsaturation of microalgal oil and its content of fatty acids with more than 4 double bonds can be reduced easily by partial catalytic hydrogenation of the oil (Jang et al., 2005; Dijkstra, 2006).

Table 14.1 Comparison of properties of biodiesel from microalgal oil and diesel fuel and ASTM biodiesel standard.

Properties	Biodiesel from microalgal oil	Fossil diesel fuel	ASTM biodiesel standard
Density (kg/m^3)	864	838	860-894
Viscosity (mm^2/s, at 40 °C)	5.2	1.9-4.1	3.5 – 5.3
Flash point (°C)	115	75	Min 100
Solidifying point (°C)	-12	-50 – 10	-
Cold filter plugging point (°C)	-11	- 3.0 (max. -6.7)	Summer max. 0; winter max. < -15
Acid value (mg KOH/g)	0.374	Max. 0.5	Max. 0.5
Heating value (MJ/kg)	41	40 – 45	-
H/C ratio	1.81	1.81	-

The chemical stability property of different biodiesel fuels against oxidation is specified as the Iodine Number or Iodine Value (IV). It is determined by measuring the number of double bonds in the mixture of fatty acid chains in the fuel by introducing iodine into 100 grams of the sample under test and then measuring the amount of iodine absorbed. Iodine absorption occurs at double bond positions - thus a higher IV number indicates a higher quantity of double bonds in the sample. The Iodine Value can be important because many biodiesel fuel standards specify an upper limit for fuel that meets the specification. For example, Europe's EN14214 specification allows a maximum Iodine Number of 120, Germany's DIN 51606 is at 115. The USA ASTM D6751 does not specify an Iodine Value. The Iodine Value does not necessarily make the best measurement for stability as it does not take into account the positions of the double bonds available for oxidation.

If the biodiesel is pure i.e. 100% biodiesel fuel, it is referred to as B100 or "neat" fuel. A biodiesel blend is a mixture of pure biodiesel and normal fossil diesel. Biodiesel blends are referred to as BYY where the YY indicates the amount of biodiesel in the blend (i.e., a B60 blend is 60% biodiesel and 40% normal fossil diesel). Biodiesel is a clear amber-yellow liquid with a viscosity (1.9-6 centistokes at 40℃) similar to that of normal diesel. Biodiesel is non-flammable and, in contrast to normal diesel, is non-explosive. But now-a-days biodiesel costs are 1.5–3 times more than fossil diesel (Demirbas, 2009).

Fuel characterization data (Table 14.2) show some similarities and differences between biodiesel and petrodiesel fuels (Shay, 1993). Sulfur content of petrodiesel is 20–50 times those of biodiesels. Several municipalities are considering mandating the use of low levels of biodiesel in diesel fuel on the basis of several studies which have found hydrocarbon (HC) and particulate matter (PM) benefits from the use of biodiesel. The use of biodiesel to reduce N_2O is attractive for several reasons. Biodiesel contains little nitrogen, as compared with petrodiesel which is also used as a reburning fuel. In addition, biodiesel contains virtually trace amount of sulfur, so SOx emissions are reduced in direct proportion to the petrodiesel replacement.

Table 14.2 Comparison of chemical properties between biodiesel and petrodiesel fuels.

Chemical property	Biodiesel	Diesel
Ash (wt%)	0.002–0.036	0.006–0.010
Sulfur (wt%)	0.006–0.020	0.020–0.050
Nitrogen (wt%)	0.002–0.007	0.0001–0.003
Aromatics (vol%)	0	28–38

For one of the most common blends of biodiesel, B20, a soybean-based biodiesel, the estimated emission impacts for percent change in emissions of NOx, PM, HC, and CO were +20%, -10.1%, -21.1%, and -11.0%, respectively (EPA, 2002). The use of blends of biodiesel and diesel oil is preferred in engines, in order to avoid problems related to the decrease of power and torque and to the increase of NOx emissions (a contributing factor in the localized formation of smog and ozone) with increasing content of pure biodiesel in the blend (Schumacher et al., 1996).

14.3 Sources of Biodiesel Raw Materials

Raw materials contribute to a major portion in the cost of biodiesel production. The main raw materials used to produce biodiesel are the vegetable oils extracted from oleaginous plants. The cost of these materials currently represents about 60-75% of the total production costs (Huang et al. 2009). The choice of raw materials depends mainly on its availability and cost. The yield/conversion of biodiesel with different oils has been listed in Table 14.3. Countries such as USA and those belonging to European community have surplus amount of edible oils to export. Hence, edible oils such as soybean and rapeseed are used in USA and European Nations, respectively. Similarly, countries such as Malaysia and Indonesia have surplus coconut oil and that is utilised for the synthesis of biodiesel (Ahouissoussi and Wetzstein, 1998). India uses unutilised and underutilised materials for biodiesel production. The raw materials used at present in India are, *Jatropha curcas* (jatropha) and *Pongamia pinnata* (karanja). The oil content from the seed of plant oil varies. Seeds of soybean contain only 20% of oil, whereas, 40% of oil can be expelled from seeds of rapeseed (Abelson, 1995). Jatropha and karanja seeds possess 40% and 33% oil, respectively (Azam et al., 2005).

Table 14.3 Biodiesel yield with different feedstock (adapted from Sharma et al., 2008).

Oil used	Yield (%)	Conversion of biodiesel (%)
Karanja (*Pongamia pinnata*)	89.5	-
Tobacco (*Nicotina tabacum*)	91	-
Polanga (*Calophyllum inophyllum*)	-	85
Jatropha (*Jatropha curcas*)	99	-
Mahua (*Madhuca indica*)	98	-
Karanja (*Pongamia pinnata*)	97-98	-
Karanja (*Pongamia pinnata*)	-	92/95
Soybean (*Glycine max*)	-	98.4
Waste cooking oil	97.02	-
Canola oil (*Brassica napus*)	90.04	98
Used frying oil	87.5	94
Sunflower oil (*Helianthus annuus*)	-	Nearly complete
Alga (*Chlorella protothecoides*)	56	-
Alga (*Chlorella protothecoides*)	-	>80

Other edible and non-edible oils, animal fats, algae, and waste cooking oils have also been investigated by researchers for the development of biodiesel (Canakci and Gerpen, 1999; Madras et al., 2004; Karmee and Chadha, 2005; Miao and Wu, 2006; Xue et al., 2006; Wang et al., 2007). Table 14.4 presents biodiesel production from various feedstocks under different conditions. Type and amount of variables such as feedstock, alcohol, molar ratio, catalyst, reaction temperature, time duration, rate and mode of stirring affect the yield and conversion of biodiesel.

Search for the new raw materials for biodiesel production is not limited to crop seeds only. Presently, researchers of United States are of the opinion that even surplus vegetable oils will not be enough to meet the future demand of biodiesel and hence have tried algae as a raw material for biodiesel production. Nowadays, high petroleum-based sourced fuel prices, the collapse of food for biodiesel initiatives and concerns about increased levels of CO_2 emissions in the atmosphere have all created awareness of the need for alternative fuel solutions. Microalgae have optimistically emerged as one of the lowest cost feedstocks for biodiesel production. The oil yield from algae is many times more than other land crops (Table 14.5). However, the production costs of high grade algae oils is, and likely will be, an obstacle in the short term. Autotrophic algae convert carbon dioxide into sugars and proteins. But, when they are starved of nitrogen, mainly oil is produced (Flam, 1994). A microalga, *Chlorella protothecoides*, has been grown under autotrophic and heterotrophic conditions to obtain lipid as raw material for biodiesel production. Lipid content in the heterotrophic cells reached 55.20% as compared to 14.57% in autotrophic cells. The microalgal oil as lipids was prepared by pulverization of heterotrophic cell powder in a mortar and extracted with n-hexane. The lipid was successfully used for biodiesel production (Miao and Wu, 2006).

Table 14.4 Biodiesel production from different feedstocks (adapted from Sharma et al., 2008).

Feedstock	Transesterification stages	Alcohol	Molar ratio (methanol to oil)	Catalyst	Reaction temp (K)	Duration	Stirring	Conversion
Sunflower oil	Single step	Supercritical methanol	40:1	No catalyst	473–673 (pressure 200 bar)	10–40 min		78–96% conversion with increase in temperature 23% conversion 27% conversion
		Supercritical ethanol Methanol Ethanol	5 :1	Supercritical CO_2 + lipase (Novozyme 435) 30 wt% of oil	318	6 h		
Pongamia pinnata	Single step	Methanol	10:1	KOH (1% by wt)	378	1.5 h		92% conversion 83% conversion
				ZnO Hb-Zeolite Montmorillonite	393	24 h		
Madhuca indica	Two step	Methanol	0.30–0.35 v/v	1% v/v H_2SO_4	333	1 h		59% conversion 47% conversion
			0.25 v/v	0.7% w/v KOH		1 h		
Rubber seed oil	Two step	Methanol	6:1	H_2SO_4 0.5% by volume	318±5	20–30 min	Magnetic stirrer	98% yield
			9:1	NaOH 0.5% by volume	318±5			
Chlorella protothecoides		Methanol	56:1	Acid catalyst	303	30 min		More than 80% conversion

Chlorella protothecoides		Methanol	56:1	H_2SO_4 (100%) on the basis of oil weight	303	4 h		63% yield
Neat canola oil	Single step	Methanol	6:1	NaOH 1.0 wt%	318	15 min	Magnetic stirring 1100 rpm in the first stage (10 min) and 600 rpm in second stage	Ester content 98 wt%
Used frying oil			7:1	NaOH 1.1 wt%	333	20 min		Ester content 94.6 wt%
Nicotiana tabacum L. (tobacco)	Two step	Methanol	18:1 6:1	H_2SO_4 (1% with lower molar ratio) (2% with higher molar ratio) KOH (1% based on the oil wt.)	333.0±0.1	25 min 30 min	Magnetic stirrer 400 rpm	Yield 91% in 30 min
Pongamia pinnata	Single step	Methanol	6:1 12:1	KOH (1% by weight)	338	2 h 1 h	Mechanical stirrer 360 rpnm	Yield 97–98%
Soybean oil		Methanol	4.5:1	TiO_2/ZrO_2 (11 wt% Ti) Al_2O_3/ZrO_2 (2.6 wt% Al) K_2O/ZrO_2 (3.3 wt% K)	448	2 h		Conversion over 95% Conversion 100%

Monosodium glutamate wastewater	Two step	Methanol	0.5 M 12.5% v/v	1 ml KOH (0.5 M) 1 ml BF_3 (12.5%, v/v)	333 353	10 min 5 min		Yield 92.54±2.00%
Soybean oil	Single step alkali catalysed	Methanol	6:1	NaOH	318	10–20 min	Mechanical stirrer 900 rpm Power ultrasonic (frequency 19.7 KHz, power 150 W) Hydrodynamic cavitation (operation pressure 0.7 MPa)	Yield 100%
Jatropha, pongamia, sunflower, soybean, palm	Single step	Methanol	3:1	NaOH/KOH (1 wt% of oil)		2–4 h	Stirring	
Calophyllum inophyllum	Three step zero catalysed Acid catalysed Alkali catalysed	Methanol	6:1 9:1	Anhydrous H_2SO_4 (98.4%) 0.65% by volume KOH 1.5% by weight	338	2 h 4 h 4 h	Mechanical stirring 450 rpm	85% yield in 90 min Complete in 4 h reaction
Jatropha curcas	Two step acid catalysed Alkali catalysed	Methanol	0.28 v/v 0.16 v/v	H_2SO_4 1.43% v/v (3.5 + acid value, w/v KOH)	333	88 min 24 min		>99% yield

Triolein	Single step	Ethanol	10:1	Anion exchange resin Cation exchange resin (heterogeneous catalyst)	323	60 min		98.8% purity
Sunflower oil	Single step	Methanol	13:1	Activated CaO (1 wt%)	333	100 min	Helix stirrer 1000 rpm	
Waste cooking oil	Two step	Methanol	10:1 6:1	Fe_2SO_4 KOH	368 338	4 h 1 h	No stirring because boiling was sufficient	97.02% conversion
Karanja	Two step	Methanol	8:1 9:1	H_2SO_4 NaOH/KOH	318±2	30 min 30 min	Magnetic/ mechanical	89.5% yield with mechanical 85% yield with magnetic

Table 14.5 Different crops' oil yield (http://www.oilgae.com/algae/oil/yield/yield.html).

Crop	Oil (L/ha)	Crop	Oil (L/ha)
Castor	1413	Soy	446
Sunflower	952	Coconut	2689
Safflower	779	Algae	100000
Palm	5950		

14.4 Microalgae and Oil Production

Microalgae or microphytes are microscopic algae generally found in marine or fresh water bodies. They are mainly free living single cellular or colonial in forms. The cultivation of the microphyta is generally termed as algaculture. Microalgae have gained worldwide attention as a food source like other macroalgae as they grow in all types of water including wastewater or industrial effluents. Microalgal products are currently available in market through some companies' brand names including biopigments and ω-3 fatty acids. The major companies such as Cognis (e.g. carotene from alga *Dunaliella salina* http://www.cognis.com), Nikken Sohonsha Corporation (e.g.,"Chlostanin Gold" Chlorella polysaccharide, http://www.chlostanin.co.jp), Algatechnology (eg. Astaxanthin, carotenoid, http://www.algatech.com), Martek (e.g. ω-3 fatty acids with *Crypthecodinium cohnii,* http://www.martek.com), and Cyanotech (e.g. BioAstin®, astaxanthin and Hawaiian Spirulina Pacifica®, http://www.cyanotech.com) have commercialized the products (Richmond, 2003; Pulz and Gross, 2004; Wijffels, 2007). Nowadays the importance of microalgae has increased due to its usefulness in biofuel as biodiesel production (Chisti, 2007). High production ability of lipids has made microalgae as a potential source for the production of biodiesel (Chisti, 2007).

It is estimated that microalgae culture can produce up to 50% oil per dry cell weight (Demitrov, 2007). Algal lipid concentration in cell for different strains is given in the Table 14.6. Recently, researchers reported a heterotrophic growth (by providing easily consumable carbon sources like dextrose) of *Chlorella protothecoides* capable of yielding as high as 55% w/w lipid content (Wu et al., 1994; Miao et al., 2004; Miao and Wu, 2004, 2006). Besides the need of expensive nutrient of thiamine hydrochloride, this alga required glucose instead of CO_2 for heterotrophic growth and hence it has almost no carbon sequestration capacity considering the severity of global warming issue. Using CO_2 as a carbon source, the strain yielded only about 15% of lipid (Widjaja et al., 2009). Allard and Templier (2000) extracted lipid from a variety of freshwater and marine microalgae and reported that lipid content varied from 1 to 26% w/w. Attention has been focused on the autotrophic green microalga, *Botryococcus braunii*, due to its high-hydrocarbon production level (Wake and Hillen, 1980; Casadevall et al., 1985; Metzger and Pouet, 1995). Sawayama et al. (1995) utilized these algae for continuously operated oil production. Despite the high-lipid content of 64%, the growth rate of this strain was reported to be very low (Sheehan et al., 1998).

Hydrocarbon from *B. braunii* has the potential to completely displace petroleum-derived transport fuels. This alga is characterized by its conspicuous ability to synthesize and accumulate a variety of lipids. These lipid substances include numerous hydrocarbons and a number of specific ether lipids. Strains of *B. braunii* isolated and grown in laboratories and wild populations of this alga both differ in the type of hydrocarbons they synthesize. Accordingly, they are sub-classified into three chemical races. Algae in race A produce essentially *n*-alkadiene and triene hydrocarbons, odd-carbon-numbered from C_{23} to C_{33}, algae in race B produce triterpenoid hydrocarbons, C_{30}–C_{37} botryococcenes (Metzger et al., 1985) and C_{31}– C_{34} methylated squalenes (Huang and Poulter 1989; Achitouv et al., 2004) and algae in race L produce a single tetraterpenoid hydrocarbon, lycopadiene (Metzger and Casadevall 1987; Metzger et al., 1990).

Utilizing marine microalgae will benefit if large pond is used for the system. As these strains can grow in brackish water, they will not compete for the land already being used by other biomass-based fuel technologies. Freshwater microalgae can compete with marine microalgae if closed photobioreactors used for culturing with lesser space and controlled conditions (Sheehan et al., 1998).

The strain improvement of promising fresh water strains to get salt tolerance may be the future aspect for biodiesel production to utilize brackish water or marine water. If we can enhance salt tolerance of *B. braunii* and if it survives in seawater, then we can reduce the demand of fresh water for the mass culture of the alga for the production of biofuels. The approaches used by microalgae to survive in environment with high and changing salinities have received much attention nowadays for the cultivation of them in an economic manner. Salt tolerance studies are reported in some bacteria and salt tolerant green alga, *Dunaliella salina*. In minimal media bacterial strains produce amounts of the compatible solutes trehalose (*Escherichia coli*) (Kaasen et al., 1992), proline (*Bacillus subtilis*) (Whatmore et al., 1990), and glucosylglycerol (GG) (*Synechocystis*) that are proportional to the stress (Reed and Stewart, 1985; Hagemann and Erdmann, 1994).

Table 14.6 Oil content of some microalgae (Sources: Chisti, 2007; http://www.oilgae.com/algae/oil/yield/yield.html; Um and Kim, 2009; Widjaja et al., 2009).

Microalga	Oil content (% dry wt)	Microalga	Oil content (% dry wt)
Botryococcus braunii	25–75	*Scenedesmus obliquus*	12-14
Chlorella sp.	28–32	*S. quadricauda*	1.9
Crypthecodinium cohnii	20	*S. dimorphus*	16-40
Cylindrotheca sp.	16–37	*Chlamydomonas rheinhardii*	21
Dunaliella primolecta	23	*Chlorella vulgaris*	14–22
D. tertiolecta	37	*C. protothecoides*	55
D. bioculata	8	*Chlorella pyrenoidosa*	2
D.salina	6	*Spirogyra* sp.	11–21
Isochrysis sp.	25–33	*Euglena gracilis*	14–20
Monallanthus salina	>20	*Prymnesium parvum*	22–39
Nannochloris sp.	20–35	*Tetraselmis maculata*	3
Nannochloropsis sp.	31–68	*Porphyridium cruentum*	9–14
Neochloris oleoabundans	35–54	*Spirulina platensis*	4–9
Nitzschia sp.	45–47	*Spirulina maxima*	6–7
Phaeodactylum tricornutum	20–30	*Synechoccus sp.*	11
Schizochytrium sp.	50–77	*Anabaena cylindrica*	4–7
Tetraselmis sueica	15–23		

Enhancement of salt tolerance in *B. braunii* can be achieved by the introduction of glycerol-3-phosphate dehydrogenase to accumulate glycerol in the chloroplasts. The enzyme that catalyzes the reversible conversion of di-hydroxyacetone phosphate (DHAP) to glycerol 3-phosphate, DHAP reductase, also known as glycerol 3-phosphate dehydrogenase (GPDH) is found in many eubacteria and eukaryotes. Nicotinamide-adenine dinucleotide (NAD+)-dependent GPDH is an important enzyme in glycerol metabolism. For glycerol synthesis, DHAP is converted to glycerol 3-phosphate catalyzed by NAD+-dependent GPDH, and then glycerol 3-phosphate is converted to glycerol catalyzed by glycerol 3-phosphatase. The roles of NAD+-dependent GPDH isoenzymes in yeast, *Saccharomyces cerevisiae*, already been detailed by several researchers earlier (Albertyn, et al., 1994; Ansell et al., 1997). It is known that heterologous expression of GPDH genes in yeast can increase glycerol production (Watanabe et al., 2004). Therefore, the NAD+-dependent GPDH genes are considered as the key genes of glycerol synthesis.

It is estimated that biodiesel from algae can reduce the cost up to five times than the conventional diesel, but all the claims and research dating from the early 1970's to date in projected algae and oil yields have been achieved (Dimitrov, 2007). The current price for microalgae production needs to be reduced much and production must be increased more. The potential productivity of microalgae is ten folds greater than that of agricultural crops and it need not require the cultivable land (Richmond, 2003; Chisti, 2007). For production of microalgal products at larger scales, such as those needed for the production of biofuels, major developments in scaling up technology including efficiency different types of bioreactors must be realized. Yields must be increased to make sustainable production of high-value compounds and economically feasible energy carriers. These developments are possible only after the improvement in the photobioreactors, with the utilization of best strain, further optimized process conditions and utilization of molecular tools to control metabolism of the desired product. The entire process should be applicable on an industrial scale, which would involve cultures of several thousands of cubic meters (Wijffels, 2007).

14.5 Biosynthetic Pathway and Enhancement of Oil Production through Strain Improvement

Algae produce lipids mainly for the synthesis of biomembranes, defense molecules against environmental stresses, storage food etc. There are different pathways of lipid biosynthesis according to the strains and types of lipids produced vary accordingly. Universal synthesis pathway of triacylglycerol (TAG) in cells is comprised of three major steps: 1) carboxylation of acetyl-coenzyme A (-CoA) to form malonyl-CoA, the committing step of fatty acid biosynthesis; 2) acyl chain elongation; and 3) TAG formation (Courchesne et al., 2009). It is generally accepted that the ability of algae to adapt with the environmental conditions is reflected in an exceptional variety of lipid patterns as well as with their ability to synthesize a number of unusual compounds. Moreover, in recent years the development of modern analytical methods combining various chromatographic techniques with sensitive detection systems and mass spectrometry, as well as new derivatization procedures, has led to significant progress in the identification of new and unusual classes of lipids and fatty acids in algae (Guschina and Harwood, 2006). The acyl lipids that make up the plastid membranes are highly enriched in polyunsaturated fatty acids in most plants and algae. In flowering plants, linoleic (18:2) and linolenic (18:3) acids are the most common unsaturated acids, whereas in many algae, highly unsaturated long-chain acids, such as arachidonic (20:4), eicosapentaenoic (20:5), and docosahexaenoic (22:6) acids, are found as major fatty acid components, depending on the species (the number of carbon atoms [*x*] and the number of double bonds [*y*], such as *x*:*y*). Unsaturated fatty acids, especially polyunsaturated ones, are important in maintaining membrane fluidity during the cold acclimation of photosynthesis. Linolenic acid was also found to be important in tolerance to high temperatures. The only known photosynthetic organisms that lack polyunsaturated acids are *Synechococcus* sp. strain PCC 6301 and its close relative PCC 7942 as well as some unicellular species of cyanobacteria (Sato and Moriyama, 2007).

Besides hydrocarbons of *B. braunii* such as n-alkadienes and trienes, triterpenoid botryococcenes and methylated squalenes, a tetraterpenoid, lycopadiene, and other classic lipids such as fatty acids, glycerolipids and sterols, algae synthesize a number of ether lipids closely related to hydrocarbons (Metzger and Largeau, 2005; Guschina and Harwood, 2006). The review of Metzger and Largeau summarizes the available information on algal biodiversity, the chemical structure and biosynthesis of hydrocarbons and ether lipids and biotechnological studies related to hydrocarbon production (Metzger and Largeau, 2005; Guschina and Harwood, 2006).

The major saturated fatty acid in algae is palmitate while oleate is much less abundant than in higher plants. In fact, in many algae, palmitoleate is the dominant monoene. 16C PUFAs may be significant, particularly in marine species while 18C PUFAs include the usual plant fatty acids (linoleate, a-linolenate) but may include octadecatetraenoate (18:4) as a major component (Harwood and Guschina, 2009).

Recently, several different molecular biology techniques have become available for genetic transformation of *Chlamydomonas*, the classical strain used for the study of many biosynthetic pathways (Fuhrmann, 2002; Guschina and Harwood, 2006). Thus, genetic engineering tools such as, insertional mutagenesis by plasmid transformation, has been used to study the biosynthesis of the sulfolipid, 2'-*O*-acyl-sulfoquinovosyldiacylglycerol (ASQD), in *C. reinhardtii* to identify genes essential for its biosynthesis (Riekhof et al., 2003; Guschina and Harwood, 2006). n-3 Octadecatetraenoic acid was the main acid attached to the head group of ASQD. Analysis of sulphoquinovosyldiacylglycerol (SQDG) molecular species indicated that there were two pools of lipid (16:0/16:0 and 18:1/16:0) of which only the second appeared to be the substrate for an acyltransferase forming ASQD. The biosynthesis study of ASQD in *C. reinhardtii* revealed that the acyltransferase responsible for acylation of SQGD could be localized in the outer envelope of the plastid, where it would have access to SQDG made in the plastid and also to fatty acids derived from the endoplasmic reticulum (Riekhof et al., 2003; Guschina and Harwood, 2006). Algae mutants defective in a certain lipid or a fatty acid have been shown to be powerful tools to investigate the functions and physiological roles of lipids. Thus, within the past decade, experiments with *C. reinhardtii* mutants deficient in SQDG or in D^3-trans-hexadecenoic acid-containing phosphatidylglycerol have led to further progress in understanding the specific role of these lipids in thylakoid membranes, namely in the functioning of Photosystem II (Dubertret et al., 2002; Sato et al., 2003; Pineau et al., 2004; Guschina and Harwood, 2006). A mutant of *C. reinhardtii* which was impaired in fatty acid desaturation of chloroplast lipids has been used to study the role of lowered unsaturation levels in such lipids with regard to high temperature tolerance of photosynthesis in this alga (Sato et al., 1996; Guschina and Harwood, 2006). Above all, phosphatidylcholine and phosphatidylserine are not present in its membranes and a major membrane lipid in this alga is a betaine lipid, diacylglyceryltrimethylhomoserine (DGTS) (Giroud et al., 1988). It makes the organism a very useful system for studying the biosynthesis of phosphatidylethanolamine which is independent of both phosphatidylcholine and phosphatidylserine metabolism, as well as the biosynthesis of DGTS itself (Moore et al., 2001; Yang et al., 2004; Guschina and Harwood, 2006). A cDNA which encodes the ethanolaminephosphotransferase protein has been sequenced from *C. reinhardtii*. Although the enzyme coded by the cDNA is clearly an ethanolaminephosphotransferase the enzyme had more activity with CDP-choline than CDP-ethanolamine as a substrate (Yang et al., 2004; Guschina and Harwood, 2006).

Not only microalgae produce oil, their accumulation is higher in microalgae compared to other organism like bacteria and fungi. Lipid accumulations differ with strain of organisms. *Rhodotorula* spp. and *Cryptococcus curvatus* can accumulate 40-70% w/w lipid, while *Saccharomyces cerevisiae* and *Candida utilis* accumulate >5-10% oil when grown in the same culture condition (Meng et al., 2009). Lipid accumulation in an oleaginous microorganism begins when it runs out a nutrient, especially nitrogenous, from the medium, but an excess of carbon (in the form of glucose) is still assimilated by the cells and is converted into triacylglycerols, while lipid is synthesized during the balance phase of growth at nearly the same rate. At nitrogen-defficient growth, cell proliferation is prevented and the lipid formed is stored within the existing cells which can no longer divide, leading to lipid accumulation. Two critical enzymes, malate enzyme and ATP-citrate lyase (ACL), have effect on lipid accumulation. There is a strong correlation between the presence of ACL activity and the ability to accumulate lipid in yeasts, fungi, and other oleaginous microorganisms (Ratledge, 2002; Meng et al., 2009). Accordingly, ACL is a prerequisite for lipid accumulation to occur, but its possession is not necessarily crucial for lipid accumulation, it cannot be the sole effecting factor (Adams et al., 2002; Meng et al., 2009). Involvement of other enzymes must be responsible for controlling the extent of lipid biosynthesis in individual organisms. Fatty acids are highly reduced materials and to achieve their synthesis a ready supply of reductant NADPH is essential (Ratledge, 2004; Meng et al., 2009). The synthesis of 1 mol of a C18 fatty acid requires 16 mols of NADPH. 2 mols NADPH are needed to reduce each 3-keto-fattyacyl group arising after every condensation reaction of acetyl-CoA with malonyl-CoA.

The strain improvement of oleaginous microorganisms for the oil production mainly focused on their ability to accumulate high amounts of intracellular lipid, their relatively high growth rates and the resemblance of their triacylglycerols fraction to plant oils. But when it comes to the production of biodiesel, only those with a high content of stearic acid (C18:0) and oleic acid (C18:1) are of potential interest because of improved oxidative stability and adaptability in the industrial production of biodiesel. The lipid composition of oleaginous microorganisms can be modified by genetic engineering methods to enable them satisfies in the actual biodiesel production. Thus, synthesis of lipids with varying fatty acid composition should be feasible by culturing oleaginous microorganism under different conditions and by metabolic regulation (Meng et al., 2009).

Three most important interdependent genetic technologies involved in strain improvement for higher oil yield are cloning genes of critical enzymes, transgenic expression of these genes aimed to achieve a fine high-product microbial oil recombination strain and modification of cloned genes in order to engineer the expressed protein. In recent years, there are successful examples of production of high oil yielding strain by recombination of oleaginous microorganisms (Meng et al., 2009). Acetyl-CoA carboxylase (ACC) was first isolated from the microalga *Cyclotella cryptica* (Roessler, 1990) and successfully transformed into the diatoms *C. cryptica* and *Navicula saprophila* (Dunahay et al., 1995; Dunahay et al., 1996; Sheehan et al., 1998). The ACC gene, *acc*1, was over expressed with the enzyme activity enhanced to two to three-folds. These experiments demonstrated that ACC could be transformed efficiently into microalgae although no significant increase of lipid accumulation was observed in the transgenic diatoms (Dunahay et al., 1995; Dunahay et al., 1996). It also suggests that over expression of ACC enzyme alone might not be sufficient to enhance the whole lipid biosynthesis pathway (Sheehan et al., 1998). But further improvement needed to generate strains having tolerance against all environmental and biological stresses for the efficient production of biodiesel.

14.6 Microalgal Biomass Production for Biodeisel

Microalgae can be cultivated autotrophically or heterotrophically for the production of oil. Generally photobioreactors are used for the cultivation as they are photosynthetic organisms. Many parameters should be taken in account for the design of photobioreactors. A prerequisite for the development of efficient photobioreactors is the efficient use of light. There are different types of such bioreactors from bottles; hanged plastic bags to big shallow tanks to trap the natural light. Algae can be grown on open settling ponds, but this approach is doubtful to provide the best yields. Raceway ponds are typical microalgal production facility with low cost and the size of the bioreactor can be varying according to the size of land or water bodies. A raceway pond is made of a closed loop recirculation channel that is typically about 0.3 m deep. Mixing and circulation are produced by a paddlewheel and it operates all the time to prevent sedimentation. Flow is guided around bends by baffles placed in the flow channel. Raceway channels are built in concrete or compacted earth, and may be lined with white plastic. During daylight, the culture is fed continuously in front of the paddlewheel where the flow begins. Broth is harvested behind the paddlewheel, on completion of the circulation loop (Chisti, 2007).

The main drawbacks of this open pond are lack of environmentally adapted organisms and contamination problem. Unfortunately, the best strains which have the tolerance against environmental and other bio-contaminants including viral, fungal and other algae borne in the atmosphere are not good productive in the case of fuel yield. Covering ponds with translucent membranes or the use of greenhouses overcomes this drawback, allowing the more productive strains to be grown free of atmospheric contamination. Even, closed pond systems enable more control not only to the contaminants but also to the other parameters as discussed earlier (http://www.renewableenergyfocus.com January/February 2009). The organism varies with their photosynthetic efficiency, the percentage of energy in the visible light spectrum that is converted into biomass, and theoretically it is 9% of visible spectrum in practice this is generally <1% (Richmond, 2003; Wijffels, 2007).

There are other advances on the design of reactors. In another approach, a series of storage tanks linked by transparent tubes that rest on support structures used for maximum sunlight trap. Algae and water are pumped through the pipes to ensure maximum exposure to sunlight. Carbon dioxide is piped into the tanks. Dissolved oxygen levels much greater than the air saturation values inhibit photosynthesis and the oxygen level may increase due to the oxygenic reaction during photosynthesis. Higher concentration of oxygen can lead to the photo oxidative damage to algal cells and inhibit the photosynthesis by photorespiration. However, under normal conditions, CO_2 loss by algal photorespiration is minimal. To prevent inhibition and damage, the maximum tolerable dissolved oxygen level should not generally exceed 400% of air saturation value. Oxygen cannot be removed from a photobioreactor tube as it negatively influences the due to the contamination problem. This limits the maximum length of a continuous run and tends to make strictness in oxygen removal (Chisti, 2007). So constant pumping is needed to the degassing center especially for the removal of oxygen.

Many different ways to produce algal oil on a commercial scale have been tried, but they face significant hurdles. There was little risk of internal contamination from inoculated strains. Productivity per hectare is also high so the equipment takes up less land than open systems. Usual way of harvesting light by algal culture is putting cylindrical (tubular) bioreactor on ground. Instead of putting down horizontally on the ground (Fig. 14.1), the tubes may be made of flexible plastic and coiled around a supporting frame to form a helical coil tubular photobioreactors (Fig. 14.2) (Chisti, 2007).

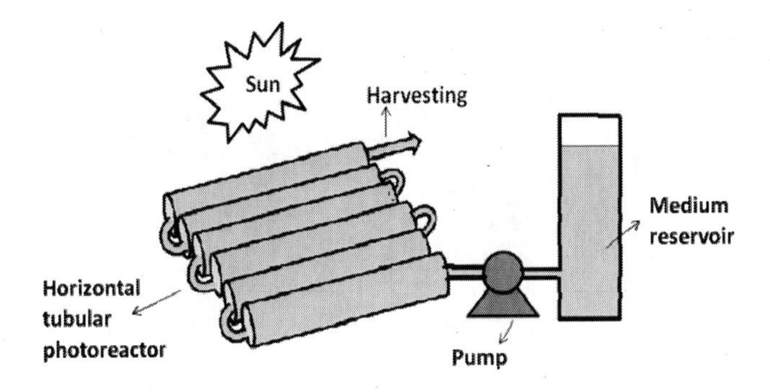

Fig. 14.1 Schematic of horizontal tubular photo-bioreactor (Chisti, 2007).

Artificial illumination of tubular photobioreactors is technically feasible (Pulz, 2001), but expensive compared with natural illumination. Nevertheless, artificial illumination has been used in large-scale biomass production (Pulz, 2001) particularly for high-value products. The solar panels for trapping solar energy and use as energy for artificial illumination can solve the problem of cost upto certain extend.

The main objective of advanced development on photobioreactor should focus on increasing the efficiencies close to theoretical value. This could be done by reducing the amount of light. At low light intensities microalgae convert light with higher efficiencies. Alternative ways for reducing the light intensity are to increase the biomass concentration in initial level and well mixing of the medium. Vonshak and Guy (1992) observed that shading *Spirulina platensis* resulted in higher productivities compared to unshaded cultures. Currently, efficient flat panel photobioreactors are used in laboratory-scale production with advantage of being thin with a shorter light path and a well-mixed culture due to gas sparging into the reactor. In these reactors it is possible to obtain high densities of microalgae with 30 to 100 times greater biomass concentration. (Qiang et al., 1998; Richmond, 2003; Richmond et al., 2003; Janssen et al., 2003). According to Grobbelaar (2007), there are at least six important factors that determine productivity in mass algal cultures: (1) the culture depth or optical cross section, (2) turbulence, (3) nutrient content and supply, (4) cultivation procedure, (5) biomass concentration and areal density, and (6) photo-acclimation.

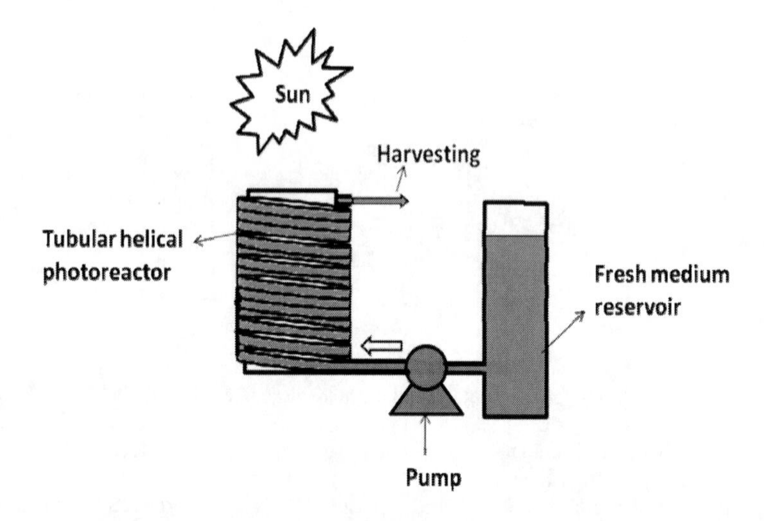

Fig. 14.2 Schematic of helical tubular photo-bioreactor (Chisti, 2007).

As a raw material CO_2 plays an important role in photosynthesis and subsequent processes. Photosynthetic cultures of *B. braunii* aerated with 0.3% CO_2-enriched air have a much shorter mass doubling time (40 h) compared to 6 days for cultures supplied with ambient air. Carbon dioxide enrichment favors the formation of lower botryococcenes (C_{30}–C_{32}), whereas cultures sparged with ambient air accumulate higher botryococcenes (C_{33}–C_{34}) (Wolf et al., 1985). In fact, methylation steps leading from C_{30} to C_{31} and C_{32} are faster in CO_2-enriched cultures than steps leading to C_{33}, C_{34} and higher homologues. Although autotrophic, *B. braunii* utilizes exogenous carbon sources for improved growth and hydrocarbon production. Various carbon sources, including C_1–C_6 compounds and disaccharides (lactose, sucrose), have been screened in attempts to decrease the mass doubling time of the alga from $\geq$1 week to less than 2 to 3 days (Weetal, 1985). Yoon et al. (2008) noted that the pH increases very rapidly at the end of the exponential growth phase. Therefore, they controlled the pH using CO_2. Carbon dioxide the source of carbon for cell growth, and it can act as both weak acid and buffer. Therefore, if CO_2 is supplied appropriately, pH can be controlled simultaneously and can be used as carbon source.

Temperature or light are considered as the two limiting factors in outdoor cultures, because nutrients are supplied in excess, and essentially only these two variables are often used to model algal productivity in mass algal cultures (Grobbelaar et al., 1990). Photo inhibition (the decline in photosynthetic rates at supra-optimal irradiancies) is a well known phenomenon. On a contrary, certain reports are available, with high biomass production at a high cell density, the light penetration distance, in the culture vessel becomes shorter. If the intensity of incident light increases, cells can absorb more light and, consequently, can grow more rapidly (Yoon et al., 2008). Attempts were done to overcome the incapability of strain to perform in adverse conditions such as high intensity of light. Entrapment of *B.*

braunii colonies in calcium alginate beads exhibits some interesting advantages to free suspension cultures. Enhancement in chlorophyll photosynthetic activity, protection against photo inhibition towards high irradiance and an increase in hydrocarbon production are some of them despite a decrease in the rate of biomass production (Baillez et al., 1985, 1986). However, the lack of stability of calcium alginate beads over a long period is not suitable for the industrial level production.

During daylight hours the temperature of the bioreactor increases and negatively influences the growth and photosynthesis of microalgae. So photobioreactors require cooling during daylight hours. Additionally, temperature control at night is also necessary as there is a loss of biomass due to the respiration in high temperature. Outdoor tubular photobioreactors are effectively and inexpensively cooled using heat exchangers. Large tubular photobioreactors have been placed within temperature controlled greenhouses (Pulz, 2001), but doing so is prohibitively expensive in larger scale.

The growth phases and nutrient components have key roles in production of hydrocarbon. For example, in the three chemical races of *B. braunii* (race A, B and L), the hydrocarbon productivity was shown to be optimal during the exponential phase of growth (Largeau et al., 1980; Metzger et al., 1985, 1990). Thus, the production of hydrocarbons appears to be a normal feature of *B. braunii*. Hydrocarbon synthesis does not occur in nitrogen- and phosphorous deficient media (Casadevall et al., 1985; Brenckman et al., 1985; Wolf et al., 1985). Similarly, ether lipid production was shown to be maximal during the exponential and early deceleration stages of growth (Villarreal-Rosales et al., 1992). Studies with nitrogen supplied as NO_3–, NO_2–, and NH_3 reveal that the primary factor regulating nitrogen metabolism in *B. braunii* is the nitrate uptake system. Nitrogen is generally supplied as nitrate salts. An initial NO_3– concentration of $\geq 0.2\ kg \cdot m^{-3}$ favours hydrocarbon production (Casadevall et al., 1983).

The culture of *B. braunii* on pre-treated domestic and piggery wastewater reduced the use of both nitrate and phosphate levels in these media (Sawayama et al., 1994; An et al., 2003). In addition, the concentration of some toxic metals, such as arsenic, chrome and cadmium, was significantly reduced (Sawayama et al., 1995). In batch culture, a biomass of 8.5 g l^{-1} dry cell weight and 0.95 g l^{-1} hydrocarbons were obtained from secondarily treated piggery wastewater (An et al., 2003).

During the growth phases there is a chance of settling down of cells. Biomass settling on the transparent surfaces will affect the penetration of light. Biomass settling on inner wall of tubes is prevented by maintaining highly turbulent flow (Chisti, 2007). Flow is produced using either a mechanical pump, or a gentler airlift pump. Mechanical pumps can smash up the microalgal biomass (Chisti, 1999; García Camacho et al., 2001, 2007; Sánchez Mirón et al., 2003; Chisti, 2007), but designing, installation and operation are easy in the case of mechanical pump. Airlift pumps have been used successfully in many experiments (Grima et al., 1999, 2000, 2001; Fernández et al., 2001; Chisti, 2007).

14.7 Microalgal Harvesting

Recovery of microalgal biomass from the broth is necessary for extracting the oil. Biomass is easily recovered from the broth by filtration, centrifugation, and other means (Chisti, 2007). High-density algal cultures can be concentrated by either chemical or physical flocculation. Products such as aluminum sulphate and ferric chloride cause cells to coagulate and precipitate to the bottom or float to the surface. Recovery of the microalgae is then accomplished by, either draining off the supernatant or skimming cells off the surface. Flotation is used in combination with flocculation for algae harvesting in waste water. It is a simple method by which algae can be made to float on the surface of the medium and removed. Dissolved Air Flotation (DAF) technique separates algae from its culture using features of both froth flotation and flocculation. It uses alum to flocculate an algae/air mixture, with fine bubbles supplied by an air compressor. Alum is a common name for several trivalent sulfates of metal such as aluminum, chromium, or iron and a univalent metal such as potassium or sodium, for example $AlK(SO_4)_2$. Flocculation using multivalent metal salts will contaminate the algal biomass while flocculation using cationic polymers is inhibited by the high ionic strength of sea water. Alternatively, bioflocculation, induced by environmental stresses such as extreme pH, temperature, or nutrient depletion, may cause cell composition changes and is generally considered too unreliable to be economical on a commercial scale. The presence of sodium in a marine environment would reduce the exocellular protein responsible for bioflocculation (Lee et al., 2008). Interrupting the carbon dioxide supply to an algal system can cause algae to flocculate on its own, which is called "autoflocculation" to reduce the cost of other chemical or biochemical flocculation.

Centrifugation of large volumes of algal culture is usually performed using a cream separator; the flow rate being adjusted according to the algal species and the centrifugation rate of the separator. Cells are deposited on the walls of the centrifuge head as a thick algal paste. However, centrifugation requires high capital, energy and running costs and is therefore only suitable for high value products (Lee et al., 2008).

Filtration is carried out commonly on membranes of modified cellulose with the aid of a suction pump. The greatest advantage of this method as a concentrating device is that it is able to collect microalgal biomass of very low density. However, concentration by filtration is limited to small volumes and leads to the eventual clogging of the filter by the packed cells when vacuum is applied. Improvement to such method involves the use of a reverse-flow vacuum in which the pressure operates from above, making the process easy that also avoids the packing of cells. This method itself has been modified to allow a relatively large volume of water to be concentrated in a short period of time. A second process uses a direct vacuum but involves a stirring blade in the flask above the filter which prevents the particles from settling during the concentration process (http://www.oilgae.com/algae/har/har.html). Filtration is suitable for filamentous or colony forming microalgae such as *Spirulina* sp. or *Micractinium* sp. (Lee et al., 2008).

Drying of biomass can reduce the volume for concentrating and easy extraction. Various drying methods differ in the extent of both investment and energy requirement. The selection of a particular drying method therefore depends on the scale of operation and final use of the dried product (Tiburcio et al., 2007).Currently, solar and oven/cabinet drying is used for small scale drying of microalgae, while large-scale producers use spray drying. Spray drying produces high quality products with high bioavailability, while solar and oven/cabinet drying frequently result in products of low and often inconsistent quality with low bioavailability (Tiburcio et al., 2007).

14.8 Extraction of Oil from Microalgae

Oil extraction from biological materials is performed by chemical means, physical means, or a combination of the two. The extraction and purification of oil can be carried out by different procedures, such as solvent extraction, urea inclusion method, HPLC, or selective enzymatic reactions. The method selected must avoid heating and oxidation of fatty acids. Free fatty acid (FFA) extracts, used to purify PUFA, were obtained by a three-step process: (a) direct saponification of biomass-oil, (b) extraction of unsaponifiable lipids, and (c) extraction of purified FFA. Solvents of low toxicity such as ethanol and hexane were used and 80-90% of saponifiable lipid extracted from this method (Fajardo, 2007).

For large scale oil extraction from microalgae, the process is usually accomplished with mechanical cell disruption after removal of maximum water followed by solvent extraction. In this case, the mechanical disruption is commonly performed with either a bead mill or ultrasonication. Bead mills work by having a vertical or horizontal cylindrical chamber that houses a series of mechanically driven agitating elements. The grinding of the cells is performed by plastic or glass bead that occupy about 80% of the chamber's volume, and this has been successfully used to disintegrate algal cells. Ultrasonication, the other main technique, uses an ultrasonic probe to disrupt wet cells. The probe uses a transducer to generate sound waves which in turn cause small bubbles to form, and it is the formation and cavitation of these bubbles that produces shock waves that rupture the cells. It has been found that at higher working volumes a higher acoustic power is required, which can cause larger bubble formation and decreased effectiveness. Thus, for larger scale use, specially designed sound disruption vessels, with a continuously-fed material being disrupted are used (Richmond, 2004).

14.8.1 Direct Liquefaction of Algae for Diesel Production

The microalgal biomass has relatively high water content (80-90%) and this is major tailback for usage in energy supply. The high water content and inferior heat require microalgal biomass pre-treatments that reduce water content and increase the energy density. As a result the energy cost increases and makes the alternative less economically attractive (Patil et al., 2008). Patil et al. (2008) mentioned that direct hydrothermal liquefaction in sub-critical water conditions is a technology that can be employed to convert wet biomass material with oil to liquid fuel. This technology is believed to mimic the natural geological processes thought to be involved in the formation of fossil fuel, but in the time scale of hours or even minutes. As moist biomass can be easily heated by microwave power and process using a novel microwave high-pressure (MHP) reactor has been developed in order to further minimize the energy consumption of the process. In addition, integrated utilization of high temperature and high pressure conditions in the process of hydrothermal liquefaction of wet biomass would significantly improve the overall thermal efficiency of the process. Suitable systems for such utilization are an internal heat exchanger network, or a combined heat and power (CHP) plant. Past research in the use of hydrothermal technology for direct liquefaction of biomass was very active. Only a few of them, however, used algal biomass as feedstock for the technology. Minowa et al. (1995) report an oil yield of about 37% (organic basis) by direct hydrothermal liquefaction at around 300°C and 10 MPa from *Dunaliella tertiolecta* with a moisture content of 78.4 wt%. The oil obtained at a reaction temperature of 340°C and holding time of 60 min had a viscosity of 150–330 mPas and a calorific value of 36 kJ g^{-1}, comparable to those of fuel oil. The liquefaction technique was concluded to be a net energy producer. In a similar study on oil recovery from *Botryococcus braunii*, a maximum yield 64% dry weight basis of oil was obtained by liquefaction at 300°C catalyzed by sodium carbonate. Also, Aresta et al. (2005a, b) have compared different conversion techniques *viz.,* supercritical CO_2, organic solvent extraction, pyrolysis, and hydrothermal technology for production of microalgal biodiesel. The hydrothermal liquefaction technique was more effective for extraction of microalgal biodiesel than using the supercritical carbon dioxide Aresta et al. (2005a). As a conclusion from among the selected techniques, the hydrothermal liquefaction is the most effective technological option for production of bio-diesel from algae. More studies are necessary to improve the hydrothermal liquefaction of algae.

14.9 Biodiesel Conversion from Oils

The extracted oil is converted to fuel by the chemical or biological conversion. There are many methods which convert into diesel with chemical catalyst and enzymes. The most usual method to transform oil into biodiesel is transesterification. This consists of the reaction between triacylglycerols (present in the oils) and an acyl-acceptor. The acyl group acceptors may be carboxylic acids (acidolysis), alcohols (alcoholysis), or another ester (interesterification).

14.10 Microalgae *vs* Carbon Sequestration and Other Environmental Pollutants

Carbon sequestration, i.e. capturing and storing carbon emitted from the global energy system, could be a major tool for reducing atmospheric CO_2 emissions from fossil fuel usage. This can be done by increasing the microalgae production that needs more CO_2 from atmosphere than conventional plants or crops and algae will in turn produce greener fuel which will reduce GHG emissions by replacing conventional fossil diesel. As shown in Fig. 14.3, CO_2 (from the fossil fuel combustion system) and nutrients can be added to a photobioreactor. Microalgae photosynthetically will convert the CO_2 into compounds of high commercial value or mineralized carbon for sequestration.

The development of CO_2-neutral fuels is one of the most urgent challenges facing our society. Achieving a 60% CO_2 emission reduction by 2020 or even a 50–85% reduction by 2050 is an enormous global challenge and which will require the development of a suite of renewable energies (Schenk et al., 2008). There is now a concerted effort to develop new biofuels, of which biodiesel and bioethanol are seen as being close to market options. The areas of biomethane, biomass-to-liquid-diesel, and biohydrogen are also developing rapidly.

Weissman and Tillett (1992) had studied the fixation of CO_2 by large pond-type systems and at optimum conditions the capture efficiency has been shown to be as high as 99% (Weissman and Tillett, 1992; Zeiler et al., 1995). It is reported algae can fix 4 g CO_2 L^{-1} day^{-1} at growth rates of 2.5 g algae L^{-1} day^{-1}, a ratio of 1.6 to 1 (Kurano et al., 1995). Taking into account the conversion of glucose into other compounds such as lipids or starch under certain conditions; the consumption of CO_2 can be as high as 2 g CO_2 to 1 g algae (Schenk et al., 2008) and based upon this rate of conversion, it is possible for one hectare of algal ponds with a growth rate of 50 g m^{-2} day^{-1} to sequester up to 1 Mg of CO_2 a day (Schenk et al., 2008). In the case of the biodiesel process, after extraction of oil, the remaining biomass can be fed into downstream carbon sequestration processes. Specifically the sequestered carbon can be converted to hard C-chips (Agrichar) through pyrolysis (Bridgwater and Maniatis, 2004). This has the additional advantage that Agrichar can be marketed to the agricultural sector; as it greatly enhances the carbon content of the soil and so its fertility. This in turn can contribute further to climate change mitigation by affecting the gas exchange of crops and soils (Marris, 2006; Lehmann, 2007). Additionally, pyrolysis acts as a sterilization process of the biomass waste, providing an environment friendly waste disposal mechanism which will increase public acceptance regarding the use of genetically modified microalgae for biodiesel production.

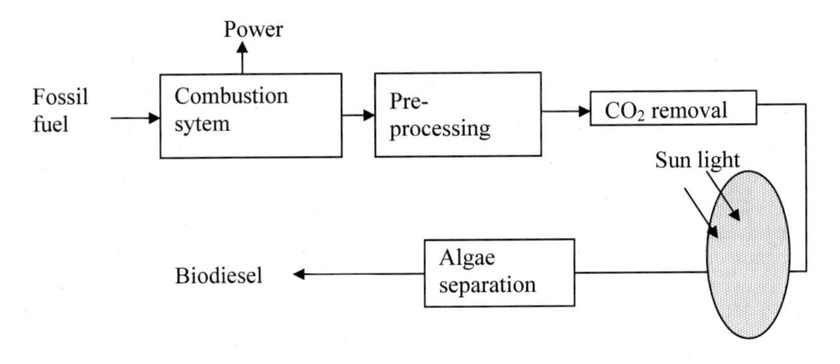

Fig. 14.3 CO_2 recovery and C-sequestration of from stationary combustion systems by photosynthesis of microalgae.

High purity CO_2 gas is not essential for algae. Direct feeding of flue gas containing 2–5% CO_2 to the photobioreactor can simplify CO_2 separation from flue gas significantly. In addition, some combustion products such as NOx or SOx can be used effectively as nutrients for microalgae. The products of the proposed process are mineralized carbon for stable sequestration, and high commercial value added products such as biodiesel. By selecting appropriate algae species for higher oil yields, either one or both can be produced. The process is a renewable cycle with minimal negative impacts on the environment.

Emissions of all pollutants except NOx appear to decrease when biodiesel is used. The fact that NOx emissions increase with increasing biodiesel concentration could be a detriment in areas that are out of attainment for ozone. Table 14.7 shows the average changes in mass emissions from diesel engines using the biodiesel mixtures relative to the standard diesel fuel (Morris et al., 2003).

Table 14.7 Average changes in mass emissions from diesel engines using the biodiesel mixtures relative to the standard diesel fuel (%) (Source: Morris et al., 2003).

Category	CO	NOx	SO_2	PM	VOC
B20	-13.1	+2.4	-20	-8.9	-17.9
B100	-42.7	+13.2	-100	-55.3	-63.2

Major disadvantages of biodiesel are higher viscosity, lower energy content, higher cloud point and pour point, higher nitrogen oxides (NOx) emissions, lower engine speed and power, injector choking, engine compatibility, high price, and higher engine wear.

14.11 Future Perspective of Microalgal Biodiesel

There is a great interest on exploiting microalgal source for the oil production. The economics of biodiesel production could be improved by advances in the production technology. In addition, extraction processes are needed that would enable the recovery of the algal oil from moist biomass pastes without the need for drying. Algal biomass production capacity (i.e. the productivity) depends on the geographical latitude where the facility is located. This is because the sunlight regimen varies with geographic location. Improved photobioreactor engineering will make predictions of productivity more reliable and enable design of photobioreactors that are more efficient. A different and complimentary approach to increase productivity of microalgae is via genetic and metabolic engineering. Genetic and metabolic engineering are likely to have the greatest impact on improving the economics of production of microalgal biodiesel.

The microalgal biodiesel costs can be reduced further by finding better microalgal strains with higher lipid contents and shorter growth cycles. Algal biodiesel becomes even more environment friendly feasible technology given the potential for greenhouse gas regulation in the near future; since for every Mg of algal biomass produced, approximately 1.6 Mg of CO_2 is fixed. Algae have immense potential as a sustainable feedstock for production of biodiesel with very low CO_2

emissions. However, there are currently serious drawbacks, it would be unjustified to overestimate the greenness of this upcoming technology. Today, algae can be grown in water bodies and bioreactors in just a few days, and oil can be extracted directly from the harvested algae. Success in microalgal culture achieved in the US is providing momentum in many other countries in that region.

There are several (more than 100,000) known strains of microalgae in the world (Sheehan et al., 1998). Researchers are looking for the strain with a combination of high oil content and a rapid growth rate. Algae reproduce by dividing their cells and generally algae with high oil contents grow relatively slowly than less oil contents. Advances in genetic transformation techniques will permit ever more sophisticated forms of metabolic network engineering to be accomplished. In addition to the development of transgenic algae, methods to protect against unintended environmental consequences will be required. These mechanisms may include control considerations in photobioreactor design and incorporation of suicide genes in the engineered organism. Ultimately, microalgae offer the potential to have a profound impact on the future welfare of the planet by addressing the pressing issues of alternative energy resources, global warming, human health, and food security.

14.12 Summary

Biodiesels produced from renewable biomass are the sustainable energy resource with greatest potential for CO_2 neutral production. The first generation biodiesel produced from corn, sugarcane, and soybeans perform poorly in many environmentally context. A sustainable and profitable biodiesel production from microalgae is possible. It is the only renewable biodiesel that can potentially completely displace liquid fuels derived from petroleum for transport requirements. This second generation biodiesel can overcome the energy and environmental needs by integrating the technologies. Large quantities of algal biomass needed for the production of biodiesel could be grown in photobioreactors combined with photonics and biotechnologies. However, more precise economic assessments of production are necessary to establish with petroleum derived fuels.

The use of the biorefinery concept and advances in photobioreactor engineering will further lower the cost of production. Photobioreactors provide a controlled environment that can be tailored to the specific demands of highly productive microalgae to attain a consistently good annual yield of oil. Achieving the capacity to inexpensively produce biodiesel from microalgae is of strategic significance to an environmentally sustainable society. Development of improved strain using genetic engineering along with highly developed farming and downstream technologies will benefit the future development of microalgal biodiesel production.

14.13 Acknowledgements

The authors are sincerely thankful to the Natural Sciences and Engineering Research Council of Canada (Grants A4984, STP235071, Canada Research Chair), and INRS-ETE for financial support. The views or opinions expressed in this article are those of the authors and should not be construed as opinions of the U.S. Environmental Protection Agency.

14.14 References

Abelson, P.H. (1995). "Renewable liquid fuels." *Science*, 268, 955.

Achitouv, E., Metzger, P., Rager, M-N., and Largeau, C. (2004). "C_{31}–C_{34} methylated squalenes from a Bolivian strain of *Botryococcus braunii*." *Phytochem.,* 65, 3159–3165.

Adams, I.P., Dack, S., and Dickinson, F.M. (2002). "The distinctiveness of ATP: citrate lyase from *Aspergillus nidulans*." *Biochim. Biophys. Acta*, 1597, 36–41.

Ahouissoussi, N.B.C., and Wetzstein, M.E. (1998). "A comparative cost analysis of biodiesel, compressed natural gas, methanol and diesel for transit bus systems." *Resour. Energy Economics*, 20, 1–15.

Albertyn, J., Hohmann, S., Thevelein, J.M., and Prior, B.A. (1994). "*GPD1*, which encodes glycerol-3-phosphate dehydrogenase, is essential for growth under osmotic stress in *Saccharomyces cerevisiae*, and its expression is regulated by the high-osmolarity glycerol response pathway." *Mol.Cell. Biol.*, 14, 4135–4144.

Allard, B., and Templier, J. (2000). "Comparison of neutral lipid profile of various trilaminar outer cell wall (TLS)-containing microalgae with emphasis on algaenan occurrence." *Phytochem.*, 54, 369.

Al-Widyan, M.I., and Al-Shyoukh, A.O. (2002). "Experimental evaluation of the transesteriWcation of waste palm oil into biodiesel." *Bioresour. Technol.*, 85, 253–256.

An, J-Y., Sim, S-J., Lee, J.S., and Kim, B.W. (2003). "Hydrocarbon production from secondarily treated piggery wastewater by the green alga *Botryococcus braunii*." *J. Appl. Phycol.*, 15, 185–191.

Ansell, R., Granath, K., Hohmann, S., Thevelein, J., and Adler, L. (1997). "The two isozymes for yeast NAD^+-dependent glycerol 3-phosphate dehydrogenase encoded by *GPD1* and *GPD2* have distinct roles in osmoadaptation and redox regulation." *EMBO. J.,* 16, 2179–2187.

Antolin, G., Tinaut, F.V., Briceno, Y., Castano, V., Perez, C., and Ramirez, A.I. (2002). "Optimisation of biodiesel production by sunflower oil transesterification." *Bioresour. Technol.*, 83, 111–114.

Aresta, M., Dibenedetto, A., and Barberio, G. (2005a). "Utilization of macro-algae for enhanced CO_2 fixation and biofuels production: Development of a computing software for an LCA study." *Fuel Process. Technol.*, 86, 1679-1693.

Aresta, M., Dibenedetto, A., Carone, M., Colonna, T., and Fragale, C. (2005b). "Production of biodiesel from macroalgae by supercritical CO_2 extraction and thermochemical liquefaction." *Environ.Chemistry Lett.*, 3, 136-139.

Azam, M.M., Waris, A., and Nahar, N.M. (2005). "Prospects and potential of fatty acid methyl esters of some non-traditional seed oils for use as biodiesel in India." *Biomass Bioen.*, 2, 293–302.

Baillez, C., Largeau, C., Berkaloff, C., and Casadevall, E. (1986). "Immobilization of *Botryococcus braunii* in alginate: influence on chlorophyll content, photosynthetic activity and degeneration during batch cultures." *Appl. Microbiol. Biotechnol.*, 23, 361–366.

Baillez, C., Largeau, C., and Casadevall, E. (1985). "Growth and hydrocarbon production of *Botryococcus braunii* immobilized in calcium alginate gel." *Appl. Microbiol. Biotechnol.*, 23, 99–105.

Belarbi, E.H., Molina, G.E., and Chisti, Y. (2000). "A process for high yield and scalable recovery of high purity eicosapentaenoic acid esters from microalgae and fish oil." *Enzyme Microb. Technol.*, 26, 516–29.

Brenckmann, F., Largeau, C., Casadevall, E., and Berkaloff, C. (1985). "Influence de la nutrition azotée sur la croissance et la production d'hydrocarbures de l'algue *Botryococcus braunii.* " In: Palz, W., Coombs, J., and Hall, D.O. (Ed.) *Energy from biomass*. Elsevier, London, pp. 717–721.

Bridgwater, A., and Maniatis, K. (2004). "The production of biofuels by thermal chemical processing of biomass." In: Archer, M., and Barber, J. (Ed.) *Molecular to global photosynthesis*. Imperial College Press, London, pp.

Campbell, C.J. (1997). *The coming oil crisis,* Multi-science Publishing Company and petroconsultants S.A, Essex, England.

Canakci, M., and Gerpen, J.V. (1999). "Biodiesel production via acid catalysis." *Trans. Am. Soc. Agric. Eng.*, 42, 1203–1210.

Casadevall, E., Dif, D., Largeau, C., Gudin, C., Chaumont, D., and Desanti, O. (1985). "Studies on batch and continuous cultures of *Botryococcus braunii*: Hydrocarbon production in relation to physiological state, cell ultrastructure, and phosphate nutrition." *Biotechnol. Bioeng.,* 27, 286.

Casadevall, E., Largeau, C., Metzger, P., Chirac, C., Berkaloff, C., and Coute, A. (1983). "Hydrocarbon production by unicellular microalga *Botryococcus braunii.*" *Biosci.,* 2, 129.

Chisti, Y. (2007). "Biodiesel from microalgae." *Biotechnol. Adv.*, 25, 294–306.

Chisti, Y. (2008). "Biodiesel from microalgae beats bioethanol. " *Trends Biotechnol.*, 26, 126–131.

Chisti,Y. (1999). "Shear sensitivity." In: Flickinger, M.C., and Drew, S.W. (Ed.) *Encyclopedia of bioprocess technology: fermentation, biocatalysis, and bioseparation*, vol. 5. Wiley, pp. 2379–2406.

Courchesne, N.M.D., Parisien, A., Wang, B., and Lan, C.Q. (2009). "Enhancement of Lipid Production Using Biochemical, Genetic and Transcription Factor Engineering Approaches." *J. Biotechnol.*, doi:10.1016/j.jbiotec.2009.02.018.

Coyle, W. (2007). *A global perspective: the future of biofuels, economic research service.* US Dept. of Agriculture, Washington, DC, 2007, Available at http://www.ers.usda.gov/AmberWaves/November07/PDF/Biofuels.pdf (accessed June, 09)

Demirbas, A. (2009). "Progress and recent trends in biodiesel fuels." *Energy Conversion Manage.*, 50, 14-34.

Dijkstra, A.J. (2006). "Revisiting the formation of trans isomers during partial hydrogenation of triacylglycerol oils." *Eur. J. Lipid Sci. Technol.*, 108(3), 249–64.

Dimitrov, K. (2007). *Green fuel technologies: a case study for industrial photosynthetic energy capture*, Brisbane, Australia. Available at http://www.nanostring.net/Algae/CaseStudy.pdf (accessed January, 2008).

Dubertret, G., Gerard-Hirne, C., and Tremolieres, A. (2002). "Importance of trans-D3 hexadecenoic acid containing phosphatidylglycerol in the formation of the trimeric light-harvesting complex in *Chlamydomonas*." *Plant Physiol. Biochem.*, 40, 829–836.

Dunahay, T.G., Jarvis, E.E., Dais, S.S., and Roessler, P.G. (1996). "Manipulation of microalgal lipid production using genetic engineering." *Appl. Biochem. Biotechnol. - Part A: Enzyme Engg. Biotechnol.*, 57-58, 223-231.

Dunahay, T.G., Jarvis, E.E., and Roessler, P.G. (1995). "Genetic transformation of the diatoms *Cyclotella cryptica* and *Navicula saprophila*." *J. Phycol.*, 31, 1004-1012.

EPA (2002) U.S. Environmental Protection Agency, *A Comprehensive Analysis of Biodiesel Impacts on Exhaust Emissions*. EPA-Draft Technical Report, EPA420-P-02-001, October 2002.

Fajardo, A.R., Cerdán, L.E., Robles-Medina, A., Fernández, F.G.A., Pedro, A., Moreno, G., and Grima, E.M. (2007). "Lipid extraction from the microalga *Phaeodactylum tricornutum*." *Eur. J. Lipid Sci. Technol.*, 109, 120-126.

Fernández, F.G.A., Sevilla, J.M.F., Pérez, J.A.S., Grima, E.M., and Chisti, Y. (2001). "Airlift-driven external-loop tubular photobioreactors for outdoor production of microalgae: assessment of design and performance." *Chem. Eng. Sci.,* 56, 2721–2732.

Flam, F. (1994). "Chemists get a taste of life at gathering in San Diego." Meeting briefs. *Science*, 264, 33.

Fuhrmann, M. (2002). "Expanding the molecular tool kits for *Chlamydomonas reinhardtii* – from history to new frontiers." *Protist.*, 153, 357–364.

García Camacho, F., Gallardo Rodríguez, J., Sánchez Mirón, A., Cerón García, M.C., Belarbi, E.H., Chisti, Y., and Molina Grima, E. (2007). "Biotechnological significance of toxic marine dinoflagellates. " *Biotechnology Advances*, 25 (2), 176-194.

García Camacho, F., Molina Grima, E., Sánchez Mirón, A., González Pascual, V., Chisti, Y. (2001). "Carboxymethyl cellulose protects algal cells against hydrodynamic stress." *Enzyme Microb. Technol.* 29, 602–610.

Giroud, C., Gerber, A., and Eichenberger, W. (1988). "Lipids of *Chlamydomonas reinhardtii*: analysis of molecular species and intracellular site(s) of biosynthesis." *Plant. Cell. Physiol.*, 29, 587–595.

Gouveia, L., and Oliveira, C.A. (2009). "Microalgae as a raw material for biofuels production." *J. Ind. Microbiol. Biotechnol.*, 36, 269–274.

Grima, E.M., Fernández, J., Fernández, F.G.A., and Chisti, Y. (2001). "Tubular photobioreactor design for algal cultures." *J. Biotechnol.,* 92, 113–131.

Grima, E.M., Fernández, F.G.A., Camacho, F.G., Rubio, F.C., and Chisti, Y. (2000). "Scale-up of tubular photobioreactors." *J. Appl. Phycol.*, 12, 355–368.

Grima, E.M., Fernández, F.G.A., Camacho, F.G., and Chisti, Y. (1999). "Photobioreactors: light regime, mass transfer, and scale-up." *J. Biotechnol.*, 70, 231–247.

Grobbelaar, J.U. (2007). "Photosynthetic characteristics of *Spirulina platensis* grown in commercial-scale open outdoor raceway ponds: what do the organisms tell us?" *J. Appl. Phycol.,* 19, 591–598.

Grobbelaar, J.U., Soeder, C.J., and Stengel, E. (1990). "Modelling algal productivity in large outdoor cultures and waste treatment systems." *Biomass,* 21, 297–314.

Guschina I.A., and Harwood, J.L. (2006). "Lipids and lipid metabolism in eukaryotic algae." *Progress Lipid Res.,* 45, 160–186.

Hagemann, M., and Erdmann, N. (1994). "Activation and pathway of glucosylglycerol biosynthesis in the cyanobacterium *Synechocystis* sp. PCC 6803." *Microbiol.,* 140, 1427–1431.

Huang, Z., and Poulter, C.D. (1989). "Tetramethylsqualene, a triterpene from *Botryococcus braunii* var. Showa." *Phytochem.,* 28, 1467–1470.

Jang, E.S., Jung, M.Y., and Min, D.B. (2005). "Hydrogenation for low trans and high conjugated fatty acids." *Comp. Rev. Food Sci. Saf.*, 4, 22–30.

Janssen, M., Tramper, J., Mur, L.R., and Wijffels, R. H. (2003). "Enclosed photobioreactors: light regime, photosynthetic efficiency, scale-up and future prospects." *Biotechnol. Bioeng.,* 81, 193–210.

Kaasen, I., Falkenberg, P., Styrvold, O.B., and Strom, A.R. (1992). "Molecular cloning and physical mapping of the *otsBA* genes, which encode the osmoregulatory trehalose pathway of *Escherichia coli*: evidence that transcription is activated by *katF* (AppR)." *J. Bacteriol.,* 174, 889–898.

Karmee, S.K., and Chadha, A. (2005). "Preparation of biodiesel from crude oil of *Pongamia pinnata*." *Bioresour. Technol*., 96, 1425–1429.

Kinast, J.A. (2003). *Production of Biodiesels from Multiple Feedstocks and Properties of Biodiesels and Biodiesel/Diesel Blends.*, Report 1 in a 6. National Renewable Energy Laboratory, Report no. NREL/SR-510-31460.

Krawczyk, T. (1996). "Biodiesel - Alternative fuel makes inroads but hurdles remain." *INFORM,* 7, 801-829.

Kurano, N., Ikemoto, H., Miyashita, H., Hasegawa, T., Hata, H., and Miyachi, S. (1995). "Fixation and utilization of carbon dioxide by microalgal photosynthesis." *Energy Convers. Manag.*, 36, 689–692.

Lang, X., Dalai, A.K., Bakhshi, N.N., Reaney, M.J., and Hertz, P.B. (2001). "Preparation and characterization of bio-diesels from various biooils." *Bioresour. Technol.,* 8, 53–62.

Largeau, C., Casadevall, E., Berkaloff, C., and Dhamelincourt, P. (1980). "Sites of accumulation and composition of hydrocarbons in *Botryococcus braunii*." *Phytochem.,* 19:1043–1051.

Lee, A.K., Lewis, D.M., and Ashman, P.J. (2008). "Microbial flocculation, a potentially low-cost harvesting technique for marine microalgae for the production of biodiesel." *J. Appl. Phycol.,* DOI 10.1007/s10811-008-9391-8.

Lehmann, J. (2007). "A Handful of carbon." *Nature*, 447, 143–144.

Madras, G., Kolluru, C., and Kumar, R. (2004). "Synthesis of biodiesel in supercritical fluids." *Fuel*, 83, 2029–2033.

Marris, E. (2006). "Black is the new green." *Nature*, 442, 624–626.

Meng, X., Yang , J., Xu , X., Zhang , L., Nie , Q., and Xian, M. (2009). "Biodiesel production from oleaginous microorganisms." *Renew. Energy,* 34, 1–5

Metzger, P., Allard, B., Casadevall, E., Berkaloff, C., and Couté, A. (1990). "Structure and chemistry of a new chemical race of *Botryococcus braunii* that produces lycopadiene, a tetraterpenoid hydrocarbon." *J. Phycol.,* 26, 258–266.

Metzger, P., Berkaloff, C., Couté, A., and Casadevall, E. (1985). "Alkadiene and botryococcene-producing races of wild strains of *Botryococcus braunii.*" *Phytochem.*, 24, 2305–2312.

Metzger, P., and Casadevall, E. (1987). "Lycopadiene, a tetraterpenoid hydrocarbon from new strains of the green alga *Botryococcus braunii.*" *Tetrahedron Lett.,* 28, 3931–3934.

Metzger, P., and Largeau, C. (2005). "*Botryococcus braunii*: a rich source for hydrocarbons and related ether lipids." *Appl. Microbiol. Biotechnol.*, 66, 486–496.

Metzger, P., and Pouet, Y. (1995). "N-Alkenyl pyrogallol dimethyl ethers, aliphatic diol monoesters and some minor ether lipids from *Botryococcus braunii*: A Race." *Phytochem.*, 40, 543.

Miao, X., and Wu, Q. (2004). "High yield bio-oil production from fast pyrolysis by metabolic controlling of *Chlorella protothecoides*." *J. Biotechnol.,* 110, 85.

Miao, X., and Wu, Q. (2006). "Biodiesel production from heterotrophic microalgal oil." *Bioresour. Technol.*, 97, 841–846.

Miao, X., Wu, Q., and Yang, C. (2004). "Fast pyrolysis of microalgae to produce renewable fuels." *J. Anal. Appl. Pyrol.*, 71, 855.

Minowa, T., Yokoyama, S.Y., Kishimoto, M., and Okakurat, T. (1995). "Oil production from algal cells of *Dunaliella tertiolecta* by direct thermochemical liquefaction." *Fuel*, 74, 1735–1738.

Moore, T.S., Du, Z., and Chen, Z. (2001). "Membrane lipid biosynthesis in *Chlamydomonas reinhardtii.* In vitro biosynthesis of diacylglyceryltrimethylhomoserine." *Plant. Physiol.*, 125, 423–429.

Morris, R.E., Pollack, A.K., Mansell, G.E., Lindhjem, C., Jia, Y., and Wilson, G. (2003). *Impact of biodiesel fuels on air quality and human health.* Subcontractor Report, NREL/SR-540-33793, National Renewable Energy Laboratory, Golden, CO.

Patil, V., Tran, K-Q., and Giselrød, H.R. (2008). "Towards sustainable production of biofuels from microalgae." *Int. J. Mol. Sci.*, 9, 1188-1195.

Pineau, B., Girard-Bascou, J., Eberhard, S., Choquet, Y., Tremolieres, A., Gerard-Hirne, C., Bennardo-Connan, A., Decottignies, P., Gillet, S., and Wollman, F.-A. (2004). "A single mutation that causes phosphatidylglycerol deficiency impairs synthesis of photosystem II cores *in Chlamydomonas reinhardtii."* *Eur. J. Biochem.*, 271, 329–338.

Pulz, O. (2001). "Photobioreactors: production systems for phototrophic microorganisms." *Appl. Microbiol. Biotechnol.,* 57, 287–293.

Pulz, O., and Gross, W. (2004). "Valuable products from biotechnology of microalgae." *Appl. Microbiol. Biotechnol.,* 65, 635–648.

Qiang, H., Zarmi, Y., and Richmond, A. (1998). "Combined effects of light intensity, light path and culture density on output rate of *Spirulina platensis* (cyanobacteria)." *Eur. J. Phycol.,* 33, 165–171.

Ratledge, C. (2002). "Regulation of lipid accumulation in oleaginous micro-organisms." *Biochem. Soc. Trans.,* 30, 1047–1050.

Ratledge, C. (2004). "Fatty acid biosynthesis in microorganisms being used for single cell oil production." *Biochimie.*, 86, 807-815.

Reed, R.H., and Stewart, W.D.P. (1985). "Osmotic adjustment and organic solute accumulation in unicellular cyanobacteria from freshwater and marine habitats." *Marine Biol.,* 88, 1–9.

Richmond, A. (2004). *Handbook of Microalgal Culture: Biotechnology and Applied Phycology,* Blackwell, Oxford, UK.

Richmond, A. (2003). *Handbook of Microalgal Cultures*, Blackwell Publishers.

Richmond, A., Zhang, C-W., and Yair, Z. (2003). "Efficient use of strong light for high photosynthetic productivity: interrelationships between the optical path, the optimal population density and cell growth inhibition." *Biomol. Eng.,* 20, 229–236.

Riekhof, W.R., Ruckle, M.E., Lydic, T.A., Sears, B.B., and Benning, C. (2003). "The sulfolipids 2`-*O*-acyl-sulfoquinovosyldiacylglycerol and sulfoquinovosyldiacylglycerol are absent from a *Chlamydomonas reinhardtii* mutant deleted in SQD1." *Plant Physiol.*, 133, 864–874.

Roessler, P.G. (1990). "Purification and characterization of acetyl-CoA carboxylase from the diatom *Cyclotella cryptica.*" *Plant Physiol.,* 92, 73-78.

Sánchez Mirón, A., Cerón García, M-C., Contreras Gómez, A., García Camacho, F., Molina Grima, E, and Chisti, Y. (2003). "Shear stress tolerance and biochemical characterization of *Phaeodactylum tricornutum* in quasi steady-state continuous culture in outdoor photobioreactors." *Biochem. Eng. J.*, 16, 287–297.

Sato, N., Aoki, M., Maru, Y., Sonoike, K., Minoda, A., and Tsuzuki, M. (2003). "Involvement of sulfoquinovosyl diacylglycerol in the structural integrity and heat-tolerance of photosystem II." *Planta*, 217, 245–251.

Sato, N., and Moriyama, T. (2007). "Genomic and biochemical analysis of lipid biosynthesis in the unicellular rhodophyte *Cyanidioschyzon merolae*: lack of a plastidic desaturation pathway results in the coupled pathway of galactolipid synthesis." *Eukaryotic cell*, 6, 1006–1017.

Sato, N., Sonoike, K., Kawaguchi, A., and Tsuzuki, M. (1996). "Contribution of lowered unsaturation levels of chloroplast lipids to high temperature tolerance of photosynthesis in *Chlamydomonas reinhardtii*." *J. Photochem. Photobiol. B: Biol.*, 36, 333–337.

Sawayama, S., Inoue, S., Dote, Y., and Yokoyama, S-H. (1995). "CO_2 fixation and oil production through microalga." *Energy Convers. Manage.*, 36, 729–731.

Sawayama, S., Inoue, S., and Yokoyama, S. (1994). "Continuous culture of hydrocarbon-rich microalga *Botryococcus braunii* in secondarily treated sewage." *Appl. Microbiol. Biotechnol.*, 41, 729–731.

Schenk, M.P., Skye, R., and Hall, T. (2008). "Second generation biofuels: high-efficiency microalgae for biodiesel production." *Bioenerg. Res.*, 1, 20–43.

Schumacher, L.G., Borgelt, S.C., Fosseen, D., Goetz, W., and Hires, W.G. (1996). "Heavy-duty engine exhaust emission test using methyl ester soybean oil/diesel fuel blends." *Bioresour. Technol.*, 57, 31–36.

Sharma, Y.C., and Singh, B. (2008). "Development of biodiesel: Current scenario." *Renew. Sustain. Energy Rev.*, in press, doi:10.1016/j.rser.2008.08.009.

Sharma, Y.C., Singh, B., and Upadhyay, S.N. (2008). "Advancements in development and characterization of biodiesel: A review." *Fuel,* 87, 2355–2373.

Shay, E.G. (1993). "Diesel fuel from vegetable oils: status and opportunities." *Biomass Bioen.*, 4, 227–242.

Sheehan, J., Dunahay, T., Benemann, J., and Roessler, P. (1998). *A look back at the U.S. Department of Energy's Aquatic Species Program—Biodiesel from Algae.* NREL Report, NREL/TP-580-24190, 1998.

Siler-Marinkovic, S., and Tomasevic, A. (1998). "Transesterification of sunflower oil in situ." *Fuel*, 77(12), 1389–1391.

Tickell, J. (2000). *From the fryer to the fuel tank, the complete guide to using vegetable oil as an alternative fuel.* Tallahasseee, USA.

Tiburcio P.C., Galvez, F.C.F., Cruz, L.J., and Gavino, V.C. (2007). "Optimization of low-cost drying methods to minimize lipid peroxidation in *Spirulina platensis* grown in the Philippines." *J. Appl. Phycol.*, 19, 719–726.

Um, B-H., and Kim, Y-S. (2009). "Review: A chance for Korea to advance algal-biodiesel" *Journal of Industrial and Engineering Chemistry*, 15(1), 1-7.

Vicente, G., Martinez, M., and Aracil, J. (2004). "Integrated biodiesel production: a comparison of different homogeneous catalysts systems." *Bioresour. Technol.*, 92, 297–305.

Villarreal-Rosales, E., Metzger, P., and Casadevall, E. (1992). "Ether lipid production in relation to growth in *Botryococcus braunii*." *Phytochem.,* 31, 3021–3027.

Vonshak, A., and Guy, R. (1992). "Photoadaptation, photoinhibition and productivity in the blue-green alga *Spirulina platensis*, grown outdoors." *Plant Cell Environ.,* 15, 613–616.

Wake, L.V., and Hillen, L.W. (1980). "Study of a ''bloom'' of the oil-rich alga *Botryococcus braunii* in the darwin river reservoir." *Biotechnol. Bioeng.*, 22, 1637.

Wang, Y., Ou, S., Liu, P., and Zhang, Z. (2007). "Preparation of biodiesel from waste cooking oil via two-step catalyzed process." *Energy Conservat. Manage.*, 48, 184–188.

Watanabe, Y., Tsuchimoto, S., and Tamai, Y. (2004). "Heterologous expression of *Zygosaccharomyces rouxii* glycerol 3-phosphate dehydrogenase gene (ZrGPD1) and glycerol dehydrogenase gene (ZrGCY1) in *Saccharomyces cerevisiae*." *FEMS Yeast Res.,* 4, 505–510.

Weetal, H.H. (1985). "Studies on nutritional requirements of the oil producing alga *Botryococcus braunii*." *Appl. Biochem. Biotechnol.*, 11,. 377.

Weissman, J.C., and Tillett, D.M. (1992). "Aquatic Species Project Report; NREL/MP-232-4174." In: Brown, L.M., and Sprague, S. (Ed.) *National Renewable Energy Laboratory.* pp. 41–58.

Whatmore, A.M., Chudek, J.A., and Reed, R.H. (1990). "The effects of osmotic upshock on the intracellular solute pools of *Bacillus subtilis*." *J. Gen. Microbiol.,* 136, 2527–2535.

Widjaja, A., Chien, C-C., and Ju, Y-H. (2009). "Study of increasing lipid production from fresh water microalgae *Chlorella vulgaris.*" *J. Taiwan Inst. Chem. Engineers,* 40, 13–20.

Wijffels, R.H. (2007). "Potential of sponges and microalgae for marine biotechnology." *Trends Biotechnol.*, 26, 26-31.

Wolf, F.R., Nonomura, A.M., and Bassham, J.A. (1985). "Growth and branched hydrocarbon production in a strain of *Botryococcus braunii* (Chlorophyta)." *J. Phycol.,* 21, 388–396.

Wu, Q.Y., Yin, S., Sheng, G.Y., and Fu, J.M. (1994). "New Discoveries in study on hydrocarbons from thermal degradation of heterotrophically yellowing algae." *Sci. China (B)*, 37, 326.

Xue, F., Zhang, X., Luo, H., and Tan, T. (2006). "A new method for preparing raw material for biodiesel production." *Process Biochem.*, 41, 1699–702.

Yang, W., Moroney, J.V., and Moore, T.S. (2004). "Membrane lipid biosynthesis in *Chlamydomonas reinhardtii*: ethanolaminephosphotransferase is capable of synthesizing both phosphatidylcholine and phosphatidylethanolamine." *Arch. Biochem. Biophys.*, 430, 198–209.

Yoon, J.H., Shin, J-H., and Park, T.H. (2008). "Characterization of factors influencing the growth of Anabaena variabilis in a bubble column reactor." *Bioresour. Technol.,* 99, 1204–1210.

Zeiler, K.G., Heacox, D.A., Toon, S.T., Kadam, K.L., and Brown, L.M. (1995). "The use of microalgae for assimilation and utilization of carbon dioxide from fossil fuel—red power plant flue gas." *Energy Convers. Manage.*, 36, 707–712.

CHAPTER 15

Heterotrophic Algal-biodiesel Production: Challenges and Opportunities

Michael J. Cooney and Christopher K.H. Guay

15.1 Introduction

Microalgae are an extremely vast, diverse group of microorganisms comprising both photoautotrophic and heterotrophic species. They are small in size (usually microscopic), uni- or multicellular (although colonies are possible with little or no cell differentiation), often colorful (due to the presence of photosynthetic pigments), and found typically - but not exclusively - in fresh and saline aqueous environments. Phylogenetically, microalgae can be prokaryotic or eukaryotic. In evolutionary terms, they can be recent or very ancient. This diversity makes microalgae, as a group, a potentially rich source of chemical products with applications in a wide array of industries (Olaizola, 2003). The value of microalgae as a source of food (blue-green and green algae), soda ash, iodine, and alginic acid (brown algae, agar, and carrageenans (red algae) has been recognized for centuries (Chen, 1996). In recent years, microalgae have begun to be exploited commercially to produce high-value molecules, such as specialty fatty acids and pigments, for nutraceuticals, cosmetics, and pharmaceuticals (Sporalore et al., 2006). Current industrial uses of microalgae also include wastewater treatment, fertilizer for agriculture, and feed for aquaculture (Sporalore et al., 2006; Mulbry et al., 2008; Muñoz et al., 2009). Large-scale commercial microalgae culture began in the early 1960s in Japan with the culture of *Chlorella* and was followed in the 1970s by the production of *Arthrospira* in Mexico (Sporalore et al., 2006). These endeavors were dedicated to food and nutraceuticals. By the end of the 1980s, over 46 large-scale production facilities throughout Asia were producing over 1000 kg of microalgae per month (mostly *Chlorella*). Concurrently, large-scale production of ß-carotene from *Duniella salina* was established in Western Australia, followed by construction of production facilities in the United States, Israel, and India (Biotechnological and Environmental Applications of microalgae [BEAM], and Australian Research Network; www.scieng.murdoch.edu.au/centres/algae/BEAM-Net-Appl0.htm, 2005). By the early 2000s, it had been estimated that over 5000 tons of microalgae (as dry matter) were produced globally per year (Pulz and Gross, 2004). Although no official statistics exist, the consensus view of experts is that the current global production of commercial autotrophic algae for nutritional products has risen to over 10,000 tons per year (personal communication, John Benneman). This includes *Spirulina* (~5,000 tons), *Chlorella* (~4,000 tons), and *Dunaliella* (~1000 tons), and *Haematococcus pluvialis* (< 100 tons), but not aquaculture feeds, for which there is no real commercial production and account for at most a few hundred tons worldwide. Roughly 40% of this production occurs in China (mostly as *Spirulina*), 30% in Japan, Taiwan, and Korea (mostly as *Chlorella*), 10% in the US (as *Spirulina* and some *Haematococcus*), and 20% other countries such as India. Photobioreactor production remains low at approximately 200 tons and occurs in Israel, Germany, and Hawaii.

Despite these successes, photoautotrophic production of microalgae is inherently constrained by the onset of light limitation at increasing cell densities. Maximum biomass concentrations achievable using advanced photobioreactors remain relatively low (< 10 grams dry weigh per liter (gdw l^{-1});(Apt and Behrens, 1999), and microalgae cultivation in open ponds (raceways) is characterized by even lower cell densities (Eriksen, 2008). Low biomass concentrations necessitate large-volume cultures, resulting in substantial water use and high costs for harvesting and drying microalgae cells. To overcome these obstacles, heterotrophic culture of microalgae was proposed in the early 1990s (Day et al., 1991; Orus et al., 1991; Barclay et al., 1994; Gladue and Maxey, 1994). Heterotrophic growth, or dark fermentation, requires organic carbon compounds, such as glucose, acetate, etc., as a source of carbon and energy. Robust, highly sophisticated fermentation technology has been previously developed and applied worldwide for industrial bacteria and yeast production (Apt and Behrens, 1999), and the adaptation of this technology provided a significant advance to commercial microalgae cultivation. Bioreactors range in volume from < 1 to over 500,000 liters, and there exists a great deal of theoretical and practical knowledge permitting the accurate correlation of pilot plant studies to performance at larger scales (Barclay et al., 2005). Preparation of inoculums, the composition and addition rate of feed media, and other important process parameters (such as pH, dissolved oxygen, and temperature) can be precisely defined and replicated over successive batches through the implementation of process control and Good Manufacturing Practice (GMP) protocols. Of particular significance is the ability to apply modeling techniques, rapid sample analysis, and fed-batch culture methodology using precisely defined media components to accurately control the carbon-to-nitrogen (C/N) ratio of the growth media (Ganuza et al., 2008). This permits the separation of the culture period into distinct phases that promote cell growth (at low C/N ratios) or lipid accumulation (at high C/N ratios). Under heterotrophic growth conditions, microalgal cultures with cell densities of 50-100 gdw l^{-1} can be achieved on a reproducible basis, assuming appropriate choices of strain, media composition, process parameters, and growth regime (Apt and Behrens, 1999). It is therefore possible to produce several thousand kilograms of dry microalgal biomass from a single industrial-scale fermentation batch, with growth rates more rapid than possible under photoautotrophic growth conditions. In this context, heterotrophic culture of microalgae can reduce the cost of production (per equivalent mass of dried microalgal biomass) by at least one order of magnitude relative to photoautotrophic culture (Radmer and Parker, 1994).

The initial focus on microalgae as an energy resource is largely attributable to work performed at the Solar Energy Research Institute, the forerunner to the National Renewable Energy Laboratory (NREL). Under the Aquatic Species Program, sponsored by the U.S. Department of Energy from 1978 through 1996, the potential of microalgae species to produce bio-oils as feedstock for industrial biodiesel production was investigated (Neenan et al., 1986; Sheehan et al., 1998). The results of this program identified promising lipid-rich strains of algae and demonstrated the technical feasibility of large-scale algal culture, but production costs were judged too high to be economically viable at the time. The scope of this program was limited, however, to photoautotrophic culture of microalgae. *Chlorella protothecoides* has been maintained at moderate cell density (15 gdw l^{-1}) and lipid content (35 – 40 wt%) under heterotrophic growth conditions in large-scale cultures (Li et al., 2007a). The extracted microalgal oil has been shown to be useful as a feedstock to produce high quality biodiesel satisfying most European fuel quality standards (Wu et al., 1994; Miao and Wu, 2005; Xu et al., 2006). *Schizochytrium* has been grown industrially to densities as high as 210 gdw l^{-1} in low D.O. fed-batch culture with direct harvesting on drum dryers (Bailey et al., 2003; Barclay et al., 2005). It has also been grown in the lab in an ammonium/pH-auxostat fed-batch system to biomass densities as high as 63 gdw l^{-1} and with lipid concentrations at 25% (w/w) (Ganuza et al., 2008). Recently, several publications have highlighted the potential of recombinant engineering to improve synthetic pathways and production efficiencies in microalgae cultured as a source of bio-oils for biodiesel production (Rosenberg et al., 2008; Song et al., 2008; Meng et al., 2009). These developments point to a revolution in technology for increasing growth rates, yields, and lipid recovery that will significantly improve the economic feasibility of biodiesel derived from heterotrophic microalgae.

15.2 Growth

15.2.1 Media

A review of the literature shows that media most commonly used to support heterotrophic microalgae growth have been variations of those originally proposed by (Grant and Hommersand, 1974; Chen and Johns, 1991). The original formulations and subsequent variants have largely been defined media – i.e., comprising known amounts of specific nutrients and devoid of complex media supplements such as peptone or yeast extract. Glucose has been used almost exclusively as the carbon source, and a general inability of microalgae to digest disaccharides (in particular, lactose) should be expected. Glycine and nitrate have been most commonly used as the nitrogen source; although ammonia and urea have also been examined (caution is required with ammonia due to its toxicity at high concentrations). Media formulations have also contained a varying assortment of salts and minerals, typically including boric acid, zinc sulfate, manganese sulfate, molybdenum trioxide, copper sulfate, and cobalt nitrate.

Heterotrophic microalgae growth on the Grant and Hommersand media was advanced by a series of manuscripts investigating the capacity of *C. protothecoides* (UTEX 25) to sequester large quantities of lipids when grown on glucose in the dark (Wu et al., 1994; Wu et al., 1996; Miao and Wu, 2004; Miao and Wu, 2005; Xu et al., 2006). The growth media used in these works contained glucose at 10-20 g l^{-1} and glycine at 0.1 g l^{-1}, yielding C/N ratios well above 200 g g^{-1}. Thiamine hydrochloride was also added to these media as a precursor for several co-enzymes. The reported total lipid content achieved in batch culture was generally above 50% (w/w), with final biomass concentrations of ~3.5 gdw l^{-1}. In the case of fed-batch culture, biomass concentrations as high as 15.0 gdw l^{-1} were achieved. Growth rates on this media were relatively slow, however, with batch growth requiring ~144 hours and fed-batch growth requiring ~72 hours to attain biomass concentrations of 3.5 gdw l^{-1}. Results from our laboratory have suggested that it is not possible to achieve high biomass densities with this media unless the initial concentration of glycine is increased significantly above 0.1 g l^{-1}. By comparison, doubling times of 6 to 9 per day have been reported for thraustochytrid strains that also showed high lipid and DHA production (Barclay et al., 2005). Thraustochytrids are generally very lightly pigmented microalgae.

The early work by Chen and Johns (1991) is noteworthy for its insightful and groundbreaking investigation into the effect of C/N ratio in the growth media on lipid production. The general consensus of this seminal work was that a C/N ratio above 22 to 25 g g^{-1} is needed to stimulate lipid concentrations above 40 g gdw^{-1}, and that lipid production generally does not begin until after the depletion of nitrogen in the presence of excess glucose (it was suggested that increased production of lipids under nitrogen limitation provides a mechanism for carbon storage). It was further observed that varying the C/N ratio in the growth media appeared to affect the fatty acid profile (i.e., distribution of carbon chain lengths and degree of saturation) as well as the bulk concentration of lipids produced by the algal cells. The importance of carbon to nitrogen variation was also verified in the development of high cell density growth of *Schizochytrium* cultures wherein low dissolved oxygen concentrations were also implemented to increase production of DHA (Barclay et al., 2005).

There is also reported evidence that varying the type of sodium salt and minimizing its concentration in marine strains dramatically increased production of bio-oils such as DHA (Barclay et al., 2005). Ganuze et al. (2008) fed-batch cultured *Schizochytrium* sp. to densities as high as 63 gdw l^{-1} on defined media that contained glucose as the carbon source, ammonium sulfate as the nitrogen source, salts (sodium chloride, magnesium sulfate, potassium phosphate and potassium chloride, calcium chloride) as well as trace element and vitamin solution (Ganuza et al., 2008). The vitamin solution contained thiamine, biotin, and cyanocobalamin. The trace element solution was similar to those commonly used in fermentation systems (see, for example, Table 15.1 below).

Table 15.1 Typical media for growth of microalgae.

	Autotrophic ($g\ l^{-1}$)	**Glucose-peptone** ($g\ l^{-1}$)	**Glucose-NO_3** ($g\ l^{-1}$)
Glucose	--	10.0	10.0
$NaNO_3$	0.125	--	0.125
Bacto peptone	0.5	0.5	
KH_2PO_4	0.0875	0.7	0.7
K_2HPO_4	0.0375	0.3	0.3
$MgSO_4*7H_2O$	0.0375	0.3	0.3
$FeSO_4*7H_2O$	--	0.003	0.003
Thiamine HCl	--	$1x10^{-6}$	$1x10^{-6}$
$CaCl_2$	--	0.015	0.015
$CaCl_2*7H_2O$	0.0125	--	--
NaCl	0.0125	0.0125	0.0125
Aaron's solution	--	2 ml	2 ml

Aaron's solution (per liter): H_3BO_3, 2.9 g; $MnCl_2*4H_2O$, 0.0125 g; $ZnSO_4*7H_2O$, 0.22 g; $CuSO_4*5H_2O$, 0.08 g; MoO_3, 0.018 g.

In summary, manipulation of media components such as the type of sodium salt and nitrogen salt used, the C/N ratio in the growth media in concert with other critical culture parameters (e.g., aeration) can maximize the bulk lipid content of the microalgae cells.

15.2.2 Cell Growth on Liquid Media in Shake Flask Culture

Microalgae cultures can be maintained on agar slants containing autotrophic media (Table 15.1) under 12:12 light cycling. Culture growth is usually obtained within one to two days, after which the slants can be used to inoculate agar plates containing the same or different growth media. Under 12:12 light cycling conditions on agar plates containing autotrophic media, the cells produce chlorophyll (among other photosynthetic pigments) and form large, dense, green colonies (Fig. 15.1A). Heterotrophic growth can be achieved in the dark on media containing a suitable complex (e.g., bacto peptone) or defined (e.g., nitrate) nitrogen source (Table 15.1). In the case of *C. protothecoides*, the cells take on a distinct "bleached" yellow color (Fig. 15.1B-C; Fig. 15.2B-C). This was first documented by Grant and Hommersand (1974), who developed heterotrophic growth methodology for *C. protothecoides* in order to produce cells devoid of pigments that could potentially interfere with spectrophotometric measurement of specific metabolic byproducts. A proposed explanation is that when cells are transferred from autotrophic to heterotrophic growth conditions, chlorophyll is catabolically degraded in concert with the simultaneous excretion of red bilin derivatives from the cells (Engel et al., 1991). The observed color change results from degradation of chlorophyll into compounds that reflect yellow light or the emergence of underlying yellow pigments to the forefront in the absence of chlorophyll. These cultures take about 3 to 4 days to develop distinct colonies.

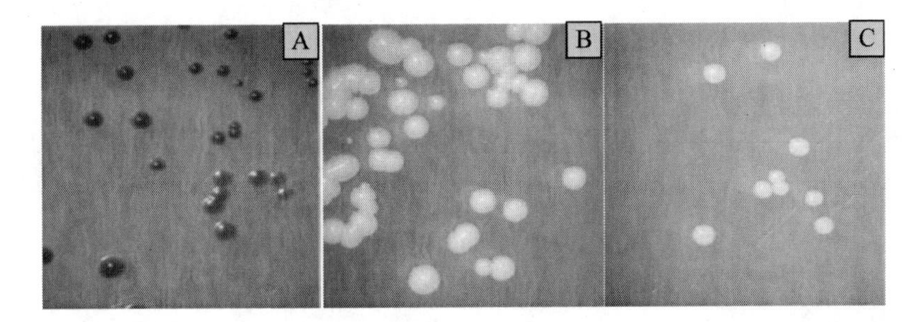

Figure 15.1 *Chlorella protothecoides* grown on (A) autotrophic agar under 12:12 light cycling, (B) glucose-peptone agar in the dark, and (C) glucose-NO_3 agar in the dark. All cultures were inoculated from a *C. protothecoides* culture grown on a photoautotrophic agar slant under 12:12 light cycling.

Colonies grown heterotrophically on glucose-NO_3 agar are visibly less intense in color, smaller, and characterized by much slower growth relative to colonies grown on glucose-peptone agar (Fig. 15.1B-C), suggesting that the nitrate reductase enzyme is either absent or less active in cells grown in the absence of light. Additional evidence is provided by observations from shake flask culture experiments using liquid growth media. A flask containing autotrophic media can be inoculated from a colony grown under autotrophic conditions (i.e., 12:12 light cycling, autotrophic agar). If grown under 12:12 light cycling, the flask culture will develop into a rich, green color characteristic of photoautotrophic growth (Fig. 15.2A). A flask containing glucose-peptone media and glucose-NO_3 media can be inoculated from a colony grown under heterotrophic growth conditions (i.e., in the dark) on agar containing glucose-peptone media. If grown under dark conditions, both flasks will possess the bleached color characteristic of heterotrophic growth of *C. protothecoides* (Fig. 15.2B-C). If a colony grown glucose-NO_3 agar in the dark is transferred into a flask containing glucose-NO_3 media, no growth will be observed (Fig. 15.2D). Such an absence of growth suggests that either *C. protothecoides* cannot produce the nitrate reductase enzyme under dark conditions or that its production is initiated and/or inhibited by chemicals yet to be identified. The growth observed in glucose-NO_3 media is likely due to the residual presence of nitrate reductase produced by the cells during their original growth on glucose-peptone agar. These results demonstrate that the method of preparing the inoculums as well as the nitrogen source used in the growth media are both critical factors that must be considered in the development of heterotrophic growth media for microalgae.

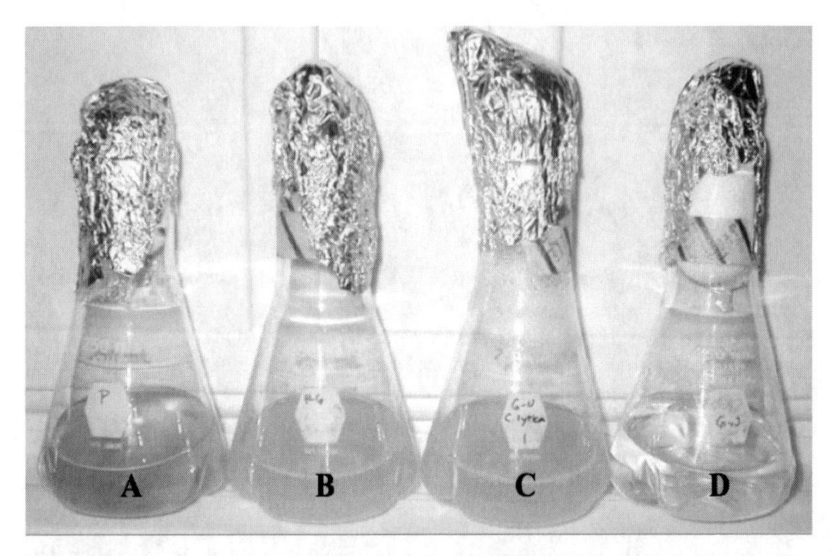

Figure 15.2 *Chlorella protothecoides* cultures grown on (A) autotrophic media under 12:12 light cycling, (B) glucose-peptone media in the dark, (C) glucose-NO_3 media in the dark, and (D) glucose-NO_3 media in the dark. Culture A was inoculated from a autotrophic agar colony grown under 12:12 light cycling. Cultures B and C were inoculated from a glucose-peptone agar colony grown in the dark. Culture D was inoculated from a glucose-NO_3 colony grown in the

15.2.3 Cell Growth on Liquid Media in Fed-batch Culture

Fed-batch culture technology using optimized feed media has been applied to improve growth rates, final biomass concentrations, and bio-oil content of heterotrophically grown microorganisms (Johnston and Cooney, 2002a; Johnston and Cooney, 2002b; Shi et al., 2002; Johnston et al., 2003; Whiffen et al., 2004; Ankerstjerne et al., 2005; Xu et al., 2006; Li et al., 2007b). Fed-batch culture is essentially a compromise between batch culture growth, in which culture components are neither removed nor added and process parameters such as temperature and pH are not tightly controlled, and continuous culture, in which feed and product are continuously added and removed at the same volumetric rate and critical process parameters are maintained at predetermined, optimum set-points (Johnston et al., 2003; Whiffen et al., 2004). In fed-batch culture, concentrated feed media (usually containing a single growth-limiting substrate -- e.g., carbon in the form of glucose or nitrogen in the form of ammonium chloride) is added at a defined rate over the culture period, but cell biomass is not harvested until the culture period has been completed. Fed-batch culture technology using optimized feed media has been applied to improve growth rates, final biomass concentrations, and bio-oil content of heterotrophically grown microorganisms (Johnston and Cooney, 2002a; Johnston et al., 2003; Whiffen et al., 2004; Ankerstjerne et al., 2005; Barclay et al., 2005; Xu et al., 2006; Li et al., 2007a; Li et al., 2007b; Ganuza et al., 2008). As with continuous culture, critical process parameters (such as temperature and pH) are maintained at predetermined, optimum set-points (Johnston et al., 2003; Whiffen et al., 2004; Li et al., 2007a; Ganuza et al., 2008).

The primary objective of fed-batch culture is the is to maximize biomass production by feeding cells large cumulative amounts of growth substrate without inoculating them into media containing concentrations of nutrients high enough to be toxic or otherwise inhibit cell growth -- e.g., some strains of *E. coli* will consume and metabolize glucose at rates beyond their ability to oxidize glucose using oxygen as the terminal electron acceptor (Whiffen et al., 2004), whereupon residual glucose is broken down via anaerobic metabolic pathways that produce growth-inhibiting metabolites, such as acetic acid or alcohols (Johnston et al., 2003). In fed-batch culture, a seed culture is typically inoculated into an initial batch media containing a moderate amount of growth-limiting substrate (e.g., glucose) capable of initiating culture growth, plus other nutrients (e.g., nitrogen, salts, minerals, vitamins, etc.) in amounts sufficient for sustaining growth to a high final biomass concentration (Ganuza et al., 2008). After the initial growth phase is complete, additional aliquots of concentrated feed media containing the growth-limiting substrate are delivered to the culture at predefined rates to maintain controlled growth over an extended period. This makes it possible to achieve significantly higher cell densities relative to batch culture or continuous culture methods.

Other advantages of fed-batch culture relative to continuous culture include a lower likelihood of contamination due to the absence of an outgoing product stream potentially open to the external environment and the ability to impose extreme nutrient conditions (e.g., high C/N ratios). Additionally, fed-batch culture is a semi-batch process, which allows for the efficient recovery and processing of large volumes of concentrated biomass in a single harvest. In contrast, continuous culture produces a product stream that is relatively low in volume and biomass concentration. The main disadvantage of fed-batch culture relative to continuous culture is that the growth rate can not be as precisely controlled. With reliable models and culture growth data, however, the delivery rate of concentrated media to fed-batch cultures can be adjusted dynamically to yield reasonably well-constrained growth rates.

Fed-batch culture presents a mechanism for producing high concentrations of biomass with high percentages of internal lipids by splitting the growth period into three distinct phases: an initial phase, during which the growth of the culture is established, an intermediate phase of controlled growth, during which the biomass concentration is steadily increased, and a final phase, in which the cells accumulate lipids. During the initial and intermediate phases, the C/N ratio of the feed media is typically kept below 20 g g^{-1} in order to support balanced growth, while it is increased above 200 g g^{-1} during the final phase to stimulate lipid accumulation. This approach is illustrated by results from fed-batch cultures of *C. protothecoides* performed in our laboratory in a 1.2-liter bioreactor using glucose as the carbon source and bactopeptone as the nitrogen source (Fig. 15.3A-C). Growth is typically slow, and residual glucose concentration is kept between 6 and 10 g l^{-1} through metered addition of concentrated glucose feed. The culture growth begins to level off sometime around 200 - 300 hr, (Fig. 15.3D), presumably due to depletion of the bacto peptone contained in the original media. Over the next 100+ hr, the culture enters the lipid accumulation phase wherein the production and storage of internal lipids by the cells is favored over cell division (Fig. 15.3D-E). Final cell lipid contents approaching 50% (w/w) were typically achieved with this strain of *C. protothecoides* (UTEX 25). While biomass concentrations were measured optically as opposed to gravimetrically (i.e., in terms of gdw l^{-1}) in our experiments, other groups performing fed-batch cultures using the same *Chlorella* strain and media have reported final biomass concentrations above 15 gdw l^{-1} (Xu et al., 2006).

Other examples of fed-batch culture include production of docosahexaenoic acid (DHA), a polyunsaturated fatty acid with food and pharmaceutical applications, by cultivation of the heterotrophic marine alga, *Crypthecodinium cohnii,* on complex media containing sea salt, yeast extract, and glucose. In a study investigating fed-batch cultivation as an alternative fermentation strategy for DHA production, glucose and acetic acid were compared as carbon sources (de Swaaf et al., 2003). For both substrates, the feed rate was adapted to the maximum specific consumption rate of *C. cohnii.* In cultures grown on glucose, this was achieved by maintaining glucose concentrations between 5 and 20 g l^{-1} throughout the fermentation period. In cultures grown on acetic acid, the feed rate was adjusted automatically in order to maintain the culture pH at 6.95. A feed containing 50% (w/w) acetic acid resulted in higher overall volumetric productivity of DHA (38 mg l^{-1} h^{-1}) than a feed containing 50% (w/v) glucose (14 mg l^{-1} h^{-1}). Using a feed consisting of pure acetic acid increased the productivity of DHA to 48 mg l^{-1} h^{-1} and resulted in final biomass, lipid, and DHA concentrations of 109 gdw l^{-1}, 61 g l^{-1}, and 19 g l^{-1}, respectively. Ganuza et al. (2008) reported fed-batch cultivation of *Schizochytrium* sp. to a cell density of 63 gdw l^{-1}. Concentrations above 200 gdw l^{-1} have been reported for *Schizochytrium* when grown in low D.O. fed-batch culture using drum dryers as the harvesting technique (Bailey et al., 2003). High cell density requires vigorous mixing to sustain desired dissolved oxygen levels. Vigorous mixing introduces large shear forces which becomes particularly troublesome if the culture releases extracellular bio-oils or polysaccharides. In the case of bio-oils serious emulsions can emerge. In the case of polysaccharides, the viscosity of the culture can increase to a point where mixing is no longer viable. For large-scale industrial production of lipids such as DHA, it may be possible to overcome this problem by adding a commercial polysaccharide-hydrolase preparation to the culture in order to reduce its viscosity (Barclay et al., 2005).

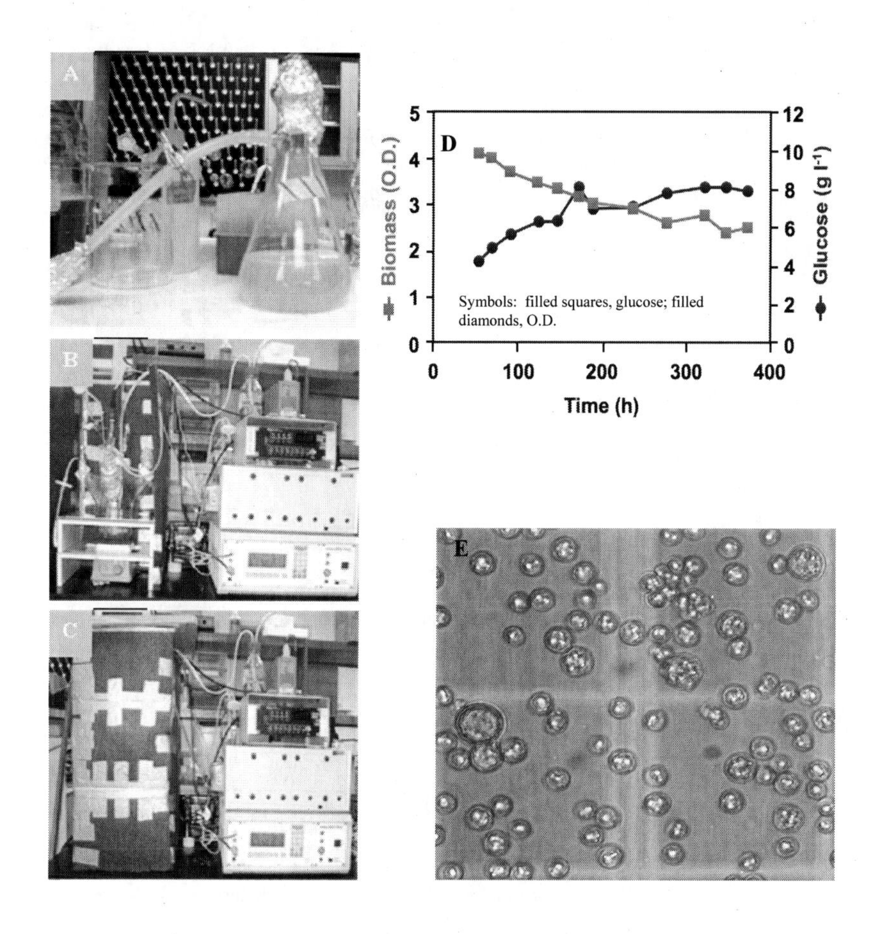

Figure 15.3 Fed-batch culture of the microalgae *C. protothecoides* using concentrated glucose feed media. The initial seed inoculum (A) is prepared for sterile addition to the culture vessel (B), which is then isolated from the light (C). Growth proceeds (D) until the cells have reached a sufficient cell density and lipid accumulation (E).

15.2.4 Summary

Although the production of biodiesel from heterotrophic microalgal bio-oils is still in an early stage of development, commercial production of lipids from heterotrophic microalgae for other industrial applications is well established (Bailey et al., 2003; Barclay et al., 2005). But as much of the data related to commercial applications are proprietary, few descriptions of large-scale production of lipid-rich heterotrophic microalgae have been reported in the open literature (Li et al., 2007a; Ganuza et al., 2008). Available data show that it has been generally feasible to cultivate heterotrophic algal cells having lipid compositions approaching 50% (w/w), with final biomass concentrations up to 14 gdw l^{-1} for *Chlorella* (Li et al., 2007a) and as high as 200 for *Schizochytrium* (Bailey et al., 2003; Barclay et al., 2005; Ganuza et al., 2008).

15.3 Product Recovery

Product recovery is overlooked when it comes to lipid production from microalgae, whether it be autotrophically or heterotrophically grown. Extraction is only one, although important, step in the process. Overall, product recovery encompasses several unit operations that include harvesting the biomass, dewatering the biomass, extraction the lipid product, concentration the lipid, and final conversion to the desire product (e.g., in this case biodiesel) (Ratledge et al., 2005). In some cases, the extraction step can be merged with some of the other unit operations. One example would be the direct transesterification reaction wherein the conversion of the lipids to fatty acid methyl esters is coupled to the extraction step (Lepage and Roy, 1984). When these sort of coupling of unit operations are executed, then the integration of the remaining down and upstream unit operations must be redesigned. In general, however, the principle design issues lie with the extraction step and the upstream harvesting unit operations are generally consistent. As heterotrophic cultures are assumed to achieve relatively high cell densities, e.g., above 15 gdw l^{-1}, the culture will likely be directly drained or pumped from the reactor at the completion of the growth phase. The dewatering is likely to occur in two phases with an initial membrane step comprised of tangential cross flow filtration followed by centrifugation to achieve solid wet paste of cells. Membrane microfiltration and ultrafiltration are possible alternatives to conventional filtration for recovering algal biomass. Microfiltration is suitable for fragile cells (Petrusevski et al., 1995), but large-scale processes for producing algal biomass do not generally use membrane filtration (Molina Grima et al., 2003). Membrane replacement and pumping are the major cost contributors to membrane filtration processes. Microfiltration can be more cost-effective than centrifugation if only small volumes (e.g., < 2 m^3 day^{-1}) are to be filtered. For larger scale of production (e.g., > 20 m^3 day^{-1}), centrifugation may be a more economic method of recovering the biomass (MacKay and Salusbury, 1988). Guidelines provided by Chisti and Moo-Young (1991) for selection and use of centrifuges are especially relevant to recovery of microalgal biomass. Centrifugal recovery can be rapid, but it is energy intensive. Nevertheless, centrifugation is a preferred method of recovering algal cells (Benemann et al., 1980; Richmond and Becker, 1986). Thereafter the biomass can be oven or vacuum dried and then stored until the extraction step is applied. Whether, or not, the dried biomass is subjected to a follow-on disruption technique (i.e. freeze-grinding, etc.) depends upon the requirements of the extraction technique.

15.3.1 Solvent Extraction

In the event a high lipid accumulating strain is obtained with rapid growth rates on a suitable carbon source, the target compound for biodiesel production – triglycerides – are accumulated intracellularly and must therefore be extracted if they are to be recovered. By the very nature of heterotrophic fermentation, the cells are immersed in aqueous media which will require that the extraction step account for or eliminate the presence of water. Since triglycerides are hydrophobic they are immiscible with water and difficult to recover in its presence. The concept that lipid rich cells can be ruptured by a simple mechanical means (such as shear stress applied by a pump or by quick pressure release) and that the "freed" lipids or bio-oils will naturally float to the surface where they will partition into a separate phase that can easily be recovered, is not realized in practice. Not only due the "freed" lipids fail to rapidly partition into their own top laying immiscible phase, problematic emulsions are usually formed. With respect to the former, the internal triglycerides quickly form small micelles in the presence of water that are not individually of sufficient density to float to the surface in any appreciable rate. With respect the latter, the intrusion of mechanical energy applied through shear forces generally creates emulsions of water, protein, and lipids. Consequently, the extraction of bio-oil from biomass requires the dissolution of the free and membrane bound lipids into a liquid solvent (system) where they can be separated and recovered. This requires that the bio-oil be partitioned from either a solid or aqueous phase within the cells to the solvent liquid phase outside the cells. The driving force behind the solubility of the lipid into the solvent phase is the introduction of attractive hydrophobic, electrostatic, or hydrogen bonds between the solvent and lipid molecules (Kates, 1988). As the chemical properties of the more polar membrane associated lipids differs from the more non-polar free lipids, in terms of their polarity and hydrophobicity, it is necessary to address this distinction when choosing extraction solvents.

Neutral triglycerides are hydrophobic molecules that may be extracted with similarly non-polar solvents, such as chloroform, ether, or benzene owing to the favorable hydrophobic interaction. By contrast, the more polar lipids, such as membrane associated free fatty acids, phospholipids, or glycolipids, require more polar solvents owing to the increased need of hydrogen bond interactions between the polar molecules. In both cases, the accumulated grouping of molecules of like polarity or hydrophobicity is an entropy driven process and consequently thermodynamically favorable (Mead et al., 1986). To address this issue (i.e., to improve the ability to extract both polar and non-polar lipids in a single solvent extraction step), the use of "miscible" organic co-solvent mixtures was proposed, comprised of a polar covalent molecule such as methanol and a more hydrophobic organic molecule like chloroform (Folch et al., 1956). Most commonly referred to as the Bligh and Dyer technique, these co-solvent mixtures were proposed because they possessed a hybrid property of both solvents, in terms of polarity and/or hydrophobic interaction and hydrogen bonding, to dissolve both the polar and non-polar lipids (Bligh and Dyer, 1959). There have also been periodic adaptations of this procedure to increase the range and total of lipids extracted. For example, in 1979 Dubinsky and Aaronson proposed the addition of HCl to the usual chloroform-methanol extraction mixture to significantly increase the yield of phospholipid fraction relative to the glycolipids and neutral lipids (Dubinsky and Aaronson, 1979). More recently, Cravotto et al. (2008) have suggested that the application of ultrasound and/or microwave can significantly enhance yields and reduce times of extraction in organic solvents. Cravotto et al. (2008) also suggested that the microwave assisted extraction technique resulted reasonable yields in the presence of the single nonpolar solvent hexane. This result is noteworthy given that in our work on extraction of bio-oils from microalgae we achieved poor yields of bio-oils and FAME product in excess hexane solvent. Table 15.2 shows representative data on the direct transesterification of microalgae biomass using stoichiometric amounts of methanol in excess hexane solvent. We found consistently low yields in part due to the immiscibility of methanol and loss of acid catalyst.

Co-solvent systems serve as both an initial extraction and subsequent partitioning medium. Initially, the extraction solvent is comprised of chloroform, methanol and water at relative ratios (1/2/0.8 (v/v %)) that yield a single solvent into which the solvents are extracted and dissolved. After the lipids have been extracted and dissolved into the single phase co-solvent, additional chloroform and water are added to create a new volume ratio (2/2/1.8 (v/v %)) that leads to a two-phase system wherein the methanol and water are partitioned to a polar phase while the lipids are partitioned to the more hydrophobic chloroform phase. Although widely referenced, the Bligh and Dyer method only works well for lipid contents smaller than 2% (Iverson et al., 2001). Although largely unnoticed, this is result of the fact that the carrying capacity of the extraction mixture is not sufficient to extract all the lipids, a fact that often leads to lower yields and a generally unrecognized limitation of the Bligh and Dyer extraction process to work with samples possessing high lipid content. This highlights one potential limitation to any solvent based system applied to the extraction of lipids for the production of biodiesel: the carrying capacity of many organic solvents may not be sufficient to extract the entire lot of lipid mass. One solution to this is to use counter current solvent extraction systems, but these may not be practical to apply due to the presence of biomass matter and the fact that such extractions process are inherently a batch process.

Table 15.2 Direct transesterification of *Nannochloropsis* microalgae in the presence of excess hexane solvent.

Experiment	1	2	3	4	5	6
Biomass (mg)	112	101	102	108	114	104
MeOH (ml)	0.04	0.005	0.005	0.005	3X 0.005	0.005
MeOH/lipids ratio (mole $mole^{-1}$)	50:1	6:1	6:1	6:1	18:1	6:1
HCl in solvent[3] (%, g g^{-1})	0.03	3	3	3	3[1]	3[2]
Reaction time	1h	1h	2h	3h	3h	2h
FAMES in biomass (g g^{-1})		0.7	1.3	1.5	3.3	0.3

[1]Every hour a new aliquot of new aliquot of catalyst was added. [2]H_2SO_4 was added instead of HCl. [3]Acetyl Chloride was initially added which reacts with the methanol to form HCl.

A more practical solution to this issue has been the proposed use of direct transesterification of the lipids contained within the dried biomass. Direct transesterification is the application of the transesterification reaction directly to dried biomass for the purpose of producing fatty acids in a one step reaction (Lepage and Roy, 1984). With this approach, the transesterification reagent (i.e., the alcohol and acid catalyst) is added directly to the dried biomass which is then stirred for 1 hour at an elevated temperature (typically between 60 to 100°C) under reflux (or sealed cap) until the reaction is complete. Afterwards the reaction product is extracted with a nonpolar solvent such as hexane or a hexane/water mixture (1/1 v/v %). Compared to the process of first extracting and then transesterifying the extracted bio-oil, direct transesterification has the advantage of minimizing the number of reaction steps and reaction time. Reducing the number of extraction steps also helps to improve the yield of bio-oil recovered. For example, Lepage and Roy (1984) reported that the direct transesterification reaction improved yields, relative to the method of Folch et al. (1956), by up to 15.8% (w/w %). A further modification of the direct transesterification approach was proposed by Rodriguez-Ruiz et al. (1998) With respect to the original work of Lepage and Roy (1984), Rodriguez-Ruiz et al. (1998) noted that their method combined both the production and extraction of FAME product into a single step without loss of yield. In fact, the authors proposed that the simultaneous transesterification and extraction even led to higher yields of recovered FAMES in dramatically shorter reaction times (i.e., 1 h vs. 10 min).

More recently, extraction techniques using single solvents at elevated temperatures and pressures have been proposed. One of these is accelerated solvent extraction (ASE), first proposed in the mid 1990's by Richter et al. (1996). Accelerated solvent extraction uses organic solvents at high pressure and temperatures above the boiling point (Richter et al., 1996). In general, the solvents used are similar to those used in standard liquid extraction techniques for Soxhlet or sonication. In general a solid sample is enclosed in a sample cartridge that is filled with an extraction fluid and used to statically extract the sample under elevated temperature (50 – 200°C) and pressure (500 – 3000 psi) conditions for short time periods (5 – 10 min). Compressed gas is used to purge the sample extract from the cell into a collection vessel. ASE is applicable to solid and semi-solid samples that can be retained in the cell during the extraction phase (using a solvent front pumped through the sample chamber at the appropriate temperature and pressure). It has been proposed for the extraction of liquid extracts (Richter et al., 1996), carotenoids (Denery et al., 2004), and lipids from microalgae (Schafer, 1998). In addition to improving yields and dramatically reducing extraction time, ASE can also be applied to remove co-extractable material from various biomass feedstocks, to selectively extract polar compounds from lipid rich samples, and to fractionate lipids from biological samples. Various absorbents (e.g., Al_2O_3 activated by placing in a drying oven at 350°C for 15 hours) can also be added to the extraction cell in order to improve the purity of the final sample (Dionex, 2007).

In most cases ASE can be an efficient technique assuming the extracting solvent, sample-solvent ratio, extraction temperature and time have been optimized. Denery et al. (2004), for example, examined these factors to optimize the extraction of carotenoids from *Dunaliella salina* and showed that equivalent extraction efficiencies (compared to traditional solvent technology) could be achieved with the use of less solvent and shorter extraction times (Denery et al., 2004). What remains unclear is the effectiveness of such an approach at large-scale in terms of how to handle large amounts of biomass as well as the energy cost. The latter is also noteworthy in the context that accelerated solvent extraction by definition uses non aqueous solvents and therefore must use dried biomass, a step that requires dewatering of the sample at high cost in terms of energy input.

Extraction steps that nullify the need for dewatering the biomass are obviously attractive in terms of their ability to reduce the energy load required to dewater the biomass. Even in high-density cell culture, the water content (on a weight basis) can be as high as 80% after centrifugation. The various unit operations that can reduce the water content to as low as 10% (w/w) (e.g., spray drying, belt drying, fluidized bed, evaporators) can be quite expensive energetically. The more economical approach of solar drying (usually using windrow drying) is unrealistic in terms of land use, pest control, and time. For these reasons, several aqueous based extraction systems have been proposed including those that mix aqueous and organic phases, as well subcritical water extraction.

Hejazi et al. (2002) proposed the two-phase system of aqueous and organic phases for the selective extraction of carotenoids from the microalgae *Dunaliella salina*. Their observation was that solvents with lower hydrophobicity reach critical concentrations more easily, and in the process broke down the cell membrane. By using solvents of higher hydrophobicity the effect of the solvent on the membrane could be decreased and the extraction efficiency for both chlorophyll and βcarotene decreased, as well. By applying a measurement of solvent hydrophobicity based on the partition coefficient of the solvent in a two-phase system of octanol and water, screening viability and activity tests of *Dunaliella salina* in the presence of different organic phases indicated that cells remained viable and active in the presence of organic solvents with a $\log P_{octanol} > 6$ and that βcarotene can be extracted more easily than chlorophyll by biocompatible solvents. This work serves as the basis for the development of follow-on technology that proposes to use solvents such as decane and dodecane in the presence of live microalgae cells that have been concentrated for the extraction of triglycerides without loss of cell viability and extraction of membrane bound free fatty acids. Conceptually, the cells can be returned to their original bioreactor for continued growth and production of triglycerides for biofuels production. Otherwise termed "cell milking", this technique has gained some attention in terms of patent and small scale-pilot applications by private companies. Long term testing of cell viability in the context of continual production, however remains to be examined. If successful, however, this method does offer the possibility of selectively extracting lipids suitable for biofuels and excluding the extraction of lipids that can not be transesterified and pigments (such as chlorophyll) that can be difficult to separate from the desired lipids and create a very viscous and tarry final product.

Subcritical water extraction is based on the use of water, at temperatures just below the critical temperature, and pressure high enough to keep the liquid state (Ayala and Castro, 2001). The technique, originally termed pressurized hot water extraction, was initially applied to whole biomass hemicellulose and a pretreatment prior to its use as a fermentation substrate (Mok and Antal Jr., 1992). More recently, however, it has been applied for the selective extraction of essential oils from plant matter (Eikani et al., 2007), the extraction of functional ingredients from microalgae (Herrero et al., 2006), and saponins from oil-seeds (Guclu-Ustundag et al., 2007). The basic premise to subcritical water extraction is that water, under these conditions, becomes less polar and organic compounds are more soluble than at room temperature. There is also the added benefit of solvent access into the biomass matrix that occurs at the higher temperatures. In principle, as the water is cooled down to room temperature, products miscible at the high temperature and pressure become immiscible at lower temperatures and are easily separated. Some of the more important advantages described for subcritical water extraction are shorter extract times, higher quality of extract, lower costs of the extracting agent, and environmental compatibility (Herrero et al., 2006). With respect to microalgae, however, whether grown phototrophically or heterotrophically, one of the more attractive aspects to subcritical water extraction is the use of water as the solvent which eliminates the need for the dewatering step. A major constraint, however, as with accelerated solvent extraction, is difficulty designing a system at large scale and the high energy load required to heat the system up to subcritical temperatures. Large scale design will require a significant cooling system to cool the product down to room temperature to avoid product degradation as well.

All solvent systems, regardless of the operating conditions, also suffer from a singular process limitation: they all require fractionation of the final product. Whether organic solvents are used at room temperature or under accelerated conditions, or whether water is used at subcritical temperatures, any compound with solubility similar to lipids (i.e., free fatty acids or triglycerides) will also be extracted. This added extraction of co-products enhances the requirements of downstream fractionation and the promise of "more selective extraction capacity" is a major driving force behind the more advanced solvent techniques such as accelerated solvent, subcritical water, and supercritical fluid extraction. Regardless of there promise, however, the reality is that any product with a miscibility or solubility similar to the target lipid will also extract. This highlights a main advantage of heterotrophic growth of microalgae over phototrophic production: many of the pigments and non transesterifiable lipids present in phototrophically grown microalgae are absent from heterotrophically grown cultures. It is also a reason why many researchers have favored direct transesterification over the process of extraction, purification, and then transesterification; the resulting FAME can be more selectively extracted using organic solvents that are selective against non transesterifiable lipids.

15.3.2 Supercritical Fluid Extraction

Although supercritical fluid extraction is technically a solvent extraction technique, it has been separated from the discussion on solvent extraction above because supercritical fluids are a unique type of solvent. Supercritical fluids are those heated above their critical point and possess "gas-like" mass transfer properties and "liquid-like" solvating properties with diffusion coefficients greater than those of a liquid (Luque de Castro et al., 1999). Supercritical fluid extraction is a relatively recent extraction process that exploits the enhanced solvating power of supercritical fluids (Luque de Castro et al., 1994). The majority of applications have used carbon dioxide because of its preferred critical properties (i.e. moderate critical temperature of 31.1℃ and pressure of 72.9 atmosphere), low toxicity, and chemical inertness, but other fluids used have included ethane, water, methanol, ethane, nitrous oxide, sulfur hexafluoride as well as n-butane and pentane (Herrero et al., 2006). The process requires a dry sample placed within a cell that can be filled with the gas before it is pressurized above its critical point. The temperature and pressure above the critical point can be adjusted as can the time of the extraction. Although often employed in batch mode, the process can also be operated continuously. One of the more attractive points to supercritical fluid extraction is that after the extraction reaction has been completed, and the extracted material dissolved into the supercritical fluid, the solvent and product can be easily separated downstream once the temperature and pressure are lowed to atmospheric conditions. Specifically, the fluid returns to its original gaseous state while the extracted product remains as a liquid or solid.

Supercritical fluid extraction has been applied for the extraction of essential oils from plants, as well as functional ingredients and lipids from microalgae (Herrero et al., 2006). Lipids have also been selectively extracted from macroalgae at temperatures between 40 to 50℃ and pressures of 241 to 379 bar (Chueng, 1999). Despite the range of products that can be extracted from microalgae using supercritical fluid extraction, its application to the extraction of lipids from microalgae (for the production of biofuels) is limited by both the high energy costs and difficulties with scale-up.

15.3.3 Thermochemical Processes

Biofuels can also be produced from microalgae using thermochemical methods, including direct combustion, heat decomposition, gasification, and thermochemical liquefaction (Yang et al., 2004). Also, pyrolysis is a related process that is generally broken into (1) direct thermal cracking and (2) a combination of thermal and catalytic cracking (Maher and Bressler, 2007). In these methods the general aim is to obtain low molecular weight liquid fuels from organic high molecular weight compounds by conversion of the biomass in water at high temperatures and high pressure in the presence of alkali catalysts without reducing gas (i.e., usually N_2 is required to remove the trace of oxygen). As with subcritical water extraction, an advantage to this approach is a relaxation on the requirement of fully dewatering the biomass. However, disadvantages, in terms of biofuel production, include the energy costs associated with the heating the water and the fact that reaction products are highly dependent upon the catalyst type and reaction products and can range from diesel like to gasoline like fractions. Despite these disadvantages, applications to algae have been pursued and many applications can be found in the literature. Yang et al. (2004), for example, examined the conversion of microalgae biomass into oil in the presence of water at high temperature (300 – 340°C) and pressure (about 20 MPa) with and without alkali catalyst. Their results produced a heavy oil largely comprised of C17-C18 hydrocarbons. Although the authors did reference a different study that suggested the liquefaction of sludge was net energy producer, they did not provide a quantitative energy balance of their process. This highlights the general limitation that these processes generally require more energy than they produce. The exception is usually when the reaction is carried out in the process of catalysts although this generally leads to a wide range of uncontrolled and unwanted final by-products. In general thermochemical techniques lead to a final heavy oil product that is generally a mix of many saturated compounds and of little use except for combustive heating. On a comparison basis, one can argue that a better use is to dry and burn the biomass through gasification. Gasification of biomass, however, is a process that is highly sensitive to the water content, composition, and texture of the feedstock which requires strict controls in order to avoid unwanted gaseous by-products.

15.3.4 Summary

The most common extraction technology applied is liquid phase solvent extraction applied to dried biomass. Assuming the cells are properly disrupted or the appropriate solvents are added (or a mixture of the two), the polar and neutral lipids can be expected to be extracted into the biomass. Thereafter, the solvent can be removed under vacuum distillation, leaving behind the extracted product. In the case of phototrophically grown microalgae the fat soluble pigments (i.e. those present in the cell, such as chlorophyll, to support light driven growth) will also carry over into the extracting solvent. In the case of heterotrophic growth, with the microalgae grown on defined or semi-defined media, this product can be relatively pure. From this perspective solvent extraction is generally the most practical approach although it can be energy intensive. Other solvent extraction techniques, such as accelerated solvent, subcritical water, or supercritical fluid extraction can be quite effective at lab scale but difficult to apply at large scale, in part because their energy requirements and in part because the biomass must be captured in tube like reactors that are held under high pressure and temperature. That said supercritical fluid extraction is perhaps the best technical approach. It uses an inert solvent and the lipid is easily recovered without an extraction step because the CO_2 will vaporize as the supercritical fluid (with lipid product) exits the reactor at atmospheric pressure. The only problems are the cost of operating supercritical reactors at high pressures and temperatures which can defeat the energy yields. Rendering is not particularly suited for microalgal cells unless the cell line used has no membrane and is therefore quite fragile. This, however, will generally not be the case for high cell density culture (which occurs for sugar based heterotrophic growth) because growth in fermenters requires the kind of mixing that induces shear stress.

15.4 Cost and Scale-up

A historical review of the scaled up production of Docosahexaeonic acid using *Schizochytrium* is well described elsewhere (Barclay et al., 2005). The process begins with a bio-rational approach to technology development. This requires identificaiton of the most desireble characteristics that an ideal prodcution strain should possess in terms of both the target production system and the target product. This is found through the implementation of an effective collection/isolation/screening program (Barclay, 1992). This generally inolves running target microalge samples through sandwhich type filters to eliminate large cells (~ > 25 um), and the plating of the collected cells on appropriate screening agar followed by analytical analysis of lipid profiles. In addition, the presence or absence of toxins should be pursued through sensitive methods such as standard high-performance liquid chromotography or sensistive bioassays. Although seemingly unimportant, these initial screens can identify slow growing strains or those that produce growth inhibiting toxins – two factors that can dramatically undermine scale up.

Laboratory experiments using a small bioreactors can be performed to develop cultivation methods for specific species of heterotrophic algae and optimize critical culture parameters (e.g., growth media compostion, culture period, mixing rate, temperature, pH, etc.). Fermentation scale up generally involves investigation into media components and their concentrations (i.e., altering forms of major nutrients such as glucose or nitrogen, adding or eliminating salts) to initially raise the cell concentration and lipid composition. Scale up follows with a focus on the processing technique. This generally involves the manipulation of the carbon and nitrogen feeds in fed-batch culture. Often a split approach wherein nitrogen starvation or limitation is imposed after high cell concentrations have been achieved is a good way to initiate the production of internal lipid stores. Thereafter investigation of processing variable such as dissolved oxygen or pH can also be important. In the case of *Schizochytrium*, for example, low dissolved oxygen concentrations can be used to improve DHA production (Barclay et al., 2005). The final step required is the development of a suitable biomass harvesting and dewatering technique. For example, the appliation of drum dryers has been applied to avoid the use of centrifugation (Barclay et al., 2005). Additional design issues to include the size and type of fermenters and other process components (i.e., equipment for sterilization, inoculum collection and storage, harvesting, dewatering, oil extraction/conversion, etc.), system instrumentation and controls, mass flows, heating/cooling loads, and power requirements.

Another often overlooked but nonetheless critical aspect of scale up is the production of preliminary process and instrumentation drawings (P&IDs), equipment lists, and mass and energy balances. These materials, along with data from the laboratory-scale fermentation experiments, provide the kind of information required to obtain quotes from equipment vendors, engineering firms, and/or design-build contractors for construction of the full-scale facility. At this stage it is then possible to estimate the capital expenditure and energy demand for the facility and construct a full life cycle cost/energy analysis for the production of algal-derived bio-oils (including other operating expenses, such as sourcing of water, nutrients for growth media, and other raw materials, labor, maintenance, waste disposal, etc.) and/or the entire process of producing biodiesel from heterotrophic algal biomass (additionally accounting for the conversion of bio-oils to biodiesel, purification of the raw biodiesel product, and distribution and use of the final biodiesel fuel and co-products). Finally, this data must be used to compare the cost of producing algal bio-oils relative to the price of conventional feedstocks (e.g., Chicago Board of Trade futures (2004-2009) for soybean oil fluctuated over a range of 0.1900-0.7418 $USD/lb).

15.5 Integrated Biorefinery Approach

The term "integrated biorefinery" refers to an industrial process in which several renewable feedstocks and conversion technologies are utilized to produce a variety of products. In our discussion the primary product is a liquid transportation fuel while additional co-products may include other fuels, chemicals, materials, heat, and/or power. The diversification of feedstocks and products and the integration of multiple processes associated with this approach can provide significant economic and sustainability advantages, including more complete utilization of raw materials, greater insulation from market volatility, shared infrastructure, waste reduction, and energy efficiency. The concept is similar to a conventional oil refinery except that biomass-derived materials are used as feedstocks instead of petroleum-based materials.

In the case of algal-derived biodiesel, an integrated biorefinery approach is particularly compelling due to the relatively low value of the finished fuel product – i.e., the profitability of the overall process to produce algal biodiesel could be greatly enhanced by simultaneously producing other products (especially high-priced specialty products for pharmaceutical and/or nutraceutical applications) from the algal biomass. As a hypothetical example, lipids could be extracted from harvested algal cells and separated/refined to produce an omega-3 fatty acid fraction for use in nutrient supplements and a bulk oil fraction for use as biodiesel feedstock. The algal cells could then be further processed to extract other useful compounds (e.g., proteins, pigments, etc.). Finally, the residual biomass could be burned or converted to digester gas for generating steam and/or electricity to sustain facility operations. The operational details for an actual biorefinery will depend on the strains of algae used as feedstock and their physical/chemical properties (e.g., lipid content, fatty acid profile, etc.), and each process step will need to be carefully evaluated in terms of the added value of its co-products relative to the additional capital and operating expenses. In spite of the additional complexity involved, incorporating algal biodiesel production as a component of an integrated biorefinery may provide the best opportunity for making the process economically feasible on an industrial scale.

15.6 Future Research Directions

Through the 1990s and 2000s, tremendous advances have been made in the number of molecular tools available to both study a variety of microalgae and to improve their productivity. These include techniques such as directed evolution, high throughput screening, metabolic engineering, and more recently synthetic biology. Individually, each of these methods and techniques provides a general class of tools to address the unique demands of strain development and optimization. Used together, they can greatly enhance the potential for commercial microbial bioproduction of valuable products. Below, the four methodologies are briefly discussed.

15.6.1 Directed Evolution

Directed evolution is a broad class of proprietary and public domain methods that can be used to optimize biological functions. From single proteins to single metabolic pathways to whole cell functions involving interrelated pathways, the directed evolution process is considered an efficient method to engineer an organism to perform a desired function. The process, at its most fundamental level, involves two steps: generating one or more genetic changes in a population of otherwise genetically homogeneous organisms or gene sequences, and determining which organism or gene from the mutated population performs the desired function better than the strain or gene before the genetic change was made. In simpler terms this can be thought of as "mutate and screen". In preferred formats, the process is iterative, where improved organisms or genes are further evolved to perform the desired function at an even higher level.

Error-prone PCR, DNA shuffling, and site directed mutagenesis are favored directed evolution tools for the "mutate" part. The "screen" aspect of the directed evolution process can be significantly automated through the use of robotic technologies that not only serve to speed up the process of assembling and testing large populations of mutated organisms of genes, but also to standardize the assay process. With robotic technology, tens of thousands of individual mutant organisms can be tested for an enhanced function in a matter of hours. The ability to perform such mass screening increases the number of improved organisms or gene sequences identified in an assay. That being said, the screening process should be performed under commercial deployment conditions that mimic, as closely as possible, the envisioned commercial production system.

Although no literature reports could be found that coupled strain improvement (through directed evolution) and bioreactor-based production strategies for heterotrophic growth, there are some useful examples for phototrophic microalgae.

Gordon and Polle (2007) described the potential for increased productivity in algal photobioreactors by an unorthodox integration between cellular engineering and photonics. Specifically, the study proposed to achieve ultrahigh productivity by combining recent progress in the cellular engineering of algal chlorophyll antenna size achieved using metabolic engineering (Gordon and Polle, 2007). Their proposed key to greater biomass yields - projected as high as 100 gdw m^{-2} h^{-1} – combined a pronounced heightening of algal flux tolerance with pulsed light-emitting diodes – which they used to tailor the photonic temporal, spectral and intensity characteristics. Specifically, a tailored photonic input was applied in concert with thin-channel ultradense culture photobioreactors with flow patterns that produce rapid light/dark algae exposure cycles. While the authors admitted that the artificial-light scheme was globally feasible only with electricity generated from renewables, their work does offer (for heterotrophic strains) promise for combining strain improvement through metabolic engineering with highly advanced dark fermentation technologies. Debus (2001) presents a good review on the introduction of site-directed, deletion, and other mutations into the photosystem II polypeptides of the cyanobacterium, *Synechocystis* sp. PCC 6803, and the green alga, *Chlamydomonas reinhardtii,* (Debus, 2001) while Xiong et al. (1998) used site directed mutagenesis to create two mutants of the unicellular green alga, *Chlamydomonas reinhardtii,* to study the effect of formate on the inhibition of electron and proton transfers in photosystem II (PSII) (Xiong et al., 1998). Jing-Jing et al. (2007) used site-directed mutagenesis to study the structural and/or functional roles of arginine in PSII complexes in *C. reinhardtii* (Jing-Jing et al., 2007). As a consequence, a molecular approach through genetic engineering has been demonstrated as a good method to achieve high activity via directed evolution (Fang et al., 2009).

15.6.2 High Throughput Screening (HTS)

HTS uses automation to run a screen of an assay against a library of candidate compounds (Liu et al., 2004). Typically, an assay will test for specific activity such as inhibition or stimulation of a biochemical or biological mechanism but it can also be used to identify strains with a particular ability to produce a target molecule, such as triglycerides appropriate for biodiesel production. Typical HTS screening libraries or "decks" can contain from 100,000 to more than 2,000,000 compounds. In some cases, the screen for toxins or growth inhibitors can be as important as the screen for lipid production (Barclay et al., 2005).

High throughput screening can be applied effectively to identify industrially suitable strains. Automating the process speeds up the optimization of strains and guarantees high quality results. A well run process comprises a number of key components: A large and growing library of microorganisms, (2) DNA libraries that are generated synthetically and from isolated organisms - industrial companies are generally active in developing a growing and diverse collection of ready-to-screen DNA libraries; (3) Bioinformatic analysis of both public and proprietary genomic data; (4) Targeted Assay Technologies - proprietary assay technologies are vital to developing the best production strains; and (5) Sensitive and application-specific assays that are applied to bioproduction targets. These highly selective assays are used to identify novel microorganisms displaying desired phenotypes. There are examples of active screening of microalgal isolates for high value compounds such as secondary carotenoids, fatty acids, polysaccharides and other active compounds (Guil-Guerrero et al., 2000; Hsiao and Blanch, 2002; Barclay et al., 2005).

Except for microalgal strains that may be found in sufficient quantities and purity in nature (e.g. cyanobacterial mats), laboratory culture of these isolates may be necessary to obtain enough material to screen. The assay must be relatively accurate and reproducible in part because the assays must be applied over a large number of measurements, and on very small sample volumes. They should also be relatively simple and quick. Recently, our lab reported modifications to the Nile Red assay (Cooksey et al., 1987) that improved its ability to screen for neutral lipids, either across multiple microalgal strains grown under identical conditions, or across varying growth conditions applied to a single strain (Elsey et al., 2007). The same assay has been applied for throughput measurement of neutral microalgal lipids (Chen et al., 2009). Fluorescent based assays are particularly useful given that they are extremely sensitive and can be applied in microliter concentrations. They are also difficult to standardize since the fluorescence emission of fluorophores is affected by the surrounding solvent and because they also suffer bleaching (which makes calibration against a known standard difficult). Still fluorescent assays using probes such as Nile Red can be extremely useful as first screens because their great sensitivity permits detection at very low amounts, which permits very small assay volumes.

15.6.3 Metabolic Engineering

Improving lipid production by microalgae has been an important metabolic engineering goal for many years, and remains so today (Fischer et al., 2008). Metabolic engineering includes the manipulation of endogenous metabolic pathways as well as the transplantation of metabolic pathways into new host organisms. In some instances the genes in a metabolic pathway are unknown, while in other cases some or all of the genes necessary and sufficient to produce a molecule are known. The massive expansion of available genomic information, particularly in the area of microbes, allows researchers to push the limits of what can be produced by a chosen organism. In addition, metabolic engineering also allows for the significant up-regulation of pathways producing desired molecules. While such productivity improvements were very uncommon five years ago, 100-fold and higher increases in production of a desired molecule in a production organism have been increasingly demonstrated in academic and commercial settings.

In one relevant example, gas chromatographic profiling of fatty acids was performed during the growth cycle of four marine microalgae in order to establish which, if any, of these could act as a reliable source of genes for the metabolic engineering of long chain polyunsaturated fatty acid (LC-PUFA) synthesis in alternative production systems (Tonon et al., 2002). Initially, a high-throughput column based method for extraction of triacylglycerols (TAGs) was used to establish how much and at what stage in the growth phase LC-PUFAs partitioned to storage lipid in the different species. Using this data, differences in the time course of production and incorporation of docosahexaenoic acid (22:6n-3, DHA) and eicosapentaenoic acid (20:5n-3, EPA) into TAGs were found in the marine microalgae *Nannochloropsis oculata, Phaeodactylum tricormutum* and *Thalassiosira pseudonana* as well as *Haptophyte Pavlova lutheri*. Differences were not only observed between species but also during the different phases of growth within a species. The first report of genetic transformation of any chlorophyll c-containing microalgal strain was presented by Dunahay et al. (1996), who introduced additional copies of the ACCase gene into diatoms in an attempt to manipulate lipid accumulation in transformed strains. More specifically, they genetically transformed two species of diatoms by introducing chimeric plasmid vectors containing a bacterial antibiotic resistance gene driven by regulatory sequences from the acetyl-CoA carboxylase (ACCase) gene from the diatom *Cyclotella cryptica*. The recombinant DNA was integrated into one or more random sites within the algal genome and the foreign protein was produced by the algal transformants.

In summary, the further metabolic engineering of the triglyceride producing pathways of strains identified, via high throughput screening, to be good candidates for successful metabolic engineering, requires some understanding of the appropriate metabolic pathways. The metabolic pathways for the synthesis of most proposed biofuels proceed through common metabolic intermediates such as acetyl-CoA or pyruvate (Fischer et al., 2008). Acetyl-CoA is the primary source of carbon for fatty acid synthesis. Long-chain fatty acids are biosynthesized through the ATP-requiring carboxylation to malonyl-CoA, followed by cycles of decarboxylative addition of malonyl-CoA to acyl units and β-reduction. Since lipids are accumulated intracellularly, which requires their extraction from crude pastes, one desired metabolic engineering strategy will be to engineer secretion of the lipid products (Fischer et al., 2008). Since many details of fatty acid and triglyceride secretion is poorly understood in model organisms such as *E. coli* and *S. cerevisiae* (Fischer et al., 2008), this area is an attractive application of metabolic engineering applied to heterotrophic microalgal strains identified as being appropriate for lipid production.

15.6.4 Synthetic Biology

Synthetic biology is the concept of re-integrating biological components into larger assemblies (Endy, 2005; Andrianantoandro et al., 2006; Heinemann and Panke, 2006). In other words, molecular parts (e.g., DNA, RNA, or protein) or devices (any combination of parts that performs desired functions) are fabricated in order to either construct new biological systems or to rewire existing cellular signaling networks (Fischer et al., 2008). In this sense various components of a biological cell, such as proteins, promoters, RNA, and DNA can be broken down into "parts" that can be re-integrated into genetic circuits that provide the desired functionality (Leonard et al., 2008). In this context, synthetic biology can be described as a "bottom-up" approach to reconstruction of artificial molecules or systems with the aim of both understanding the dynamic behavior of biological systems and controlling them (Fischer et al., 2008). In the long run, studies along this line may provide new tools for designing and constructing synthetic systems that can produce liquid phase fuels such as biodiesel from an number of substrates, whether they be byproducts of food production (e.g., glucose) or as compounds in waste streams (e.g., acetate). Although a complex human-made biological system is yet to exist, a number of reviews on the potential of this field are available (Endy, 2005; Andrianantoandro et al., 2006; Heinemann and Panke, 2006; Leonard et al., 2008) and task specific literature reports are beginning to appear. For example, based on the synthetic biology pSB1C3 platform an *Escherichia coli* expression vector that is particularly useful for construction and production of fusion proteins has been presented (Škrlj et al., 2008). In another recent report it was shown, for the first time, that membrane proteins can be functionally synthesized inside vesicles of proper lipid composition, and that such proteins can synthesize lipidic components at the aim of stimulating liposome growth and self-division (Kuruma et al., 2008).

15.7 Summary

While humans have taken limited advantage of natural populations of microalgae for centuries (e.g., *Nostoc* in Asia and *Spirulina* in Africa and North America for sustenance), it is only recently that we have begun to realize the full potential of microalgal biotechnology. The potential exists for producing a vast array of materials, including foodstuffs, industrial chemicals, compounds with therapeutic applications, and bioremediation solutions. Modest success has been achieved for a relatively small number of high-value microalgal products such as whole cell microalgae (e.g., spirulina) or nuetraceuticals (e.g., DHA). In the case of heterotrophic biodiesel production, commercial viability will be harder to attain because of the cost differential of the feedstock (glucose) and the product (biodiesel). While technological tools exist to develop and culture strains of microalgae at high cell density and lipid content (Wen et al., 2002; Barclay et al., 2005; Xu et al., 2006), the major driving force will be the identification of a cheap sugar source that can sustain microalgae growth without substantial pretreatment. One example might be raw sugar juice from the first press of sugar cane. Without this advance it will be difficult for large-scale commercial production of heterotrophic production of biodiesel in microalgae to succeed. Also, the emergence of new technologies in extraction will be a primary factor governing the future feasibility of heterotrophic algal biodiesel production (Cooney et al., 2009).

15.8 Acknowledgements

The authors would like to thank Dr. Kirk Apt for a series of helpful discussions during the preparation of this material.

15.9 References

Andrianantoandro, E., Basu, S., Karig, D. K., and Weiss, R. (2006). "Synthetic biology: new engineering rules for an emerging discipline." *Molecular Systems Biology,* 2, 0028.

Ankerstjerne, R., Wiebe, M. G, and Eriksen, N. T. (2005). "Heterotrophic high-cell density fed batch cultures of the phycocyanin-producing red alga *Galdieria sulphuraria*." *Biotechnology and Bioengineering,* 90(1), 77 - 84.

Apt, K. E., and Behrens, P. W. (1999). "Commercial developments in microalgal biotechnology." *Journal of Phycology,* 35, 215 - 226.

Ayala, R. S., and Castro, L. (2001). "Continuous subcritical water extraction as a useful tool for isolation of edible essential oils." *Food Chemistry*, 75, 109 - 113.

Bailey, R. B., DiMasi, D., Hansen, J. M., Mirrasoul, P. J., Rucker, C. M., Veeder, G. M., Kaneko, T., and Barclay, W. (2003). U.S. patent 6607900.

Barclay, W. (1992). U.S. patent 5130242.

Barclay, W., Weaver, C., and Metz, J. (2005). "Development of a Docasahexanenoic Acid Production Technology Using *Schizochytrium*: A Historical Perspective." In: Cohen, Z., and Ratledge C. (Ed.) *Single Cell Oils*. AOCS Publishing, pp. 37 - 52.

Barclay, W. R., Meager, K. M., and Abril, J. R. (1994). "Heterotrophic production of long chain omega-3 fatty acids utilizing algae and algae-like microorganisms." *Journal of Applied Phycology,* 6, 123 - 129.

Benemann, J. R., Kopman, B. L., Weissman, D. E., Eisenberg, D. E., and Goebel, R. P. (1980). "Development of microalgae harvesting and high rate pond technologies in California." In: Shelef, G., and Soeder, C. J. (Ed.) *Algal biomass.* Amsterdam: Elsevier, pp. 457.

Bligh, E. G., and Dyer, W. J. (1959). "A rapid method for total lipid extraction and purification." *Canadian Journal of Biochemistry and Physiology,* 37, 911 - 917.

Chen, F. (1996). "High Density culture of microalge in heterotrophic growth." *Trends in Biotechnology,* 14, 421 - 426.

Chen, F., and Johns, M. R. (1991). "Effect of C/N ratio and aeration on the fatty acid composition of heterotrophic *Chorella sorokiniana*." *Journal of Applied Phycology,* 3, 203 - 209.

Chen, W., Zhang, C., Song, L., Sommerfield, M., and Hu, Q. (2009). "A high throughput Nile red metho for quantitative measurement of neutral lipids in microalgae." *Journal of Microbial Methods,* 77, 41-47.

Chisti, Y., and Moo-Young, M. (1991). "Fermentation technology, bioprocessing, scale-up and manufacture." In: Moses M., and C.R.E. (Ed.) *Biotechnology: the science and the business.* New York: Harwood Academic Publishers, pp. 167-209.

Chueng, P. C. K. (1999). "Temperature and pressure effects on supercritical carbon dioxide extraction of n-3 fatty acids from red seaweed." *Food Chemistry,* 65, 399 - 403.

Cooksey, K. E., Guckert, J. B., Williams, S., and Callis, P. R. (1987). "Fluorometric determination of the neutral lipid content of microalgal cells using Nile Red." *Journal of Microbiological Methods*, 6, 333 - 345.

Cooney, M. J., Young, G. L., and Nagle, N. (2009). "Extraction of Bio-oils from Microalgae." *Separation and Purification Reviews,* in press.

Cravotto, G., Boffa, L., Mantegna, S., Perego, P., Avogadro, M., and Cintas, P. (2008). "Improved extraction of vegetable oils under high-intensity ultrasound and/or microwaves." *Ultrasonics Sonochemistry*, 15, 898 - 902.

Day, J. G., Edwards, A. P., and Rodgers, G. A. (1991). "Development of an industrial-scale process for the heterotrophic production of a micro-algal mollusc feed." *Bioresource Technology*, 38, 245 - 249.

de Swaaf, M. E., Sijtsma, L., and Pronk, J. T. (2003). "High-cell-density fed-batch cultivation of the docosahexaenoic acid producing marine alga *Crypthecodinium cohnii*." *Biotechnology and Bioengineering*, 81(6), 666-672.

Debus, R. J. (2001). "Amino acid residues that modulate the properties of tyrosine YZ and the manganese cluster in the water oxidizing complex of photosystem II." *Biochimica et Biophysica Acta*, 1503, 164-186.

Denery, J. R., Dragull, K., Tang, C. S., and Li, Q. X. (2004). "Pressurized Fluid Extraction of carotenoids from *Haematococcus pluvialis* and *Dunaliella salina* and kavalactones from *Piper methysticum*." *Analytica Chimica Acta*, 501, 175 - 181.

Dionex, C. (2007). "Accelerated solvent extraction techniques for in-line selective removal of interferences." Technical Note **210**(LPN 1931): Sunnyvale, CA, USA.

Dubinsky, Z., and Aaronson, S. (1979). " Increase of lipid from some algae by acid extraction." *Phytochemistry (Oxford)*, 18, 51 - 52.

Eikani, M. H., Golmohammad, F., and Rowshanzamir, S. (2007). "Subcritical water extraction of essentail oils from coriander seeds (*Coriandrum sativium L*)." *Journal of Food Engineering*, 80, 735 - 740.

Elsey, D., Jameson, D., Raleigh, C. B., and Cooney, M. J. (2007). "Fluorescent measurement of microalgal neutral lipids." *Journal of Microbial Methods,* 68, 639 - 642.

Endy, D. (2005). "Foundations for engineering biology." *Nature*, 438, 449 - 453.

Engel, N., Jenny, T. A., Mooser, V., and Gossauer, A. (1991). "Chlorophyll catabolism in Chlorella protothecoides." *FEBS Letters*, 293(1,2), 131 - 133.

Eriksen, N. T. (2008). "The Technology of microalgal culturing." *Biotechnology Letters*, 30, 1525 - 1536.

Fang, Y., Lu, Y., Lv, F., Bie, X., Zhao, H., Wang, Y., and Lu, Z. (2009). "Improvement of alkaline lipase from *Proteus vulgaris* T6 by directed evolution." *Enzyme and Microbial Technology,* 44, 84-88.

Fischer, C. R., Kleini-Marcuschamer, D., and Stephanopolous, G. (2008). "Selection and optimization of microbial hosts for biofuels production." *Metabolic Engineering*, 10, 295 - 304.

Folch, J., Lees, M., and Sloane Stanley, G. H. (1956). "A simple method for the isolation and purification of total lipids from animal tissues." *Journal of Biological Chemistry*, 226, 497 - 509.

Ganuza, E., Anderson, A. J., and Ratledge, C. (2008). "High-cell-density cultivation of *Schizochytrium* sp. in an ammonium/pH-auxostat fed-batch system." *Biotechnology Letters*, 30, 1559 - 1564.

Gladue, R. M., and Maxey, J. E. (1994). "Microalgal feeds for aquacultrue " *Journal of Applied Phycology*, 6, 131 - 134.

Gordon, J. M., and Polle, J. E. W. (2007). "Ultrahigh bioproductivity from algae." *Applied Microbiology and Biotechnology*, 76(5), 969-975.

Grant, N. G., and Hommersand, M. H. (1974). "The repiratory chain of Chlorella prototheocoides." *Plant Physiology*, 54: 50 - 56.

Guclu-Ustundag, O., Balsevich, J., and Mazza, G. (2007). "Pressurized low polarity water extraction of sponins from cow cockle seed." *Journal of Food Engineering,* 80, 619 - 630.

Guil-Guerrero, J. L., Belarbi, E. H., and Rebolloso-Fuentes, M. M. (2000). "Eicosapentaenoic and arachidonic acids purification from the red microalga *Porphyridium cruentum*." *Bioseparation*, 9(5), 299-306.

Heinemann, M., and Panke, S. (2006). "Synthetic biology - putting engineering into biology." *Bioinformatics*, 22, 2790 - 2799.

Herrero, M., Cifuentes, A., and Ibanez, E. (2006). "Sub- and supercritical fluid extraction of functional ingrediants from different natural sources: Plants, food-by-products, algae and microalgae. A review." *Food Chemistry*, 98, 136 - 148.

Hsiao, T. Y., and Blanch, H. W. (2002). *Physiological study of algae: growth characteristics and fatty acid synthesis in Glossomastix chrysoplastos.,* The 4th Asia_/Pacific Marine Biotechnology Conference, Honolulu, USA.

Iverson, S. J., Lang, S. L. C., and Cooper, M. H. (2001). "Comparison of the bligh and dyer and folch methods for total lipid determination in a broad range of marine tissue." *Lipids*, 36(11), 1283 - 1287.

Jing-Jing, M., Liang-Bi, L., Yu-Xiang, J., and Ting-Yun, K. (2007). "Mutation of residue Arginine(18) of cytochrome b(559) alpha-subunit and its effects on photosystem II activities in *Chlamydomonas reinhardtii*." *Journal of Integrative Plant Biology*, 49(7), 1054-1061.

Johnston, W. A., Stewart, M., Lee, P., and Cooney, M. J. (2003). "Tracking the acetate threshold using DO transient control in medium and high cell density cultivation of recombinant E.coli in complex media." *Biotechnology and Bioengineering*, 84(3), 314 - 323.

Johnston, W. J., and Cooney, M. J. (2002a). "Industrial control of recombinant E. coli fed batch culture: new perspectives on traditional controlled variables." *Bioprocess and Biosystems Engineering*, 25, 111 - 120.

Johnston, W. J., and Cooney, M. J. (2002b). "Biomass recycling to reduce process noise resulting from concentrated substrate addition in fed-batch cultivation of *E. coli*." *Biotechnology Progress*, submitted.

Kates, M. (1988). *Techniques of lipidology*, Elsevier, New York, USA.

Kuruma, Y., Stano, P., Ueda, T., and Luisi, P. L. (2008). "A synthetic biology approach to the construction of membrane proteins in semi-synthetic minimal cells." *Biochimica et Biophysica Acta*, 1788(2), 567-574.

Leonard, E., Nielsen, D., Solomon, K., and Prather, K. J. (2008). "Engineering microbes with synthetic biology frameworks." *Trends in Biotechnology*, 26(12), 674 - 681.

Lepage, G., and Roy, C. C. (1984). "Improved recovery of fatty acid through direct transesterfication without prior extraction or purification." *Journal of Lipid Research,* 25, 1391 - 1396.

Li, X., Xu, H., and Wu, Q. (2007a). "Large-scale biodiesel production from microalga *Chlorella protothecoides* through heterotrophic cultivation in bioreactors." *Biotechnology and Bioengineering*, 98(4), 764 - 771.

Li, Y., Zhao, Z., and Bai, F. (2007b). "High-density cultivation of oleaginous yeast Rhodosporidium toruloides Y4 in fed-batch culture." *Enzyme and Microbial Technology,* 41, 312 - 317.

Liu, B., Li, S., and Hu, J. (2004). "Technological advances in high-throughput screening." *American Journal of Pharmacogenomics,* 4(4), 263-276.

Luque de Castro, M. D., Jimenez-Carmona, M. M., and Fernandez-Perez, V. (1999). "Towards more rational techniques for the isolation of valuable essential oils from plants." *Trends in Analytical Chemistry*, 18(11), 708 - 715.

Luque de Castro, M. D., Valcarcel, M., and Tena, M. T. (1994). *Supercritical fluid extraction,* Springer Verlag, Heidelberg.

MacKay, D., and Salusbury, T. (1988). "Choosing between centrifugation and crossflow microfiltration." *Chemical Engineering (London)*, 477, 45- 50.

Maher, K. D., and Bressler, D. C. (2007). "Pyrolysis of triglycerie materials for the production of renewable fuels and chemicals." *Bioresource Technology*, 98, 2351 - 2368.

Mead, J. M., Alfin-Slater, R. B., Howton, D. R., and Popjak, G. (1986). *Lipids Chemistry, Biochemistry, and Nutrition,* Plenum Press, New York.

Meng, X., Yang, J., Xu, X., Zhang, L., Nie, Q., and Xian, M. (2009). "Biodiesel production from oleaginous microorganisms." *Renewable Energy*, 34, 1 - 5.

Miao, X., and Wu, Q. (2004). "High yield bio-oil production from fast pyrolysis by metabolic controlling of *Chlorella prototheocoides*." *Journal of Biotechnology,* 110, 85 - 93.

Miao, X., and Wu, Q. (2005). "Biodiesel production from heterotrophic microalgal oil." *Bioresource Technology*, 97, 841 - 846.

Mok, W. S. L., and Antal Jr, M. J. (1992). "Uncatalyzed solvolysis of whole biomass hemicellulose by hot compressed liquid water." *Industrial & Engineering Chemistry Research*, 31(4), 1157 - 1161.

Molina Grima, E., Belarbi, E. H., Acien Fernandez, F. G., Robles Medina, A., and Chisti, Y. (2003). "Recovery of microalgal biomass and metabolites: process options and economics." *Biotechnology Advances*, 20, 491 - 515.

Mulbry, W., Kondrad, S., Pizarro, C., and Kebede-Westhead, E. (2008). "Treatment of dairy manure effluent using freshwater algae: Algal productivity and recovery of manure nutrients using pilot-scale algal turf scrubbers." *Bioresource Technology,* 99(17), 8137-8142.

Muñoz, R., Köllner, C., and Guieysse, B. (2009). "Biofilm photobioreactors for the treatment of industrial wastewaters." *Journal of Hazardous Materials*, 161(1), 29-34.

Neenan, B., Feinberg, D., Hill, A., McIntosh, R., and Terry, K. (1986). *Fuels from microalgae : Technology, Status, Potential, and Research Requirements,* Solar Energy Research Institute, Golden Colorado, pp. 149.

Olaizola, M. (2003). "Commercial development of microalgal biotechnology: from the test tube to the marketplace." *Biomolecular Engineering,* 20, 459 - 466.

Orus, M. I., Marco, E., and Marinez, F. (1991). "Suitability of Chlorella vulgaris UAM 101 for heterotrophic biomass production." *Bioresource Technology,* 38, 179 - 184.

Petrusevski, B., Bolier, G., Van Bremen, A. N., and Alaerts, G. J. (1995). "Tangential flow filtration: a method to concentrate freshwater algae." *Water Resources*, 29, 1419-1424.

Pulz, O., and Gross, W. (2004). "Valuable products from biotechnology of microalgae." *Applied Microbiology and Biotechnology,* 65, 635–648.

Radmer, R. J., and Parker, B. C. (1994). "Commercial applications of algae: opportunities and constraints." *Journal of Applied Phycology*, 6, 93 - 98.

Ratledge, C., Streekstra, H., Coeh, Z., and Fichtali, J. (2005). "Down-Stream processing, extraction, and purification of single cell oils." In: Cihen, Z., and Ratledge, C. (Ed.) *Single Cell Oils*. AOCS Publishing, pp. 202 - 219.

Richmond, A., and Becker, E. W. (1986). "Technological aspects of mass cultivation, a general outline." In: Richmond, A. (Ed.) *CRC handbook of microalgal mass culture*. Boca Raton, CRC Press, pp. 245 - 263.

Richter, B. E., Jones, B. A., Ezzell, J. L., Porter, N. L., Avdalovic, N., and Pohl, C. (1996). "Accelerated solvent extraction: a technique for sample preparation." *Analytical Chemistry*, 68, 1033 -1039.

Rodriguez_Ruiz, J., Belarbi, E.H., Sanchez, J.L.G., and Alonso, D.L. (1998). "Rapid simultaneous lipid extraction and transesterification for fatty acid analysis." *Analytical Chemistry,*68, 1033-1039.

Rosenberg, J. N., Oyler, G. A., Wilkinson, L., and Betenbaugh, M. J. (2008). "A green light for engineered algae: redirecting metabolism to fuel a biotechnology revolution." *Current Opinion in Biotechnology*, 19, 430 - 436.

Schafer, K. (1998). "Accelerated solvent extraction of lipids for determining the fatty acid composition of biological material." *Analytica Chimica Acta*, 358, 69 - 77.

Sheehan, J., Dunahay, T., Benemann, J., and Roessler, P. (1998). A Look Back at the U.S. Department of Energy's Aquatic Species Program: Biodiesel from Algae. Golden Colorado, National renewable Energy Laboratory.

Shi, X. M., Jian, Y., and Chen, F. (2002). "High-Yield production of Lutein by the green microalga Chlorella protothecoides in heterotrophic fed-batch culture." *Biotechnology Progress*, 18, 723 - 727.

Škrlj, N., Ercˇulj, N., and Dolinar, M. (2008). "A versatile bacterial expression vector based on the synthetic biology plasmid pSB1." *Protein Expression and Purification*, 64(2), 198-204.

Song, D., Fu, J. M., and Shi, D. (2008). "Exploitation of Oil-bearing Microalgae for Biodiesel." *Chinese Journal of Biotechnology,* 24(3), 341-348.

Sporalore, P., Joannis-Cassan, C., Duran, E., and Isambert, A. (2006). "Commercial applicaitons of microalgae." *Journal of Bioscience and Bioengineering*, 101, 87 - 96.

Tonon, T., Harvey, D., Larson, T. R., and Graham, I. A. (2002). "Long chain polyunsaturated fatty acid production and partitioning to triacylglycerols in four microalgae." *Phytochemistry (Oxford),* 61(1), 15-24.

Wen, Z. Y., Jian, Y., and Chen, F. (2002). "High cell density culture of the diatom Nitzchia laevis for eicosapentaenoic acid production: fed batch development." *Process Biochemistry,* 37, 1447-1453.

Whiffen, V., Cooney, M. J., and Cord-Ruwisch, R. (2004). "On-line detection of feed demand in high cell density cultures of *Escherichia coli* by measurement of changes in dissolved oxygen transients in complex media." *Biotechnology and Bioengineering,* 85(4), 423 - 433.

Wu, Q. Y., Yin, S., Sheng, G. Y., and Fu, J. M. (1994). "New Discoveries in study on hydrocarbons from thermal degradation of heterotrophically yellowing algae." *Science in China (Series B)*, 37(3), 326 - 335.

Wu, Q., Zhang, B., and Grant, N. G. (1996). "High yield of hydrocarbon gases resulting from pyrolysis of yellow heterotrophic and bacterially degraded *Chlorella protothecoides*." *Journal of Applied Phycology*, 8, 181 - 184.

Xiong, J., Minagawa, J., Crofts, A., and Govindjee (1998). "Loss of inhibition by formate in newly constructed photosystem II D1 mutants, D1-R257E and D1-R257M, of *Chlamydomonas reinhardtii*." *Biochimica et Biophysica Acta,* 1365, 473-491.

Xu, H., Miao, X., and Wu, Q. (2006). "High quality biodiesel production from a microalga *Chlorella protothecoides* by heterotrophic growth in fermenters." *Journal of Biotechnology,* 126, 499 - 507.

Yang, Y. F., Feng, C. P., Inamori, Y., and Maekawa, T. (2004). "Analysis of energy conversion characteristics in liquefaction of algae." *Resources Conservation & Recycling*, 43, 21 - 33.

CHAPTER 16

Microalgal Ethanol Production: A New Avenue for Sustainable Biofuel Production

Rojan P. John, Puspendu Bhunia, S. Yan, R.D. Tyagi, and R.Y. Surampalli

16.1 Introduction

A small fraction of the solar energy (~0.1–0.5%) is captured by photosynthesis and drives most living systems. Life on earth is carbon-based and the energy is used to fix atmospheric carbon dioxide into biological material (biomass), indeed fossil fuels that we consume today are a legacy of mostly algal photosynthesis. Marine and freshwater algae have higher photosynthetic efficiencies than terrestrial plants and are more efficient in capturing carbon (Packer, 2009).

The earliest internal combustion engines were run on biologically-derived fuels (e.g. plant oils for diesel engines and bioethanol for spark-ignition engines) until plentiful cheap petroleum-derived fuels took over (Packer, 2009). Several researchers reported the potential of algal biomass to generate biofuels and most of these are based on the fact that algae utilize the CO_2 emitted from petroleum-based power stations or other industrial sources (Benemann, 1993, 1997; Hughes and Benemann, 1997; Vunjak-Novakovic et al., 2005; de Morais and Costa, 2007; Ratledge and Cohen, 2008). Some of the microalgae and macroalgae are oleaginous in nature and can be exploited for the production of biodiesel.

Algae can be easily cultivated and can be used as a feedstock for the production of biofuel such as bioethanol. Currently food grain, such as corn and wheat, or cane sugar and cane juice products are used for ethanol production. With increasing utilization of grains for ethanol production, it can lead to fuel versus food concern. Lignocellulosic biomass and starchy wastes are alternative feedstocks, which can be hydrolyzed by chemical or biological methods to fermentable sugars for subsequent biofuel production.

Some leading companies like Algenol Biofuels Inc. had developed technology to utilize microalgal cell as a tiny biorefinary for ethanol production using the energy from sunlight. Using specialized bioreactor the algal cells are allowed to produce and accumulate sugar by photosynthesis and these photosynthate are directly turn to produce ethanol using the enzymes presents in the same organisms. Algenol's technology produces industrial-scale and low-cost ethanol using algae, sunlight, CO_2 (from air or industrial sources), and seawater. Besides the system purify air by removing CO_2 (a green house gas) and pumping oxygen to the environment, it can produce pure water while sea water is used as source for photosynthesis. The technology can utilize marginal or desert land instead of agricultural land. They claimed that the system can produce 6,000 gallons of ethanol per acre per year, far greater than corn at a rate of 400 gallons of ethanol per acre per year (Algenol Biofuels, 2009).

Microalgae like cyanobacteria have simple growth requirement, grow to high densities, and use light, carbon dioxide, and other inorganic nutrients efficiently (Deng and Coleman, 1999), so they could be attractive source for ethanol production. This chapter focuses on the different strategies of direct and multi stepped ethanol production using microalgae.

16.2 Algal Bioethanol Production

The interests in the production of oil from microalgae are due to their higher productivity and yield. Chisti (2007) reported up to 127 Mg/ha/year of microalgae in high-rate raceway ponds while Carlsson et al. (2007) suggested an algal yield of 50 to 60 Mg/ha/year could be achievable. In photobioreactor, productions of up to 150 Mg/ha/year microalgae have already been obtained (Carlsson et al., 2007). Chisti (2007) has suggested an algal yield of up to 263 Mg/ha/year. Application of carbohydrate–rich marine microalgae can serves as an alternative biomass resource for ethanol production by providing huge amount of carbohydrate year around (Matsumoto et al., 2003). Some of the microalgae which are minute photosynthetic organisms like *Chlorella, Dunaliella, Chlamydomonas, Scenedesmus,* and *Spirulina* are known to contain a large amount (> 50% of the dry weight) of starch and glycogen, useful as raw materials for ethanol production (Ueda et al., 1996). Table 16.1 summarises the starch content of some microalgae. The accumulated starch can be extracted from algae by physico-chemical or biological methods and can be fermented into ethanol following enzyme hydrolysis. The accumulated starch can also be converted directly to ethanol by the same microalgae under anaerobic condition during dark. Generalized steps in the conversion of microalgal starch or biomass to ethanol is shown in Figure 16.1. The direct utilization of microalgal biomass may open a new avenue for the cost effective production of ethanol.

Table 16.1 Starch content of some microalgae belonging to *Chlorophyceae* and *Cyanophyceae*.

Algae types	**% (g starch / dry weight)**	**References**
Green algae		
Green alga NKG 121701	>50.0	Matsumoto et al., 2003
Chlamydomonas reinhardtii UTEX 90	53.0	Kim et al., 2006
C. reinhardtii	17.0	Spolaore et al., 2006
Chlorella vulgaris	12.0-17.0	Spolaore et al., 2006
Dunaliella salina	32.0	Spolaore et al., 2006

Scenedesmus obliquus	10.0-17.0	Spolaore et al., 2006
Synechococcus sp.	15.0	Spolaore et al., 2006
Chlorella sp.TISTR 8262	21.5	Rodjaroen et al., 2007
Chlorella sp.TISTR 8485	27.0	Rodjaroen et al., 2007
Chlorella sp. TISTR8593	22.0	Rodjaroen et al., 2007
Chlorococcum sp.TISTR8583	26.0	Rodjaroen et al., 2007
Chlorococcum sp.TISTR 8973	16.8	Rodjaroen et al., 2007
Scenedesmus sp.TISTR 8579	20.4	Rodjaroen et al., 2007
Scenedesmus sp.TISTR 8982	13.3	Rodjaroen et al., 2007
S. acuminatus TISTR 8457	7.3	Rodjaroen et al., 2007
S. acutiformis TISTR 8495	16.4	Rodjaroen et al., 2007
S. acutus TISTR 8447	18.6	Rodjaroen et al., 2007
S. arcuatus TISTR 8587	12.9	Rodjaroen et al., 2007
S. armatus TISTR 8591	15.4	Rodjaroen et al., 2007
S. obliquus TISTR 8522	23.7	Rodjaroen et al., 2007
S. obliquus TISTR 8546	23.4	Rodjaroen et al., 2007
Blue green algae		
Nostoc sp. TISTR 8872	30.7	Rodjaroen et al., 2007
Nostoc sp. TISTR 8873	32.9	Rodjaroen et al., 2007
N. maculiforme TISTR 8406	30.1	Rodjaroen et al., 2007
N. muscorum TISTR 8871	33.5	Rodjaroen et al., 2007
N. paludosum TISTR 8978	32.1	Rodjaroen et al., 2007
N. piscinale TISTR 8874	17.4	Rodjaroen et al., 2007
Oscillatoria sp. TISTR 8869	19.3	Rodjaroen et al., 2007
O. jasorvensis TISTR 8980	9.7	Rodjaroen et al., 2007
O. obscura TISTR 8245	12.6	Rodjaroen et al., 2007
O. okeni TISTR 8549	8.1	Rodjaroen et al., 2007
Phormidium angustissimum TISTR 8979	28.5	Rodjaroen et al., 2007
Spirulina fusiformis	37.3 - 56.1	Rafiqul et al., 2003
Spirulina maximas	44.0	Canizares-Villanueva et al., 1995

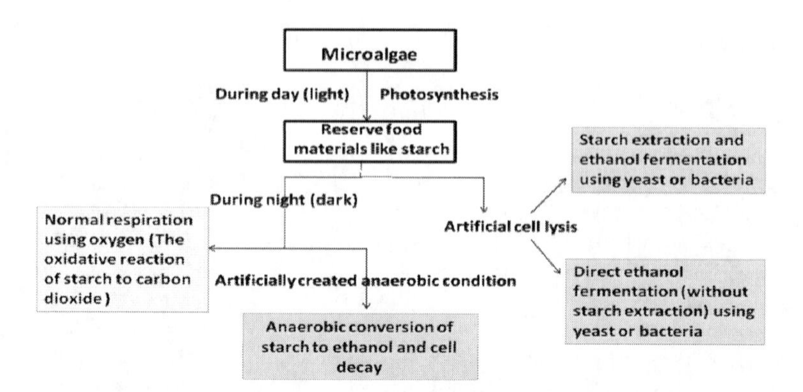

Figure 16.1 Utilization of microalgal biomass for ethanol production.

16.3 Algae and Their Culturing Advantage

Algae, including macro and microalgae, can be either fresh water or marine and some grow optimally at intermediate saline levels while some grow in hyper saline conditions. Macroalgae are multicellular algae that have defined tissues containing specialized cells. Microalgae are microscopic algae and may be unicellular. They can be motile or non-motile depending on the presence of locomotory structures. Mostly microalgae fall under algal groups like dinoflagellates, Chlorophyceae, Chryosophyceae and diatoms (Packer, 2009). Green algae include about 8000 species, covering both marine and fresh water environments and contain complex polysaccharides in their cell walls. It is easy to provide optimal nutrient levels for culturing of microalgae. This is due to the well-mixed aqueous environment as compared to soil and requirement of fewer nutrients. Absence of non-photosynthetic supporting structures (roots, stems, etc.) also favors the microalgal cultivation in aquaculture. Most of the microalgae under consideration are single-celled organisms that are self-contained and are productive. Microalgae do not have to spend energy towards storage molecules like starch between tissues. Asexual reproduction of microalgae like fragmentation helps to obtain biomass from very low levels to maximum under optimal conditions during continuous production. In the continuous culture systems such as raceway ponds and bioreactors, harvesting efforts can be controlled to match productivity. Due to their high cell division rate, handling is often simpler in research application and it can be performed several times faster with microalgae than that of the terrestrial crop species (Packer, 2009). There is evidence that small-scale experiments can be effectively translated into a large-scale facility for carbon dioxide capturing and biofuel production (Sheehan et al., 1998). In regard to strain improvement through genetic engineering, these species are acquiescent to 1) nuclear transformation for control of metabolic pathways; 2) chloroplastic transformation for high levels of protein expression; and 3), more clear-cut approaches to genetic alteration compared to higher plants (Rosenberg et al., 2008).

16.4 Application of Starch-Accumulating Microalgal for Bioethanol Production

16.4.1 Culturing of Starch Accumulating Microalgae

Microalgae are easily growing and they do not require fertile land or irrigation like terrestrial plants. Use of marine algae even provides an alternative to freshwater use (Rosenberg et al., 2008). The most cost-effective way to cultivate microalgae is in large, circulating ponds; however, a serious disadvantage in the use of outdoor ponds is the risk of contamination by other microbes or indigenous algal species (Benemann and Oswald, 1996). Nutrient composition may vary depending on the evaporation of water or by rain in open ponds and need continuous checking. Closed photobioreactors provide sterility and control over culture parameters such as light intensity, carbon dioxide, nutrient levels, and temperature (Rosenberg et al. 2008). Under optimal conditions, microalgal populations are capable of doubling within hours and achieving high cell densities, corresponding to as much as 60 g of heterotrophic biomass per liter and 5 g of photoautotrophic biomass per litre. Mixotrophic conditions can be provided for high-density algal growth (Rosenberg et al. 2008). Mixotrophic condition is achieved by the accumulation of starch by regular photosynthesis (autotrophic) in day time and accumulation of starch by utilizing provided sugars or organic acids (heterotrophic) during night time. Heterotrophic condition can be made economic by providing agro-industrial sugary waste or hydrolysate. As microalgae are hydrophilic microbes, they can easily absorb nutrients from water that can be effectively utilized for the metabolic activities. Checking the concentration of each nutrient and manipulation in water can be easily done for further enhancement of growth. Starch accumulating microalgae trap sunlight effectively using inexpensive water and CO_2 for sugar synthesis via light and dark reaction and store excess food in the form of starch. The duration, availability and intensity of light can be controlled by different type bioreactors and mixing of media.

16.4.2 Extraction of Starch from Microalgae

Photo bioreactors or shallow ponds like race way can be used for the production of microalgal biomass or by culturing it heterotrophically in the dark and in the presence of organic materials such as sugars and organic acids (Ueda et al., 2001; Chisti, 2007). Since, the grown microalgae store starch mainly in the cells, biomass can be harvested at regular intervals for the extraction of starch. The starch can be extracted from the cells with the aid of mechanical means (e.g., ultrasonic, explosive disintegration, mechanical shear, etc.) or an enzyme by dissolution of cell walls. The starch is then separated by extraction with water or an organic solvent.

16.4.3 Ethanol Production from Microalgal Starch

The production of ethanol by fermentation of algal-derived carbohydrates composes of two processes, saccharification and fermentation (Matsumoto et al.,

2003). Starch is mainly composed of two fractions of high molecular weight, amylose and amylopectin. The minor fraction, amylose (20–30%), is mainly a linear glucose polymer formed by α-1,4-glucosidic linkage and some α-1,6-branching points. Amylopectin represents the major fraction of starch (70–80%) and is highly branched (Yamada et al., 2009). Usually, organisms, which do not produce amylase, need additional hydrolysis step to utilize complex starchy substrate. Once the intracellular microalgal starch is extracted, the starch can be fermented to ethanol using the process similar to other starch-based feedstocks. Fig. 16.2 shows different steps in microalgal starch to ethanol.

The enzymatic hydrolysis of starch poses no problem while using agro-industrial wastes or agricultural products like wheat or corn starch as raw materials. However, utilization of marine biomass would require desalinization if a terrestrially-derived enzymes were to be used. Therefore, development of a system utilizing amylase from a marine source would be beneficial (Matsumoto et al., 2003).

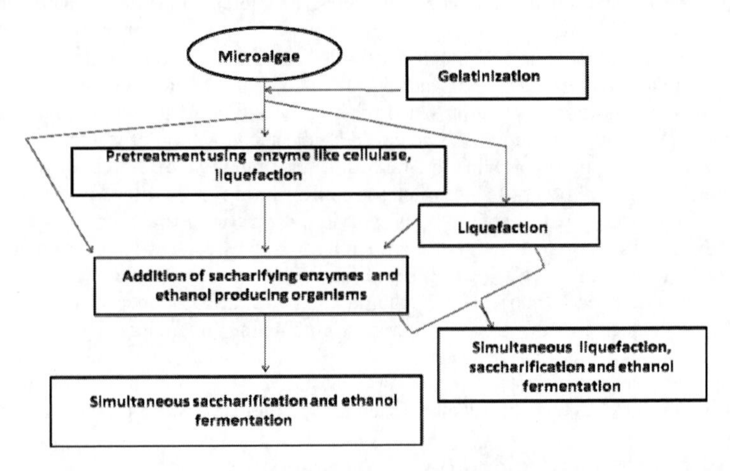

Figure 16.2 Possible ways of saccharification and ethanol fermentation of microalgal biomass.

Matsumoto et al. (2003) hydrolyzed marine microalgal starch using amylase from marine bacteria in saline conditions after mechanical breakdown of algal cell and gelatinization. The authors isolated an amylase-producing bacterium identified as *Pseudoalterimonas undina* NKMB 0074. The green microalgal strain NKG 120701 was determined to have the highest concentration of intracellular carbohydrate (> 50% of dry cell weight) and was selected from their algal culture stocks. *P. undina* NKMB 0074 was inoculated into suspensions containing NKG 120701; the suspended sugars were increasingly reduced with incubation time. The activity of terrestrial amylase and glucoamylase was found to be inactive in saline suspension. Thus marine amylase is necessary in saline conditions for successful saccharification of marine microalgae.

Saccharification of microalgal starch can also be conducted through acid hydrolysis. However, neutralization step generates extra cost and salt concentration in saccharified broth may inhibit fermentation process. Microalgal-derived carbohydrates contain not only starch but also other polysaccharides like cellulose, hemicelluloses, etc. For the complete utilization of the algal biomass, cellulose- and hemicellulose-degrading enzymes are also needed.

One of the methods of enhancing the ethanol productivity is the use of very high gravity (high dissolved solids) technology. It is usually used for the conversion of grain and tuber-based starch to ethanol after hydrolysis. The technology involves the preparation and fermentation of mash containing high dissolved solids (>300 g/L) to yield a high ethanol concentration (Thomas et al., 1993). It is of interest for fuel ethanol production because of its potential to increase fermentor throughput, and reduce processing costs (Thomas et al., 1996). Very high gravity technology has been intensively conducted in ethanol production from tubers by Srichuwong et al. (2009). It is interesting to workout the economy of microalgal starch for ethanol production by very high gravity hydrolysis and fermentation. This method may be employed as the biomass after harvesting is highly concentrated.

Higher viscosity of the substrate causes several handling difficulties during processes, and may lead to incomplete hydrolysis of starch to fermentable sugar (Ingledew et al., 1999; Wang et al., 2008; Srichuwong et al., 2009). The addition of water can reduce the mash viscosity. This, however, causes dilution of fermentable sugars thereby requiring more energy for evaporation. On the other hand, required viscosity can be accomplished by the enzymatic dissociation of cell-wall matrix (Srichuwong et al., 2009). Enzymes cellulase and pectinase have been reported in reducing the viscosity of waste residues from sugar beet (Beldman et al., 1984), potato (Slominska and Starogardzka, 1987) and cassava (Ahn et al., 1995; Sriroth et al., 2000). Similar approach i.e. application of enzymatic digestion of polysaccharides can overcome the viscosity problem and facilitates the use of very high gravity technology in microalgal bioethanol production.

16.5 Microalgal Ethanol Production under Anaerobic Conditions

Microalgae fix CO_2 during photosynthesis under illumination with sunlight (autotrophic nutrition) and thereby accumulate starch in their cells. Moreover, some microalgae can also grow under dark conditions in the presence of organic nutrients such as sugars and thereby accumulate starch (heterotrophic nutrition) (Ueda et al., 1996). It is necessary to separate intracellular starch by extraction as mentioned above under starch extraction section. However, since many microalgae have strong cell walls, significant power is consumed during mechanical cell disintegration or an expensive enzyme is needed for the dissolution of cell walls. Moreover, a large amount of organic solvent is required in the starch extraction step. Since the starch separated by extraction is raw, the starch must be subjected to a heat treatment for gelatinization before being hydrolyzed to glucose. A large amount of heat energy is required for this purpose. Usually, this heat energy for gelatinization accounts 20 to 30% of the total energy consumed in the ethanol production process (Ueda et al., 1996).

In the absence of light and organic nutrients, microalgae usually maintain their life by consuming starch or glycogen stored in the cells and decomposing them oxidatively to carbon dioxide (aerobic respiration). Under these conditions, the production of ethanol does not take place as there is enough stored food and can be used for metabolic purposes utilizing oxygen (normal catabolism) (Figure 16.1). However, it has been found that, if dark and anaerobic conditions are established, the oxidative reaction of starch to carbon dioxide does not proceed to completion. Thus, depending on the type of the microalga, hydrogen gas, carbon dioxide, ethanol, lactic acid, formic acid, acetic acid, and other products are produced in varying proportions (Ueda et al., 1996).

The inventors in the United States (Patent No. 5578472) used microalgae as starting materials for the production of ethanol (Ueda et al., 1996). The algal cells contained a large amount of polysaccharides composed of glucose (more than 50% of the dry weight), such as starch, glycogen etc., in the cells, which were metabolized rapidly under dark and anaerobic conditions to ethanol. Microalgae undergoing these pathways fall under classes Chlorophyceae, Prasinophyceae, Cryptophyceae and Cyanophyceae. Typical genera belonging to the class Chlorophyceae include *Chlamydomonas* and *Chlorella*, and typical genera belonging to the class Cyanophyceae include *Spirulina, Oscillatoria and Microcystis* (Ueda et al., 1996).

Production of ethanol by starch accumulating filament-forming or colony forming algal biomass has been reported (Bush and Hall, 2006). The starch-accumulating filament-forming or colony-forming algae were selected from Zygnemataceae, Cladophoraceae, Oeologoniales, or a combination and the starch-accumulating filament-forming or colony-forming algae selected from *Spirogyra, Cladophora, Oedogonium*, or a combination. The digested biomass was fermented with *Saccharomyces cerevisiae* and *Saccharomyces uvarum* for alcohol production. It had described the process for forming ethanol from algae, comprising: (a) growing starch-accumulating, filament-forming or colony-forming algae by aqua culture (can be open raceway pond or closed photobioreactor); (b) harvesting the grown algae to form a biomass (can be flocculation, sedimentation, filtration, centrifugation, etc.); (c) placing the biomass in a dark and anaerobic aqua environment to initiate decay of the biomass; (d) contacting the decaying biomass with a yeast capable of fermenting it to form a fermentation solution; and, (e) separating the resulting ethanol from the fermentation broth. This technology has been claimed superior to another patented technology (United States Patent No. 5578472) which describes a different source of fermentable sugars, single-cell free floating algae. The latter technology is not industrially scalable due to the inherent limitations of single-cell free floating algae (Bush and Hall, 2006).

Genetically engineered microalgae, such as *Synechocystis* can produce ethanol using their own sugar produced during photosynthesis. There are numerous benefits from producing ethanol using photosynthetic microorganisms such as *Synechocystis*, such as, economic ethanol production, positive environmental impacts, reduction in global warming, and improved food security. There is no requirement of extra supplements of sugars and other complex nutrients because it can utilize atmospheric nitrogen and can trap light. These methods for producing ethanol from solar energy and CO_2 using cyanobacteria offer significant savings in both capital and operational costs, in comparison to the biomass-based ethanol production facilities. A decreased expenditure is achieved by factors such as: simplified production processes, absence of agricultural crops and residues, no solid wastes to be treated, no enzymes needed, etc. The cyanobacteria fermentation involves no hard cellulose or hemicellulose which is difficult to treat. As a result, there will be no emissions of hazardous air pollutants and volatile organic compounds from cyanobacterial ethanol production plants (Fu and Dexter, 2006). The major pathway for ethanol synthesis is catalyzed by two enzymes, pyruvate decarboxylase (PDC) and alcohol dehydrogenase (ADH). PDC catalyzes the nonoxidative decarboxylation of pyruvate, which produces acetaldehyde and CO_2. Acetaldehyde is then converted to ethanol by ADH. This fermentation pathway plays a role in the regeneration of NAD^+ for glycolysis under anaerobic conditions in fungi, yeasts, and higher plants (Deng and Coleman, 1999). Deng and Coleman (1999) reported the ethanol production in aerobic photosynthetic cyanobacteria through transformation of the *pdc* and *adh* genes. Ethanol production in the medium was quite low compared to industrial ethanol production process. The rate of ethanol released into the medium and the final concentration could be increased by using a greater cyanobacterial cell density (Deng and Coleman, 1999). Usually ethanol fermentation can be done only in dark anoxygenic conditions due to the inhibition of light and oxygen. Deng and Coleman (1999) reported a genetically engineered strain which could produce ethanol in oxygenic photosynthesis.

Many unicellular, colonial and filamentous microalgae can accumulate starch as a major portion of their biomass. Exploiting these microalgal strains to accumulate starch and directly utilize their enzymatic or anaerobic digestion systems to produce ethanol can provide a cost effective bioethanol production process. Further research is required on screening of high starch accumulating microalgae from corresponding water bodies or to generate efficient microalgae with conventional mutagenesis or using modern tools like genetic engineering. It can increase the production efficiency of fuel ethanol.

The possibility of competition between different pathways for carbon metabolism, including carbohydrate biosynthesis and storage, may limit ethanol production (Deng and Coleman, 1999). There must be a condition to produce ethanol simultaneously with photosynthesis and avoid the steps of accumulation of starch and conversion back to sugar for ethanol. In this approach, there may be an inhibition by accumulated ethanol towards the metabolic activity of algae and hence could decrease the productivity. This necessitates the need of ethanol tolerant algae for effective ethanol production. Development of high salt and temperature tolerant microalgae is necessary for a better utilization of marine water and trapping sunlight in elevated temperature area for getting higher growth and productivity. Akin to Algenol microalgae, the ethanol release in the form of vapour can reduce the risk of ethanol concentration and inhibition in the medium. Therefore, it is necessary to use process engineering approach for the development of effective bioreactor for the simultaneous production and recovery of ethanol.

16.6 Summary

The microscopic nature of microalgae is paradoxical considering their gigantic benefits for application in industry. The utilization of microalgal biomass, either starchy or cellulosic, for bioethanol production is undoubtedly a sustainable and eco-friendly approach in the field of renewable biofuel production. As the importance of microalgae in biodiesel production is growing, an equal or more attention is needed for the efficient use of easily cultivable microorganism to generate green fuel in the form of bioethanol. There are many microalgal strains, both true algae and blue green algae, which have the capability to grow rapidly even in saline water and have the capacity to accumulate starch or oil in a large quantity. It will help culturing the high productive strain in marine or other waste waters for the production of algal biomass and thereby increase in the bioethanol production too. Genetic engineering of selected strain to survive in adverse conditions and development of new bioreactor for the effective production and recovery of ethanol is necessary to generate fuel for the exploding consumption.

16.7 Acknowledgements

The authors are sincerely thankful to the Natural Sciences and Engineering Research Council of Canada (Grants A4984, STP235071, Canada Research Chair), and INRS-ETE for financial support. The views or opinions expressed in this article are those of the authors and should not be construed as opinions of the U.S. Environmental Protection Agency.

16.8 References

Ahn, Y.H., Yang, Y.L., and Choi, C.Y. (1995). “Cellulases as an aid to enhance saccharification of cassava root.” *Biotechnol. Lett.*, 17, 547–550.

Algenaol Biofuels . (2009). *Harnessing the Sun to Fuel the World*™ . Available at http://www.algenolbiofuels.com/Algenol%20101%20PUBLIC%20WEBSITE .pdf (accessed on June, 2009).

Beldman, G., Rombouts, F.M., Voragen, A.G.J., and Pilnik, W. (1984). “Application of cellulase and pectinase from fungal origin for the liquefaction and saccharification of biomass.” *Enzyme Microb. Technol.*, 6, 503–507.

Benemann, J. (1993). “Utilization of carbondioxide from fossil fuel-burning power plants with biological systems II.” *Energy Convers. Manage.*, 34, 999–1004.

Benemann, J. (1997). “CO_2 mitigation with microalgal systems.” *Energy Convers. Manage.*, 38, S475–S479.

Benemann J.R., and Oswald W.J. (1996) “Algal mass culture systems, Systems and Economic Analysis of Microalgae Ponds for Conversion of CO_2 to Biomass.” *US Department of Energy*, Pittsburgh Energy Technology Center., 42–65.

Bush, R.A., and Hall, K.M. (2006). “Process for the production of ethanol from algae” U.S. patent 7135308.

Canizares-Villanueva, R.O., Dominguez, A.R., Cruz, M.S., and Rios-Leal, E. (1995). "Chemical composition of cyanobacteria grown in diluted, aerated swine wastewater." *Bioresour. Technol.*, 51, 111-116.

Carlsson, A.S., Van Bilein, J.B., Möller, R., Clayton, D., and Bowles, D. (2007). *Outputs from EPOBIO Project: micro- and macro-algae utility for industrial application,* CPL Press, York, UK.

Chisti, Y. (2007). "Biodiesel from microalgae." *Biotechnol. Adv.*, 25, 294–306.

Deng, M., and Coleman, J. (1999). "Ethanol Synthesis by Genetic Engineering in Cyanobacteria." *Appl. Environ. Microbiol.*, 65(2), 523-428.

De Morais, G. M., and Costa, J.A. (2007). "Biofixation of carbondioxide by *Spirulina* sp. and *Scenedesmus obliquus* cultivated in a three-stage serial tubular photobioreactor." *J. Biotechnol.,* 129, 439–445.

Fu, P.P., and Dexter, J. (2006). "Methods and compositions for ethanol producing cyanobacteria." U.S. patent (http://www.faqs.org/patents/app/20090155871)

Hughes, E., Benemann, J. (1997). "Biologicalfossil CO_2 mitigation." *Energy Convers. Manage.*, 38, 467–473.

Ingledew, W.M., Thomas, K.C., Hynes, S.H., and McLeod, J.G. (1999). "Viscosity concerns with rye mashes used for ethanol production." *Cereal Chem.*, 76, 459–464.

Kim, M-S., Baek, J-S., Yun, Y.-S., Sim, S. J., Park, S., and Kim, S-C. (2006). "Hydrogen production from *Chlamydomonas reinhardtii* biomass using a two-step conversion process: Anaerobic conversion and photosynthetic fermentation." *Int. J. Hydrogen Ener.*, 31,812-816.

Matsumoto, M., Yokouchi, H., Suzuki, N., Ohata, H., and Matsunaga, T. (2003). "Saccharification of marine microalgae using marine bacteria for ethanol production." *Appl. Biochem. Biotechnol.*, 105–108, 247-254

Packer, M. (2009). "Algal capture of carbon dioxide; biomass generation as a tool for greenhouse gas mitigation with reference to New Zealand energy strategy and policy." *Energy Policy,* in press, doi:10.1016/j.enpol.2008.12.025.

Rafiqul, I.M., Hassan, A., Sulebele, G., Orosco, C.A., Roustaian, P., and Jalal, K.C.A. (2003). "Salt stress culture of blue-green algae *Spirulina fusiformis*." *Pakistan J. Biol. Sci.*, 6, 648-650.

Ratledge, C., and Cohen, Z. (2008). "Microbial and algal oils: do they have a future for biodiesel or as commodity oils?" *Lipid Technol.*, 20, 155–160.

Rodjaroen, S., Juntawong, N., Mahakhant, A., and Miyamoto, K. (2007). "High Biomass production and starch accumulation in native green algal strains and cyanobacterial strains of *Thailand Kasetsart.*" *J. Nat. Sci.*, 41, 570–575.

Rosenberg, J.N., Oyler, G. A., Wilkinson, L., and Betenbaugh, M. J. (2008). "A green light for engineered algae: redirecting metabolism to fuel a biotechnology revolution." *Current Opinion Biotechnol.,* 19, 430-436.

Slominska, L., and Starogardzka, G. (1987). "Application of a multi-enzyme complex in the utilization of potato pulp." *Starch-Starke*, 39, 121–125.

Spolaore, P., Joannis-Cassan, C., Duran E., and Isambert, A. (2006). "Review: Commercial application of microalgae." *J. Biosci. Bioeng.*, 101, 87-96.

Srichuwong, S., Fujiwara, M., Wang, X., Seyama, T., Shiroma, R., Arakane, M., Mukojima, N., and Tokuyasu K. (2009). "Simultaneous saccharification and fermentation (SSF) of very high gravity (VHG) potato mash for the production of ethanol." *Biomass Bioener.*, doi:10.1016/j.biombioe.2009.01.012.

Sriroth, K., Chollakup, R., Chotineeranat, S., Piyachomkwan, K., and Oates, C.G. (2000). "Processing of cassava waste for improved biomass utilization." *Bioresour. Technol.*, 71, 63–69.

Sheehan, J., Dunahay, T., Benemann, J., and Roessler, P. (1998). "A Look Back at the U.S. Department of Energy's Aquatic Species Program- Biodiesel from Algae." NREL/TP-580-24190. National Renewable Energy Laboratory. U.S. Department of Energy.

Thomas, K.C., Hynes, S.H., and Ingledew, W.M. (1996). "Practical and theoretical considerations in the production of high concentrations of alcohol by fermentation." *Process Biochem.*, 31, 321–331.

Thomas, K.C., Hynes, S.H., Jones, A.M., and Ingledew, W.M. (1993). "Production of fuel alcohol from wheat by VHG technology: effect of sugar concentration and fermentation temperature." *Appl. Biochem. Biotechnol.*, 43, 211–226.

Ueda, R., Hirayama, S., Sugata, K., and Nakayama, H. (2001). "Process and system for the production of ethanol from microalgae." European patent EP0645456.

Ueda, R., Hirayama, S., Sugata, K., and Nakayama, H. (1996). "Process for the production of ethanol from microalgae." U.S. patent 5578472.

Vunjak-Novakovic, G., Kim, Y., Wu, X., Berzin, I., and Merchuk, J.C. (2005). "Air-lift bioreactors for algal growth on flue gas: mathematical modeling and pilot-plant studies." *Ind. Eng. Chem. Res.*, 44, 6154–6163.

Wang, D., Bean, S., McLaren, J., Seib, P., Madl, R., Tuinstra, M,. Shi Y, Lenz M, Wu X, and Zhao R. (2008). "Grain sorghum is a viable feedstock for ethanol production." *J. Ind. Microbiol. Biotechnol.*, 35(5), 313–320.

Yamada, R., Bito, Y., Adachi, T., Tanaka, T., Ogino, C., Fukuda, H., and Kondo, A. (2009). "Efficient production of ethanol from raw starch by a mated diploid *Saccharomyces cerevisiae* with integrated α-amylase and glucoamylase genes." *Enzyme Microb. Technol.*, 44, 344–349.

CHAPTER 17

Value-Added Processing of Residues from Biofuel Industries

Prachand Shrestha, Marry L. Rasmussen, Saoharit Nitayavardhana, Samir Kumar Khanal, and J. (Hans) van Leeuwen

17.1 Introduction

Human society has become, and is still, exceedingly and increasingly, dependent on energy. Energy is the mainstay of the economy and is used in agriculture, manufacturing, domestic appliances and for transportation. The latter, in particular, requires liquid fuels as the safest and most convenient source of energy to propel goods and people in vehicles on land, sea, and in the air. Currently, these liquid fuels are almost exclusively made from fossil sources. Not only is this undesirable from a strategic point of view for industrial nations, which are generally net importers of petroleum, but petroleum is a non-renewable resource that cannot be relied on to provide future fuel needs.

The ultimate source of all energy on Earth is sunlight. The sun drives all atmospheric and oceanic processes and is ultimately the source of energy bound up in fossil fuels. Wind power, as one manifestation of solar power, is increasingly used as a source of energy, but it is still relatively expensive and certainly not convenient for modern transportation. Solar power is the most essential part for plant growth and thereby provides the main source of energy for all life on Earth. Plants present an easy way to capture and convert solar power. Plants are also the most economical route to produce biofuels.

Plants can be used as biofuels either directly as solid fuel for heating and power generation or converted into various liquid biofuels. Direct combustion leads to solid residues such as fly ash, while production of liquid biofuels, such as ethanol, butanol, and biodiesel, leads to solid and liquid co-products of relatively low value. Some co-products lead to disposal problems, but there are also many opportunities to make good use of these. This chapter reviews various options for conversion of biofuel residues into high-value products.

17.2 Types and Characteristics of Biofuel Industrial Residues

17.2.1 Starch-based Feedstocks

Corn Ethanol Residues. Ethanol is mainly produced from corn grain in the United States. The corn grain, or kernel, contains approximately 72% starch, 10% protein, 10% fiber, and 4% fat, on a dry wt. basis (White and Johnson, 2003). Dry-grind processing converts primarily the starch in corn to ethanol by milling, cooking, enzymatic hydrolysis of starch to fermentable sugars, yeast fermentation, and distillation. The POET BPX process, or raw starch hydrolysis, skips the cooking step by using raw starch hydrolyzing enzymes (POET, 2009). Approximately one-third of the corn is converted to ethanol, one-third to carbon dioxide, and one-third remains as organic solids in the distillation leftovers, known as stillage. The dry-grind corn process is by far the most commonly used for ethanol production, generating ~ 6 gallons stillage per gallon ethanol in a typical 50 MGY plant (Khanal, 2008). Most solids – wet distillers grains – are removed by centrifugation and the liquid centrate – thin stillage – is partially recycled for in-plant use. Recycling thin stillage as backset is limited to 50% to avoid build-up of solids and fermentation byproducts. Thin stillage is high in organic content, measured as 80–100 g chemical oxygen demand (COD)/L and 7-11% total solids, and low in pH, 3.3–4.5 (Khanal, 2008). The remaining thin stillage is currently concentrated by flash evaporation, blended with distillers dried grains to produce distillers dried grains with solubles (DDGS).

A modern dry-grind corn ethanol plant produces 2.8 gallons of ethanol, 17 pounds of carbon dioxide, and 16 pounds of distillers grains per bushel of corn (RFA, 2008). The co-product distillers grains returns protein and other non-starch nutrients in corn kernels back to feed markets. Distillers grains are sold as livestock feed, but are low in essential amino acids and digestibility for non-ruminants, limiting usage for poultry and swine. U.S. ethanol plants produced 14.6 million metric tons of distillers grains in 2007, of which cattle/dairy consumed 84%; poultry and swine consumed only 5 and 11%, respectively (RFA, 2008). Approximately a quarter of the distillers grains was shipped wet locally, reducing energy inputs for drying and transportation costs.

Cassava Ethanol Residues. Industrial processing of cassava involves separation of tuber peels and subsequent washing and screening steps to collect starch slurry and a solid waste, which is known as cassava bagasse (Pandey et al., 2000a). On dry-weight basis, cassava bagasse contains about 50 % starch. Jaleel et al. (1988) evaluated the fibrous cassava residue for ethanol fermentation via simultaneous solid-phase fermentation and saccharification. Addition of 2.5 to 4% (v/w) of concentrated sulfuric acid to sun-dried cassava residue was followed by pH adjustment (to 4.5), glucoamylase and yeast (*Saccharomyces cerevisiae*) addition to acid hydrolyzed substrate. The substrate-enzyme-yeast mixture was allowed to ferment for 72 hours.

Cassava bagasse has been intensively studied as substrate suitable for producing organic acids, flavors, aromas, and enzymes (Pandey et al., 2000a). However, limited research work has been published in utilization of cassava waste products into value-added products like ethanol. High starch and fiber contents make the solid residue like cassava bagasse a very suitable substrate for ethanol production.

17.2.2 Sugar-based Feedstocks

Sugarcane, sugar beets, and cane/beet molasses are easily fermented to ethanol compared to starch-based feedstocks (Khanal et al., 2008). Starch is a more complex carbohydrate that requires hydrolysis to yield fermentable sugars. The dimeric sugar in sugarcane, for example, is directly fermentable to ethanol. Sugarcane is one of the traditional sugar crops cultivated for ethanol production. Another common feedstock is molasses, the syrup byproduct from processing sugarcane into sugar. The main byproducts from sugarcane-to-ethanol production are bagasse, the fibrous residue from cane stalks, and vinasse, the liquid distillation leftovers. Vinasse may also be referred to as stillage and distillery spent wash. Processing one metric ton of crude sugarcane generates 100 kg of raw sugar, 35 kg of molasses and 270 kg of dry bagasse (Drummond and Drummond, 1996; Garcìa-Pèrez et al., 2002).

Sugarcane bagasse contains approximately 50% cellulose, 25% hemicellulose, and 25% lignin (Pandey et al., 2000a). Bagasse is often burned to generate energy for sugar mill operation and to sell excess electric power. Oven-dried sugarcane bagasse (12% moisture) has a heating value of 18.4 MJ per kg (Zandersons et al., 1999). The cellulose fiber is also suitable for papermaking, which uses 10% of global bagasse production (Khanal et al., 2008). Bagasse accounts for 20% of paper production in South America and India.

Ethanol production from sugarcane juice and molasses yields 13–20 L vinasse per L ethanol, respectively (Wilkie et al., 2000). Vinasse produced from cane sources is high in organic strength and nutrient variability. Literature compiled by Wilkie et al. (2000) indicated that cane juice vinasse has a COD of 30.4 ± 8.2 g/L, biochemical oxygen demand (BOD) of 16.7 ± 3.4, and a low pH of 4.0 ± 0.5. Cane molasses vinasse has a still higher, more variable organic content, with a COD of 84.9 ± 30.6 g/L, BOD of 39.0 ± 10.8 g/L, and pH, 4.5 ± 0.4.

17.2.3 Biodiesel Residues

The transesterification reaction of oil and alcohol produces fatty acid methyl ester (FAME) and glycerol (1,2,3–propanetriol). The former product is also called biodiesel. About 10 % (w/w) of the feedstock oil is converted into a byproduct commonly known as biodiesel glycerol (Kemp, 2006). The production of biodiesel has been tremendously increased in recent years. At present, there are 149 biodiesel plants in the United States, with annual production capacity of 2.5 billion gallons. Various types of feedstock (e.g. soybean oil, corn oil, canola oil, palm oil, rapeseed oil, animal fats, yellow grease, etc.) are used to produce biodiesel (Biodiesel Magazine, 2009). For each gallon of biodiesel produced, approximately 0.35 kg (0.76 lb) of crude glycerol is also produced (Thompson and He, 2006). Glycerol is a colorless, odorless, and viscous co-product. Glycerol is considered as a versatile molecule due to three hydroxyl groups (1°, 2°, and 3°) and therefore, can undergo different reactions ultimately leading to various value-added products (Pagliaro et al., 2007).

17.2.4 Second Generation Biofuel Residues

Second generation biofuels production generally uses ubiquitous lignocellulosic materials, such as crop residues, grasses, forestry, and sawmill wastes as raw materials for liquid biofuel production. The material is then either subjected to a high heat process to pyrolyze or gasify the organic matter (Brown et al., 2000) or pretreated under milder conditions to prepare for subsequent enzymatic hydrolysis to sugars. The former process has a high yield, but is a complex process with many co-products, not all useful.

The enzymatic process uses some form of pretreatment to either remove lignin or cause the lignin bonds between cellulosic fibers to separate. The material is then treated further to break cellulose and hemicellulose into smaller oligosaccharides. These are then hydrolyzed to six- and five-carbon sugars (hexoses and pentoses) respectively using enzymes. Shrestha et al. (2008 and 2009) have developed methodology to use wood-rot and soft-rot fungi to produce in-situ enzymes for hydrolysis. The sugars can be fermented with yeasts to produce ethanol. A better approach would be to convert these sugars into yeast oil (Van Leeuwen et al., 2009a) or into fungal oil using filamentous fungi in the Mycofuel™ process (Van Leeuwen et al., 2009b). All these processes leave residues of unused cellulose, hemicellulose, and lignin.

Lignocellulosic ethanol plants also generate liquid residue, derived after distillation process, called stillage. Ethanol production from lignocelllulosic material yields approximately 11.1 L stillage per L ethanol (Wilkie et al., 2000). A substantial quantity of stillage poses a considerable pollution potential as it has a relatively low pH ((5.35 ± 0.53) and contains high organic matters (with COD 61.3 ± 40 g/L and BOD, 27.6 ± 15.2 g/L) (Wilkie et al., 2000). The following characteristics of stillage: total nitrogen, 2787 ± 4554 mg/L; total phosphorus, 28 ± 30 g/L; sulfate, 651 ± 122 g/L, were shown in the literature compiled by Wilkie et al. (2000).

The solid residues, including unused cellulose, hemicellulose, and lignin, are often burned as boiler fuel for biofuel plants. However, significant increases in ethanol production will require an effective solution for residue management to improve sustainability and profitability of ethanol plants.

17.3 Potential Value-Added Products from Residues

17.3.1 Biofuels

Bioethanol . The main components of co-products/residues from corn-based ethanol industries (dry-grind or wet milling plants) or sugarcane ethanol industries are primarily residual starch and complex polysaccharides – cellulose and hemicellulose as shown in the Table 17.1.

Table 17.1 Important co-products composition.

Co-product	% dry mass components				
	Residual starch	**Cellulose**	**Hemicellulose**	**Lignin**	**Reference**
DDGS	ND	17	19	11*	Miron et al., 2001
Corn fiber	20	18	35	1.3	Abbas et al., 2004; Shrestha et al., 2009
Sugarcane bagasse	ND	50	25	25	Pandey et al., 2000a

* as acid detergent lignin (ADL). ND : not determined

The recalcitrant nature of these co-products impedes enzymatic hydrolysis of complex polysaccharides to fermentable sugars. Physical, chemical, or biological pretreatment methods are necessary unit step(s) in processing of cellulosic biomass to produce biofuels. NREL (2004) reported that fibrous fraction of DDG can increase net ethanol production (per bushel of corn) by 13%. Physical and chemical pretreatment options like liquid hot water treatment, steam explosion, hot dilute acid, ammonia fiber expansion (AFEX), lime, etc. are employed to break down cellulose-lignin interaction or dissolve lignin (See Chapter 9 on pretreatment). Upon pretreatment, enzymatic hydrolysis of cellulosic biomass using commercial enzymes like Accellerase^{TM}, Spezyme^{CP}, and βglucosidase would become faster and more economical.

Bacteria and fungi are also capable of converting cellulosic feedstock to sugars, which can be fermented to ethanol. Cloete and Malherbe (2002) reported bacterial and fungal species that degraded cellulose and hemicellulose to monomeric sugars. *Trichoderma reesei* has been widely studied in producing cellulase enzymes for biodegradation of cellulosic feedstock (Schulein, 1988). Wood-rot fungi such as white- and brown-rot fungi have also been studied for enzymes responsible for lignocellulose degradation (Glenn et al., 1983; Highley and Dashek, 1998) using different substrates and fermentation conditions (Pandey et al., 2000a; Hongzhang et al., 2001; Rodriguez-Vazquez et al., 2003).

Agricultural and Industrial by/co-products like rice husks, corn cobs, sugarcane bagasse, corn fiber, etc. have been extensively studied for their potential application in cellulosic biofuel development. Gutierrez-Correa et al. (1999) and Dien et al. (2006), respectively reported enzymatic saccharificaiton of alkali pretreated sugarcane bagasse and hot water pretreated corn fiber. Wood-rot fungal treatment and subsequent saccharification of cellulosic feedstock for ethanol fermentation have been reported (Abd El-Nasser et al., 1997; Shrestha et al., 2008) for ethanol fermentation. Several yeasts and bacteria have also been evaluated (Lee, 1997) for fermenting both pentoses and hexoses. *Zymomonas mobilis*, *E. coli* (K 011), *Clostridium thermocellum*, and *Klebsiella planticola* are some of the bacteria studied for fermentation of C-5 and C-6 sugars to ethanol. *Pichia stipitis, Candida tropicalis*, and *Candida shehatae* are the yeast species frequently studied for ethanol fermentation from xylose. In a pilot-scale study, fermentation of dilute acid pretreated corn fiber yielded ethanol at 25g/L (Schell et al., 2004).

Formation of inhibitory compounds following biomass pretreatment has also been studied extensively. The performance of microbial fermentation depends on concentration and variation of these compounds (Klinke et al., 2004). Therefore, characterization and detoxification of such products are necessary prior to saccharificaiton and fermentation (Palmqvist and Hahn-Hagerdal, 2000).

Biobutanol. There have been several studies in producing acetone-butanol-ethanol (ABE) from cellulosic feedstock. Chronological development of acetnone-butanol process has been reported by Jones and Woods (1986). The fermentation of cellulosic feedstock is primarily carried out by species of Clostridium like *Clostridium pasteurianum, C. beijerinckii, C. acetobutylicum* etc. Hemi/cellulolytic enzymes like cellulases and xylanase have also been attributed to these bacterial species capable of solvent (acetone and butanol) production.

ABE fermentation of xylan and corn fiber using *C. acetobutylicum* (US patent # 4,649,112) has been reported (Datta and Lemmel, 1987). Qureshi et al. (2006, 2007) studied ABE fermentation of corn fiber and DDGs. ABE yield of 0.35 to 0.39 g ABE/ g sugar from corn fiber using *C. beijerinckii* fermentation was reported. Large-scale fermentation (50 m^3) of pretreated corn cobs using *C. acetobutylicum* produced acetone-butanol, carbon dioxide, and hydrogen (Marchal et al., 1992).

Biodiesel. The unmanageable volume of crude glycerol has become a great concern for biodiesel industries. Pure glycerol has some limited utilization in food, cosmetic, and pharmaceutical industries (Voegele, 2009). Low cost and saturation of demand for crude glycerol in these industries also impede purification steps of crude glycerol. Scientific communities have lately been evaluating other usage of biodiesel glycerol as low cost feedstock.

Pagliaro et al. (2007) reported on the possibility of catalytic conversion of glycerol into products like drugs, fuels, and polymers. Glycerol serves as functional molecules due to its three hydroxy groups (1°, 2°, and 3°) and therefore, can undergo selective oxidation, esterification, hydrogenolysis, dehydration, reforming, or fermentation to produce wide varieties of intermediates or final products. Researchers at Washington State University and Virginia Tech. are exploring the potential of converting glycerol (as carbon source) to omega-3 (ω3) fattyacid like docosahexaenoic acid (DHA) and eicosapentaenoinc acid (EPA) using microalgae (Voegele, 2009). Ashby et al. (2008) reported production of bacterial polyesters-polyhydroxyalkanoates (PHA) by feeding glycerol to different strains of *Pseudomonas*. Da Silva et al. (2009) reviewed on application of glycerol for industrial microbiology. The authors reported microbial conversion of glycerol to valuable organic acids (succinic, propionic, and citric acids), ethanol, pigments, polyesters, and biosurfactants.

Similar research work has also been conducted at Iowa State University, where researchers have been evaluating a process to convert biodiesel glycerol into lipids via feeding crude biodiesel glycerol to oleaginous yeasts. These yeasts can convert a carbon source like glycerol into high lipids, as much as 60 % of their body weight. The lipids can then be extracted and converted into biodiesel while the remaining yeast cells can serve as single cell protein (SCP) (Van Leeuwen et al., 2009a,b). Further research in this area has led to using oleaginous filamentous fungi,

Mucor circinelloides. These fungi grow into pellets of several mm in diameter, and are readily separated and converted to biodiesel using catalyzed ultrasonication for trans-esterification (Van Leeuwen et al., 2009b). This Mycofuel process was granted a 2009 R&D 100 award.

17.3.2 Bio-Oil

Pyrolysis is thermal cracking in the absence of oxygen to produce solid and gas products (Briens et al., 2008). Liquid bio-oil, typically the most valuable product, is recovered by rapid cooling of the condensable vapors in the gas product stream. The endothermic pyrolysis process requires a continuous, high rate of heat transfer to the ground biomass feed. Product gas is usually burned for energy to drive the reaction. Conventional pyrolysis is used for making charcoal. Fast pyrolysis, on the other hand, achieves high yields of bio-oil by rapid heating to around 500℃ and short residence times for the product gas (Bridgwater and Peacocke, 2000). The vapors produced condense to form bio-oil, a dark brown mobile liquid; the higher heating value (17 MJ/kg) is less than half that of conventional fuel oil (42-44 MJ/kg). Bio-oil yields of up to 80% (w/w) on a dry feed basis have been reported.

Generating energy from biofuel residues, such as distillers grains, would greatly improve the energy efficiency of ethanol production (Briens et al., 2008). Pyrolysis of distillers grains to bio-oil may be an appealing option, particularly if bio-oil fractions can be converted to transportation fuels and used as an energy source for ethanol plant operation. Direct combustion of distiller grains to produce heat is more energy efficient, but produces twice the heat requirements of ethanol production. Most existing plants, which are set up for natural gas combustion, would need a major retrofit to directly combust distiller grains. Natural gas burners, however, can be easily retrofitted to use bio-oil. Preliminary pilot-plant pyrolysis tests of distiller grains demonstrated large yields of bio-oil containing aqueous and oily (organic) phases. Distiller grains pyrolysis at 10% moisture resulted in approximately 42.0% organics, 22.4% water, 19.3% solids, and 16.3% losses/gases. After valuable chemicals are extracted, the aqueous phase could be fermented to produce more ethanol. The oily phase could be burned to as a fuel in the ethanol process or converted into a syngas.

Bagasse conversion into high-density fuel sources, like charcoal and bio-oil, can significantly improve the profitability of sugarcane plantations (Drummond and Drummond, 1996). These solid and liquid fuels offer opportunities to export power and can be easily transported to distant consumers. Pyrolysis of sugarcane bagasse has been reported in literature using a variety of reactor types and conditions. Asadullah et al. (2007) produced bio-oil from bagasse in a fixed-bed reactor at temperatures ranging from 300 to 600°C; the maximum yield of bio-oil was 66.0% (w/w) (Table 17.2). Non-condensable, carbon-based gases, including CO, CO_2, methane, ethane, ethene, propane, and propene, were produced as byproducts. The bio-oil density and viscosity were within the range of proposed specifications for the different grades of pyrolysis oils. The pH of the bio-oil was between 3.5 and 4.5 due to the presence of organic acids. Zandersons et al. (1999) also performed experiments on bagasse in a fixed bed reactor (Table 17.2). Product yields and properties were investigated in vacuum pyrolysis tests of bagasse particles (>450 μm) on bench and pilot plant scales by Garcìa-Pèrez et al. (2002). The pyrolysis tests performed on laboratory scale yielded more liquid and less charcoal than achieved on pilot scale (Table 17.2). The bio-oil produced had a low ash content (0.05%), relatively low viscosity, and high calorific value (22.4 MJ/kg) – improving its potential value as a liquid fuel. The charcoal yield was between 19 and 26% (w/w) (bagasse anhydrous basis). This solid byproduct could be used as fuel in boilers where bagasse is currently burned. It is also a potential feedstock for activated carbon production (Xia et al., 1998). Brossard Perez and Cortez (1997) pyrolyzed bagasse in a 50 kg/h continuous, slow pyrolysis reactor, with yields of approximately 30% charcoal, 50% liquid, and 20% gas (on wt basis).

Table 17.2 Research on bagasse thermal conversion.

Reactor type	**Temperature (°C)**	**Yield (wt. %)**			**Reference**
		Liquid	**Char**	**Gas**	
Batch, fixed bed (lab-scale)	500	66.1	24.9	9.0	Asadullah et al., 2007
Batch, vacuum (lab-scale)	500	62.0	19.4	17.6	Garcìa-Pèrez et al., 2002
Batch, vacuum (pilot-scale)	530	51.3	25.6	22.0	Garcìa-Pèrez et al., 2002
Fixed bed (pilot-scale)	530	44.5	35.1	20.4	Zandersons et al., 1999
Thermoreactor with paddle stirrer (lab-scale)	520	57.7	20.7	21.6	Zandersons et al., 1999

In addition to distiller grains and bagasse, pyrolysis of the cassava rhizomes has been studied by Pattiya et al. (2008), who pyrolyzed cassava rhizomes in a laboratory-scale reactor using pyrolysis-gas chromatography/mass spectrometry (Py-GC/MS) to investigate the effect of different catalysts on the bio-oil quality.

17.3.3 Biobased Products

Corn Ethanol. DDGS is an inexpensive, abundant co-product that contains valuable corn components, but is not currently used for industrial applications. Therefore, in addition to increasing consumption among ruminant and non-ruminant animals, new markets and uses (fertilizer, pet litter, packaging materials, food, etc.) are under investigation (RFA, 2009). The high fiber levels in DDGS enable incorporation into biopolymers, an option that could earn greater economic returns. Tatara et al. (2009) evaluated the feasibility of using DDGS as a biofiller in molded, phenolic resins to produce a new biomaterial. DDGS was blended with phenolic resin at four levels (0-75%, by weight) and compression molded at different pressures and temperatures. Pressure and temperature had little effect on the mechanical and physical properties of the molded specimens. DDGS content greatly affected all of the properties, including tensile yield strengths, water absorption, biodegradability, and surface hardness. Results were similar to those obtained in other studies on biofillers. Cellulose was extracted by Xu et al. (2009) from corn kernels and DDGS to evaluate its suitability for films and absorbents. DDGS contains about 9–16% cellulose on dry wt. basis. Alkali and enzymes were used to extract cellulose, with resulting minimum yields (crude cellulose content) of 1.7% (72%) and 7.2% (81%) from corn kernels and DDGS, respectively. The crude cellulose obtained after extraction were made into films using water only. The cellulose recovered could also be used as an absorbent; it was found to hold water up to 9 times its weight. Other potential applications include paper, composites, lubricants, and nutritional supplements.

Destarched corn fiber, a co-product of corn wet milling, was alkali-pretreated and the dissolved hemicellulose was precipitated with ethanol for the recovery of a valuable co-product in a study by Gáspár et al. (2007). Hemicelluloses, such as xylans, have some important applications. The highly-branched heteroxylan from corn is used as a new food gum and is an excellent additive to increase the life of thermoplastic-starch composites. Hemicelluloses also gained attention recently as polymers for chemical and pharmaceutical applications, e.g. cationic biopolymers, hydrogels, and long-chain alkyl ester derivatives. The remaining material, primarily cellulose, in the study by Gáspár et al. (2007) was hydrolyzed with enzymes, and the sugars were fermented to ethanol. Another option for hydrolysates of corn fiber or DDGS, other than production of more ethanol, is fermentation to produce the solvents such as acetone and butanol (Ezeji and Blaschek, 2008).

Cassava Ethanol. Organic acids are important products of microbial cultivation on cassava bagasse (Pandey et al., 2000b). Citric acid production on cassava bagasse, in particular, has been well studied using solid-state fermentations. Kolicheski et al. (1995) studied citric acid production by *Aspergillus niger* on three cellulosic supports – cassava bagasse, sugarcane bagasse, and vegetable sponge. Cassava bagasse was found to be a good substrate, resulting in 13.6 g citric acid per 100 g dry substrate (42% yield based on sugars consumed) and up to a 70% yield after improving fermentation conditions. Vandenberghe et al. (1999) investigated citric acid production by *A. niger* on sugarcane bagasse, coffee husk, and cassava bagasse. Cassava bagasse had the best fungal growth, giving the highest citric acid yield (8.8 g/100 g dry matter) among the three substrates. An alternative approach is to hydrolyze cassava bagasse using amylolytic enzymes and then to use the hydrolysate for microbial cultivations. Fumaric acid is an important organic acid produced by *Rhizopus* strains on cassava bagasse hydrolysate (Carta et al., 1999). Fumaric acid has a wide range of applications, including use as an acidulant in the food and pharmaceutical industries.

Sugarcane Ethanol. Researchers have studied utilization of sugarcane bagasse as a raw material for production of a variety of value-added products, such as protein-enriched animal feed, enzymes, pharmaceutical compounds, and biomaterials (Pandey et al., 2000a). Cellulosic biomass, such as bagasse, is a potentially inexpensive renewable feedstock for the production of biomaterials. To synthesize biopolyesters, for example, sugarcane bagasse first requires pretreatment to hydrolyze the cellulose and release sugars (Yu and Stahl, 2008). In research by Yu and Stahl

(2008), bagasse was subjected to a dilute acid pretreatment, releasing sugars and inhibitory hydrolysates. An aerobic bacterium, *Ralstonia eutropha*, was employed to remove the organic inhibitors, primarily formic acid, acetic acid, furfural, and acid-soluble lignin, for potential recycling of the process water. Simultaneous biosynthesis of polyhydroxyalkanoates (PHAs) for the production of bioplastics was also achieved; PHA biopolyesters accumulated to 57% of cell mass. Poly(3-hydroxybutyrate) was the major biopolyester produced on the bagasse hydrolysates.

Sugarcane bagasse has also been investigated for the production of organic and amino acids. Bagasse can serve as an inert support material or be pre-treated to release fermentable sugars (Pandey et al., 2000a). Researchers produced lactic acid (Soccol et al., 1994), citric acid (Lakshminarayana et al., 1975; Manonmani and Sreekantiah, 1987) and glutamic acid (Nampoothiri and Pandey, 1996) using bagasse as an inert carrier during solid-state fermentations. Soccol et al. (1994), for instance, impregnated bagasse with a liquid medium of glucose and calcium carbonate for lactic acid production by a strain of *Rhizopus oryzae*. Yields of approximately 94 and 137 g/L of L(+)-lactic acid were obtained with 120 and 180 g/L of glucose added in liquid- and solid-state fermentations, respectively.

Second Generation Biofuel. Lignocellulosic residues serve as low-cost materials for microbe cultivation for producing valuable enzymes and organic acids. A study from Iowa State University has suggested that the fungal processing is more effective in degrading complex carbohydrates than bacterial cultivation (Van Leeuwen et al., 2003). Fungal process is not a new approach. Teunissen et al. (1992) showed that the fungus, *Piromyces* sp., was able to utilize several substrates including cellobiose, cellulose, fructose, mannose, xylan, and xylose, which remain in ligocellulosic residues (unused cellulose and hemicellulose). In anaerobic environment, metabolic pathway of cellulolytic and xylanolytic enzymes was induced resulting in the production of formate and acetate as main fermentation products. Bacterial cultivation on cellulosic material has also been investigated for the ability of organic acid production. Co-culture of *Ruminococcus albus.* and a hydrogen-using acetogen (HA),which is a gram-negative coccobacillus isolated from horse feces, on cellulosic material resulted in bioconversion of cellulose to acetate (Miller and Wolin, 1995). The end product, acetate, could be used in the production of de-icing salt, calcium magnesium acetate, and potassium acetate. Lactic acid is another organic acid, which could be generated by bioconversion of cellulosic material. Lactic acid can be used in multiple applications as it potentially serves as a raw material for industrial chemicals, such as ethyl lactate, propylene glycol, 2,3-pentandione, propanoic acid, acrylic acid, acetaldehyde, dilactide, and biodegradable plastic (polylactic acid, PLA) (Datta et al., 1995; Litchfield, 1996; Varadarajan and Miller, 1999; John et al., 2009). The possibility of lactic acid production from 2nd generation biofuel residues has confirmed by Wee and Ryu (2009). This study showed an economically feasible lactic acid production by using lignocellulosic hydrolysate as an inexpensive raw material.

Several high value biochemical compounds could be generated by microbial cultivation on 2nd generation biofuel residues. Xylitol, natural sweetener in food product, could be produced from the utilization of hemicellulose hydrolysate by metabolically engineered *Escherichia coli* (Sakakibara et al., 2009). Vanillin, an aromatic flavor molecule, is another high value biochemical, which is widely used in food, pharmaceutical, and cosmetic industries (Fragues et al., 1996; Hocking, 1997; Muheim and Lerch, 1999; Laufenberg et al., 2003). Although few studies have investigated the biological production of vanillin directly from lignin, it has been believed that lignin can be biologically converted to vanillin. Muheim and Lerch (1999) investigated the ability of biological vanillin production from ceugenol or ferulic acid. This research suggested the use of bacteria, *Pseudomonas putida* and *Streptomyces setonii*, for vanillin production. The biological conversion of lignin into vanillin could be supported by the ability of these bacteria to degrade lignin (Haider, 1978; Pettey and Crawford, 1984). Therefore, microbial cultivation in the lignin residue from 2nd generation biofuel process is one of the possible approaches for vanillin production.

17.3.4 Enriched Feed Products

Joshi and Sandhu (1996) studied the feed quality of residue from solid-state fermentation of apple pomace. Increase in crude protein, fats, and vitamins have been reported in the dried fermentation residue. Fermentation with *Saccharomyces cerevisiae* had comparatively higher soluble protein yield. Protein enriched fermentation residue can serve as better feed for monogastric animals, including poultry. Essential amino acids, vitamins and trace elements are some of the essential parameters to evaluate the feed quality of solid residues from fermentation.

Corn-based ethanol industries in the United States produces huge quantities of fibrous co-products like DDG and DDGS, corn gluten meal (CGM), corn gluten feed (CGF) etc. The two former products (DDG and DDGS accounted up to 14.6 million metric tons in 2007), along with thin stillage are produced from corn dry-grind processes while CGM and CGF are produced in corn wet-milling processes. DDG, DDGS, and CGF are basically included as greater proportion in cattle ration. CGM is used as swine and poultry feed because of higher protein (gluten) fraction. The usage of distillers grains in the animal feed sector is as follows (RFA, 2009): cattle (42%), dairy (42%), swine (11%), and poultry (5%).

Feed trial studies performed by Ham et al. (1994) using DDGS and thin stillage implicated the necessity of combining distillers grains, fiber, and thin stillage for better performance of cattle. Inclusion of DDGS, high protein DDG, and corn germ, by reducing quantities of soybean meal and soybean oil, in swine diet did not have a negative impact on overall growth performance, carcass composition and pork palatability (Widmer et al., 2008). Swiatkiewicz and Koreleski (2008) also reported that inclusion of 5-8% DDGS in starter diets and 12-15% DDGS in grower-finisher diets for broilers, layers and turkeys is acceptable. Kleinschmit et al. (2006) conducted series of experiments on performance of lactating cows with a diet that contained some DDGS from various sources. Low lysine content in corn co-products has always been a great concern in animal nutrition. One way of increasing the lysine content would be to increase the total protein content of co-products like DDGS. POET's BFRAC^{TM} process separates corn into fiber, germ, and endosperm. Liquefaction and fermentation of endosperm produces ethanol and high protein residue, also called as high protein DDGS. High protein DDGS has comparatively higher amino acid contents (including lysine lysine content: soybean meal – 2.76, DDGS – 0.88, and High Protein DDGS – 1.11).

Researchers at Iowa State University are conducting experiments to grow feed grade fungus: *Rhizopus oligosporus* in thin stillage from corn dry-grind industry. Growing fungus in organically rich stillage could solve multiple problems: energy saving, water-nutrient-enzyme recycling and producing fungal protein (Rasmussen et al., 2007; Van Leeuwen et al., 2008). Fungal protein rich in essential amino acid like lysine can supplement a solid co-product like DDG. The success of such a study not only improves the feed value of the co-product, but also helps hundreds of ethanol industries to save expenses on energy and water usage.

Single-cell protein (SCP) could also be produced from microbe cultivation on lignocellulosic residues. Several fungi are capable of using lignin and cellulosic materials as substrates for their cultivation. White-rot fungi are a more attractive culture for animal feed production because they can degrade lignin and preserve cellulose fibers thereby improving the quality of lignocellulosic substrates for animal feed (Karunanandaa and Varga, 1996).

17.4 Process Options for Recovery of Products

17.4.1 Corn Ethanol

The corn ethanol industry is pursuing innovative processes to recover new value-added products, to enhance the protein content of feed co-products, and to reduce energy consumption. Raw starch fermentation, corn fractionation, and corn oil extraction are a few of the processing options utilized by ethanol companies (RFA, 2008). Several corn fractionation methods, both dry and wet, have been investigated to recover the germ and fiber prior to fermentation of the endosperm fraction in dry grind corn ethanol production (Murthy et al., 2009). In dry fractionation, the main unit operations added are corn tempering, size reduction, sieving, aspiration, and density separation, similar to conventional corn dry milling. Wet fractionation methods are similar to conventional corn wet milling. An aqueous medium is used to separate and recover the corn germ, pericap fiber, and/or endosperm fiber prior to fermentation. The main unit operations are soaking/steeping, size reduction, density separation, and sieving. Corn fractionation prior to fermentation helps reduce the amount of DDGS produced by up to 66% depending on the fractionation method (Wang et al., 2005). Three commercial dry grind ethanol plants (BFRAC^{TM}, POET, Sioux Falls, SD) dry fractionate corn to recover the germ and bran fiber (Clark, 2006; Nilles, 2006).

Technology is also being developed to remove corn oil from the syrup prior to mixing with wet distillers grains in the dryer (RFA, 2007a). The oil recovered could serve, for instance, as a feedstock for biodiesel production. Removing the oil enhances the value of distillers grains as a feed co-product by concentrating the protein. A high-protein, animal feed co-product, for swine and poultry in particular, could also be produced from the thin stillage stream as shown in research at Iowa State University (Van Leeuwen et al., 2008). Thin stillage contains biodegradable organics and sufficient micronutrients at pH 4 – an ideal fungal cultivation feedstock. A fungal treatment process for thin stillage can provide energy savings by avoiding thin stillage evaporation, recovery of nutritious fungal biomass (rich in highly-digestible essential amino acids), and potential for water and enzyme recycling. The high-protein fungus could be fed to non-ruminants.

A newer process for fungal beneficiation of thin stillage is under development by MycoInnovations and Iowa State University. This process uses an oleaginous fungus, *Mucor circinelloides,* to convert part of the organic content into an intracellular oil that can readily be converted into a form of biodiesel. This process, Mycofuel, has been awarded an R&D 100 award in 2009 (Van Leeuwen et al, 2009b).

Another processing option is to use ethanol co-products to supply part of the energy demands of the ethanol plant. Corn Plus, a Minnesota-based ethanol producer, installed a fluidized bed reactor to generate steam for running the facility from biomass, such as syrup (RFA, 2007a). The resulting reduction in natural gas usage was more than 50%. In addition, Corn Plus has two wind turbines to provide about 45% of the electric needs.

The production of more ethanol from corn by hydrolysis and fermentation of agricultural/ethanol processing residues, like corn stover, corn cobs, corn fiber, and distillers grains, is an attractive possibility; it will help bridge the industry to the conversion of other cellulosic feedstocks and energy crops, such as switchgrass and miscanthus (RFA, 2008).

17.4.2 Cassava Ethanol

Cassava bagasse is an ideal substrate for microbial processing to recover value-added products (Pandey et al., 2000b); it is rich in organic materials (30-50% starch on dry weight basis) and low in ash content. Researchers have investigated the production of several products, including single-cell protein (SCP), organic acids, flavor and aroma compounds, and mushrooms, from cassava bagasse. Most bioconversion processes attempted have been based on solid-state fermentation.

17.4.3 Sugarcane Ethanol

Processes involving microbial cultivations on agro-industrial residues, such as bagasse, can be classified into two broad groups: submerged (liquid) fermentation and solid-state fermentation (Pandey et al., 2000a). Submerged fermentation can be sub-divided based on the substrate utilized: the whole residue or the liquid hydrolysate from hydrolyzed residue. Solid-state fermentations can be divided based on the use of residue as the source of carbon/energy or as an inert solid support. The production of enzymes, ethanol, single-cell protein (SCP), and other value-added products have been reported using submerged fermentation of whole or hydrolyzed bagasse. Cellulolytic enzymes production using bagasse has been widely studied. The fungi from basidiomycetes family are commonly used for these fermentations; isolates contain extracellular cellulases, such as exoglucanase, endoglucanase, β glucosidase, and ligninases (Kalra et al., 1984; Sidhu et al., 1983; Sarkar and Prabhu,

1983; Nigam and Prabhu, 1991; Nigam et al., 1987, 1988).

17.5 Disposal or Reuse Options of Residues from Recovery Processes

Usage of residues from various biofuel industries are generally dependent on their (i) organic content (carbohydrates, protein and lipids), (ii) successful research studies for their application as substrate for microbial growth and ancillary benefits from microbes (like single cell proteins – SCPs, pharmaceuticals, organic products, enzymes, etc.), (iii) possibility of their reuse as secondary (complex) substrate for biofuel production (like ethanol, biodiesel, biobutanol, etc.), (iv) end use as feedstock for energy (for boilers, anaerobic digester etc) generation, and (v) proper manageable disposal (landfill, soil amendments, fertilizer, etc.) without causing long term environmental harms.

The reuse or proper disposal of these residues is also governed by technological cost and maturity (and adaptability) of existing technologies. The biological, chemical, and physical natures of the residues from processes like fermentation, nutrient recycle and reuse have to be considered prior to opt disposal or reuse options of the residue. Liquid streams (from microbial fermentation) have to be characterized for pH, organic acids, residual nutrients, enzymes, and potency of microbial propagules (spores, hyphae, etc.). These variabilities are to be treated (pH adjustment, nutrient removal, sterilization, etc.) prior to their treatment at municipal waste treatment facilities or disposal into natural water bodies.

Adjustment of nutrient levels and acidity-alkalinity for reuse can substitute quantitative volume of fresh water and medium in upstream process in the same or near by industries. High organic liquid waste can be feedstock for anaerobic digesters to produce energy-rich methane gas. On-site treatment of liquid waste can provide irrigation water in surrounding fields or parks. On the other hand, solid residue can be used as soil amendments or fertilizer. Solid residue may still be reused as support for microbes for fermentation. Following proper treatment: pH adjustment, sterilization, drying, etc., the solid residue from fermentation can be used as an animal feed because of availability of residual but essential and readily available residual proteins along with enzymes to further hydrolyze the organics. There have been wide applications of microbes (bacteria and fungi) in solid-substrate and submerged fermentation. Feed studies on fermentation residues (mixed with microbes and their byproducts) need greater evaluation.

17.6 Energy Implications of Producing Further Co-Products

The energy requirement is usually the largest operational expense in processing biofuels. The same often applies to processing co-products. It would be good policy to select options for producing further co-products on the basis of minimizing energy requirements. This will be illustrated by an example.

Dry-grind ethanol plants typically dispose of their excess thin stillage by concentrating this to a syrup, where about 80% of the water is evaporated. Typically, about 3 gallons of water is evaporated per gallon ethanol produced. Using a fungal process to purify the water to a quality suitable for recycling would avert the need for evaporation. Not evaporating 3 gal water/gal ethanol saves around 3 x 5,490 BTU (multiple effect evaporation) = 16.5 kBTU/gal ethanol. With an energy recovery of up to 50%: 8.25 kBTU/gal ethanol, or 0.91 million MBTU on 110 Mgal/year scale, the expected net energy saving would be 0.91 x 10^{12} BTU – 42 GWh x (3414/0.4) x10^6 = 0.55 x10^6 MBTU/year on a 110 Mgal/year plant. Industry-wide net savings in the US would be 50 x 10^6 MBTU/year = 50 billion ft^3 natural gas (Eq. ~1,600 MW power plant).

The fungal cultivation process also needs energy, though. The fungi require oxygen to grow and the process must be kept aerobic. Van Leeuwen et al. (2008) calculated the energy requirements for a fungal water recovery process on a 110 Mgal/year plant as 21 GWh/year. When it is assumed that power plants operate at a 40% efficiency, the equivalent energy consumption for aeration and other operational power for fungal cultivation amounts to about 25% of the energy saved by not evaporating.

Fungal harvesting makes the process more attractive as the fungi contain high levels of lysine and methionine. These make the fungal biomass a valuable co-product.

17.7 Summary

A clear weakness in the biofuel industry is that there are various residues with little or even a negative value. The biofuel main product is subject to severe fluctuation in tandem with the price of oil. Finding opportunities in value-adding to reduce disposal costs co-products in the biofuel industry and it is essential to increase and stabilize the income.

Biodiesel production makes glycerol, saturating the market for this co-product, so that conversion of this co-product to other substances, for which there is a strong demand, would be preferable. However, the separation and purification of glycerol also add to the cost of many such schemes. Using the glycerol as feedstock for oleaginous fungi and converting the intracellular oil to more biodiesel may be a much simpler solution, as this does not require prior separation of the glycerol. Microbial processes are generally very useful to remove and convert dissolved organic substances. Microbes can also be used to convert solid residues into useful co-products. Furthermore, the microbes can be harvested as yet another valuable co-product.

Other solid residues from biofuel production may require other approaches. Burning bagasse to provide the energy needs for sugar extraction, fermentation and, ethanol production makes sense, but making paper or other products from bagasse would obviously add more value, although it would require further investment in equipment for paper making.

There are many challenges remaining in the biofuels industry. Research into co-product beneficiation is of utmost importance to improve the profitability and longer term sustainability in this important emerging industrial sector.

17.8 References

Abbas, C., Beery, K., Dennison, E., and Corrington, P. (2004). Thermochemical treatment, separation and conversion of corn fiber to ethanol. In *Lignocellulose Biodegradation.* Eds., Saha, B. C., and Hayashi, K. pp. 84–97.American Chemical Society, Washington, D.C.

Abd El-Nasser, N.H., Helmy, S.M., and El-Gammal, A.A. (1997). Formation of enzymes by biodegradation of agricultural wastes with white rot fungi. *Polym. Deg. Stab.*, 55, 249 - 255.

Asadullah, M., Rahman, M.A., Ali, M.M., Rahman, M.S., Motin, M.A., Sultan, M.B., and Alam, M.R. (2007). Production of bio-oil from fixed bed pyrolysis of bagasse. *Fuel,* 86, 2514 - 2520.

Ashby, R.D., Wyatt, V.T., Foglia, T., and Solaiman, D. (2008). Industrial products from biodiesel glycerol. In: *Biocatalysis and Bioenergy*. Eds., Hou, C.T., and Shaw, J-F. pp 131-154.John Wiley & Sons, Inc., Hoboken, New Jersey.

Biodiesel magazine. (2009). Retreieved February 11, 2009 from http://www.biodieselmagazine.com/plant-list.jsp

Bridgwater, A.V., and Peacocke, G.V.C. (2000). Fast pyrolysis processes for biomass. *Renew. Sustain. Energ. Rev.,* 4, 1 - 73.

Briens, C., Piskorz, J., and Berruti, F. (2008). Biomass valorization for fuel and chemicals production – a review. *Int. J. Chem. Reactor Eng.* 6, 1 - 49.

Brossard Perez, L.E., and Cortez, L.A.B. (1997). Potential for the use of pyrolytic tar from bagasse in industry. *Biomass and Bioenergy*, 12, 363 - 366.

Brown, R.C., Liu, Q., and Norton, G. (2000). Catalytic effects observed during the co-gasification of coal and switchgrass. *Biomass and Bioenergy,* 18, 499 - 506.

Carta, F.S., Soccol, C.R., Ramos, L.P., and Fontana, J.D. (1999). Production of fumaric acid by fermentation of enzymatic hydrolysates derived from cassava bagasse. *Bioresour. Technol.*, 68, 23 - 28.

Clark, P.J. (2006). Ethanol growth inspires advances in cereal milling. *Food Technology Magazine,* 60, 73 - 75.

Cloete, T.E., and Malherbe, S. (2002). Lignocellulose biodegradation: Fundamentals and applications. *Rev. Environ. Sci. Biotechnol.*, 1, 105 - 114.

Da Silva, G.P., Mack, M., and Contlero, J. (2009). Glycerol: a promising and abundant carbon source for industrial microbiology. *Biotechnol. Adv.*, 27(1), 30 - 39.

Datta, R., and Lemmel, S.A. (1987). Utilization of xylan and corn fiber for direct fermentation by *Clostridium acetobutylicum*, U.S. patent 4649112.

Datta, R., Tsai, S.P., Bonsignore, P., Moon, S.H., and Frank, J.R. (1995). Technological andeconomic potential of poly (lactic acid) and lactic acid derivatives. *FEMS Microbiol. Rev.*, 16, 221 - 231.

Dien, B.S., Li, X.L., Iten, L.B., Jordan, D.B., Nichols, N.N., O'Bryan, P.J., and Cotta, M.A. (2006). Enzymatic saccharification of hot-water pretreated corn fiber for production of monosaccharides. *Enzym. Microb. Tech.*, 39, 1137 - 1144.

Drummond, A.R.F., and Drummond, I.W. (1996). Pyrolysis of sugar cane bagasse in a wire-mesh reactor. *Ind. Eng. Chem. Res.*, 35, 1263 - 1268.

Ezeji, T., and Blaschek, H.P. (2008). Fermentation of dried distillers' grains and solubles (DDGS) hydrolysates to solvents and value-added products by solventogenic clostridia. *Bioresour. Technol.*, 99, 5232 - 5242.

Fragues, C., Mathias, A., and Rodrigues, A. (1996). Kinetics of vanillin production from kraft lignin oxidation. *Ind. Eng. Chem. Res.,* 35, 28 - 36.

Garcìa-Pèrez, M., Chaala, A., and Roy, C. (2002). Vacuum pyrolysis of sugarcane bagasse. *J. Anal. Appl. Pyrol.*, 65, 111 - 136.

Gáspár, M., Kálmán, G., and Réczey, K. (2007). Corn fiber as a raw material for hemicellulose and ethanol production. *Process Biochem.* 42, 1135 - 1139.

Glenn J.K., Morgan M.A., Mayfield, M.B., Kuwahara, M., and Gold, M.H. (1983). An extracellular H_2O_2-requiring enzyme preparation involved in lignin biodegradation by the white rot basidiomycete *Phanerochaete chrysosporium. Biochem. Biophys. Res. Comm.*, 14, 1077–1083.

Gutierrez- Correa, M., Portal, L., Moreno, P., and Tengerdy, R. P. (1999). Mixed culture solid substrate fermentation of *Trichoderma reesei* with *Aspergillus niger* on sugar cane bagasse. *Bioresour. Technol.*, 68, 173 - 178.

Haider, K., Trojanowski, J., and Sundman, V. (1978). Screening for lignin degrading bacteria by means of 14C-labelled lignins. *Arch. Microbiol.*, 119, 103 - 106.

Ham, G.A., Stock, R.A., Klopfenstein, T.J., Larson, E.M., Shain, D.H., and Huffman, R.P. (1994). Wet corn distillers byproducts compared with dried corn distillers grains with solubles as a source of protein and energy for ruminants. *J. Anim. Sci.*, 72, 3246 - 3257.

Highley, T.L., and Dashek, W.V. (1998). Biotechnology in the study of brown- and white-rot decay. In: *Forest Products Biotechnology*. pp. 15–36. Taylor and Francis, London, United Kingdom.

Hocking, M.B.J. (1997). Vanillin: synthetic flavoring from spent sulfite liquor. *Chem. Educ.,* 74, 1055 - 1059.

Hongzhang, C., Fujian, X., and Zuohu, L. (2001). Solid-state production of lignin peroxidase (LiP) and manganese peroxidase (MnP) by *Phanerochaete chrysosporium* using steam-exploded straw as substrate. *Bioresour. Technol.*, 80, 149 - 151.

Jaleel, S.A., Srikanta, S., Ghildyal, N.P., and Lonsane, B.K. (1988). Simultaneous solid-phase fermentation and saccharificaiton of cassava fibrous residue for production of ethanol. *Starch,* 40(2), 55 - 58.

John, R.P., Anisha, G.S., Nampoothiri, K.M., and Ashok P. (2009). Direct lactic acid fermentation: Focus on simultaneous saccharification and lactic acid production. *Biotechnol. Adv.,* 27, 145 - 152.

Jones, D.T., and Woods, D.R. (1986). Acetone-butanol fermentation revisited. *Microbiol. Revi.*, 50(4), 484 - 524.

Joshi, V.K., and Sandhu, D.K. (1996). Preparation and evaluation of an animal feed byproduct produced by solid-state fermentation of apple pomace. *Bioresour. Technol.,* 56(2), 251 - 255.

Kalra, M.K., Sidhu, M.S., Sandhu, D.K., and Sandhu, R.S. (1984). Production and regulation of cellulases by *Trichoderma harzianum. Appl. Microbiol. Biotechnol.,* 20, 427 - 429.

Karunanandaa, K., and Varga, G.A. (1996). Colonization of rice straw by white-rot fungi (*Cyathus stercoreus*): effect on ruminal fermentation pattern, nitrogen metabolism, and fiber utilization during continuous culture. *Anim. Feed Sci. Technol.,* 61, 1 - 16.

Kemp, W.H. (2006). *An introduction to biodiesel. Biodiesel basics and beyond: a comprehensive guide to production and use for the home and farm,* Aztext Press, Ontario, Canada.

Khanal, S.K. (2008). *Anaerobic Biotechnology for Bioenergy Production: Principles and Application*, Wiley-Blackwell Publishing, Ames, IA, USA.

Khanal, S.K., Rasmussen, M., Shrestha, P., van Leeuwen, J., Visvanathan, C., and Liu, H. (2008). Bioenergy and biofuel production from wastes/residues of emerging biofuel industries. *Water Environ. Res.*, 80, 1625 - 1647.

Kleinschmit, D.H., Schingoethe, D.J., Kalscheur, K.F., and Hippen, A.R. (2006). Evaluation of various sources of corn dried distillers grains plus solubles for lactating dairy cattle. *J. Dairy Sci.*, 89, 4784 - 4794.

Klinke, H.B., Thomsen, A.B., and Ahring, B.K. (2004). Inhibition of ethanol-producing yeast and bacteria by degradation products produced during pre-treatment of biomass. *Appl. Microbiol. Biotechnol.*, 66, 10 - 26.

Kolicheski, M.B., Soccol, C.R., Marin, B., Medeiros, E., and Raimbault, M. (1995). Citric acid production on three cellulosic supports in solid state fermentation. In *Advances in Solid State Fermentation.* Eds., Roussos, S., Lonsane, B.K., Raimbault, M., and Viniegra-Gonzalez, G. pp. 447 - 460. Kluwer Academic Publishers, Dordrecht, The Netherlands.

Lakshminarayana, K., Chaudhary, K., Ethiraj, S., and Tauro, P. (1975). A solid state fermentation method for citric acid production using sugar cane bagasse. *Biotechnol. Bioeng.*, 7, 291 - 293.

Laufenberg, G., Kunz, B., and Nystroem, M. (2003). Transformation of vegetable waste into value added products: (A) the upgrading concept; (B) practical implementations. *Bioresour. Technol.,* 87, 167 - 198.

Litchfield, J.H. (1996). Microbiological production of lactic acid. *Adv. Appl. Microbiol.,* 42, 45 - 95.

Lee, J. (1997). Biological conversion of lignocellulosic biomass to ethanol. *J. Biotech.*, 56, 1 - 24.

Manonmani, H.K., and Sreekantiah, K.R. (1987). Studies on the conversion of cellulase hydrolysate into citric acid by *Aspergillus niger*. *Process Biochem.*, 22, 92 - 94.

Marchal., R., Ropars, M., Pourquie, J., Fayolle, F., and Vandecasteele, J.P. (1992). Large-scale enzymatic hydrolysis of agricultural lignocellulosic biomass. Part 2: Conversion into acetone-butanol. *Bioresour. Technol.*, 42(3), 205 - 217.

Miller, T.L., and Wolin, M.J. (1995). Bioconversion of cellulose to acetate with pure cultures of *Ruminococcus albus* and a hydrogen-using acetogens. *Appl. Environ. Microbiol.,* 61, 3832 - 3835.

Miron, J., Yosef, E., and Ben-Ghedalia, D. (2001). Composition and in vitro digestibility of monosaccharide constituents of selected byproduct feeds. *J.Agric. Food Chem.,* 49(5), 2322 - 2326.

Muheim, A., and Lerch, K. (1999). Towards a high-yield bioconversion of ferulic acid to vanillin. *Appl. Microbiol. Biotechnol.,* 51, 456 - 461.

Murthy, G.S., Sall, E.D., Metz, S.G., Foster, G., and Singh, V. (2009). Evaluation of a dry corn fractionation process for ethanol production with different hybrids. *Ind. Crops Prod.*, 29, 67 - 72.

Nampoothiri, K.M., and Pandey, A. (1996). Solid state fermentation for L-glutamic acid production using *Brevibacterium* sp. *Biotechnol. Lett.*, 18, 199 - 204.

National Renewable Energy Laboratory. (2004). *Getting Extra "Corn Squeezins." Technology Brief-11/1993*. Retrieved November 21, 2004 from http://www.nrel.gov/docs/gen/old/5639.pdf

Nigam, P., and Prabhu, K.A. (1991). Effect of cultural factors on cellulase biosynthesis in submerged bagasse fermentation by basidiomycetes cultures. *J. Basic Microbiol.*, 31, 285 - 292.

Nigam, P., Pandey, A., and Prabhu, K.A. (1987). Cellulase and ligninase production by basidiomycete culture in solid-state fermentation. *Biol.Waste.,* 20, 1 - 9.

Nigam, P., Pandey, A., and Prabhu, K.A. (1988). Fermentation of bagasse by submerged fungal cultures: Effect of nitrogen sources. *Biol. Waste.,* 23, 313 - 317.

Nilles, D. (2006). On the edge of innovation. *Ethanol Prod. Mag.*, 12, 42 - 43.

Pagliaro, M., Ciriminna, R., Kimura, H., Rossi, M., and Pina, C.D. (2007). From glycerol to value-added products. *Angewandte Chemie Inter. Edit.*, 46 (24), 4434 - 4440.

Palmqvist, E., and Hahn-Hagerdal, B. (2000). Fermentation of lignocellulosic hydrolyzates. II: inhibitors and mechanisms of inhibition. *Bioresour. Technol.*, 74, 25 - 33.

Pandey, A., Soccol, C.R., Nigam, P., and Soccol, V.T. (2000a). Biotechnological potential of agro-industrial residues. I: sugarcane bagasse. *Bioresour. Technol.*, 74, 69 - 80.

Pandey, A., Soccol, C.R., Nigam, P., Soccol, V.T., Vandenberghe, L.P.S., and Mohan, R. (2000b). Biotechnological potential of agro-industrial residues. II: cassava bagasse. *Bioresour. Technol.*, 74, 81 - 87.

Pettey, T.M., and Crawford, D.L. (1984). Enhancement of lignin degradation in *Streptomyces* spp. by protoplast fusion. *Appl. Environ. Microbiol.,* 47, 439 - 440.

POET. (2009). POET Research. Retrieved February 11, 2009 from http://www.poetenergy.com/about/showDivision.asp?id=6

Pattiya, A., Titiloye, J.O., and Bridgwater, A.V. (2008). Fast pyrolysis of cassava rhizome in the presence of catalysts. *J. Anal. Appl. Pyrol.*, 81, 72 - 79.

Qureshi, N., Li, X.L., Hughes, S., Saha, B.C., and Cotta, M.A. (2006). Butanol production from corn fiber xylan using *Clostridium acetobutylicum*. *Biotechnol. Prog*., 22, 673 - 680.

Qureshi, N., Ezeji, T.C., Ebener, J., Dien, B.S., Cotta, M.A., and Blaschek, H.P. (2007). Butanol production by *Clostridium beijerinckii*. Part I: use of acid and enzyme hydrolyzed corn fiber. *Bioresour. Technol*., 99 (13), 5915 - 5922.

Rasmussen, M., Khanal, S.K., Pometto III, A.L., Van Leeuwen, J. (2007). Bioconversion of thin stillage from corn dry-grind ethanol plants into high-value fungal biomass. Annual International Meeting (AIM) – Association of Agricultural and Biological Engineers (ASABE), Minneapolis, MN. *Paper number: 077030.*

Renewable Fuels Association (RFA) (2007a) *Building New Horizons – Ethanol Industry Outlook 2007*. Washington, D.C.

Renewable Fuels Association (RFA) (2007b). How ethanol is made. Retrieved September 21, 2007 from www.ethanolrfa.org/resource/made/

Renewable Fuels Association (RFA) (2008) *Changing the Climate – Ethanol Industry Outlook 2008*. Washington, D.C.

Renewable Fuels Association (RFA). (2009). *Industry resources: co-products.* Retrieved January 21, 2009 from http://www.ethanolrfa.org/industry/resources/co-products/

Rodriguez-Vazquez, R., Roldan-Carrillo, T., Diaz-Cervantes, D., Vazquez-Torres, H., Manzur-Guzman, A., and Torres-Dominiguez, A. (2003). Starch-based plastic polymer degradation by the white rot fungus *Phanerochaete chrysosporium* grown on sugarcane baggase pith: enzyme production. *Bioresour. Technol.*, 86, 1 - 5.

Sakakibara, S., Saha, B.C., and Taylor P. (2009). Microbial production of xylitol from L-arabinose by metabolically engineered *Escherichia coli. J. Biosci. Bioeng.,* 107, 506 - 511.

Sarkar, C., and Prabhu, K.A. (1983). Studies on cellulolytic enzymes production by *Trichoderma* sp. utilizing bagasse. *Agricultural Wastes,* 6, 99 - 113.

Schell, D.J., Riley, C.J., Dowe, N., Farmer, J., Ibsen, K.N., Ruth, M.F., Toon, S.T., and Lumpkin, R.E. (2004). A bioethanol process development unit: initial operating experience and results with a corn fiber feedstock. *Bioresour. Technol.*, 91, 179 - 188.

Schulein, M. (1988). Cellulases of *Trichoderma reesei* in methods in enzymology. In: *Biomass Part A: Cellulose and Hemicellulose Vol 160*. Eds., Wood, W.A., and Kellog, S.T. pp. 234 – 242.Academic Press, San Diego, California.

Shrestha, P., Rasmussen, M., Khanal, S. K., Pometto, A.L., and van Leeuwen, J (Hans). (2008). Solid-state fermentation of corn fiber by *Phanerochaete chrysosporium* and subsequent fermentation of hydrolysate into ethanol. *J. Agric. Food Chem.*, 56 (11), 3918 - 3924.

Shrestha, P., Khanal, S.K., Pometto, A.L., and van Leeuwen, J (Hans). (2009). Enzyme Production by Wood-Rot and Soft-Rot Fungi Cultivated on Corn Fiber Followed by Simultaneous Saccharification and Fermentation. *J. Agric. Food Chem.*, 57 (10), 4156 - 4161.

Sidhu, M.S., Kalra, M.K., and Sandhu, D.K. (1983). Production, purifcation and characterization of cellulase from *Trichoderma longibranchiatum. Process Biochem.*, 18, 13 - 15.

Soccol, C.R., Marin, B., Raimbault, M., and Lebeault, J.M. (1994). Potential of solid state fermentation for production of L(+)-lactic acid by *Rhizopus oryzae. Appl. Biochem. Biotechnol.,* 41, 286 - 290.

Swiatkiewicz, S., and Koreleski, J. (2008). The use of distillers dried grains with solubles (DDGS) in poultry nutrition. *World's Poultry Science Journal*, 64, 257 -266.

Tatara, R.A., Rosentrater, K.A., and Suraparaju, S. (2009). Design properties for molded, corn-based DDGS-filled phenolic resin. *Ind. Crops Prod.,* 29, 9 - 15.

Teunissen, M.J., de Kort, G.V.M., op den Camp, H.J.M., Huis, J. H. J., and Huis in't Veld, J.H. J. (1992). Production of cellulolytic and xylanolytic enzymes during growth of anaerobic fungus *Piromyces* sp. on different substrates. *J. Gen. Microbiol.,* 138, 1657 - 1664.

Thompson, J.C., and He, B.B. (2006). Characterization of crude glycerol from biodiesel production from multiple feedstocks. *Appl. Eng. Agr.*, 22(2), 261 - 265.

Vandenberghe, L.P.S., Soccol, C.R., Pandey, A., and Lebeault, J.M. (1999). Solid state fermentation for the synthesis of citric acid by *Aspergillus niger. Bioresour. Technol.*, 74, 175 – 178.

Van Leeuwen, J. (Hans), Hu, Z., Yi, T., Pometto, A.L., and Jin, B. (2003). Kinetic model for selectivecultivation of microfungi in a microscreen process for food processing wastewater treatment and biomass production. *Acta Biotechnol.*, 23, 289 - 300.

Van Leeuwen, J. (Hans), Rasmussen, M.L., Khanal , S.K., and Pometto, A.L. (2008). R&D 100 Award: The MycoMax process for value-adding to the corn-to-ethanol process. Fungal Research Group, Iowa State University and MycoInnovations; Ames, IA.

Van Leeuwen J. (Hans), Beattie, S., Kim, T.H., Grewell, D., Johnson, L., Hammond, E., Pometto III, A., Iassonova, D., Shrestha, P., and Vincent, M. (2009a). Single cell bio-oil from an integrated fungal lignocellulosic biorefinery. Recipient of Grand Prize for University Research, American Academy of Environmental Engineers, Washington DC.

Van Leeuwen J. (Hans), Beattie, S., Kim, T.H., Grewell, D., Mitra, D., Ziel, C., Chintareddy, V.R., Chand, P., Montalbo-Lomboy, M., and Verkade, J. (2009b). Mycofuel – an integrated fungal process to produce biodiesel from wastes. Recipient of R&D 100 – 2009 Award.

Varadarajan, S., and Miller, D.J. (1999). Catalytic upgrading of fermentation-derived organic acids. *Biotechnol. Prog.,* 15, 845 - 854.

Voegele, E. (2009). Glycerin: research turns up new uses. *Biodiesel Magazine.* Retreived March 28, 2009 from http://www.biodieselmagazine.com/article.jsp?article_id=3237

Wang, P., Singh, V., Xu, L., Johnston, D.B., Rausch, K.D., and Tumbleson, M.E. (2005). Comparison of enzymatic (e-mill) and conventional dry grind corn processes using a granular starch hydrolyzing enzyme. *Cereal Chem.*, 82, 734 - 738.

Wee, W.J., and Ryu, H.W. (2009). Lactic acid production by *Lactobacillus* sp. RKY2 in a cell-recycle continuous fermentation using lignocellulosic hydrolyzates as inexpensive raw materials. *Bioresour. Technol.,* 100, 4262 - 4270.

White, P.J., and Johnson, L.A. (2003). *Corn Chemistry and Technology*, American Association of Cereal Chemists, St. Paul, MN, USA.

Wilkie, A.C., Riedesel, K.J., and Ownes, J.M. (2000). Stillage characterization and anaerobic treatment of ethanol stillage from conventional and cellulosic feedstocks. *Biomass and Bioenergy*, 19, 63 - 102.

Widmer, M.R., McGinnis, L.M., Wulf, D.M., and Stein, H.H. (2008). Effects of feeding distillers dried grains with solubles, high-protein distillers dried grains, and corn germ to growing-finishing pigs on pig performance, carcass quality, and the palatability of pork. *J. Anim. Sci.*, 86, 1819 - 1831.

Xia, J., Noda, K., Kagawa, S., and Wakao, N. (1998). Production of activated carbon from bagasse grown in Brazil. *J. Chem. Eng. Jap.*, 6, 987 - 990.

Xu, W., Reddy, N., and Yang, Y. (2009). Extraction, characterization and potential applications of cellulose in corn kernels and distillers' dried grains with solubles (DDGS). *Carbohydr. Polym.*, 76(4), 521 – 552.

Yu, J., and Stahl, H. (2008). Microbial utilization and biopolyester synthesis of bagasse hydrolysates. *Bioresour. Technol.*, 99, 8042 - 8048.

Zandersons, J., Gravitis, J., Kokorevics, A., Zhurinsh, A., Bikovens, O., Tardenaka, A., and Spince, B. (1999). Studies of the Brazilian sugarcane bagasse carbonization process and products properties. *Biomass and Bioenergy,* 17, 209 - 219.

CHAPTER 18

Energy Life Cycle Analysis of a Biofuel Production System

Dev S. Shrestha, and Anup Pradhan

18.1 Introduction

Energy Life Cycle Analysis (ELCA) is a branch of general life cycle analysis (LCA) which accounts for all the energy that goes into making a biofuel and compares it with energy contained in the produced fuel. ELCA is relevant for the products that is used as a fuel or used for energy. Net Energy Ratio (NER) is a measure of the efficiency of making a biofuel and generally expressed as:

$$\text{NER} = \frac{\text{Biofuel energy output}}{\text{Biofuel share of total energy input}} \tag{18.1}$$

Biofuel is usually produced in conjunction with other co-products, for example soybean meal and glycerol in case of biodiesel and distillers dried grains with solubles (DDGS) in case of corn ethanol. Therefore, only biofuel share of total energy input is used in denominator of equation 18.1.

Much of the attention directed toward biofuels production focuses on the perception that they are renewable and have superior environmental attributes compared to their petroleum counterparts (EPA, 2002; Knothe et al., 2005). Biofuels are desirable because they are derived from renewable biological material, whereas petroleum fuels come from a finite resource that is depleting rapidly. According to Energy Information Administration, proven reserve of crude oil in the United States to be about 22 billion bbl in year 2007 and the production was about 8.5 million bbl per day (EIA, 2008). With this rate of crude oil production, the reserves-to-production (R/P) ratio for US is estimated to be just 7 years. The R/P ratio is the number of years for which the current level of production of fuel can be sustained by its reserves and is calculated by dividing proved reserves at the end by the production in that year (Feygin and Satkin, 2004). Compared to 8.5 million bbl of daily production, US consume about 20 million bbl of crude petroleum. The deficit is fulfilled through imported fuels. Biofuels have potential to reduce the nation's dependence on imported fuels and conserve fossil fuels for the alternative uses. Therefore, developing renewable biofuels is a need of current time.

A significant amount of nonrenewable energy is used in farm operations, transportation, and to convert biological materials into biofuels. The fossil energy requirement for biofuel production is a key to understand the extent to which the biofuel is a renewable energy source. The amount of fossil energy used is measured over the entire life cycle of biofuel to determine its renewability. The renewability can range from completely renewable (if no fossil energy input is needed), to nonrenewable (if the fossil energy required is as much or more than the energy content in the biofuel).

It is important to measure the renewability of a biofuel for two reasons. Firstly, it is useful to know how much a biofuel relies on fossil energy. The less a biofuel depends on fossil energy, the larger is the contribution it can make towards energy security in a long run. Secondly, the renewability of different biofuels can be compared by policymakers and others to favor or discourage a particular biofuel production.

This raises an important question on how to define the degree of renewability. One way is to express the result obtained from the ELCA as the Fossil Energy Ratio (FER), which is defined as the ratio of the energy output from the final biofuel to the fossil energy required to produce the biofuel (Spath and Mann, 2000). According to Sheehan et al. (1998), FER is expressed as:

$$\text{FER} = \frac{\text{Biofuel energy output}}{\text{Biofuel share of fossil energy input}} \tag{18.2}$$

Please note that equation 18.2 uses only fossil energy use in denominator instead of total energy as in equation 18.1. If no renewable fuel is used in biofuel production, FER is equal to NER defined in eq. (18.1). FER is commonly used to measure and compare the advantages of different biofuels. Although the authors believe that equation 18.2 is a good definition of FER, there are other definitions used in literature to define FER. Other definitions and their logical explanations are discussed in the section 18.2.4, later in this chapter. The results can arbitrarily vary depending on how the FER is defined. Therefore, when the value of FER is interpreted, the reader should be aware of the definition of FER is being used and the meaning of that number.

Another common way to express the advantage of biofuel is Net Energy Value (NEV). NEV is simply the difference between energy in biofuel and fossil energy it uses usually expressed as energy per unit area of land. The advantage of using NEV is that it quantifies the absolute energy gain, which is not obviously quantified in FER. Both FER and NEV should be used as complimentary indicators of the benefits from a biofuel. To illustrate, let us assume two cropping systems; one crop system produces 1000 units of biofuel and consumes 200 units of fossil fuel. Another crop system produces 2000 units of biofuel and uses 500 units of fossil fuel

(Figure 18.1). If we look at the FER, obviously the first system is better as it has FER of 5 compared to 4 for the latter; but if we look at NEV, the second system has NEV of 1500 units compared to only 800 units for the first system. This makes the second system look better over the first.

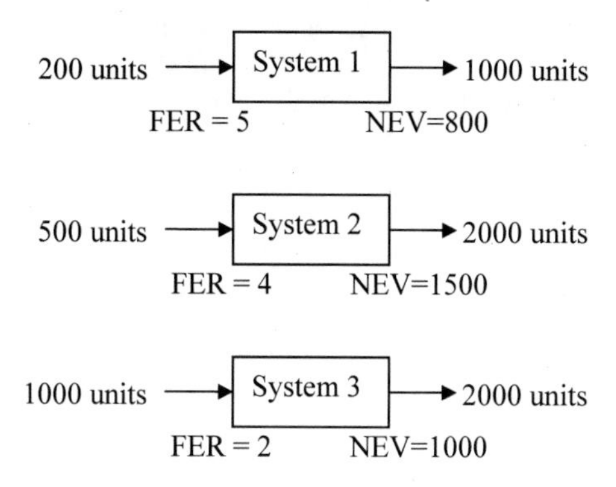

Figure 18.1 FER and NEV are both indicators of energy gain. Which system is more desirable, with higher FER or higher NEV? The answer depends on the opportunity cost.

Since system 2 is producing more energy per unit area, one might argue that system 2 is more desirable than system 1 and NEV is a better indicator. Although NEV seemed a better indicator for energy gain in above discussion, it has its own pitfall. To see this, let's change the scenario for the second cropping system with 2000 units of energy output and 1000 units of fossil energy input (system 3 in Figure 18.1), is this system still favorable over system 1? NEV for system 3 is still higher than system 1, but FER has reduced to 2. In order to get advantage of one thousand additional units of energy, 800 additional units of fossil energy needs to be spent. The return on additional energy input is marginally low; is that justifiable? This is exactly the same question farmers ask themselves in economic sense as well that what level of productivity will give them the maximum benefit on their investment? They definitely want to maximize the return rate on their investment but that does not mean having the highest revenue to investment ratio, which would occur at the lower investment range according to the law of diminishing return. The answer to optimum level depends on opportunity cost on additional investment. However, when it comes to biofuel production, it is hard to assign opportunity cost for additional use of fossil fuel to increase NEV. There is no definitive single answer to this question. Several other factors need to be considered other than energy such as carbon footprint and environmental impacts. This makes the use of LCA analysis more useful and at the same time more complicated.

18.2 ISO Standards for LCA

ISO standards (ISO 14040, 2006a; ISO 14044, 2006b) provide guidelines for conducting LCA. According to these ISO standards, LCA should be conducted in four phases (Figure 18.2). The first phase of the standard, "Goal and scope definition" requires a precisely defined system boundary and level of detail of an LCA study, which depends primarily on the subject and the intended use. The second phase, inventory analysis is the detailed accounting of inventory that enters and leaves the system boundary. The input and output data for each individual process are collected to meet the goals of the defined study. The third phase, impact assessment evaluates the environmental impacts from each individual process and provides assessment of a product's life cycle inventory results. The fourth and the final phase is interpretation in which the results of the study are interpreted and significant issues are identified in accordance with the goal and scope definition.

Even though ISO standards have been set forth to standardize the procedures, FER calculations are still difficult to compare as experts use different system boundaries, data citation and assumptions based on their specific objective.

18.2.1 System Boundary

A reasonable system boundary should be defined and justified in life cycle analysis based on the objectives of the study (ISO, 1998). System boundary defines what is to be included or excluded in the life cycle analysis of a specific biofuel production. For a biofuel system, LCA usually starts from crop production, which is considered as the cradle and ends after the fuel is used in a vehicle and this is considered as the grave. For a meaningful LCA, the entire production system must be well defined, for instance biodiesel production from soybean oil includes five distinct processes: (i) soybean farming, (ii) soybean transportation to the processing facility, (iii) oil extraction and purification, (iv) conversion of oil into biodiesel (or transesterification), and (v) transportation of biodiesel for distribution (Sheehan et al., 1998). The system boundary for soybean agriculture proposed by Sheehan et al. (1998) in the life cycle study of biodiesel production is illustrated in Figure 18.3. The figure considers all processes of the entire life cycle of soybean agriculture system that have significance for producing soybeans.

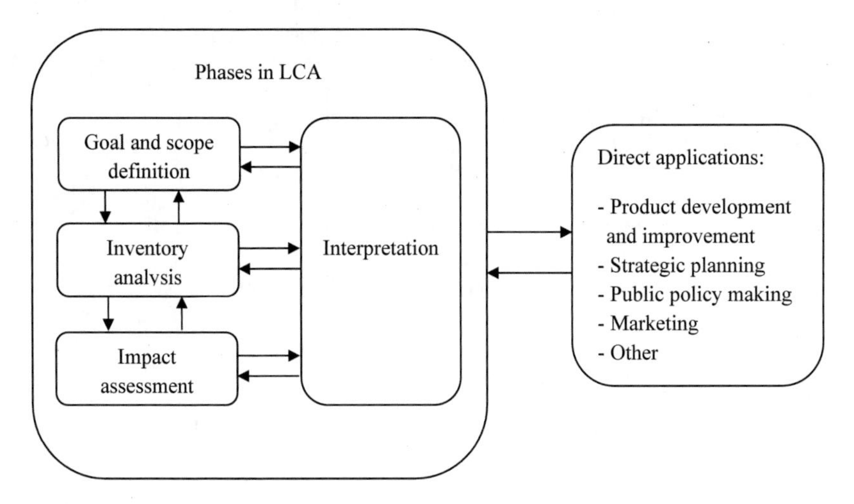

Figure 18.2 Phases in LCA (Source: ISO 14040, 2006a). This material is reproduced from ISO 14040:2006 with permission of the American National Standards Institute (ANSI) on behalf of the International Organization for Standardization (ISO). No part of this material may be copied or reproduced in any form, electronic retrieval system or otherwise or made available on the Internet, a public network, by satellite or otherwise without the prior written consent of the ANSI. Copies of this standard may be purchased from the ANSI, 25 West 43rd Street, New York, NY 10036, (212) 642-4900, http://webstore.ansi.org

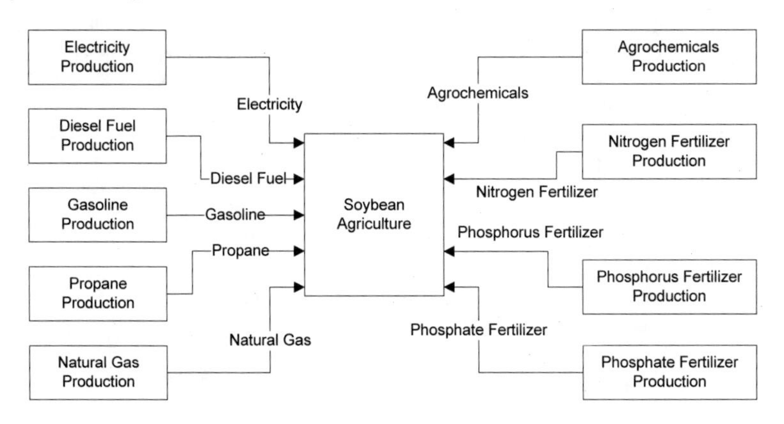

Figure 18.3 Soybean agriculture system boundary (Source: Sheehan et al., 1998).

It is nearly impossible to track all the energy used over the life cycle of a product because each input has a life cycle of its own. For instance, machinery used in agriculture has its own life cycle that may involve factory building to computer software. In turn, each of these has its own life cycle and the chain continues endlessly. Therefore, a researcher must limit the system boundary used in the analysis yet providing enough information for making the analysis meaningful (Pradhan et al., 2008). All significant sources of energy must be included in the inventory, such as the liquid fuel and electricity used powering the equipment in the system. The energy needed for the production of materials such as fertilizers, pesticides and other petrochemicals should be included. Researchers should also carefully exclude the subsystems that are hard to estimate such as labor and can be ignored.

18.2.2 Calorific Value or Life Cycle Energy?

In general, there are two approaches of assigning energy equivalence to an input. One way is to consider the absolute chemical or physical energy contributed by an input and referred to as "calorific value." For instance, diesel fuel has a calorific value of about 36 MJ/L of fuel, which is the amount of heat released when 1L of diesel is burned (Usually lower calorific value). Even though this approach is simpler and direct, it does not tell about renewability and environmental impacts because producing 1L of diesel requires additional processing energy. The second method uses the energy consumed in producing a specific input as an energy equivalence of that input and referred to as "life cycle energy." For instance, it takes some energy to extract and refine the mineral oil into diesel fuel. Usually this is the energy used in LCA to reflect the ultimate impact of using the product. The disadvantage of using the second approach is that it can be extremely complex and creates ambiguity in defining the system boundary.

Usually, the life cycle energy of an input is higher than its calorific value. However, there are some exceptions such as seed used in planting. It takes 3.16 MJ of fossil energy (Sheehan et al., 1998) to produce one kg of soybean seed; whereas soybeans contain 16.8 MJ/kg (estimated from the equivalent energy of protein, carbohydrate and fat in the seed). The rest of the energy in seed is coming from solar energy trapped by the crop.

Therefore, whether to use calorific energy or life cycle energy for an input depends on the original goal and the scope definition. If the purpose of the study is to assess the renewability of biofuel production, then it makes sense to consider the life cycle energy of all inputs. However, if the objective of the model is to calculate the system efficiency, it is generally recommended to consider the bigger of the calorific or the life cycle energy.

18.2.3 Co-product Credit

Usually more than one final product is generated during biofuel production. For instance, soybean meal and glycerin are co-products of soybean biodiesel production and distillers dried grains with soluble (DDGS) is obtained in corn ethanol production. These co-products share the input energy in some way and hence the input energy must be allocated among co-products. Since detailed information is often unavailable to measure the energy requirements for an individual co-product, an appropriate method needs to be used to assign co-product energy share. There are primarily two ways to estimate the co-products share of energy, namely allocation and displacement methods (Huo et al., 2008). In general, no method has absolute advantage and hence the appropriate method should be chosen based on the goal of the study.

An allocation method systematically allocates the material use, energy use, and emissions between the primary product and co-products based on either mass, energy content, or economic revenue. Among these allocation methods, mass allocation method is the most frequently used as it is easy to apply and provides reasonable and reproducible results (Vigon et al., 1993). This method simply allocates energy co-products by their proportionate weight fraction. The economic value is the most practical in that the biofuel ultimately must make profit to be a viable. This method shares input energy among co-products based on their economic value. Economic value-based method effectively differentiates among various forms of energy. In reality, a unit of energy from coal does not cost as much as a unit of energy from electricity. Economic based co-product allocation takes account of this. The drawback of this method is that the market dynamics make one product's FER a moving target and harder to comprehend. Energy based allocation method assigns input energy to different co-product based on their relative energy content. This method is also as simple to apply as mass allocation and have underlying assumption that value of a product is nearly proportional to its energy content. However, it does not takes into account of the fact that not all co-products are valued in terms of its energy content and not all co-products are used for fuel.

The displacement method determines an equivalent product replaced by each co-product and energy equivalent of those replaced co-products are subtracted from the total energy input to calculate the bioenergy share of energy input (Shapouri et al., 2006). This method is argued as the scientifically preferred methods (Kim and Dale, 2002; Farrell et al., 2006). The difficulties with the displacement method are accurate determination of the displaced products and identification of the approach to obtain their life cycle energy use and emissions (Huo et al., 2008). For example, DDGS or canola meals are animal feed products similar to soybean meal, but they are not exact substitutes for soybean meal. Therefore, it is impossible to calculate a precise comparable energy value even for comparable substitutes.

18.2.4 NER/FER Definitions

Eqs. (18.1 and 18.2) define NER and FER, respectively, as the ratio of renewable fuel output to its share of total (for NER) or fossil (for FER) fuel input. However, this is not the universal definition used in the literature. Based on the definition used, the NER or FER value could be quite different. The following example illustrates for biodiesel production.

Soybean meal and glycerin are the co-products of biodiesel production. Thus, the energy required to produce biodiesel is the fraction of total energy assigned to biodiesel in proportion to the mass fraction of the biodiesel output stream. E_1, E_2 and E_3 are the energy inputs. In case of biodiesel, the total and the fossil energy inputs are not very different except for the part of electricity used for processes shown in Figure 18.4.

Assume that E_1 is the fossil energy used in soybean agriculture, bean transport and crushing. The products after these processes are meal and oil. Only oil product is used for biodiesel production. If f_1 is the mass fraction for oil production, $E_1 \times f_1$ is the energy used for oil production.

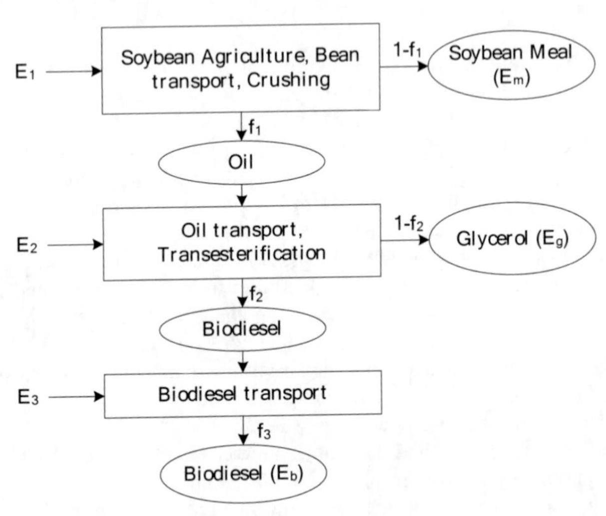

Figure 18.4 Energy allocation model: $E_{1...3}$ are the energy inputs, $f_{1...3}$ are the fractions of the energy inputs attributed to biodiesel, and E_b, E_m and E_g are the energy equivalents (calorific values) of biodiesel, meal and glycerol, respectively.

The sum of the energy splits is equal to the total fossil energy input. The allocation rule separates the energy used to produce the soybean oil and biodiesel from the energy used to produce the soybean meal and glycerin respectively in the following manner:

$$\text{Biodiesel share of energy input} = ((E_1 \times f_1 + E_2)\, f_2 + E_3)\, f_3 \qquad (18.3)$$

where E_1 is the combined energy input for agriculture, bean transport and crushing, f_1 is the fraction of energy to oil, E_2 is energy input for oil transport and transesterification, f_2 is the fraction of energy to biodiesel; E_3 is energy input for biodiesel transport, f_3 is biodiesel fraction of energy. For mass based allocation Sheehan et al. (1998) used $f_1 \cong 0.183$, $f_2 \cong 0.824$ and $f_3 = 1$.

To determine FER, E_1, E_2 and E_3 are considered as fossil energy inputs. Sheehan et al. (1998) and Argonne National Laboratory (2006) used eq. (18.3) to estimate the biodiesel share of FER as the ratio of the energy content in biodiesel to the fossil energy required to produce the biodiesel and defined FER as:

$$FER_1 = \frac{E_b}{((E_1 \cdot f_1 + E_2)f_2 + E_3)f_3} \quad (18.4)$$

where, E_b is the calorific value of biodiesel (≅37 MJ/kg). The $E_{1...3}$ of the eq. (18.3) indicates fossil energy input. In a similar way NER can also be defined according to eq. (18.4) in which$E_{1...3}$ would be total energy instead of fossil energy.

Pimentel and Patzek (2005) calculated the biodiesel share of energy differently. Instead of using eq. 18.3 to calculate biodiesel share of energy, the authors subtracted the calorific value of co-products (E_c) from total energy input. It should be noted in Figure 18.4 that $E_m \neq E_1$ $(1\text{-}f_1)$, i.e. energy contained in meal is not equal to the meal's share of energy input. This is because solar energy trapped during photosynthesis is not accounted in ELCA and usually meal will have higher energy than its share of total input energy. In this case, the FER is defined as:

$$FER_2 = \frac{E_b}{E_1 + E_2 + E_3 - E_c} \quad (18.5)$$

The disadvantage of using this definition for FER is that E_c has a potential to be equal or exceeding the sum of total fossil energy input. E_c contains stored solar energy that is not included in ELCA and hence E_c could be greater than the sum of the fossil energy inputs. As E_c approaches the sum of the total fossil energy inputs, the ratio approaches infinity and if it exceeds then eq. (18.5) yields an absurd negative result for energy ratio (Van Gerpen and Shrestha, 2005).

Ahmed et al. (1994) calculated the co-product energy by multiplying the energy input by the mass fraction of the co-products, and its value was added to E_b in the numerator instead of subtracting it from the denominator, as in the Pimentel's study. Mathematically, the FER was defined as:

$$FER_3 = \frac{E_b - E_c}{E_1 + E_2 + E_3} \quad (18.6)$$

Ahmed's FER definition is logical if the co-products are also used for fuel. Since, soybean meal is not generally used for energy production; this definition carries little meaning in evaluating the renewability of the fuel. Hill et al. (2006) defined the FER as a combination of Pimentel and Ahmed's definition. The authors calculated E_c as the calorific value of the co-products following (Pimentel and Patzek, 2005), but added it to the numerator just like in equation 18.6.

Soybean meal co-product is not used for energy, and using the calorific value of the co-product would make the number harder to interpret. If a co-product such as glycerol is actually used as a source of energy, then it should be taken into account.

Nevertheless, in practice, co-products such as glycerol or meal are not used as a source of energy they should not be accounted for the estimation of biodiesel's renewability. Therefore, the FER definition used by Sheehan et al. (1998) (eq. 18.4) is the most appropriate one as it only accounts for the fossil energy inputs shared by biodiesel fraction at each stage throughout the biodiesel production steps.

18.3 Corn Ethanol ELCA

US ethanol production has grown from just a few thousand gallons in the mid 1970s to 6.5 billion gallons in 2007 (Shapouri et al., 2002; NCGA, 2008). The US Department of Agriculture estimates the ethanol production to grow to 10.8 billion gallons in 2009 (USDA, 2007). Ethanol is primarily produced from corn in the United States. Currently, about 95% of ethanol in the United States is produced from corn. In addition to an alternative to the gasoline, ethanol has been established as an octane enhancer and blending of ethanol with gasoline has become a common practice to meet the oxygen requirements mandated by the Clean Air Act Amendments of 1990 (Shapouri et al., 2002). Energy Policy Act (EPACT) of 2005 includes a mandate for up to 7.5 billion gallons of biofuels to be used in gasoline by 2012. A new Energy Independence and Security Act established in 2007 (EISA, 2007) expands the renewable fuel standard to 9 billion gallons in 2008, and increase to up to 36 billion gallons by 2022.

The cornstarch conversion into ethanol through yeast is a well-established technology. In the United States, either dry corn milling or wet corn milling is adopted for ethanol production (Figure 18.5). The difference between these two methods is the initial treatment of grain. In dry corn milling, corn mash is gelatinized and enzyme alpha-amylase is added to produce liquefied cornstarch, which is converted to sugars using enzyme, glucoamylase. The produced sugar is then fermented to ethanol by yeast. DDGS and carbon dioxide (CO_2) are obtained as co-products (Khanal, 2008). Wet milling involves multiple steps in which corn germ (oil), corn gluten feed and corn gluten meals are separated from starch before fermenting starch into ethanol (Davis, 2001). The co-products include corn steep liquor (CSL), corn oil, gluten feed, gluten meal, high fructose corn syrup (HFCS) and carbon dioxide.

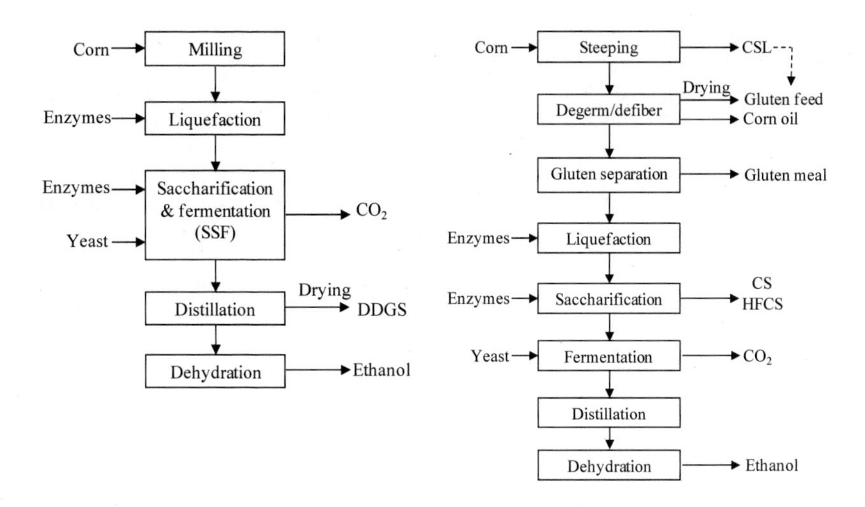

Figure 18.5 Corn ethanol production processes. (Left) dry corn milling process; (right) wet corn milling process (Adapted from Drapcho et al., 2008, reproduced with permission).

The energy balance issue was first surfaced in the mid 1970s when ethanol began to receive attention as a substitute to petroleum gasoline (Shapouri et al., 2002). Several studies analyzed the energy benefits of replacing gasoline with ethanol. The net energy value (NEV) of corn ethanol was found slightly negative (Chambers et al., 1979). The energy balance studies of ethanol resurfaced in the late 1980s due to the environmental concerns (Shapouri et al., 2002). The energy balance studies on corn ethanol production take into account the entire life cycle of ethanol production: (i) the energy used to produce and transport corn, (ii) the energy used to produce ethanol, and (iii) the energy used in distribution of ethanol in gasoline (NCGA, 2008). Co-products, such as DDGS in dry milling process, are also produced during ethanol production, which share the fraction of energy inputs. Thus, not all the energy used by an ethanol plant is directed towards ethanol production.

As discussed earlier, it is difficult to estimate the energy input from some sources due to large variability. Some of these include farm machinery and energy use to build the infrastructure. These inputs are used over a course of several years and some kind of assumption need to be made as how the input should be amortized over time. Depending on the farm size and assumptions made, the outcome may be quite different. If the anticipated impact from such an input is deemed small, it is not uncommon to exclude them. Unless an extensive data is available for highly variable and considerably small input, and does not adequately represent the industry practice, it is justifiable to leave such input out.

Shapouri et al. (2002) updated the energy balance of corn ethanol in 2002 using 1996 survey data. The summary result of the study is shown in Table 18.1. The energy inputs required for the production of corn include direct energy used on farms (such as gasoline, diesel, natural gas, LP gas, and electricity), energy used to produce fertilizers (nitrogen, phosphorus and potassium), and energy used to produce pesticides (herbicides and insecticides). The study used data from 1996 Farm Costs and Returns Survey (FCRS). The farm energy input estimates are weighted average of nine states and the corn yield used in the estimate was 125 bushels per acre (average corn yield in 1995-1997). The study did not include the energy input to farm machinery, which may have been considered negligible. The energy inputs associated with ethanol production include electricity (used for grinding and running electric motors) and thermal energy from natural gas or coal (used for fermentation, ethanol recovery and dehydration). The study did not include secondary energy inputs, such as farm machinery in corn production and building materials (cement, steel, etc.) in ethanol plants.

The table also shows how the co-product allocation method could change the NER value. The energy content of ethanol considered in the study was higher heating value (HHV) of 83,961 Btu/gal, which is the standard heat of combustion reference to water in combustion exhaust as liquid water (Shapouri et al., 2002). Either energy value can be used but it should be consistent throughout the study. The result of this study shows that corn ethanol has a positive NER in all cases. Although several studies showed an overall net positive energy gain, there are claims that the use of ethanol does not reduce gasoline use.

Table 18.1 further shows that that the minimum energy ratio of ethanol is 1.30 for wet milling plants with replacement method. This means ethanol yields 1.30 units of energy for every unit of primary energy consumed over its life cycle. A study conducted by the researchers at the University of Minnesota found that corn ethanol provides 25% more energy than required for its production (Hill et al., 2006). A considerable amount of variation was observed in these studies stemming from various assumptions including corn production and ethanol conversion, co-products allocation, and data citation.

Table 18.1 NER with different energy allocation processes.

Allocation method	Energy allocation		Energy use	Co-product credit	Energy use with co-product credit	NER
	Ethanol	Co-product				
	Percent		Btu/gal			
Mass based						
Wet mill	48	52	79,503	40,516	38,987	2.15
Dry Mill	49	51	74,447	37,158	37,289	2.25
Energy content						
Wet mill	57	43	79,503	33,503	46,000	1.83
Dry Mill	61	39	74,447	28,415	46,032	1.82

Market value						
Wet mill	70	30	79,503	23,374	56,129	1.50
Dry Mill	76	24	74,447	17,486	56,961	1.47
Replacement						
Wet mill	81	19	79,503	14,804	64,699	1.30
Dry Mill	82	18	74,447	13,115	61,332	1.37

Source: Shapouri et al. (2002).

Ten independent studies since 1995 found that ethanol has a positive net energy balance, while only two studies by the same author reported negative net energy return (Table 18.2). Ten studies showing positive energy balance have NER more than one, which indicates that a corn ethanol has a favorable renewability. Pimentel and Patzek (2005), however, reported that corn ethanol requires 29% more fossil energy to produce than its energy content. These two studies showing negative energy balance is mainly attributed to the use of outdated data unrepresentative of current processes and not giving any credit to ethanol co-products (Farrell et al., 2006).

It is worth noting that even if energy balance is negative (requiring more fossil energy than fuel has) it still may be desirable to make ethanol because ethanol replaces liquid transportation fuel which is one of the hardest to replace among all forms of energy. If we look at the LCA of gasoline fuel itself, it yields only about 0.81 units of energy per unit of fossil energy consumed (Wang, 2005).

The ELCA study is dynamic; hence, the study should be revised frequently to reflect changes in technology and industrial practices. An ELCA study update in 2004 (using 2001 survey data) yielded NER of 1.67 (Shapouri et al., 2004) compared to NER of 1.24 (Shapouri et al., 1995 using 1991 survey data) and 1.34 (Shapouri et al., 2002 using 1996 survey data). The major factor influencing increased energy balance of corn ethanol is the improved production efficiency of ethanol plants. Since 2002, ethanol plants produce 15 percent more ethanol from a bushel of corn using 20 percent less energy in the process (NCGA, 2008). Furthermore, increase in corn yield resulting from better corn varieties and improved production practices add to the increased energy balance.

Table 18.2 Comparative results of twelve corn ethanol energy balance studies (1992 – 2006).

Reference	Year	NEV BTU/gal	Energy Ratio
Morris and Ahmed	1992	25139	1.33
Lorenz and Morris	1995	30589	1.38
Shapouri et al.	1995	16193	1.24
Agri-Food Canada	1999	30465	1.56
Wang and Santini	2000	20400	1.37
Shapouri et al.	2002	21105	1.34
Graboski	2002	13332	1.21

Pimentel	2003	-22119	0.78
Shapouri et al.	2004	30528	1.67
Pimentel and Patzek	2005	-21993	0.78
Farrell et al.	2006	16122	1.2
Hill et al.	2006	860	1.25

The ELCA study is dynamic; hence, the study should be revised frequently to reflect changes in technology and industrial practices. An ELCA study update in 2004 (using 2001 survey data) yielded NER of 1.67 (Shapouri et al., 2004) compared to NER of 1.24 (Shapouri et al., 1995 using 1991 survey data) and 1.34 (Shapouri et al., 2002 using 1996 survey data). The major factor influencing increased energy balance of corn ethanol is the improved production efficiency of ethanol plants. Since 2002, ethanol plants produce 15 percent more ethanol from a bushel of corn using 20 percent less energy in the process (NCGA, 2008). Furthermore, increase in corn yield resulting from better corn varieties and improved production practices add to the increased energy balance.

18.3.1 Cellulosic Ethanol ELCA

Majority of the ethanol currently produced in the United States is derived from cornstarch. However, lignocellulosic sources such as switchgrass, woody plants, agri-residues and prairie grasses are being studied as potential ethanol feedstocks because these feedstocks require less energy to grow and harvest than annual crops and can be grown on agriculturally marginal lands (Schmer et al., 2007). Besides, these are non-food feedstocks and do not contribute to food versus fuel concern. The authors reported that on average 13.1 MJ of cellulosic ethanol was produced from switchgrass for every MJ input of petroleum fuel, i.e. FER of cellulosic ethanol produced from switchgrass is 13.1. Schmer et al. (2007) found that NEV varied with year of production and ethanol yield but always exceeded 14.5 MJ per liter. The authors reported that switchgrass produces more than 500 percent renewable energy than energy consumed to produce it.

In contrast to the above positive results, Pimentel and Patzek (2005) reported an unfavorable FER for ethanol from both switchgrass and wood biomass. They reported that ethanol production using switchgrass required 50% more fossil energy than the ethanol fuel produced and ethanol production using wood biomass required 57% more fossil energy than the ethanol fuel produced. The careful study on the sources of discrepancy revealed the fact that the system boundary and other assumptions were different for the two studies. For one, Pimentel and Patzak (2005) included human consumption of food and energy working in bioenergy production process as inputs, which are not included in other life cycle analyses.

Von Blottnitz and Curran (2007) made a comparative study of ethanol production from various feedstocks and reported FER of 7.9 for sugarcane in Brazil, 2 for sugar beet in UK and 5.2 for wheat straw in UK. Wu et al. (2006) reported 0.16 Btu fossil energy input per Btu of energy output with ethanol from wood residues, and 0.09 Btu of input per Btu of ethanol from corn stover. This corresponds to FER of 6.25 for ethanol produced from woody biomass and about 11 for the ethanol from corn stover. Their report further showed 0.76 Btu of fossil energy input for ethanol from corn grain. They argued that conventional corn ethanol plants receive process heat and power from NG and coal. In contrast, cellulosic ethanol plant is self-sufficient in heat and power supply as the lignin residue from stover can be burned to meet the internal energy demands.

18.4 Soybean Biodiesel ELCA

Similar to corn ethanol, biodiesel production in the United States is also rapidly growing. The US production has increased from under half a million gallons in 1999 to about 700 million gallons in 2008 (NBB, 2009). Soybean oil is the primary feedstock for biodiesel production in the United States. The first comprehensive LCA for biodiesel produced in the United States from soybean oil, also known as NREL study, was completed in 1998 (Sheehan et al., 1998).

A summary of energy ratios for different processes compared to biodiesel energy from NREL study is shown in Table 18.3 (Sheehan et al., 1998). The energy inputs required for the production of soybean include direct energy used in farms (such as diesel, gasoline, LP gas, natural gas, and electricity), energy used to produce fertilizers (nitrogen, phosphorus, and potassium) and energy used to produce pesticides. The study used data from 1990 USDA Farm Costs and Returns Survey (FCRS). The farm energy input estimates are the weighted average of 14 states and the soybean yield used in the estimate was 36 bushels per acre (3-year average soybean, 1989-91; 1990-92; 1992-94). Oil extraction plants included energy inputs associated with electricity, steam and hexane. The energy inputs included in biodiesel conversion plants were energy associated with steam, electricity, alcohol and chemicals (such as sodium hydroxide, and hydrochloric acid).

The study did not include labor and farm machinery in soybean production, and building materials in biodiesel plants.

The NREL study assigned energy allocation to co-products (soy meal and glycerin) in proportion to the mass fraction of output. The result of the study shows that soybean biodiesel has a positive energy balance with FER of 3.2. It should be noted that this report used FER and not NER to report ELCA results.

Table 18.3 Energy balance of soybean biodiesel production (Source: Sheehan et al., 1998).

Process	MJ/MJ of fuel
Soybean agriculture	0.0656
Soybean transport	0.0034
Soybean crushing	0.0796
Soy oil transport	0.0072
Soy oil conversion	0.1508
Biodiesel transport	0.0044
Total	0.311

All studies show that biodiesel has a favorable renewability factor (FER), except one study that claimed the use of biodiesel does not reduce petroleum use (Table 18.4). The variation in results published by different researchers is attributed to input data variability, conflicting system boundaries, and differences in energy ratio definitions (Pradhan et al., 2008).

A report published in 2005 claims that the energy output from soybean biodiesel is less than the fossil energy inputs (Pimentel and Patzek, 2005). It claims that soybean biodiesel required 27% more fossil energy than the energy contained in the biodiesel. The wide discrepancy in this result showing negative energy balance comes from the co-products credits and lime use application rate. This study assigned only 19.3% of the total input energy to the soybean meal, but in reality, 82% of the soybean mass goes into meal (Van Gerpen and Shrestha, 2005). In addition, the study assigned 4,800 kg lime ha-1 year -1 for the average soybean crop, whereas according to the source used by the study (Kassel and Tidman, 1999), lime use was recommended for only acidic soil to correct pH once in several years (usually 5 to 10 years).

Table 18.4 Net energy ratios reported for soybean biodiesel.

FER	Source	FER	Source
2.51	Ahmed et al. (1994)	0.79	Pimentel and Patzek (2005)
3.20	Sheehan et al. (1998)	1.93	Hill et al. (2006)

Pimentel and Patzek (2005) also included the lifetime energy consumption of biodiesel labor as energy input. The authors of this article argue against the inclusion of energy from labor. Not including labor as an energy input in biodiesel LCA is justified by four reasons: (i) the average per capita energy consumption may not represent a person living on a farm compared to a person who commutes 50 miles (80 km) daily to work, (ii) consideration of annual energy consumption by human laborers does not aid in evaluating the renewability of biodiesel, as human food consumption is independent of soybean agriculture, (iii) people from other sectors of the economy use the service provided by people involved in biodiesel production through use of the biodiesel, thus reducing the consumption of services from the competitive fuel industry (i.e., regular diesel fuel), and (iv) people are hired primarily for their ability to perform a task and not for their energy output. The physical energy input from human labor makes up a negligible fraction of the total energy input (Pradhan et al., 2008).

The biodiesel produced from yellow grease is shown to have even higher FER than raw vegetable oil (Mittelbach and Remschmidt, 2004). This is because the energy inputs involved in feedstock production and transportation are not relevant to recycled frying oil and tallow, commonly known as yellow grease. Piedmont Biofuels reported FER of the biodiesel produced from waste vegetable oils to be 7.8, which proves the biodiesel produced from waste oils to be highly renewable (Hoover, 2005). The higher energy balance is mainly because biodiesel production from waste oil only requires transporting from the source of waste oil and conversion into biodiesel.

The energy balance of the biodiesel production from soybean is expected to increase with increasing soybean yield and improved processing facilities. The use of genetically modified seeds contributes to the reduction in pesticide uses and increase soybean yield. The stand-alone biodiesel plants without crushing capability have to purchase soybean oil or some other feedstock to produce biodiesel. Separating the crusher from the biodiesel plant requires more energy because an extra step is added when the oil is hauled to the biodiesel plant. Hence, larger plants with more capital investment are expected to be more energy efficient, which will affect life cycle energy use positively.

18.4.1 Biodiesel from Jatropha

There is a rapidly growing interest in using jatropha (*Jatropha curcas*) as a feedstock for the production of biodiesel, particularly because jatropha is a wild plant that can grow in dry and marginal lands without irrigation (Prueksakorn and Gheewala, 2008). Jatropha is also reported to have few pests and diseases, is inedible source of oil and does not compete with food. The oil from jatropha can as easily converted into biodiesel as any other oil feedstock that meets the American and European standards (Achten et al., 2008). The energy balance of jatropha biodiesel or jatropha methyl esters (JME) is generally positive, but the NER value will depend on how efficiently the co-products are used. Prueksakorn and Gheewala (2008) reported the net energy value of 236 GJ per hectare/year (Equivalent of about 770 gallons of biodiesel per acre per year) and net energy ratio (NER) of 6.03 from the life cycle energy balance of JME and co-products. They also reported the margin of error of about 50% of the reported value. In their report, the agriculture phase consumed the highest energy followed by transesterification and transportation process. The oil extraction phase required the lowest energy input.

As discussed earlier, it makes a big difference on NER depending on how it was defined. Prueksakorn and Gheewala (2008) defined their NER as the ratio of energy from all co-products to the total energy input as in equation 18.6. They have reported the actual biodiesel production of 2700 kg/ha/yr (about 330 gallon of biodiesel per year). In their paper, although they had a report saying the jatropha yield varies from 0.88 to 12.50 metric tonne per year toward the end of third year, they assumed the higher end yield potential of 12.5 metric tonne per year for 20 years. In reality, the yield from jatropha depends on amount of water availability, soil characteristic, plant spacing and several other climatic factors. Openshaw (2000) reported that in Mali, jatropha seed yield varied from 2.5-3.5 metric tonne/ha/yr. This is equivalent of about 100 gallon of biodiesel per acre per year. Achten et al. (2008) pointed out that jatropha seed yield is still a difficult issue and the actual mature seed yield per ha per year is not known, since systematic yield monitoring only started recently. Earlier reported figures exhibit a very wide range (0.4–12 metric tonne/ha/yr) and are not coherent. It was indicated in their paper that the yield from a single tree cannot be extrapolated to estimate yield as planting distance greatly affect the seed yield.

Palm, rapeseed, and waste getable oil are other major sources of oil to produce biodiesel. FER for biodiesel made from these feedstocks is summarized in Table 18.5.

Table 18.5 FER in making biodiesel from various feedstocks.

Feedstock	Study Applicable to	FER	Source
Rapeseed Oil	UK	2.29	Horne et al. (2003)
Rapeseed Oil	Europe	3.0	Scharmer and Gosse, 1995
Jatropha Oil	India	1.9	Whitaker and Heath (2008)
Palm Oil	Brazil	7.8 – 10.3	Costa and Lora (2006)
Palm Oil	Colombia	5.9 – 6.9	Costa and Lora (2006)
Recycled frying Oil	Germany	5.51	Süß (1999)
Waste vegetable Oil	USA	7.8	Hoover (2005)
Waste vegetable Oil	Spain	7.96	Rúa et al. (2006)

18.5 Analysis of Uncertainty

There is always some uncertainty associated with collected data. This is especially true with LCA data where most of the data comes from survey. Data variability should be kept in mind while interpreting the results. Data variation occurs due to geographical and technological differences, difference in production scale, and estimation error. Geographical difference causes variation in crop production practices, yield variation and variation in transportation mode. For example, soybean grown in Iowa may have a different energy per unit of production than soybean grown in North Dakota. Technological differences cause variation in input stream such as base catalyzed reaction for transesterification versus heterogeneous catalyst reaction. The difference in production scale makes the unit energy input different per unit production of biofuel. Bigger production plants tend to have a better efficiency and hence lower energy input and cost per unit of production than smaller producers.

In that regard, NER or FER calculation should be conducted with some kind of probability distribution in mind. Monte Carlo simulation is used if enough data is available to estimate the probability distribution of parent population and if the data are not adequate to estimate the parent population distribution, boot-strapping technique could be used to study the most likely range of FER values (Pradhan et al., 2008). In any case, the mean reported FER value should be taken an estimate and keep in mind that variation exists from that likely number.

18.6 Summary

Renewable fuels have a clear advantage as a substitute of the petroleum fuels. Renewability is measured using fossil energy ratio (FER) which is the ratio of energy produced from biofuel to biofuel share of fossil energy input. It is important to notice that the definition of FER may from this generic definition. Major variations in definition are subtracting co-product energy from numerator or adding co-product energy to denominator. These variations in definitions can make a big difference in the outcome. Another big factor than can vary the result is the way the input energy is allocated among co-products. A well-defined goal and objectives, as required by ISO 14040 (2006a), is essential to achieve a meaningful energy balance results.

The results from the majority of the energy balance studies on biodiesel and ethanol are positive. The positive energy balance displayed by biofuels is mainly due to the contribution of solar energy collected by the crop from which the fuel is made. Solar energy captured during photosynthesis is not included as input energy. The variation in the energy balance result among different studies is coming from variation is data, conflicting system boundaries, co-product energy allocation, and differences definition of the metric.

Comparing corn ethanol and soybean biodiesel's renewability based on FER, soybean biodiesel showed more positive energy balance compared to corn ethanol. The main feedstock for biodiesel production in the United States is soybean oil and for ethanol is cornstarch. Efforts are being made to reduce the use of edible food feedstock such as corn and soybean oil to non-food feedstock such as cellulosic biomass and non-edible oil like jatropha. FER for the alternative non-food feedstocks are shown even higher than their counterpart food feedstock, albeit these new technologies needs to be matured more to make them economically viable.

The net energy ratios of corn ethanol and soybean biodiesel have increased over time due to the use of genetically engineered seeds (increased yield), improved farm production practices (reduced pesticide use), and improved biofuel production facilities. Furthermore, establishment of larger plants with more capital investment saves energy required in hauling soybean oil to the biodiesel conversion plants.

18.7 References

Achten, W.M.J., Verchot, L., Franken, Y.J., Mathijs, E., Singh,. V.P., Aerts, R., and Muys, B. (2008). "Jatropha bio-diesel production and use." *Biomass and Bioenergy*. 32(12), 1063-1084.

Agri-Food Canada. (1999). Assessment of net emissions of greenhouse gases from ethanol-gasoline blends in Southern Ontario. Prepared by Levelton Engineering Ltd. and (S & T)^2 Consulting Inc.

Ahmed, I., Decker, J., and Morris, D. (1994). "How much energy does it take to make a gallon of soydiesel?" Jefferson City, Mo.: National Soydiesel Development Board.

Argonne National Laboratory. (2006). "The greenhouse gases, regulated emissions, and energy use in transportation (GREET) model." *Version 2.7. Argonne, Ill: Argonne National Laboratory.* Available at http://www.transportatoin.anl.gov/modeling_simulation/GREET/index.html. (accessed October, 2008).

Chambers, R.S., Herendeen, R.A., Joyce, J.J., and Penner, P.S. (1979). "Gasohol: Does it or doesn't it produce positive net energy?" *Science*, 206, 790-795.

Costa, R.E., and Lora, E.E.S. (2006). "The energy balance in the production of palm oil biodiesel – Two case studies: Brazil and Colombia." *Federal University of Itajubá/Excellence Group in Thermal and Distributed Generation NEST (IEM/UNIFEI), Oil Palm Research Center CENIPALMA/Colombia and Bahia Federal University-UFBA.* Available at: http://www.svebio.se/attachments/22/295.pdf.

Davis, K. (2001). "Corn milling, processing and generation of co-products." *Technical Symposium, Minnesota Nutrition Conference, September 2011.* Available at http://www.ddgs.umn.edu/articles-proc-storage-quality/2001-Davis-%20Processing.pdf. (accessed October, 2008).

Drapcho, C., Nghiem, J., and Walker, T. (2008). *Biofuels Engineering Process Technology.* McGraw-Hill.

EIA. (2008). "United States energy profile." *Energy Information Administration.* Available at http://tonto.eia.doe.gov/country/country_energy_data.cfm?fips=US. (accessed on January, 2009).

EISA. (2007). "Independence and Security Act." *Public Law 110 – 140, 110th Congress.* Available at http://frwebgate.access.gpo.gov/cgi-bin/getdoc.cgi?dbname=110_cong_public_laws&docid=f:publ140.110. (accessed on January, 2009).

EPA. (2002). "A comprehensive analysis of biodiesel impacts on exhaust emissions." Report no. EPA420-P-02-001.

Farrell, A.E., Plevin, R.J., Turner, B.T., Jones, A.D., O'Hare, M., and Kammen, D.M. (2006). "Ethanol can contribute to energy and environmental goals." *Science*, 311(5760): 506 – 508.

Feygin, M., and Satkin, R. (2004). "The oil reserves-to-production ratio and its proper interpretation." *Natural Resources Research.*, 13(1), 57-60.

Graboski, M.S. (2002). "Fossil energy use in the manufacture of corn ethanol." Prepared for National Corn Growers Association. Available at http://www.ncga.com/ ethanol/pdfs/energy_balance_report_final_R1.PDF

Hill, J., Nelson, E., Tilman, D., Polasky, S., and Tifanny, D. (2006). "Environmental, economic, and energetic costs and benefits of biodiesel and ethanol biofuels." *PNSS.*, 103(30), 11206-11210.

Hoover, S. (2005). "Energy balance of a grassroots biodiesel production facility." *School of Engineering Science, Division of Science and Engineering, Murdoch University*. Available at http://www.biofuels.coop/education/energy-balance/. (accessed on October, 2008).

Horne, R.E., Mortimer, N.D., and Elsayed, M. A. (2003). "Energy and carbon balances of biofuels production: biodiesel and bioethanol." *Proceedings. The International Fertiliser Soceity.,* April, 3, 1 – 56.

Huo H., Wang M., Bloyd C., and Putsche V. (2008). "Chapter 4: Co-product credits for biofuels in Life-cycle assessment of energy and greenhouse gas effects of soybean-derived biodiesel and renewable fuels." *Argonne, IL: Argonne National Laboratory.*, pp. 26-33.

ISO. (1998). "Environmental management - Life cycle assessment - Goal and scope definition and inventory analysis. ISO 14041." Geneva, Switzerland: International Standardization Organization.

ISO. (2006a). "Environmental management – Life cycle assessment – Principles and framework. ISO 14040." Geneva, Switzerland: International Standardization Organization.

ISO. (2006b). "Environmental management – Life cycle assessment – Requirements and guidelines. ISO 14044." Geneva, Switzerland: International Standardization Organization.

Kassel, P., and Tidman, M. (1999). Ag lime impact on yield in several tillage systems. Ames, Iowa: Iowa State University, Integrated Crop Management.

Kim, S., and Dale, B. E. (2002). "Allocation procedure in ethanol production system from corn grain." *Intl. J. Life Cycle Assessment*, 7(4), 237 – 243.

Khanal, S. K. (2008). *Anaerobic Biotechnology for Energy Production: Principles and Applications.* Wiley-Blackwell Publishing, Ames, USA.

Knothe, G., Van Gerpen, J.H., and Krahl, J. (2005). *The biodiesel handbook.* Champaign, Ill: AOCS Press.

Lorenz, D., and Morris, D. (1995). "How much energy does it take to make a gallon of ethanol?" Institute for Local Self-Reliance.

Mittelbach, M., and Remschmidt, C. (2004). *Biodiesel: The Comprehensive Handbook. 2nd ed.* Vienna, Austria: Boersedruck.

Morris, D., and Ahmed, I. (1992). "How much energy does it take to make a gallon of ethanol?" Institute for Local Self-Reliance.

National Corn Growers Association (NCGA). (2008). "Ethanol and coproducts." Available at http://ncga.com/ethanol-coproducts. (accessed on October, 2008).

NBB. 2009. "Estimated US biodiesel production by fiscal year." *National Biodiesel Board.*
Available at http://www.biodiesel.org/pdf_files/fuelfactsheets/Production_ Graph_Slide.pdf. (accessed on January, 2009).

Openshaw, K. (2000). "A review of Jatropha curcas: an oil plant of unfulfilled promise." *Biomass and Bioenergy*., 19(1), 1 – 15.

Pimentel, D. (2003). "Ethanol fuels: Energy balance, economics, and environmental impacts are negative." *Natural Resources Research*., 12(2), 127 – 134.

Pimentel, D., and Patzek, T.W. (2005). "Ethanol production using corn, switchgrass, and wood; biodiesel production using soybean and sunflower." *Natural Resources Research*., 14(1), 65 – 76.

Pradhan, A., Shrestha, D.S., Van Gerpen, J.H., and Duffield, J. (2008). "The energy balance of soybean oil biodiesel production: a review of past studies." *Transactions of the ASABE*., 51(1), 185 – 194.

Prueksakorn, K., and Gheewala, S.H. (2008). "Full chain energy analysis of biodiesel from Jatropha curcas L. in Thailand." *Environ. Sci. Technol*., 42, 3388 – 3393.

Rúa, C. de la, Lechón, Y., Cabal, H., Lago, C., Izquierdo, L., and Sáez, R. (2006). "Life cycle environmental benefits of biodiesel production and use in Spain". CIEMAT, Energy Department. 8th Environment Symposium, Nicosia (Cyprus). Available at http://www.ciemat.es/recursos/doc/Areas_Actividad/Energia/ASE/2102996515_1522007122853.pdf

Scharmer, K., and Gosse, G. (1995). "Energy balance, ecological impact and economics of vegetable oil methyl ester production in Europe as substitute for fossil diesel." EU – Study Altener 4.1030/E/94-002-1.

Schmer, M.R., Vogel, K.P., Mitchell, R.B., and Perrin, R.K. (2007). "Net energy of cellulosic ethanol from switchgrass." *Proceedings of the National Academy of Sciences of the United States of America (PNAS).*, 105(2), 464 – 469.

Shapouri, H., Duffield, J. A., and Graboski, M. S. (1995). "Estimating the net energy balance of corn ethanol." *USDA – Agricultural Economic.*, Report no. 721.

Shapouri, H., Duffield, J.A., and Wang, M. (2002). "The energy balance of corn ethanol: An update." *Agricultural Economic*, Report No. 813, USDA.

Shapouri, H., Duffield, J.A., McAloon, A., and Wang, M. (2004). "The 2001 net energy balance of corn-ethanol." *USDA*., Available at http://www.ncga.com/ethanol/pdfs/netEnergyBalanceUpdate2004.pdf. (accessed on October, 2008).

Shapouri, H., Wang, M., and Duffield, J. (2006). "Net energy balance and fuel-cycle analysis." In: Dewulf J., and Van Langenhove H. (Ed.) *Renewables-Based Technology*. Chichester, U.K.: John Wiley and Sons. pp. 73 – 86.

Sheehan, J., Camobreco, V., Duffield, J., Graboski, M., and Shapouri, H. (1998). "Life cycle inventory of biodiesel and petroleum diesel for use in an urban bus." NREL/SR-580-24089 Golden, CO: National Renewable Energy Laboratory. U.S. Department of Energy.

Spath, P.L., and Mann, M.K. (2000). "Life cycle assessment of a natural gas combined-cycle power generation system." NREL/TP-570-27715. Golden, Colo.: National Renewable Energy Laboratory.

Süß, A.A.A. (1999). "Wiederverwertung von gebrauchten Speiseölen/-fetten im energetisch-technischen Bereich. Fortschritt-Berichte VDI." *Reihe* 15:

Umwelttechnik Düsseldorf: VDI-Verlag 219. (from book – Biodiesel: the comprehensive handbook by Mittelbach and Remschmidt)

USDA. (2007). "An analysis of the effects of an expansion in biofuel demand on US agriculture. Economic Research Service and Office of the Chief Economist." *US Department of Agriculture.* Available at http://www.greenlandsbluewaters.org/2007USDAbiofuel_demand.pdf. (accessed on January, 2009).

Van Gerpen, J.H., and Shrestha, D. (2005). "Biodiesel energy balance." *Moscow, Idaho: University of Idaho, Department of Biological and Agricultural Engineering.* Available at www.uidaho.edu/bioenergy/NewsReleases/Biodiesel%20Energy%20Balance_v2a.pdf. (accessed on February, 2008).

Vigon, B.W., Tolle, D.A., Cornaby, B.W., Latham, H.C., Harrison, C.L., Boguski, T.L., Hunt, R.G., and Sellers, J.D. (1993). "Life cycle assessment: Inventory guidelines and principles." Washington D.C. & Cincinnati: United States Environmental Protection Agency, Office of Research and Development. (EPA/600/R-92/245).

Von Blottnitz, H., and Curran, M.A. (2007). "A review of assessments conducted on bio-ethanol as a transportation fuel from a net energy, greenhouse gas, and environmental life cycle perspective." *Journal of Cleaner Production.*, 15(7), 607 – 619.

Wang, M. (2005). "Updated energy and greenhouse gas emission results of fuel ethanol." *The 15th International Symposium on Alcohol Fuels, San Diego, CA. September 2005.*
Available at http://www.transportation.anl.gov/pdfs/TA/375.pdf. (accessed on October, 2008).

Wang. M., and Santini, D. (2000). "Corn-based ethanol does indeed achieve energy benefits." *Environmental and Energy Study Institute.*

Whitaker, M., and Heath, G. (2008). "Life cycle assessment of the use of jatropha biodiesel in Indian locomotives." *Technical report*, NREL/TP-6A2-44428. Available at http://www.nrel.gov/docs/fy09osti/44428.pdf (accessed on December, 2008).

Wu, M., Wang, M., and Huo, H. (2006). "Fuel-cycle assessment of selected bioethanol production pathways in the United States." Reference No. ANL/ESD/06-7 Argonne, IL: Argonne National Laboratory.

CHAPTER 19

Comprehensive Study of Cellulosic Ethanol Using Hybrid Eco-LCA

Anil Baral, and Bhavik R. Bakshi

19.1 Introduction

Life cycle assessment (LCA) is emerging as an important tool in systems analysis to quantify resource consumption, and environmental and societal impacts of products and services, and assist in sustainable decision-making. LCAs can be used to compare alternative processes, products, and services that currently exist or will be available in the future. Conducting an LCA of future technologies or products assumes added significance since an LCA can identify the potential impacts early in the development process and provides us with opportunities for undertaking preventive or mitigating measures before they are widely used. Cellulosic ethanol is one such a product that is likely to shape the future transportation fuels. Cellulosic ethanol refers to ethanol derived from lignocellulosic feedstocks such as solid waste, agriculture residues, grasses, and dedicated energy crops. Cellulosic ethanol is emerging as a leading alternative to corn ethanol.

Several reasons can be attributed to growing commercial and research interests on cellulosic ethanol. First and foremost, it avoids food vs. fuel dilemma associated with the use of food crops, e.g. corn, sorghum, cassava, sugarcane, and soybean for biofuel production. Feedstocks such as agriculture residues (rice straw, corn stover, wheat straw, etc.) and municipal solid waste do not require additional croplands because they are the byproducts of existing economic systems. It has been argued that the drive for crop-based biofuels may lead to clearing of forests to make room for biofuel production making them a net carbon source (Fargione et al., 2008). Moreover, lignocellulosic feedstocks can be grown in areas not suitable for crop production such as marginal lands, rangeland, and prairies. Farming requirements including land preparations, fertilizers, pesticides, diesel, and machinery are less for agriculture residues or zero for feedstocks such as municipal solid waste and recycled newsprint. Intensive use of fertilizers in crop farming such as corn, wheat, and soybean among others has been linked to the dead zone in the Gulf of Mexico. The dead zone has caused million dollars of damage to fishery industry in the Gulf Coast and loss of biodiversity. There is also a concern about downwind impact of pollen from genetically modified Bt corn (Rosi-Marshall et al., 2007). The Bt corn is genetically modified corn that has the toxin-producing gene from *Bacillus thuringiensis*. On the other hand, cellulosic ethanol may meet part or all of its process energy requirements from renewable electricity produced from lignin and residues and hence relies less on fossil fuels. Lignin is usually produced as a byproduct in cellulosic ethanol production. Due to reduced energy inputs from the economy, fossil fuel consumption for cellulosic ethanol can be small than that for corn ethanol. The push for cellulosic ethanol is further accentuated by the provision in the Energy Independence and Security Act of 2007 that requires that16 billion gallons of biofuels be derived from lignocellulosic feedstocks by 2022.

Despite many desirable attributes of cellulosic ethanol, its overall environmental performances remain uncertain mainly because cellulosic ethanol technology is still in infancy. A number of life cycle assessment (LCA) studies have been conducted in the past few years to quantify its environmental performances and determine how they compare with those of gasoline and grain-based bioethanol. The environmental performances included in the studies pertain to energy, greenhouse gas (GHG) emissions, and criteria air pollutants. There has been a limited effort in quantifying ecological footprints of cellulosic ethanol in particular and of biofuels in general. Ecological footprints refer to quantification of ecological goods and services such as water, sunlight, soil, minerals, ores, biomass, pollination, and photosynthesis, which are essential for the sustenance of economic activities. Ecological footprints include natural resources which are outside the boundary of the economic system and enter the economy as free resources. Inclusion of ecological footprints in LCA is necessary to make decision-making process more holistic and avoid a decision that may lead to "tragedy of commons." Not all natural resources are renewable and plentiful and free. It is a customary practice in LCA of biofuels to use a handful of metrics to assess the alternatives such as energy and GHG emissions. Considering the overwhelming emphasis bestowed on energy security and climate change by the governments around the globe, common use of energy and GHG metrics in LCA is understandable. However, in addition to these important metrics, there can be several metrics which are useful in comparing alternatives and eliciting additional insights about the alternatives that may inform the policy debate.

In this regard, this study aims to provide a comprehensive analysis of cellulosic ethanol by (1) considering the contributions of ecological goods and services using exergy analysis, and (2) utilizing a number of aggregate metrics that are developed from life cycle resource consumption and emissions data obtained from a hybrid model of Eco-LCA analysis. Five different lignocellulosic feedstocks were selected to capture the variability in feedstock compositions and methods of production. Three of the feedstocks chosen, newsprint, municipal solid waste (MSW), and corn stover, represent waste and agriculture residue, and the remaining two, switchgrass and yellow poplar, represent dedicated energy crops. Gasoline is included as a reference fuel for the purpose of comparison. This chapter provides an overview of existing literature on LCA of cellulosic ethanol, and describes a hybrid Eco-LCA model used to study cellulosic ethanol followed by discussions on aggregation methods based on energy, exergy analysis, and mid-point environmental impact analysis. It also provides disaggregated data of ecological goods and services in terms of exergy to quantify their relative contributions to cellulosic ethanol and gasoline production. Moreover, a comparative analysis of cellulosic ethanol and gasoline with respect to various aggregate metrics is also provided. The goal is to identify desirable attributes that may serve as the basis for decision making for future biofuel options.

19.2 Review of LCA Studies on Cellulosic Ethanol

Theoretically, cellulosic ethanol can be produced from any lignocellulosic sources. Research on cellulosic ethanol production has been directed mostly at agricultural residues (corn stover, rice straw), and dedicated energy crops, and forest thinnings, (Levelton Engineering Ltd. and S&T Consultants, 1999; Wooley et al., 1999; Aden et al., 2002; Sheehan et al., 2004; Kemppainen and Shonnard, 2005; Kalogo et al., 2007). Few studies relate to waste products such as recycled newspaper (Kemppainen and Shonnard, 2005) and municipal solid waste (Kalogo et al., 2007). Due to variations in the chemical composition of the feedstock, municipal solid waste (MSW) poses an additional challenge for sustaining predictable ethanol yield. To expand the volume of ethanol production, reliance on only one type of feedstock may not be commercially viable in specific cases and a plant may need to employ a mixed feedstock. As a result, ethanol production from a mixed feedstock has recently become a motivating factor in cellulosic ethanol research. Most life cycle studies of lignocellulosic feedstocks follow biochemical route with acid pretreatment. In the case of ethanol production from MSW, a patented technology known as Gravity Pressure Vessel (GVP) Process was used as a benchmark for developing the life cycle inventory (Kalogo et al., 2007).

A detailed and comprehensive process and economic analysis of cellulosic ethanol has been conducted by the National Renewable Energy Laboratory (NREL) (Wooley et al, 1999; Aden et al., 2002, Sheehan et al., 2004). They have been the source of several life cycle studies conducted to date for various feedstocks. For example, Spatari et al. (2005) and Kemppainen and Shonnard (2005) projected process level energy consumption and emissions based on chemical composition and ethanol yield relative to the modeling data from NREL studies on poplar and corn stover.

The metrics used in life cycle assessment of biofuels relate to energy consumption, emissions, mostly greenhouse gases (GHG); and land use efficiency (ha per km traveled). Cellulosic ethanol is invariably associated with better energy return on investment (r_E), however, variations r_E are found across the studies due to differences in feedstocks, boundaries, energy credits, and assumptions. Electricity produced from lignin combustion can influence the amount of energy credits depending on how much electricity is sold to the grid. The only exception that reports an r_E smaller than 1 for cellulosic ethanol comes from a study of Pimentel and Patzek (2005). A comparative review of LCA studies by Fleming et al. (2006) found that r_E of cellulosic ethanol varies from 3.84 (General Motors North America, 2001) to 33.3 (European Council for Automotive R&D, 2004).

Reductions in GHG emissions were reported in almost all studies (Wang et al., 1999; Sheehan et al., 2004; Wang, 2004; Brinkman et al., 2005; Delucchi, 2006; Farrell et al., 2006; Fleming et al., 2006) for substituting cellulosic ethanol for gasoline. For example, well-to-tank GHG emissions range from -1020 to -1610 g equivalent CO_2 per liter of ethanol for switchgrass, and -1085 to -1404 g equivalent CO_2 per liter for corn stover (Spatari et al., 2005). An inter-study of well-to-wheel GHG emissions showed that an average of 86% reduction could be achieved from cellulosic ethanol substitution (Fleming et al., 2006). Wang et al., (2004) estimated that E85 could reduce GHG emissions by 68-102% for the near term cellulosic ethanol case. Likewise, Delucchi (2006) projected that replacing gasoline with E90 would result in GHG reductions of 53% and 81% for the years 2010 and 2020, respectively. The larger reduction in the year 2020 is due to assumption of improved cellulosic ethanol technology whereas there would be little improvement in refining technology since the technology is already mature. Among four lignocellulosic feedstocks studied by Levelton Engineering (1999) - corn stover, switchgrass, wheat straw, and hay, ethanol production from switchgrass and hay emitted lower GHG emissions despite their low ethanol yields per ha, followed by corn stover and wheat straw. This is due to the assumption of substantial soil carbon sequestration in the cases of switchgrass and hay.

For pollutants such as volatile organic carbon (VOC), carbon monoxide (CO), NO_x, and PM10, there are net increases in emissions over the life cycle for E85 in comparison to reformulated gasoline (Brinkman et al., 2005). For SO_x, conflicting results have been published with Spatari et al. (2005) reporting an increase and Brinkman et al. (2005) reporting a net decrease in SO_x emissions. Brinkman et al. (2005) assumed that 50% of ethanol comes from herbaceous biomass and 50% of ethanol comes from woody biomass. Existing LCA studies mainly focus on criteria air pollutants and GHG. Very few cellulosic ethanol LCA studies report land use efficiencies even though it is an important component of bioethanol production from wood or grasses. Land use efficiency as measured in kilometer traveled/ha has been reported by Watson et al. (1996) for a number of biofuels and hydrogen. According to this study, cellulosic ethanol has a better land use efficiency than corn ethanol. It is even better than sugar cane ethanol which has a higher yield per ha than corn ethanol. Only hydrogen fuel derived from biomass gasification has a better land use efficiency than cellulosic ethanol (Watson et al., 1996).

An environmental impact assessment of cellulosic ethanol using midpoint or end-point methods is not common in existing LCA studies. Most studies focus on developing inventories for energy consumption and emissions. A study by Kemppainen and Shonnard (2005) introduced a life cycle composite index from a number of environmental indices such as human health and fish toxicity to compare cellulosic ethanol derived from upper Michigan timber and newsprint. Using the Environmental Fate and Risk Assessment Tool, the authors showed that timber has a higher environmental composite index indicating that it exerts more environmental impact. However, timber is associated with a lower global warming index.

Recently, exergy analysis has been used in LCA of biofuels to account for contributions of materials and energy. The majority of such studies focus on starch-based ethanol and biodiesel derived from soybean, rapeseed, and castor bean

(Berthiaume et al., 2001; Patzek, 2004; Dewulf et al., 2005; Sciubba and Ulgiati, 2005). Exergy, by definition, is the maximum theoretical work that can be extracted from a system as it approaches a thermodynamic equilibrium. Exergy analysis has been incorporated in LCA as a scientifically rigorous tool to integrate materials and energy and identify areas of inefficiencies for improvements (Ayres et al., 1998; Rosen, 2002).

Traditional LCA of biofuels usually focuses on energy consumption and emissions. It ignores the fact that it is possible to improve the overall system efficiency by minimizing material consumption. Exergy analysis allows for the quantification of contributions of materials and fuels to the overall system efficiency. Besides exergetic efficiency, metrics such as exergy breeding factor (Dewulf et al., 2005) and renewability indicator (Berthiaume et al., 2001) were discussed in the literature to derive insight with regard to biofuels. The exergy breeding factor provides information about how much renewable exergy is generated by expending 1 joule of non-renewable exergy. A renewability indicator introduced by Berthiaume et al. (2001) exhibits the degree of renewability of biofuels by taking into account the non-renewable exergy used in environmental restoration work. The larger the environmental restoration work, the lower will be renewability of biofuels.

One possible criticism of exergy based analysis is that integrating materials and fuels together would lead to loss of information about their qualities. For example, the quality of 1 J of coal and 1 J of steel is not the same. Coal can be combusted to generate electricity but steel cannot be combusted to produce electricity. Therefore, when we state that a system consumes 2 J of exergy, it is not clear if the consumed exergy refers to exergy of coal or exergy of steel. Emergy analysis (Odum, 1996) attempts to deal with quality differences of resources but it is the beyond the scope of this book chapter. A similar criticism also applies to energy analysis. Exergy analysis is better than energy analysis due to its focus on useful energy and its capacity to aggregate fuel and non-fuel (materials) resources.

19.3 Methodology

An LCA can be conducted using traditional process LCA, economic input-output LCA (EIOLCA), and hybrid LCA, which is a combination of process LCA and economic input-output LCA. A process LCA is normally an attributional LCA where emissions, energy and resource consumption are assigned to a product or co-products in an isolated system. Process LCAs utilize refined process level data but considers major inputs to the processes since it is not feasible to include all the inputs in the supply chains. This leads to truncation errors. An EIOLCA is a consequential LCA which quantifies the resource consumption and environmental impact caused by a change in the product system. The EIOLCA model can tract energy and natural resource consumption and associated environmental impacts across the economic sectors. There is no artificial boundary in EIOLCA and, hence it reduces the truncations errors. However, the EIOLCA model uses the coarse data at the economy scale and has a high degree of product and service aggregation. To overcome this limitation, a hybrid LCA method has been proposed (Suh et al., 2004). When we combine a process LCA with an EIOLCA, a hybrid LCA model results. There are three main types of hybrid LCA models: (1) tiered hybrid LCA (2) input-output hybrid LCA, and (3) integrated hybrid LCA.

19.3.1 Hybrid Model

A hybrid ecologically-based life cycle assessment (Eco-LCA) model developed by Baral and Bakshi (2009) was used to estimate life cycle resource consumption and emissions. The Eco-LCA model was developed using economic input-output analysis (Zhang et al., 2009) which considers a wide range of provisioning and supporting services and some regulating services by representing their flows in physical units besides emissions. The Eco-LCA model estimates resource consumption and emissions in physical unit (*R*) to money (*$*) ratios (Zhang et al., 2009). Physical units can be mass, energy, emissions, industrial cumulative exergy consumption (ICEC), and ecological cumulative exergy consumption (ECEC) of industry sectors. The equation used in Eco-LCA is:

$$R/\$ = M\hat{X}^{-1}(I - A)^{-1} \tag{19.1}$$

where, *M* is the resource consumed by or emissions from corresponding economic sectors, X is the diagonal matrix of the total economic throughput, *I* is the identity matrix, and *A* is the direct requirements matrix. If *R/$* is multiplied by the cost (*$*) of each product or service under consideration, total resource consumption or emissions in various physical units of the product or service is obtained.

The Eco-LCA model uses the 1997 US economy model with the 490 × 490 commodity sector matrix (Zhang et al., 2009). Once the production price of a commodity belonging to a particular industry sector is known, it is possible to find out cumulative resource consumption and emissions associated with the commodity by multiplying a physical unit/$ ratio of the particular industry sector with the price of a commodity. For example, if one wishes to calculate GHG emissions for producing X dollars worth of gasoline, GHG/$ ratio of the petroleum refineries sector is multiplied by price (X dollars) of gasoline to obtain GHG emissions.

For a product which is not adequately represented by industry sectors, either due to aggregation or novelty, one may need to use a hybrid Eco-LCA model by combining process level inventory with ecologically based life cycle inventory (Eco-LCI), i.e., physical unit/$ ratios. Also hybrid model is useful if use phase and disposal phase are to be included for cradle- to-grave LCA. A hybrid Eco-LCA model can be constructed as follows for gasoline with GHG emissions as an example. First GHG emissions associated with inputs used in production of X dollars worth of gasoline are calculated by finding out inputs required for gasoline production and their production costs.

$$\text{GHG} = \sum_{i=1}^{i=n} \frac{GHG_i}{\$} \times \$_i \tag{19.2}$$

where, $GHG_i/\$$ represents GHG to money ratio associated with the ith input in gasoline production and $\$_i$ represent the production cost of the ith input.

The amount obtained from Eq. 19.2 does not include GHG emissions resulting from the direct use of inputs in gasoline production process and the final use of the end product. When GHG emissions from the direct use of inputs (such as coal combustion) and the final use of the end product are calculated separately using emission factors and process level data, and added with GHG emissions obtained from Eq. 19.2, then total life cycle GHG emissions are obtained and can be shown as:

$$GHG_{total} = GHG + GHG_{production} + GHG_{use} \qquad (19.3)$$

where, $GHG_{production}$ and GHG_{use} refer to GHG emissions from gasoline production and its use in a vehicle.

The model (Eq. 19.3) that combines Eco-LCI (Eq. 19.2) with process level inventory is called hybrid Eco-LCA. The hybrid Eco-LCA model is a mixture of process LCA and economic input-output LCA. As a result, it is more precise than Eco-LCA. Also due to inclusion of upstream processes, it eliminates truncation errors that results from selecting an artificial system boundary in LCA. The steps included for LCA of cellulosic ethanol and gasoline are shown in Figure 19.1. For yellow poplar, switchgrass, corn stover, and newsprint, cellulosic ethanol was assumed to be produced via acid pretreatment followed by simultaneous saccharification and co-fermentation (SSCF) due to availability of model data. In the case of MSW, cellulosic ethanol was assumed to be produced from a patented technology called Gravity Pressure Vessel Process (Kalogo et al., 2007). It was assumed that lignin, a byproduct, would fully provide steam and electricity for in-plant use for SSCF process. Any excess electricity was assumed to be sold to the grid. For Gravity Pressure Vessel Process, lignin was not used as the source of energy. The life cycle studies of cellulosic ethanol is based on the assumed yields of 340 liters, 352 liters, 327 liters, 257 liters, and 405 liters per dry ton of corn stover, yellow poplar, switchgrass, MSW and newsprint, respectively.

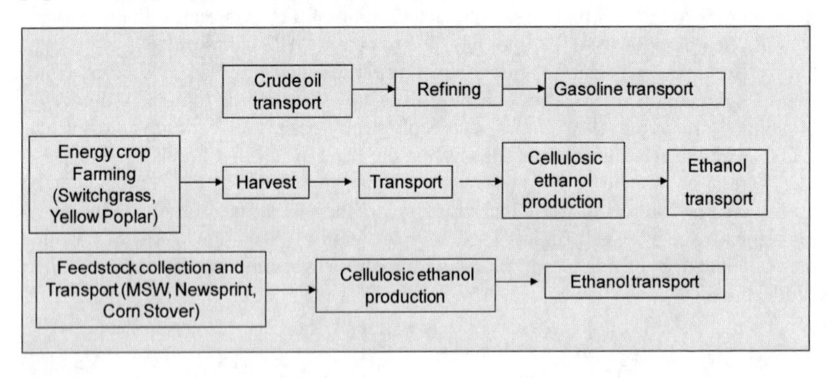

Figure 19.1 Steps included in LCA of cellulosic ethanol and gasoline (system boundary).

19.3.2 Aggregation

Aggregation of resources and emissions were performed to reduce a large number of data to meaningful and easily interpretable metrics for assessing biofuels. Natural resources are aggregated using energy and exergy analyses. Whereas emissions that cause similar impacts are aggregated using characterization factors. Since aggregation of resource is accompanied by the loss of information about their qualities, disaggregate resource consumption data in terms of per km travelled are also provided to maintain the integrity of the study.

Energy Analysis. Energy analysis permits aggregation of only those resources that have heating values (enthalpy) such as fossil fuels, biomass and electricity. Materials which do not release heat upon combustion such as steel are excluded by energy analysis. Therefore, energy analysis is only restricted to inputs which provide heat upon combustion and mechanical energy (electricity), and provides insight only about the energy efficiency or energy competiveness.

$$E = \sum x_i \tag{19.4}$$

where, x_i represents energy in Joule of the ith resource and E is the total energy consumed in a process. This aggregation assumes that x_i of the ith resource is substitutable for x_{i+1} of the $(i+1)^{th}$ resource. However, energy of all fuels is not of the same quality. For example, energy of electricity is of high quality than energy of natural gas. Therefore, 1 Joule of electricity cannot be substituted by 1 Joule of natural gas. In this study total energy and non-renewable energy consumption required to drive a Chevrolet Impala 2006 (passenger car) by a kilometer are chosen as metrics for comparing well-to-wheel energy efficiencies of various fuels. Energy breeding factor, which is defined as the energy return per unit J of non-renewable energy consumption, was also calculated. This metric is also known as energy return on investment (r_E).

ICEC Analysis. In recent years, there has been a growing interest in exergy analysis as part of LCA (Ayres et al., 1998; Berthiaume, 2001; Rosen, 2002; Sciubba and Ulgiati, 2005). Exergy is the maximum work that can be done as the system reaches equilibrium with its surroundings. It includes Gibbs free energy, gravitational potential energy and kinetic energy. Since exergy is a measure of useful energy, it represents the quality of resources, more so than energy analysis does. For example, water at 30°C can do little work than water at 100°C even if the total energy of these two systems is the same. This ability to distinguish energy quality makes exergy analysis appealing in LCA. Other advantage lies in its capacity to integrate energy and material resources together on a common thermodynamic unit. While energy analysis enables us to add together different energy types such as biomass, natural gas, electricity, and coal based on calorific values, it cannot take into account materials that cannot produce heat upon combustion such as steel. Therefore, energy metrics can only provide insight about energy efficiency. On the other hand, exergy analysis can tell about the overall system efficiency. The recognition that energy is as much as important as materials from the efficiency point of view is what separates exergy analysis from energy analysis. Many of the economic products and services are derived from non-renewable natural resources. Therefore, quantification of efficiency that reflects both materials and energy consumption is important to identify avenues of inefficiencies for targeting resource conservation and cost reductions. A production system may use less energy but more water; and energy analysis may show the process to be more energy efficient while ignoring the fact that the overall efficiency can be also improved by reducing water consumption. When exergy of all natural resources and economic products required to make a product are added, it results in industrial cumulative exergy consumption (ICEC). Let's suppose, y (J) is the exergy of iron ore and z (J) is the exergy of energy inputs required to process iron ore into steel, then ICEC of steel is given as $(y+z)$ J. Similarly, ICEC of coal delivered at power plant includes exergy of coal plus exergy invested in coal mining and transportation.

$$ICEC = \sum Ex_i \tag{19.5}$$

where, ICEC is industrial cumulative exergy consumption and Ex_i represents the exergy of the i-th resource. One limitation of ICEC analysis is that it assumes that Ex_i of the ith resource is substitutable for x_{i+1} of the $(i+1)^{th}$ resource. ICEC metrics that can be of interest in LCA of biofuels for providing unique insight in LCA are exergy breeding factors, exergy return investment and well-to-wheel ICEC. Exergy breeding factors and exergy return on investment are similar for biofuels and hence only exergy breeding factors are discussed here. Exergy breeding factor is defined as the ratio of exergy of biofuel to total non-renewable exergy consumed.

For biofuels, resource intensity can be expressed in terms of ICEC per kilometer traveled.The metric ICEC/km is straightforward and shows how much cumulative exergy needs to be invested to drive a vehicle by kilometer. A higher ICEC/km signals inefficiency and a large exergy loss. From a resource conservation point of view, the main goal would be to lower ICEC/km. Alternatively, it is desirable to choose the fuel with less nonrenewable ICEC/km. Several other forms of ICEC metrics can also be formulated for use in LCA such as % renewable ICEC, renewability indicator, and environmental loading ratio for the purpose of comparison.

Aggregation of Emissions. Emissions that exert similar environmental effects can be aggregated using equivalency factors. The equivalency factors, also known as characterization factors, enable us to express environmental impact in equivalent units. For example, methane is 25 times more potent than carbon dioxide as far as its contribution to global warming is concerned. Therefore, 1 unit of methane equals 25 units of carbon dioxide, and 25 is known as characterization factor. The indices developed by aggregating emissions that cause similar effects are known as mid-point impact indictors. The mid-point indicators included in the study are global warming potential, ecotoxicity, photochemical oxidation, human toxicity and eutrophication.

Aggregation scheme for mid-point impact analysis can be represented as:

$$E_{impact} = \sum Ai \times Bi \quad (19.6)$$

where, A_i refers to mass (kg) of the ith emission and B_i refers to a characterization factor of the ith emission. Characterization factors used in this study come from the Institute of Environmental Sciences (CML)-Leiden University impact assessment methods and characterization factors (CML, 2004).

Allocation. Except for MSW, lignin (a byproduct of cellulosic ethanol production) was assumed to be combusted to generate steam and electricity for in-plant use and selling excess electricity to the grid. This excess electricity was taken as co-product. For newsprint and MSW, energy saved from the avoidance of land filling was also taken as a co-product for the purpose of allocation. Resource consumption and emissions were allocated based on the market values of co-products as follows:

$$X = M \times (Z/Y) \quad (19.7)$$

where, X is the amount of resource consumption or emissions allocated to the co-product A, M is the total inputs (resources) or emission of the process, Y is the total market value of all co-products, and Z is the market value of the co-product A.

19.4 Results

19.4.1 Natural Resource Consumption

Since most resources have exergy, it is possible to quantify natural resource consumption in terms of industrial cumulative exergy consumption (ICEC). Material flow analysis concerns only with resources that have mass and excludes resources such as sunlight, electricity, etc. In Figures 19.2 to 19.5, ICEC of individual ecological goods and services needed for producing 3.79×10^6 liters (l) of cellulosic ethanol and the km-equivalent amount of gasoline are provided. The idea behind providing disaggregated ICEC data of individual resources is to determine the overall composition of natural resource mix required for fuel production, and to identify if critical resources, particularly non-renewable, are consumed more by one alternative versus the other. ICEC analysis provides ecological footprints in terms of exergy for economically derived products. Figures 19.2-19.5 depict not only what resources are involved in biofuel production pathways but in what quantities. It is recommended that both disaggregated and aggregated data are provided in LCA for eliciting complementary information. Aggregation of data reduces large number of data into a handful of metrics that can tell about the attributes of the overall system under investigation but it comes with information loss about qualities of resources being consumed. On the other hand, disaggregated data show relative contribution of each individual resource to the attribute that is being measured.

Figures 19.2-19.5 show how the consumption of individual natural resources for cellulosic ethanol obtained from five different feedstocks compares with that of gasoline over their life cycle. The consumption data are based on the mileage of a Chevrolet Impala 2006, a standard-size car. Due to a disparate scale of consumption, the figures are plotted on a log scale. As seen from Figure 19.2, cellulosic ethanol from all feedstocks consumes less crude oil compared to gasoline. Crude oil consumption by cellulosic ethanol is lower than that of gasoline by factors ranging from 8 to 27 depending on the method of cellulosic ethanol production and the types of feedstock used. It suggests that cellulosic ethanol can be an effective strategy for reducing dependence on crude oil. For other fossil fuels, the results are mixed. For example, corn-stover derived cellulosic ethanol consumes more of and yellow poplar cellulosic ethanol consumes less of coal, natural gas, and nuclear energy compared to gasoline.

When it comes to metallic ores, gasoline is associated with lower ICEC of these resources than cellulosic ethanol (Figure 19.3). Similarly, gasoline consumes less non-metallic minerals than cellulosic ethanol (Figure 19.4). Note that metallic ores and non-metallic mineral consumption refers to indirect consumption since these resources are not utilized directly in biofuel production. For example, if steel is used in equipment, its consumption appears as iron ore. Cellulosic ethanol derived from feedstocks directly dependent on farming (corn stover, switchgrass, yellow poplar) relies more on renewable natural resources than gasoline including soil (Figure 19.5). Note that although soil is classified here as renewable, it is a critical resource since it cannot be replenished quickly. Only cellulosic ethanol derived from newsprint and MSW consumes less soil (causes less soil erosion) than gasoline since it is not directly dependent on agriculture. If corn stover is considered as waste, one may not allocate soil erosion to corn stover. In that case, soil erosion of corn stover-derived cellulosic ethanol can be small than shown in Figure 19.5 but still larger than that of gasoline. ICEC of soil erosion is two-three orders of magnitude larger for cellulosic ethanol directly dependent on agriculture than that of gasoline. Overall, cellulosic ethanol consumes more of ecological resources considered in the study with a few exceptions such as crude oil. Fortunately, some of the resources are renewable such as sunlight, detrital matter (dead organic matter), nutrients from mineralization, and may not be a cause of concern.

19.4.2 Aggregate Metrics

In this section, metrics developed using energy and exergy analyses are analyzed and discussed. Moreover, environmental impact potentials of cellulosic ethanol and gasoline are assessed using mid-point impact indicators.

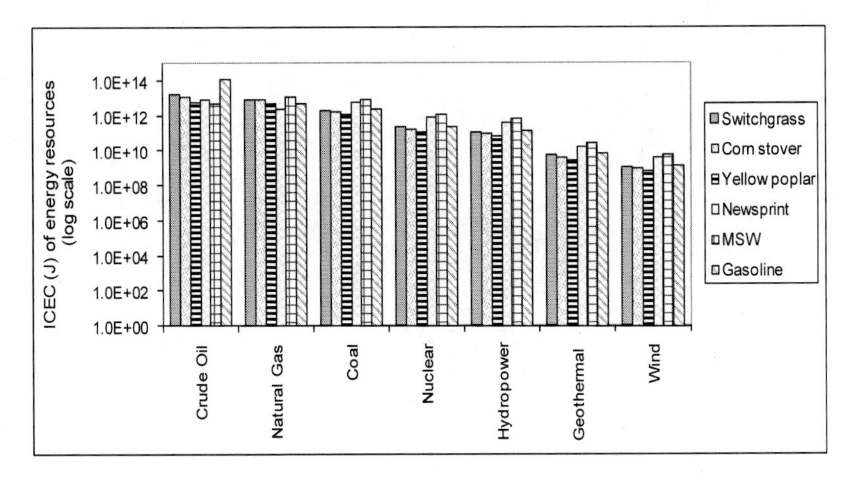

Figure 19.2 ICEC contribution of energy sources for cellulosic ethanol (3.79 E+06 liters) and km-equivalent amount of gasoline (2.61E+06 liters).

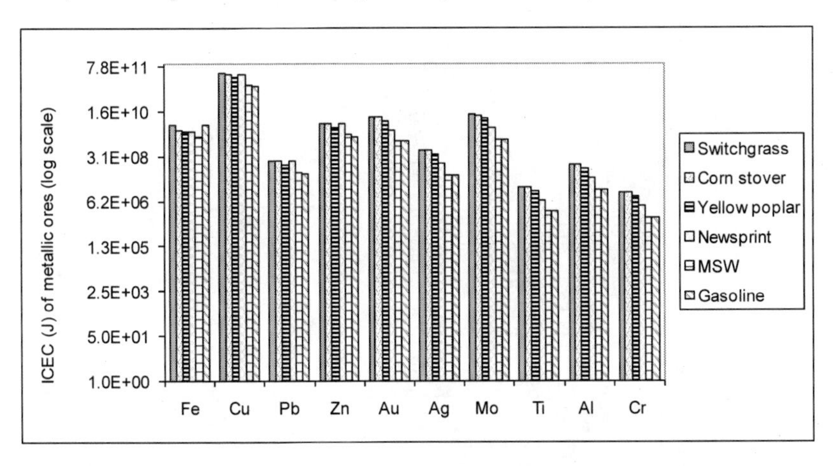

Figure 19.3 ICEC contribution of metallic ores for cellulosic ethanol (3.79 E+06 liters) and km-equivalent amount of gasoline (2.61E+06 liters).

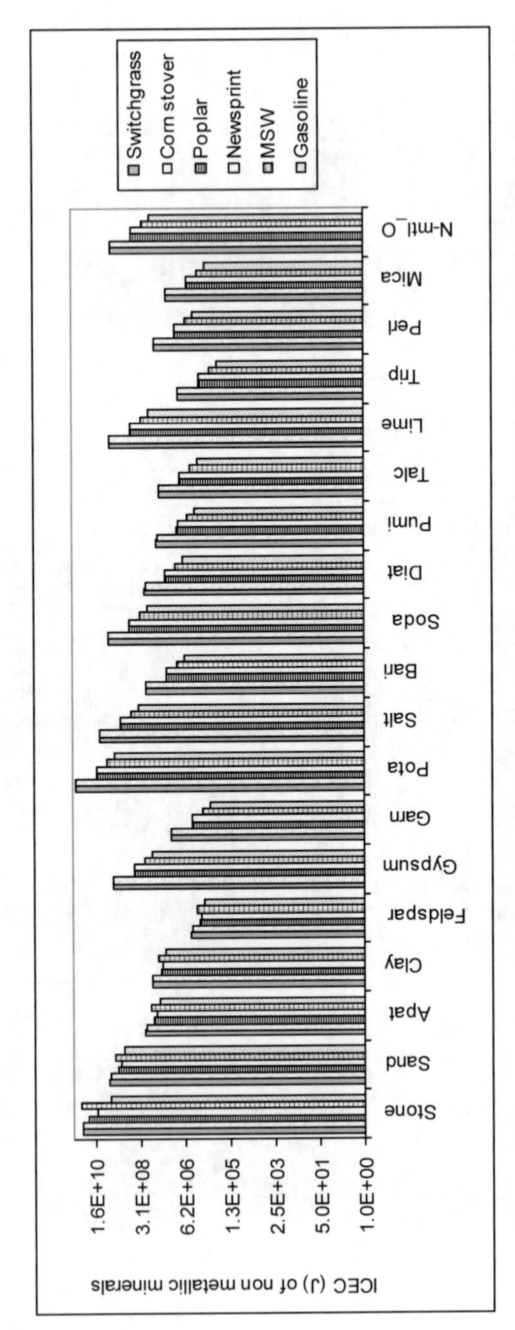

Abbreviations: Apat -Apatite, Garn -Garnite, Pota -Potassium, Salt -NaCl, Bari- Barium, Soda - Na_2CO_3, Pumi -Pumice, Trip-Triplite, Perl- Perlite, N_mtl_O – Other non-metallic ores

Figure 19.4 ICEC contribution of non-metallic minerals for cellulosic ethanol (3.79 E+06 liters) and km-equivalent amount of gasoline (2.61E+06 liters).

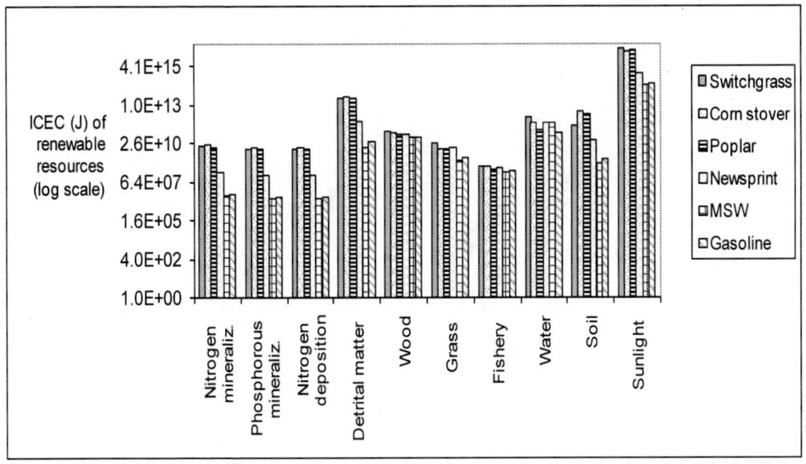

Figure 19.5 ICEC contribution of renewable resources for cellulosic ethanol (3.79 E+06 liters) and km-equivalent amount of gasoline (2.61E+06 liters).

19.4.2.1 Well-to-Wheel Energy Consumption

Well-to-wheel (WTW) energy consumption is a measure of how efficiently energy is utilized per kilometer traveled when both direct and indirect energy consumption is taken into account. Figure 19.6 provides well-to-wheel total energy consumption as well as non-renewable energy consumption for E85 obtained from cellulosic ethanol and gasoline which is modeled after Chevrolet Impala, 2006. Except for yellow poplar and newsprint, E85 from the rest of feedstocks has higher WTW energy consumption than gasoline, hence, is less efficient and less energetically competitive. This shows that there exists room for increasing energy efficiency by improving process parameters for ethanol production. Note that a large portion of energy required in ethanol production comes from lignin.

Since the primary purpose of using biofuels is to minimize the dependence on fossil fuels, WTW non-renewable energy consumption provides a better indication of how much non-renewable energy can be saved by switching to E85 obtained from cellulosic ethanol. The non-renewable energy savings can vary from 56% for E85 of MSW to as high as 70% for E85 of newsprint compared to gasoline. The savings largely result from the use of renewable energy (electricity and steam) produced from lignin combustion except in the case of MSW. If the future technology development allows for the use of 100% ethanol in a passenger vehicle in the USA, non-renewable energy savings can be even higher. It is to be noted that pure ethanol obtained from sugarcane is currently being used in vehicles in Brazil.

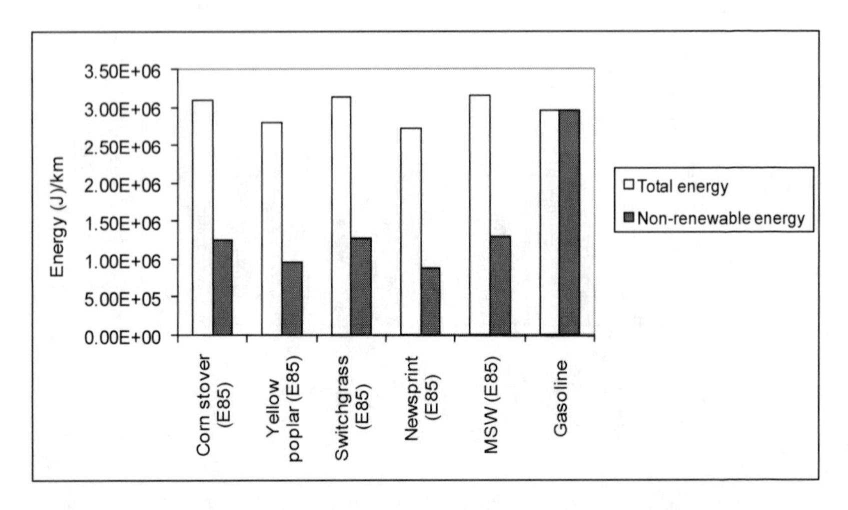

Figure 19.6 WTW energy consumption for E85 derived from various lignocellulosic feedstocks.

19.4.2.2 Well-to-Wheel ICEC

When all of the resources consumed are aggregated on a common unit of exergy (J), well-to-wheel ICEC can be calculated (Figure 19.7). ICEC/km, also known as WTW exergetic efficiency, estimates useful energy of various resources including materials that are consumed in the production and use of an economically derived product. A higher ICEC/km indicates inefficient resource utilization. Since the major focus in sustainability debate has been on conservation of non-renewable resources, non-renewable ICEC may better reflect the intensity of non-renewable resource consumption per km. Ideally the future choice of alternative sustainable transportation fuel should be the one with lower non-renewable ICEC per km.

As indicated in Figure 19.7, cellulosic ethanol (E85) consumes more ICE than gasoline per km indicating its relative resource intensity. It is recalled here that cellulosic ethanol consumes more of all of resources considered in the study (Figures 19.2-19.5) which when added together results in a large ICEC. More than 99 percent of exergy consumption comes from sunlight for cellulosic ethanol since sunlight is inefficiently utilized by plants. Therefore, by improving photosynthetic efficiency or biomass yield per hectare, it is possible to improve the WTW exergetic efficiency of cellulosic ethanol. Since sunlight is renewable and plentiful, large consumption of sunlight is not much of a concern since it does not represent a resource constraint from a large scale production point of view. It is clear from Figure 19.7 that E85 derived from lignocellulosic feedstocks consumes less non-renewable resources in aggregate. In this study, soil is classified as renewable resource. Even if it is classified as non-renewable, cellulosic ethanol still consumes less non-renewable ICE per km. Reduction of non-renewable ICEC per km when E85 substitutes for gasoline varies from 43% for MSW to 60% for newsprint. The lower non-renewable ICEC for newsprint and yellow poplar is attributed to the utilization of lignin as process energy and lower farm inputs for yellow poplar production.

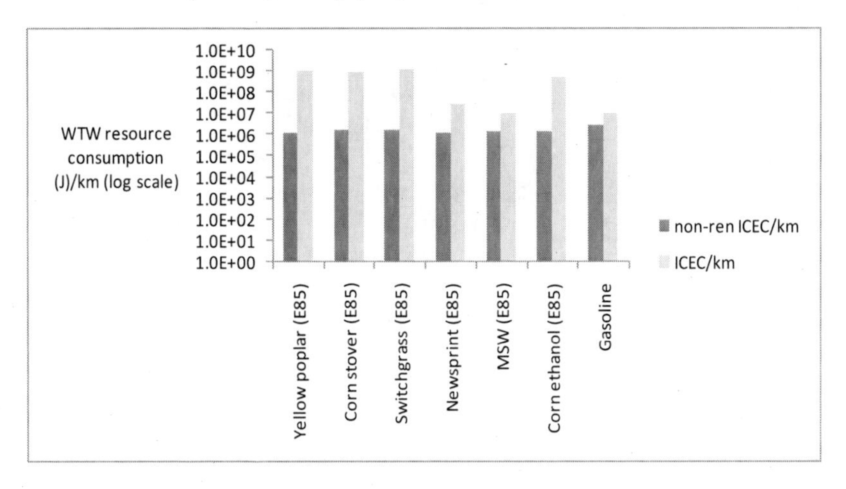

Figure 19.7 ICEC/km and non-renewable ICEC/km for E85 derived from five lignocellulosic feedstocks.

19.4.2.3 Energy and Exergy Breeding Factors

Alternatively, when comparing biofuels, it may be of interest to quantify how much renewable exergy can be generated per joule of non-renewable exergy. This term is referred to as exergy breeding factor (Dewulf et al., 2005). The energy version of it is called energy breeding factor. Energy breeding factor is also a return on energy investment (r_E) which shows the return per joule of non-renewable energy consumed. The goal for future transportation fuels is to obtain more renewable output with little non-renewable inputs. Non-renewable exergy inputs include not only energy but also materials such as metal ores, non-metallic minerals, etc. Cellulosic biofuels from newsprint and yellow poplar have excellent energy and exergy breeding factors and cellulosic ethanol from MSW has the lowest energy and exergy breeding factors (Figure 19.8). The higher breeding factors for newsprint- and yellow poplar-derived cellulosic ethanol are attributed to the utilization of renewable lignin as a source of process energy. Yellow poplar also requires less farm inputs and therefore requires less fossil fuel. In contrast, the lower breeding factors of MSW-derived cellulosic ethanol is due to the lack of utilization lignin for electricity generation in GPV process; although by process modification it is possible to use lignin as a source of energy. Therefore, MSW represents the conservative estimates of breeding factors for cellulosic ethanol. Nonetheless, breeding factors of MSW-cellulosic ethanol are still higher than those of corn ethanol (Figure 19.8). Overall, newspaper and yellow poplar appear to be the feedstocks of choice if more renewable output is to be generated from the least non-renewable exergy input.

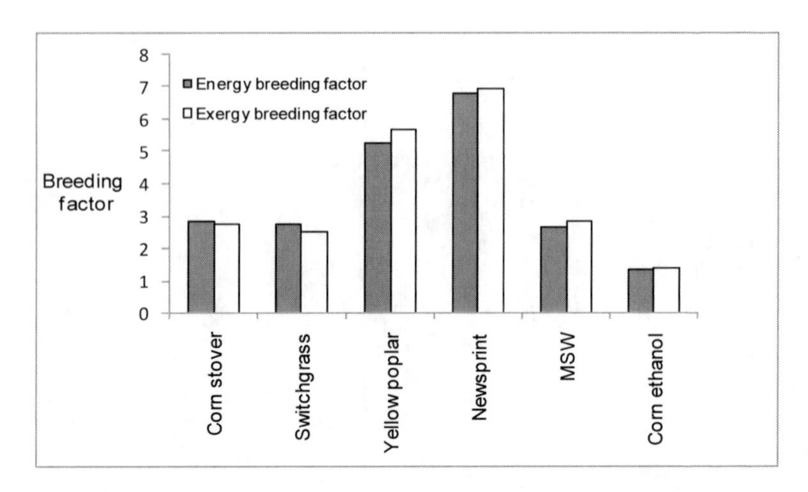

Figure 19.8 Exergy breeding factors for cellulosic ethanol obtained from different lignocellulosic feedstocks.

If the reduced reliance on non-renewable resource is taken as an indication of sustainability, newsprint- and yellow poplar-derived cellulosic biofuels are more sustainable than the cellulosic ethanol derived from other feedstocks and gasoline. Besides non-renewable consumption, there are other factors that can affect sustainability such as life cycle environmental impacts which are discussed below.

19.4.2.4 Mid-Point Environmental Indicators

Life cycle environmental impact analysis of cellulosic ethanol and gasoline was conducted using CML's impact assessment method and characterization factors (CML, 2004). A midpoint indicator provides a measurement of potential to cause damage in a particular impact category. It is possible to further aggregate mid-point indicators to arrive at single end point indicator that indicates overall impact. Such a simplification is not recommended since it can lead to misleading interpretation. There are a number of impact categories that have been developed to assess the environmental impacts of alternatives. Examples of mid-point indicators in CML's impact assessment are ozone depletion, human toxicity, photochemical oxidation, eutrophication, abiotic depletion, land use depletion, freshwater ecotoxicity, marine aquatic ecotoxicity, terrestrial ecotoxicity, etc. Since only air emissions including SO_x, VOC, NO_x, N_2O, CO, PM10, CO_2, CH_4, and HFCs were considered in the life cycle study, relevant mid-point indicators which include photochemical oxidation, acidification, eutrophication, human toxicity, and global warming were calculated and are provided in Figure 19.9.

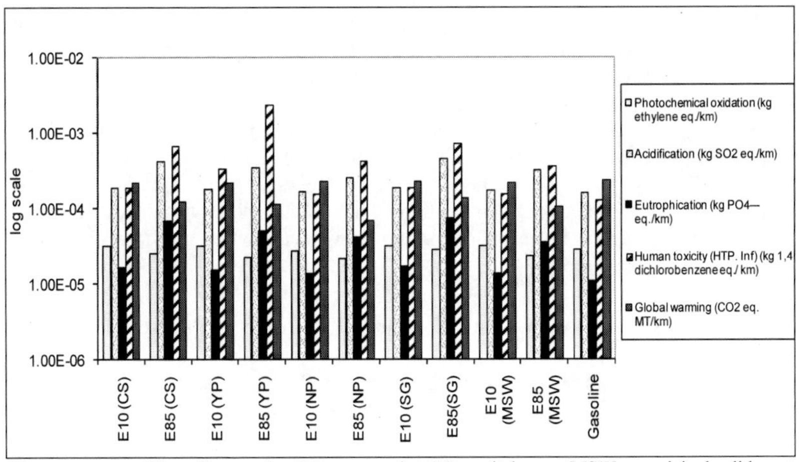

CS= corn stover, YP= yellow poplar, NP = newsprint, SG= switchgrass, MSW= municipal solid waste.

Figure 19.9 Potential environmental impacts from the use of E10, E85 and gasoline per km.

As expected, global warming potential of cellulosic ethanol (E10 and E85) is lower than that of gasoline per kilometer traveled. This is due to less fossil fuel consumption in cellulosic ethanol production pathways. Even though, nitrous oxide is released in agricultural farming from nitrogen fertilizer use, it is not large enough to offset lower GHG emissions from fossil fuel use in biofuel production. Overall, E85 can reduce global warming potential over gasoline by 40-71% depending on the feedstocks employed. The lowest reduction is observed for switchgrass and the largest reduction for newsprint. Note that GHG emissions from indirect land uses are not included in this study due to large uncertainties associated with land use models, data, and economic factors of large scale biofuel production.

In contrast E10 and E85 derived from five feedstocks cause higher acidification, eutrophication and human toxicity than gasoline. Increases in acidification from E85 substitution can range from 64% to 186% for various feedstocks which are linked to increase in SO_x emissions. Similarly, substitution of E85 for gasoline can increase eutrophication by 223% for MSW and 580% for switchgrass. Eutrophication in question is contributed by N_2O since N and P releases into surface waters were not considered in this study. Inclusion of N and P release into surface waters further increases eutrophication for yellow poplar, switchgrass, and corn stover. Likewise, human toxicity can increase from the use of E85 over gasoline which is contributed by increases in NO_x, SO_x, and PM10 emissions. The increase can range from 178% for MSW to 1728% for yellow poplar. The higher human toxicity of yellow poplar is due to PM10, NO_x, and SO_x emissions from lignin combustion. It is to be noted that lignin content of yellow poplar is higher (27.7%) among the feedstocks considered.

With respect to photochemical oxidation, conflicting results are observed for E10 and E85. Substitution of E10 for gasoline increases photochemical oxidation whereas substitution of E85 decreases the photochemical oxidation. This is because E10 from a Chevrolet Impala 2006 releases more CO than gasoline does. The results from environmental impact analysis suggest that utilization of cellulosic ethanol contributes to climate change mitigation but may worsen other environmental impacts. However, it needs to be determined whether increases in other emissions are large enough to cause significant effects on human health and the environment. One way of doing this would be to normalize emissions with regional emissions or national emissions.

19.5. Land Requirements and Ethanol Production Potential

Other attributes that are of interest in comparative analysis of biofuels are land requirements and production potential. If the alternative fuel possesses desirable characteristics but cannot be produced in large amounts, its potential to be economically viable and provide the significant improvements will be severely limited. Consequently, cellulosic ethanol production potential from lignocellulosic feedstocks should be assessed. Considering the availability of land that does not compete with food crops, yellow poplar can potentially produce 61 billion gallons of cellulosic ethanol. Switchgrass can also provide 30 billion gallons of ethanol. Even MSW, newsprint, and corn stover can each produce nearly as much or more ethanol than what is produced from corn (6.5 billion gallons in 2007). Since land can be a limiting resource, land use efficiency can also provide a valuable guidance about the choice of future transportation fuels. Although, potential ethanol production from yellow poplar is large, it requires a large land area. Cellulosic ethanol from corn stover is associated with the largest land area per million gallons produced but it is possible to argue that cropland should not be allocated to corn stover since it is primarily dedicated for corn. On the other hand, MSW and newsprints need less land per million gallons which refer to indirect land use.

Table 19.1 Land area requirements and ethanol production potential for various feedstocks.

Feedstock	Potential supplies (dry MT)	Potential ethanol production (gallons)	Land requirements (ha)/million gallons[a]
MSW	1.38E+08	9.35E+09	3
Newsprint	7.72E+07	8.26E+09	91
Corn stover	6.80E+07	6.11E+09	1190
Yellow Poplar	6.55E+08	6.09E+10	848
Switchgrass	3.42E+08	2.95E+10	862

[a] Land area refers to direct and embodied (indirect) land use in the supply chains. 1 gallon = 3.7854 liters.

19.6 Challenges and Future Research Direction

Since cellulosic ethanol technology is yet to be commercialized, large uncertainties exist with regard to process parameters and assumptions about yields due to lack of rigorous data. The present study does not take into account GHG emissions from indirect land use changes. Indirect land use changes (ILUC) refers to land conversions that occur in response to diversion of crops and feedstocks from their primary use to biofuel production. Although GHG emissions from ILUC are likely to be small for dedicated energy crops such as yellow poplar and negligible for agriculture residues and MSW as compared to crops, it may be necessary to estimate GHG emissions from ILUC since low carbon fuel-related standards that are in various stages of development may require the inclusion of ILUC in LCA of biofuels. Therefore, ILUC is going to be an important future research field in LCA of biofuels. Moreover, aggregation of resources can be further expanded by incorporating emergy or ecological cumulative exergy consumption (ECEC) analysis. Emergy or ECEC analysis can be appealing since, unlike exergy analysis, it enables aggregation of resources by taking into account quality of resources and may provide additional insights about biofuels.

19.7 Summary

This study makes the first attempt to quantify contributions of important ecological goods and services in LCA of cellulosic ethanol. The ecological goods and services considered here can be broadly classified as provisioning, regulating and supporting services. Provisioning services includes minerals, ores, renewable and non-renewable energy, biomass, water, land, etc. Examples of regulating services include pollution dissipation, mitigation, water purification, climate regulation, pollination, etc. For example, carbon sequestration by plants is one such regulating service considered in this study. Supporting services included in this study are soil formation via detrital matter and nutrients mineralization and photosynthesis. Compared to gasoline, cellulosic ethanol consumes less crude oil but more of other renewable and non-renewable resources indicating its resource intensity. As indicated by breeding factors, the overall dependence of cellulosic ethanol on non-renewable resources is lower than that of corn ethanol indicating the possibility of conserving appreciable amounts of non-renewable resources, particularly crude oil, if cellulosic ethanol is substituted for gasoline. As far as environmental impact is concerned, the main advantage of cellulosic ethanol lies in its capacity to contribute to climate change mitigation. In the rest of impact categories considered here, E85 from cellulosic ethanol will increase eutrophication, human toxicity, and acidification while it may lower photochemical oxidation. The land use requirements for cellulosic ethanol derived from MSW and newsprint are significantly less because of the avoidance of direct land use. Theoretical ethanol production potential from agriculture residue and municipal solid waste is at par with what is currently obtained from corn ethanol. The reduced land requirements for some of the feedstocks, newsprint and MSW, avoidance of food fuel conflicts, lower reliance on non-renewable resources and contribution to GHG mitigation make cellulosic ethanol an appealing choice as the future transportation fuel.

19.8 Acknowledgments

This work is supported by the National Science Foundation (Grant No. ECS-0524924)) and US EPA.

19.9 References

Aden, A., Ruth, M., Ibsen, K., Jechura, J., Neeves, K., Sheehan, J., Wallace, B., Montague, L., Slayton, A., and Lukas, J. (2002). "*Lignocellulosic Biomass to Ethanol Process Design and Economics Utilizing Co-Current Dilute Acid Prehydrolysis and Enzymatic Hydrolysis for Corn Stover*." National Renewable Energy Laboratory, Golden Colorado, NREL/TP-510-32438.

Ayres, R.U., Ayres, L.W., and Martinas, K. (1998). "Exergy, waste accounting, and life-cycle analysis." *Energy,* 25 (3), 355-363.

Baral, A., and Bakshi, B.R. (2009). "Thermodynamic metrics in aggregation of natural resources in life cycle analysis: Insight via application to some transportation fuels." Submitted to *Environ. Sci. Technol.*

Berthiaume, R., Bouchard, C., and Rosen, M.A. (2001). "Exergetic evaluation of the renewability of a biofuel." *Exergy Int. J.,* 1(4), 256-268.

Brinkman, N., Wang, M., Weber, T., and Darlington, T. (2005). "*Well-to-Wheel Analysis of Advanced Fuel/Vehicle Systems – A North American Study of Energy Use, Greenhouse Gas Emissions, and Criteria Pollutant Emissions.*" General Motors Corporation and Argonne National Laboratory.
Available at www.transportation.anl.gov/pdfs/TA/339.pdf (accessed on May 27 2007).

Delucchi, M.A. *(2006). "Lifecycle Analyses of Biofuels*." Draft Report, UCD-ITS-RR-06-08, Institute of Transportation Studies, UC-Davis, CA.
Available at http://www.its.ucdavis.edu/people/faculty/delucchi/index.php (accessed on March 2007).

Dewulf, J., Van Langenhove, H., and Van De Velde, B. (2005). "Exergy-based renewability assessment of biofuel production." *Environ. Sci. Technol.*, 39(10), 3878-3882.

European Council for Automotive R&D. (2004). "*Well-to-Wheels Analysis of Future Automotive Fuels and Powertrains in the European Context*." Joint report by European Council for Automotive R&D (EUCAR), Conservation of Clean Air and Water in Europe (CONCAWE) and the Joint Research Council (JRC). Available at http://ies.jrc.cec.eu.int/Download/eh.

Fargione, J., Hill, J., Tilman, D., Polasky, S., and Hawthorne, P. (2008). "Land clearing and the biofuel carbon debt." *Science*, 319, 1235-1238.

Farrell, A.E., Plevin, R.J., Turner, B.T., Jones, A.D., O'Hare, M., and Kammen, D.M. (2006). "Ethanol can contribute to energy and environmental goals." *Science*, 311(5760), 506-508.

Fleming, J.S., Habibi, S., and MacLean, H.L. (2006). "Investigating the sustainability of lignocellulose-derived fuels for light-duty vehicles." *Transport. Res. Part D 11*, 146–159.

General Motors North America. (2001). "*Well-to-Wheel Energy Use and Greenhouse Gas Emissions of Advanced Fuel/Vehicle Systems –North American Analysis*." General Motors Corporation Report.
Available at http://greet.anl.gov/publications.html (accessed on June 2006).

Institute of Environmental Sciences (CML). (2004). *CML-IA - CML's Impact Assessment Methods and Characterisation Factors*. Version 2.7, Leiden University. Available at http://www.leidenuniv.nl/interfac/cml/ssp/index.html (accessed on November 2006).

Kalogo, Y., Habibi, S., MacLean, H.L., and Joshi, S. (2007). "Environmental implications of municipal solid waste-derived ethanol." *Environ. Sci. Technol.*, 41(1), 35-41.

Kemppainen, A.J., and Shonnard, D.R. (2005). "Comparative life-cycle assessments for biomass-to-ethanol production from different regional feedstocks." *Biotechnol. Prog.*, 21, 1075-1084.

Levelton Engineering Ltd. and S&T Consultants. (1999). "*Assessment of Net Emissions of Greenhouse Gases from Ethanol-Blended Gasolines in Canada: Lignocellulosic Feedstocks*." Report prepared for Agriculture and Agrifood Canada, Ottawa, ON, File 499-0893.

Odum, H.T. (1996). *Emergy and Environmental Decision Making*. John Wiley & Sons, New York.

Patzek, T.W. (2004). "Thermodynamics of the corn-ethanol biofuel cycle." *Crit. Rev. Plant Sci.,* 23 (6), 519-567.

Pimentel, D., and Patzek, T. (2005). "Ethanol production using corn, switchgrass, and wood; biodiesel production using soybean and sunflower." *Nat. Resour. Res.*, 14(1), 65-76.

Sciubba, E., and Ulgiati, S. (2005). "Emergy and exergy analyses: Complementary methods or irreducible ideological options?" *Energy*, 30, 1953–1988.

Rosi-Marshall, E.J., Tank, J.L., Royer, T.V., Whiles, M.R., Evans-White, M., Chambers, C., Griffiths, N.A., Pokelsek, J., and Stephen, M. L. (2007). "Toxins in transgenic crop byproducts may affect headwater stream ecosystems." *Proc. Natl. Acad. Sci.,* 104(41), 16204-16208.

Sheehan, J., Aden, A., Paustian, K., Kendrick, K., Brenner, J., Walsh, M., and Nelson, R. (2004). "Energy and environmental aspects of using corn stover for fuel ethanol." *J. Indust. Ecol.*, 7(3-4), 177-146.

Spatari, S., Zhang, Y., MacLean, H.L. (2005). "Life cycle assessment of switchgrass- and corn stover-derived ethanol-fueled automobiles." *Environ. Sci. Technol.*, 39, 9750-9758.

Suh, S., Lenzen, M., Treloar, G.J., Hondo, H., Horvath, A., Huppes, G., Jolliet, O., Klann, U., Krewitt, W., Moriguchi, Y., Munksgaard, J., and Norris, G. (2004). "System boundary selection in life-cycle inventories using hybrid approaches." *Environ. Sci. Technol.,* 38(3), 657-664.

Rosen, M.A. (2002). "Clarifying thermodynamic efficiencies and losses via exergy." *Exergy*, 2(1), 3-5.

Wang, M. (2004). "*Greenhouse Gas, Regulated Emissions, and Energy Use in Transportation (GREET) Model*." Argonne National Laboratory, version 1.6.

Available at http://www.transportation.anl.gov/software/GREET/index.html (accessed on April 2006).

Wang, M., Saricks, C., and Santini, D. (1999). *Effects of Fuel Ethanol Use on Fuel-Cycle Energy and Greenhouse Gas Emissions.* Center for Transportation Research, Energy Systems Division, Argonne National Laboratory, 9700 South Cass Avenue, Argonne, Illinois 60439.

Watson, R.T., Zinyowera, M.C., Moss, R.H., and Dokken, D.J., (eds.) (1996). *Climate Change 1995: Impacts, Adaptations and Mitigation of Climate Change,* Cambridge University Press, Cambridge, UK.

Wooley, R., Ruth, M., Sheehan, J., Ibsen, K., Majdeski, H., and Galvez, A. (1999). "*Lignocellulosic Biomass to Ethanol Process Design and Economics Utilizing Co-Current Dilute Acid Prehydrolysis and Enzymatic Hydrolysis, Current and Futuristic Scenarios.*" National Renewable Energy Laboratory, Golden, CO, NREL/TP-580-26157. Available at http://www.nrel.gov/docs/fy99osti/26157.pdf.

Zhang, Y., Baral. A., and Bakshi, B.R. (2009). "Toward accounting for ecological resources in life cycle assessment." Submitted to *Environ. Sci. and Technol.*

CHAPTER 20

Biobutanol Production from Agri-Residues

S. Yan, Bala subramanian S., R.D. Tyagi, R.Y. Surampalli, and Tian C. Zhang

20.1 Introduction

Butanol is an alcohol which can be used as a transport fuel with properties very similar to gasoline. Biobutanol refers to biofuel produced from renewalable resources. Production of butanol by fermentation utilizes a number of organisms including *Clostridium acetobutylicum* or *Clostridium beijerinckii*. Butanol can be produced by the traditional acetone-butanol-ethanol (ABE) fermentation - the anaerobic conversion of carbohydrates by clostridia into acetone, butanol and ethanol (Ezeji et al., 2007a; Qureshi et al., 2007; Lee et al., 2008a; Qureshi et al., 2008a). The ABE process, formerly a large biotechnology industry, declined due to increasing substrate costs and the competition with the petrochemically-derived butanol (Durre, 1998; Ezeji et al., 2007b).

In recent years, there has been renewed interest in biobutanol production for transportation fuel. Compared to ethanol, butanol offers several unique advantages as a substitute for gasoline because of higher energy content and higher hydrophobicity. Economic analyses have demonstrated that the fermentation substrate is one of the most important factors that influenced the price of butanol, therefore, many studies have been focused on the use of renewable and low-cost substrates such as agricultural residues as substrates for ABE production (Qureshi and Blaschek, 2000a; Atsumi et al., 2008). The total solvents produced from these alternative renewable resources ranged from 14.8 to 30.1 g/L (Ezeji et al., 2004a). This chapter provides comprehensive review of biobutanol production from agri-residues. The microbiology and important considerations of butanol fermentation, the strain improvement by metabolic engineering and advanced fermentation technologies are also discussed. Some challenges and future research directions are also outlined.

20.2 Microbiology of Butanol Fermentation

20.2.1 Clostridia

Butanol (and acetone, ethanol, and isopropanol) are naturally formed by a number of clostridia. Clostridia are rod-shaped, spore-forming gram positive bacteria and typically strict anaerobes. Solventogenic (solvent producing) clostridia can utilize a large variety of substrates from monosaccharides including many pentoses and hexoses to polysaccharides (Jones and Woods, 1986; Ezeji et al., 2004a).

A large number of solventogenic clostridia have been studied over the years (Zverlov et al., 2006). In spite of the high number of isolates, only about 40 solventogenic clostridia have survived in public strain collections (Zverlov et al., 2006). A number of clostridia for ABE fermentation are listed in review papers (Volesky and Szczesny, 1983). The development of molecular techniques for solventogenic clostridia (i.e., *C. acetobutylicum* and *C. beijerinckii*) in combination with recent advances in fermentation techniques has resulted in the development of an integrated ABE fermentation system for simultaneous production and removal of ABE from the fermentation vessel. These improvements have resulted in reduction of toxicity, improved productivity, increased solvent yield, and improved carbohydrate utilization (Ezeji et al., 2004a). The yield of butanol of these strains will be discussed in detail in latter section.

20.2.2 Enteric Bacteria

Butanol production by *Clostridium* in mixed-product fermentation is well known. However, in this fermentation, it is difficult to control butanol yield, and furthermore, due to the relatively unknown genetic system and complex physiology of the microorganism, it is also difficult to achieve optimal biobutanol production (Ezeji et al., 2004a; Lee et al., 2008). Thus, there is strong interest to produce butanol from a user-friendly organism. *Escherichia coli* is a well-characterized microorganism with a set of readily available tools for genetic manipulation and its physiological regulation is well-studied (Atsumi et al., 2008; Cann and Liao, 2008). However, *E. coli* does not produce butanol as a fermentation product. Atsumi et al. (2008) engineered a synthetic pathway in *Escherichia coli* and demonstrated the production of butanol from this non-native user-friendly host. Alternative genes and competing pathway deletions were evaluated for butanol production. The results showed promise for using *E. coli* for butanol production (Atsumi et al., 2008).

20.3 Butanol Production Pathway

The fermentation of carbohydrates to acetone, butanol, and ethanol by solventogenic clostridia is well known. The butanol production pathway or the metabolic pathway from acetyl-CoA to butanol in *Clostridium* spp. is illustrated in Figure 20.1 (Ezeji et al., 2007b; Lee et al., 2008a).

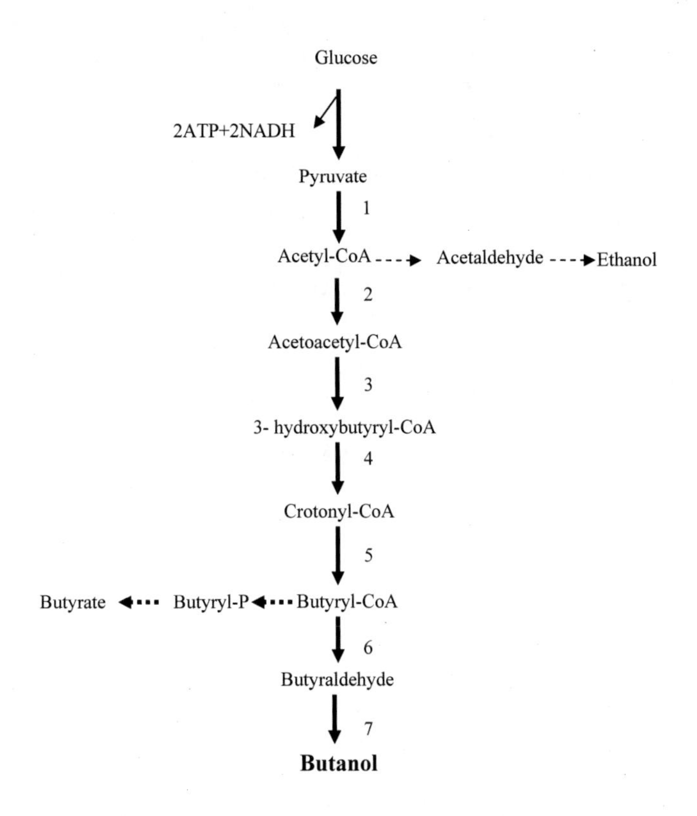

Figure 20.1 Metabolic pathway for butanol production by clostridia. 1, pflB (pyruvate-ferrodoxin oxidoreductase); 2, thl (thiolase); 3, hbd (3- hydroxybutyryl-CoA dehydrogenase); 4, crt (crotonase); 5, bcd (butyryl-CoA dehydrogenase); 6, adhE butyraldehyde dehydrogenase); and 7, bdhAB (butanol dehydrogenase).

The metabolic pathways of ABE-producing clostridia consist of two distinct characteristic phases, namely, acidogenesis and solventogenesis (Shinto et al., 2008). Clostridia produce butanol by conversion of a suitable carbon source into acetyl-CoA. Substrate acetyl-CoA then enters into the solventogenesis pathway to produce butanol using six concerted enzyme reactions. The formation of butanol requires the conversion of acetyl-CoA into acetoacetyl-CoA by acetyl transferase. This reaction is followed by the conversion of acetoacetyl-CoA into 3-hydroxylbutyryl-CoA by 3-hydroxyl-CoA dehydrogenase, which is followed by the conversion of 3-hydroxylbutyryl-CoA into crotonyl-CoA by 3-hydroxybutyryl-CoA dehydratase (also named crotonase) and the conversion of crotonyl-CoA into butyryl- CoA by butyryl-

CoA dehydrogenase and followed by the conversion of butyryl-CoA to butyraldehyde by butyraldehyde dehydrogenase, with the final conversion of butyrylaldehyde to butanol by butanol dehydrogenase (Jones and Woods, 1986; Shinto et al., 2008).

20.4 Bioreactor Configuration and Butanol Yield

The batch process is the most commonly studied method of fermentation for butanol production due to simple operation and reduced risk of contamination (Ezeji et al., 2004a; Lee et al., 2008). However, the productivity achievable in a batch reactor is low due to the lag phase, product inhibition as well as down time for cleaning, sterilizing, and filling. Under batch operation conditions, *C. beijerinckii* BA101 was able to produce 18–33 g/L ABE in 72 h of fermentation using glucose or starch as the substrate (Husemann and Papoutsakis, 1990; Avcibasi Guenilir and Deveci, 1996; Ishizaki et al., 1999; Maddox et al., 2000; Ezeji et al., 2004b; Abdel-Hakim et al., 2006). Table 20.1 summarises butanol (ABE) production from a number of carbohydrates using different clostridia in various reactors.

Fed-batch fermentation is applied to the butanol production to avoid substrate inhibition and to increase cell mass (Fond et al., 1984; Srivastava and Volesky, 1991). In fed-batch process, the reactor is initiated in a batch mode with a low substrate concentration (noninhibitory to the culture) and a low medium volume, and then the culture volume increases in the reactor over time. Since butanol is toxic to *C. acetobutylicum* or *C. beijerinckii* cells, the fed-batch fermentation technique cannot be applied unless one of the novel product recovery techniques is applied for simultaneous separation of product. As a result of substrate reduction and reduced product inhibition, greater cell growth occurs and reactor productivity is improved. This process was also employed for butanol fermentation (Table 20.1).

The continuous culture technique is also used in butanol production(Eckert and SchuÌˆgerl, 1987; Soni et al., 1987; Godin and Engasser, 1988; Holt et al., 1988; HuÌˆsemann and Papoutsakis, 1989; Meyer and Papoutsakis, 1989; Qureshi et al., 2000;

Table 20.1 Production of ABE from different fermentation and production removal system.

Strain	Fermentation process	Substrate type	Butanol conc. (gL^{-1})	ABE conc. (gL^{-1})	ABE Productivity ($gL^{-1}h^{-1}$)	ABE yield (gg^{-1})	Reference
C. acetobutylicum	Batch	Glucose	19.6	24.2	0.34	0.42	(Evans and Wang, 1988)
C. beijerinckii BA101	Batch	Glucose	-	24.2	0.34	0.42	(Qureshi and Blaschek, 1999a)
C. beijerinckii BA101	Batch with pervaporation	Glucose	-	51.5	0.69	0.42	(Qureshi and Blaschek, 1999a)
C. acetobutylicum	Batch with pervaporation	Glucose	-	32.8	0.50	0.42	(Evans and Wang, 1988)
C. beijerinckii BA101	Batch with gas stripping	Glucose	-	75.9	0.60	0.47	(Ezeji et al., 2003a)
C. beijerinckii BA101	Fed-batch with gas stripping	Glucose	-	233	1.16	0.47	(Ezeji et al., 2004b)
C. beijerinckii LMD	Fed-batch with pervaporation	Glucose	-	165.1	0.98	0.43	(Groot et al., 1984)
C. beijerinckii BA101	Continuous immobilized reactor	Glucose	-	7.9	15.8 Highest productivity	0.38	(Qureshi et al., 2000b)
C. acetobutylicum	Continuous membrane cell recycle reactor	Glucose	-	13	6.5	0.35	(Pierrot et al., 1986)
C. beijerinckii BA101	Continuous with gas stripping	Glucose	-	460	0.91	0.4	(Ezeji et al., 2004a)
C. acetobutylicum	Continuous membrane bioreactor with liquid–liquid extraction	Glucose	-	14.5	3.08	0.3	(Eckert and Schulˆgerl, 1987)
C. beijerinckii	Batch	Liquefied corn starch; LCS	13.4	18.4	0.15	0.41	(Ezeji et al., 2007b)
C. beijerinckii	Fed-batch with product recovery by gas stripping	Saccharified liquefied corn starch (SLCS)	56.2	81.3	0.59	0.36	(Ezeji et al., 2007b)

Table 20.1 Cont'd.

C. beijerinckii BA101	Batch	Glucose	19.6	24.2	0.34	0.42	(Evans and Wang, 1988)
C. beijerinckii BA101	Batch	Corn starch	15.8	24.7	9.34	0.44	(Blaschek et al., 2002)
C. beijerinckii BA101	Batch	Maltodextrins	18.6	26.1	0.37	0.50	(Parekh et al., 1999)
C. beijerinckii BA101	Batch	Soy molasses	18.3	22.8	0.19	0.39	(Qureshi and Blaschek, 2000c)
C. beijerinckii BA101	Batch	Agricultural waste	9.8	14.8	0.22	-	(Blaschek et al., 2002)
C. beijerinckii BA101	Batch	Packaging peanuts	15.7	21.7	0.20	0.37	(Blaschek et al., 2002)
C. beijerinckii NCIMB 8052 parent strain	Batch	Glucose/corn steep water (CSW) 20-l fermentation using 5% glucose/CSW medium	8.5 12.7	19.2	-	-	(Parekh et al., 1999)
C. beijerinckii BA101	Batch	Glucose/corn steep water (CSW) 20-l fermentation using 5% glucose/CSW medium	16 17.8	23.6	-	-	(Parekh et al., 1999)

Ezeji et al., 2003a,b; Huang et al., 2004; Qureshi et al., 2004; Tashiro et al., 2004; Blaschek, 2005; Ezeji et al., 2005a,b; Ezeji et al., 2007a; Lee et al., 2008a). In this process, the reactor is initiated in a batch mode and cell growth until the cells are in the exponential phase. While the cells are in the exponential phase, the reactor is fed continuously with the medium and a product stream is withdrawn at the same flow rate as the feed, thus keeping a constant volume in the reactor. This mode of fermentation runs much longer than in a typical batch process. This technique of butanol production by *C. beijerinckii* BA101 was investigated and resulted in a productivity rate of 1.74 g/L h^{-1} (Formanek et al., 1997). In a single-stage continuous system, high reactor productivity may be obtained, but this occurs at the expense of low product concentration when compared to that achieved in a batch process. The use of a single-stage continuous reactor does not yet seem practical on an industrial-scale due to the fluctuating solvent level and the complexity of butanol fermentation. Two or more multistage continuous fermentation systems have been investigated to reduce fluctuation and increase solvent concentration in the product stream (Bahl et al., 1982; Ezeji et al., 2004a). This is done by allowing growth, acid production, and solvent production to occur in separate bioreactors. In a two-stage system, Bahl et al. (1982) reported a solvent concentration of 18.2 g/L when using *C. acetobutylicum* DSM 1731. Gapes et al. (1996) reported on the continuous cultivation of *C. beijerinckii* NRRL B592 with a two-stage chemostat with on-line solvent removal. This strain maintained the ability to produce ABE at an overall dilution rate of 0.13 h^{-1} and achieved an overall ABE productivity of 1.24 $gL^{-1}h^{-1}$. This compares to a volumetric ABE production rate of 1.74 $gL^{-1}h^{-1}$ for *C. beijerinckii* BA101 (a 40% increase) grown in a single-stage chemostat at a dilution rate of 0.20 h^{-1} (Gapes et al., 1996).

Immobilized cell reactors and cell recycle reactors have also been applied to butanol production (Afschar et al., 1985; Vijjeswarapu et al., 1985; Yang and Tsao, 1995; Lienhardt et al., 2002; Qureshi et al., 2005a,b; Tashiro et al., 2005) in order to increase productivity 40–50 times as compared to batch reactors (Ezeji et al., 2004a; Lee et al., 2008). Reactor productivity can be improved by increasing the cell concentration in the reactor. *C. beijerinckii* BA101 was immobilized onto clay brick particles and the fermentation was performed for ABE production (Qureshi et al., 2000). The material chosen to immobilize cells was clay brick which is readily available, and is an inexpensive material. At a dilution rate of 2 h^{-1}, ABE productivity of 15.8 $gL^{-1}h^{-1}$ was achieved with a yield of 0.38 gg^{-1} and concentration of 7.9 gL^{-1} (Table 20.1). It was found that the yield and concentration increased with lower dilution rate (Qureshi et al., 2000). Therefore, it is suggested that sporulation should be blocked to achieve higher productivity. Nevertheless, an immobilized cell continuous reactor is a strong candidate for an industrial fermentation process.

Membrane cell recycle reactor is another option to improve fermentation productivity. Pierrot et al. (1986) studied a hollow-fiber ultrafiltration to separate and recycle cells in a continuous system. At a dilution rate of 0.5 h^{-1}, cell mass, solvent concentration, and ABE productivity of 20 gL^{-1}, 13 gL^{-1}, and 6.5 $gL^{-1}h^{-1}$, respectively, were obtained (Table 20.1). However, fouling of the membrane with the fermentation broth was a major obstacle of this system. Lipnizki et al. (2000) suggested a way to overcome this problem by allowing only the fermentation broth to undergo filtration by using the immobilized cell system (Ezeji et al., 2004a; Lee et al., 2008). However, fermentation must be combined with a suitable product removal technique (Table 20.1).

20.5 Agri-Residues as Substrate

In order to reduce the butanol production cost, various renewable and economically feasible substrates such as agricultural residues have been used as the fermentation substrates including wheat straw, corn waste, or rice straw (Nout and Bartelt, 1998; Parekh et al., 1998; Parekh and Blaschek, 1999; Parekh et al., 1999; Qureshi and Blaschek, 2001; Campos et al., 2002; Qureshi et al., 2004; Qureshi, 2006; Ezeji et al., 2007a; Ezeji et al., 2007c; Li and Chen, 2007; Liu et al., 2007; Zhang et al., 2007; Fan et al., 2008; Qureshi et al., 2008b). The ABE produced from these alternative renewable resources ranged from 14.8 to 30.1 gL^{-1} (Table 20.1). Other than the use of corn, liquefied corn meal and corn steep liquor (a byproduct of corn wet milling process that contains nutrients leached out of corn during soaking) for acetone–butanol production have been tested. In the batch process with recovery, 60 gL^{-1} of liquefied corn meal and corn steep liquor yielded 26 gL^{-1} of solvent. It showed that the use of liquefied corn meal and corn steep liquor had great potential for the bioconversion of starch to acetone–butanol. However, the presence of sodium metabisulfite in the liquefied starch and corn steep liquor may be a major problem in long-term fermentation by *C. beijerinckii* BA101 (Ezeji et al., 2007b; Parekh et al., 1998; Qureshi 2006; Qureshi et al., 2008a).

Ezeji et al. (2007b) studied the potential industrial substrate (liquefied corn starch; LCS) employed for successful ABE production. Fermentation of LCS (60 gL^{-1}) in a batch process resulted in the production of 18.4 gL^{-1} ABE, a batch fermentation of LCS integrated with product recovery resulted in 92% utilization of sugars present in the feed. It showed that a combination of fermentation of this novel substrate (LCS) to butanol with product recovery may be the economical approach.

20.6 Important Considerations in Biobutanol Fermentation

20.6.1 Pretreatment

The clostridia are not able to efficiently hydrolyze fiber-rich agricultural residues (Ezeji et al., 2007a). Therefore, agricultural biomass must be hydrolyzed to simple sugars using economically developed methods. Dilute sulfuric acid pretreatment can be applied to agricultural residues for hydrolysis. Another pretreatment method for wheat straw is the use of alkali pretreatment and enzyme hydrolysis (Marchal et al., 1984). There are enzymes available that can hydrolyze cellulose to glucose that can, in turn, be fermented to butanol (Ezeji et al., 2007a). In a study using corn stover as substrate (Parekh et al., 1988), SO_2-catalysed prehydrolysis and enzyme hydrolysis were conducted to obtain fermentable sugars. However, during acid hydrolysis, a complex mixture of microbial inhibitors is generated. Examples of the inhibitory compounds include furfural, hydroxymethyl furfural (HMF), and acetic, ferulic, glucuronic, and r-coumaric acids (Zaldivar et al., 1999; Varga et al.,2004). *C.beijerinckii* BA101 can utilize the individual sugars present in lignocellulosic (e.g., corn fiber, distillers dry grains with solubles (DDGS)) hydrolysates such as cellobiose, glucose, mannose, arabinose, and xylose. It was found that some of the lignocellulosic hydrolysate inhibitors associated with *C. beijerinckii* BA101 growth and ABE production. When 0.3 gL^{-1} r-coumaric and ferulic acids were introduced into the fermentation medium, growth and ABE production by *C. beijerinckii* BA101 decreased significantly. Furfural and HMF are not inhibitory to *C. beijerinckii* BA101; rather they have stimulatory effect on the growth of the microorganism and ABE production (Ezeji et al., 2007a).

Generally corn fiber hydrolysis process is hydrolyzed using dilute (0.5% v/v) sulphuric acid (Ebener et al., 2003). Detoxification of the corn fiber hydrolysate (CFH) is needed through overliming (Martinez et al., 2001).

Soni et al. (1982) reported bioconversion of agro-wastes into acetone and butanol by *C. saccharoper butylacetonicum*. Following a 2.25% alkali autoclaving pretreatment of bagasse or rice straw, a 30-h enzymatic hydrolysis (*A. wentii* and *T. reesei* culture filtrate) resulted in 6% reducing sugar content of 10% slurry. Poor solvents production was improved by ammonium sulphate precipitation and activated carbon treatment of the hydrolyzate. Addition of ferrons sulphate into the culture medium resulted in further solvents yield improvement to maximum of 33.5% in 60 h (Volesky and Szczesny, 1983).

There is a group working on bioconversion of lignocellulosic waste into butanol (Angenent, 2008). The first stage of the process used a mixed microbial culture in converting fractionated corn fiber into butyrate followed by optimizing a mixed culture for a continuous process in converting the resulting butyrate-rich mixture into butanol. The study is focus on the effects of pretreatment on molecular structure and microbial break down efficiency and manipulating a diverse community bioprocess in developing predominantly butyrate. This process may yield some good data in laboratory-scale experiment, the process may be too complex to scale-up.

In a study conducted by Qureshi et al. (2008c), *C. beijerinckii* P260 was used to produce butanol (acetone–butanol–ethanol, or ABE) from wheat straw (WS) hydrolysate in a fed-batch reactor. It was found that simultaneous hydrolysis of WS to achieve 100% hydrolysis to simple sugars and fermentation to butanol is possible. In addition to WS, the reactor was fed with a sugar solution containing glucose, xylose, arabinose, galactose, and mannose. The culture utilized all of the above sugars. A productivity of 0.36 $g\ L^{-1}.h^{-1}$ was observed and a productivity of 0.77 $g\ L^{-1}.h^{-1}$ was obtained when the culture was highly active. The fed-batch fermentation was operated for 533 h (Qureshi et al., 2008c).

20.6.2 Operating pH

The pH of the medium is very important for butanol fermentation. In acidogenesis, rapid formation of acetic and butyric acids causes a decrease in pH. Solventogenesis starts when pH reaches a critical point, beyond which acids are reassimilated and butanol and acetone are produced. Therefore, low pH is a prerequisite for solvent production (Kim et al., 1984; Lee et al., 2008). However, if the pH decreases below 4.5 before enough acids are formed, solventogenesis will be brief and unproductive. Increasing the buffering capacity of the medium is a simple way to increase growth and carbohydrate utilization as well as butanol production (Bryant and Blaschek, 1988; Lee et al., 2008).

It was reported that ABE was produced by *C. acetobutylicum* using glucose on a large scale only at an acidic pH (Beesch 1953; Ross 1961). When *C. acetobutylicum* was grown in continuous culture under glucose limitation at neutral pH and varying dilution rates the only fermentation products formed were acetate, butyrate, carbon dioxide and molecular hydrogen. Acetone and butanol were produced when the pH was decreased below 5.0 (optimum at pH 4.3). The addition of butyric acid (20 to 80 mM) to the medium with a pH of 4.3 resulted in a shift of the fermentation from acid to solvent formation.

ABE production was not observed when *C. acetobutylicum* was grown in continuous culture with glucose or nitrogen limitation at pH values of 6.5 or 5.7 (Gottschal and Morris 1981). Chemostat cultures growing at neutral pH and forming exclusively acids as fermentation products over weeks could be changed to producing solvents by a decrease of the pH and by an addition of butyrate to the medium. The optimum pH was found to be 4.3. Values in the range of 4.2 to 5.8 for batch cultures were reported in the literature (Lee et al., 2008). In a study (Bahl et al., 1982), it was found that butyrate had no effect at pH values above pH 5.0; it also could not be replaced by acetate. The reason might be that, at low pH, butyric acid enters the cells by diffusion where it could interfere with the formation of butyrate from butyryl-coenzyme A (Bahl et al., 1982; Gavard et al., 1957; Lee et al., 2008; Twarog and Wolfe, 1962).

Geng and Park (1993) reported that *C. acetobutylicum* B18 produced a large amount of butanol over a wide range of pH (4.5-6.0). At pH 6.0 fermentation and cell growth were most active at pH 6.0, and the highest values of glucose consumption rate (4.37 g $L^{-1}.h^{-1}$), butanol productivity (1.0 g $L^{-1}.h^{-1}$), butyric acid recycle rate (0.31 g $L^{-1}.h^{-1}$), and cell growth rate (0.2 h^{-1}) were obtained. There existed a critical pH between 6.0 and 6.5 above which cells switched to acid producing mode. Clostridial stage appeared essential for solvent production by strain B18 but sporulation was not necessary for solvent formation (Geng and Park, 1993).

20.6.3 Nutrients

The additions of various organic materials to the basic medium for production ABE by clostridia have been widely practiced by supplementing organic nitrogen, amino acids, vitamins and other compounds. The main substance for the large-scale process was usually a very complex mixture of carbohydrates and other carbonaceous and inorganic compounds which differed greatly from one operation to another (Volesky and Szczesny, 1983).

Complex nitrogen sources such as yeast extract are generally required for good growth and ABE production, but otherwise the nutrient requirements for the growth of clostridia are simple (Monot et al., 1982). Clostridia require high negative redox potential to produce butanol (and ethanol) and the supply of additional reducing power results in increased butanol and ethanol formation with reduced acetone formation (Mitchell, 1998).

Acetone and butanol can be produced by batch and continuous culture of *C. acetobutylicum* on a nitrogen limited synthetic medium. When cells were grown on a high glucose concentration, conversion yields close to 0.3 g solvents per g glucose were obtained at pH 5.0. (Monot and Engasser, 1983). When *C. acetobutylicum* was grown in continuous culture under phosphate limitation (0.74 mM) at a pH of 4.3, glucose was fermented to butanol, acetone and ethanol as the major products. The results showed that high yields of butanol and acetone were obtained in batch culture under phosphate limitation (Bahl et al., 1983).

20.6.4 Operating Temperature

The effect of temperature on the ABE production by *C. acetobutylicum* has been studied in the range 25 to 40°C. It was found that the solvent yield decreased with increasing temperature; seemingly because of a reduction in acetone production. It appeared that the yield of the other major solvent, butanol, was not affected by the temperature. Considering total ABE yield and productivity only, the optimum fermentation temperature is 35°C for C. acetobutylicum (McNeil and Kristiansen, 1985).

20.6.5 Biokinetics of Butanol Fermentation

A kinetic simulation model of ABE fermentation of xylose (ModelXYL) was proposed by substituting Embden–Meyerhof–Parnas (EMP) pathway equations in the glucose model (ModelGLC) by pentose phosphate (PP) pathway equations of xylose utilization. The results with the developed model suggested that *C. saccharoperbutylacetonicum* N1-4 has a robust metabolic network in acid- and solvent-producing pathways. Furthermore, sensitivity analysis revealed that slow substrate utilization would be effective for higher butanol production, coinciding with the experimental results. Therefore, it was considered that the proposed model to be one of the best kinetic simulation candidates describing the dynamic metabolite behavior in ABE production. These results were consistent with the experimental results. By using these models, it can successfully create an optimal design for bioreactors and elucidate metabolic networks in detail, and consequently propose the genetic manipulation strategy for higher butanol production in ABE fermentation (Shinto et al., 2008).

20.7 Butanol Recovery Technologies

High product recovery cost is another problem in biobutanol production. Besides the traditional distillation process, several other processes including gas stripping, liquid–liquid extraction, pervaporation, adsorption, and reverse osmosis have been developed to improve recovery performance and reduce costs (Qureshi and Blaschek, 2000b; Fadeev and Meagher, 2001; Huang and Meagher, 2001; Qureshi et al., 2001; Ezeji et al., 2003a, 2004a; Ezeji et al., 2005b; Qureshi et al., 2005a; Ezeji et al., 2007c; Thongsukmak and Sirkar, 2007; Lee et al., 2008b; Vane, 2008). The traditional recovery process employing distillation has a high operation cost due to the low concentration of butanol in the fermentation broth. To solve this problem as well as that of the solvent toxicity problem simultaneously, *in situ* recovery systems have been employed (Table 20.1). Gas stripping is a simple technique for recovering butanol (acetone or ethanol) from the fermentation broth. Liquid–liquid extraction is another technique that can be used to remove solvents (acetone, butanol, ethanol) from the fermentation broth. Pervaporation is a membrane-based process that is used to remove solvents from fermentation broth by using a selective membrane. From an economic point of view, reverse osmosis is most preferable. However, it has disadvantages of membrane clogging or fouling. In contrast, liquid–liquid extraction has high capacity and selectivity, although it can be expensive to perform (Durre, 1998). These techniques have been described in details in the literature (Maddox, 1989; Groot et al., 1992; Ezeji et al., 2004a; Lee et al., 2008).

20.8 Techno-Economic Analysis of Butanol Fermentation

Qureshi and Blaschek (2000b) estimated the cost to produce butanol from corn using the hyperbutanol producing strain of *C. beijerinckii* BA101. It was found that substrate price was the most influential factor affecting the price of butanol. A butanol price of US$ 0.55/kg is based on corn price of US$71/ton. An increase in corn price to US$118/ton would result in butanol price of US$0.66/kg. An inflated corn price of US$237/ton would result in butanol price of US$1.15/kg. ABE yield is another important factor which influences the butanol production cost. In the reactors containing *C. beijerinckii* BA101, an ABE yield of 0.40-0.50 was obtained. If ABE yield of 0.42 was increased to a yield of 0.45 (average of 0.40 and 0.50), the butanol price will reduce from US$0.55/kg to US$0.51/kg. It was found that there was an inverse linear relationship between yield and butanol price. As different strains show varied yield, there is different butanol production cost using different strains. *C. beijerinckii* BA101 was found superior to other butanol producing cultures as it resulted in higher production of butanol and improved yield. *C. acetobutylicum* results in total ABE yield of 0.30 to 0.33. If ABE is produced using *C.acetobutylicum* with a yield of 0.30 then the price of butanol would be US$0.73/kg compared to US$0.55/kg when using *C. beijerinckii* BA101.

There are other costs, for example: interest on the borrowed capital, rate of return (profit), steam, electricity, depreciation, maintenance and repair, federal taxes, and insurance are also the major factors contributing to the price of butanol.

During the last decade, a significant amount of research has been conducted on the development of alternative technologies designed to remove the butanol continuously from the fermentation broth (e.g., gas stripping, ionic liquids, liquid–liquid extraction, pervaporation, aqueous two-phase separation, supercritical extraction, perstraction, etc.). These recovery techniques can reduce the product inhibition which results in a reduction in the process streams, higher productivity, and lower distillation costs. According to Qureshi and Blaschek (2001), when using corn for butanol production using batch fermentation process, employing distillative recovery showed the butanol cost of US$0.55/kg, if pervaporative recovery was used, it significantly reduced the butanol production cost to US$0.11–0.36/kg.

20.9 Limitations of Butanol Fermentation and Future Perspective

Biobutanol has a number of clear-cut advantages over bioethanol as a renewable alternative fuel; it is still not produced commercially as a biofuel for economic reasons. The current method of biobutanol production, by ABE fermentation and distillation, is too complex and energy-intensive. The major problems of ABE fermentation include product toxicity to the producing bacteria, substrate-to-product conversion efficiency, the ability to utilize an inexpensive substrate, and the potential for culture degeneration. Anyway, further improvements in butanol production are possible based on the development of molecular genetic systems for the manipulation of the solventogenic clostridia, the improved upstream and downstream engineering- based approaches for production and recovery process. These improvements will undoubtedly affect the economics of fermentation-derived butanol production (Ezeji et al., 2004a; Lee et al., 2008).

On the other hand, a process using extracellular hydrolysis by chemical or by physical or by enzymes for pretreatment (is presently under development for production of bioethanol) could be applied to furnish lignocellulosic substrates for microbial conversion to butanol (Zverlov et al., 2006).

Furthermore, the new membrane technology should enable economic production of butanol, a green fuel alcohol, from renewable resources. Existing and future corn-to-ethanol plants could be used for butanol production if the technology were proven to be economical. In addition, the technology could be adopted to replace petroleum-derived butanol and acetone (Huang et al., 2008).

Future research on biobutanol production shall focus to achieve simultaneously high yield of products and recovery. By using agri-residues as the substrate for economic biobutanol production, the study should focus on mixed sugar fermentation and inhibitory effects on different soluble chemicals produced during pretreatment. Optimization of fermentation conditions shall be considered for higher butanol yield. Furthermore, further research and development, especially in strain selection and genetic modification, as well as in the downstream processing of the ABE are essential for cost-effective butanol production.

20.10 Summary

The conversion of agricultural residues into butanol is a feasible technology, which could contribute considerably to make a national economy less dependent on imported fossil oil. The technology of fermenting biomass hydrolyzates to solvents is a promising link in the chain of advances toward the use of lignocellulosic biomass for fuel production. Necessary improvements for a modern process would include breeding suitable plant varieties, growing the energy plants in large scale, pretreatment and hydrolysis, fermentation, downstream processing, and marketing. Although the production of butanol and acetone is technically more challenging than that of ethanol but also more advantageous if the substitution of gasoline as well as diesel is the target.

The process economics can be further improved when on-line product removal is used. The ABE process using the advanced product recovery such as gas stripping, liquid–liquid extraction, pervaporation, adsorption, and reverse osmosis has dramatically increased the efficiency and productivity of fermentation-based butanol production to the point that this process is expected to become competitive with petrochemically-derived butanol. ABE fermentation has a promising future for the production of biosolvents from renewable resources. Continued advances in upstream and downstream processes will result in improved efficiency of butanol production.

20.11 Acknowledgements

The authors are sincerely thankful to the Natural Sciences and Engineering Research Council of Canada (Discovery Grants A4984, Canada Research Chair). The authors are also thankful to NSERC for providing PDF scholarship to S. Yan. The views or opinions expressed in this chapter are those of the authors and should not be construed as opinions of the USEPA.

20.12 References

Abdel-Hakim, E., El-Ardi, O., El-Zanati, E., and Fahmy, M. (2006) "Modeling and simulation of batch butanol fermentation." CHISA 2006 - 17th International Congress of Chemical and Process Engineering.

Afschar, A. S., Biebl, H., Schaller, K., and Schugerl, K. (1985). "Production of acetone and butanol by Clostridium acetobutylicum in continous culture with cell recycle." *Applied Microbiology and Biotechnology*, 22(6), 394-398.

Angenent, L.T. (2008) "Butyrate from lignocellulose for butanol." *Industrial Bioprocessing* 30: 2.

Atsumi, S., Cann, A. F., Connor, M. R., Shen, C. R., Smith, K. M., Brynildsen, M. P., Chou, K. J. Y., Hanai, T., and Liao, J. C. (2008). "Metabolic engineering of Escherichia coli for 1-butanol production." *Metabolic Engineering*, 10(6), 305-311.

Avcibasi Guenilir, Y., and Deveci, N. (1996). "Production of acetone-butanol-ethanol from corn mash and molasses in batch fermentation." *Applied Biochemistry and Biotechnology - Part A Enzyme Engineering and Biotechnology,* 56(2), 181-188.

Bahl, H., Andersch, W., and Gottschalk, G. (1982). "Continuous production of acetone and butanol by Clostridium acetobutylicum in a two-stage phosphate limited chemostat." *European Journal of Applied Microbiology and Biotechnology,* 15(4), 201-205.

Beesch, S. C. (1953) "Acetone-butanol fermentation of starches." *Appl. Microbiol.* 1, 85–95.

Blaschek, H. P. M. (2005). "Continuous butanol fermentation of starch." *Industrial Bioprocessing,* 27(1), 2.

Blaschek, H. P., Formanek, J., and Chen, C. K. (2002). "Method of producing butanol using a mutant strain of Clostridium beijerinckii." University of Illinois Urbana-Champaign. U.S. Patent No. 6,358,717. March 19, 2002.

Bryant, D. L., and Blaschek, H. P. (1988). "Buffering as a means for increasing growth and butanol production by Clostridium acetobutylicum." *Journal of Industrial Microbiology,* 3(1), 49-55.

Campos, E. J., Qureshi, N., and Blaschek, H. P. (2002). "Production of acetone butanol ethanol from degermed corn using Clostridium beijerinckii BA101." *Applied Biochemistry and Biotechnology - Part A Enzyme Engineering and Biotechnology,* 98-100, 553-561.

Cann, A.F., and Liao, J.C. (2008) "Production of 2-methyl-1-butanol in engineered *Escherichia coli*." *Applied Microbiology and Biotechnology*. 81(1): 89-98.

Durre, P. (1998). "New insights and novel developments in clostridial acetone/butanol/isopropanol fermentation." *Applied Microbiology and Biotechnology,* 49(6), 639-648.

Ebener J, Ezeji TC, Qureshi N, Blaschek HP. (2003). "Corn fiber hydrolysis and fermentation to butanol using *Clostridium beijerinckii* BA101." Proceedings of the 25th Symposium on Biotechnology for Fuels and Chemicals, Breckenridge, CO, USA.

Eckert, G., and Schuì^gerl, K. (1987). "Continuous acetone-butanol production with direct product removal." *Applied Microbiology and Biotechnology*, 27(3), 221-228.

Evans, P. J., and Wang, H. Y. (1988). "Enhancement of butanol formation by Clostridium acetobutylicum in the presence of decanololeyl mixed extradants." *Appl. Environ. Microbiol.*, 54, 1662-1667.

Ezeji, T. C., Groberg, M., Qureshi, N., and Blaschek, H. P. (2003a). "Continuous production of butanol from starch-based packing peanuts." *Applied Biochemistry and Biotechnology - Part A Enzyme Engineering and Biotechnology,* 106(1-3), 375-382.

Ezeji, T. C., Qureshi, N., and Blaschek, H. P. (2003b). "Production of acetone, butanol and ethanol by Clostridium beijerinckii BA101 and in situ recovery by gas stripping." *World Journal of Microbiology and Biotechnology*, 19(6), 595-603.

Ezeji, T. C., Qureshi, N., and Blaschek, H. P. (2004a). "Acetone butanol ethanol (ABE) production from concentrated substrate: Reduction in substrate inhibition by fed-batch technique and product inhibition by gas stripping." *Applied Microbiology and Biotechnology,* 63(6), 653-658.

Ezeji, T. C., Qureshi, N., and Blaschek, H. P. (2004b). "Butanol fermentation research: Upstream and downstream manipulations." *Chemical Record*, 4(5), 305-314.

Ezeji, T. C., Karcher, P. M., Qureshi, N., and Blaschek, H. P. (2005a). "Improving performance of a gas stripping-based recovery system to remove butanol from Clostridium beijerinckii fermentation." *Bioprocess and Biosystems Engineering*, 27(3), 207-214.

Ezeji, T. C., Qureshi, N., and Blaschek, H. P. (2005b). "Continuous butanol fermentation and feed starch retrogradation: Butanol fermentation sustainability using Clostridium beijerinckii BA101." *Journal of Biotechnology*, 115(2), 179-187.

Ezeji, T. C., Qureshi, N., and Blaschek, H. P. (2007a). "Bioproduction of butanol from biomass: from genes to bioreactors." *Current Opinion in Biotechnology*, 18(3), 220-227.

Ezeji, T. C., Qureshi, N., and Blaschek, H. P. (2007b). "Production of acetone butanol (AB) from liquefied corn starch, a commercial substrate, using Clostridium beijerinckii coupled with product recovery by gas stripping." *Journal of Industrial Microbiology and Biotechnology*, 34(12), 771-777.

Ezeji, T., Qureshi, N., and Blaschek, H. P. (2007c). "Butanol production from agricultural residues: Impact of degradation products on Clostridium beijerinckii growth and butanol fermentation." *Biotechnology and Bioengineering*, 97(6), 1460-1469.

Ezeji, T., Qureshi, N., and Blaschek, H. P. (2007d). "Production of acetone-butanol-ethanol (ABE) in a continuous flow bioreactor using degermed corn and Clostridium beijerinckii." *Process Biochemistry*, 42(1), 34-39.

Fadeev, A. G., and Meagher, M. M. (2001). "Opportunities for ionic liquids in recovery of biofuels." *Chemical Communications*(3), 295-296.

Fan, Y. T., Xing, Y., Ma, H. C., Pan, C. M., and Hou, H. W. (2008). "Enhanced cellulose-hydrogen production from corn stalk by lesser panda manure." *International Journal of Hydrogen Energy*, 33(21), 6058-6065.

Fond, O., Petitdemange, E., Petitdemange, H., and Gay, R. (1984). "Effect of glucose flow on the acetone butanol fermentation in fed batch culture." *Biotechnology Letters,* 6(1), 13-18.

Formanek, J., Mackie, R., and Blaschek, H. P. (1997). "Enhanced butanol production by Clostridium beijerinckii BA101 grown in semidefined P2 medium containing 6 percent maltodextrin or glucose." *Applied and Environmental Microbiology*, 63(6), 2306-2310.

Gapes, J. R., Nimcevic, D., and Friedl, A. (1996). "Long-term continuous cultivation of Clostridium beijerinckii in a two- stage chemostat with on-line solvent removal." *Applied and Environmental Microbiology*, 62(9), 3210-3219.

Gavard R, Hautecoer B, Descourtieux H (1957) "Phosphotransbu-tyrylase *in Clostridium acetobutylicum*." *CRH Acad Sci* , 244: 2323–2326

Geng, Q., and Park, C. H. (1993). "Controlled-pH batch butanol-acetone fermentation by low acid producing *Clostridium acetobutylicum* B18." *Biotechnology Letters,* 15(4), 421-426.

Godin, C., and Engasser, J. M. (1988). "Improved stability of the continuous production of acetone-butanol by Clostridium acetobutylicum in a two-stage process." *Biotechnology Letters*, 10(6), 389-392.

Gottschal JC, Morris JG (1981) "Non-production of acetone and butanol by *Clostridium acetobutylicum* during glucose-and ammonium-limitation in continuous culture." *Biotech Lett,* 3:525–530

Groot, W. J., van den Oever, C. E., and Kossen, N. W. F. (1984). "Pervaporation for simultaneous product recovery in the butanol/isopropanol batch fermentation." *Biotechnology Letters*, 6(11), 709-714.

Groot, W. J., van der Lans, R. G. J. M., and Luyben, K. C. A. M. (1992). "Technologies for butanol recovery integrated with fermentations." *Process Biochemistry*, 27(2), 61-75.

Holt, R. A., Cairns, A. J., and Morris, J. G. (1988). "Production of butanol by Clostridium puniceum in batch and continuous culture." *Applied Microbiology and Biotechnology*, 27(4), 319-324.

Huang, J., and Meagher, M. M. (2001). "Pervaporative recovery of n-butanol from aqueous solutions and ABE fermentation broth using thin-film silicalite-filled silicone composite membranes." *Journal of Membrane Science*, 192(1-2), 231-242.

Huang, W. C., Ramey, D. E., and Yang, S. T. (2004). "Continuous production of butanol by Clostridium acetobutylicum immobilized in a fibrous bed bioreactor." *Applied Biochemistry and Biotechnology - Part A Enzyme Engineering and Biotechnology*, 115(1-3), 887-898.

Huang HJ, Ramaswamy S, Tschirner UW, Ramarao BV. (2008) "A review of separation technologies in current and future biorefineries." *Separation and Purification Technology*; 62(1):1-21.

Husemann, M. H. W., and Papoutsakis, E. T. (1989). "Enzymes limiting butanol and acetone formation in continuous and batch cultures of Clostridium acetobutylicum." *Applied Microbiology and Biotechnology*, 31(5-6), 435-444.

Husemann, M. H. W., and Papoutsakis, E. T. (1990). "Effects of propionate and acetate additions on solvent production in batch cultures of Clostridium acetobutylicum." *Applied and Environmental Microbiology*, 56(5), 1497-1500.

Ishizaki, A., Michiwaki, S., Crabbe, E., Kobayashi, G., Sonomoto, K., and Yoshino, S. (1999). "Extractive acetone-butanol-ethanol fermentation using methylated crude palm oil as extractant in batch culture of Clostridium saccaharoperbutylacetonicum N1-4 (ATCC 13564)." *Journal of Bioscience and Bioengineering*, 87(3), 352-356.

Jones, D. T., and Woods, D. R. (1986). "Acetone-butanol fermentation revisited." *Microbiological Reviews*, 50(4), 484-524.

Kim, B. H., Bellows, P., Datta, R., and Zeikus, J. G. (1984). "Control of carbon and electron flow in Clostridium acetobutylicum fermentations: Utilization of carbon monoxide to inhibit hydrogen production and to enhance butanol yields." *Applied and Environmental Microbiology*, 48(4), 764-770.

Lee, S. Y., Park, J. H., Jang, S. H., Nielsen, L. K., Kim, J., and Jung, K. S. (2008a). "Fermentative butanol production by clostridia." *Biotechnology and Bioengineering,* 101(2), 209-228.

Lee, S. M., Cho, M. O., Park, C. H., Chung, Y. C., Kim, J. H., Sang, B. I., and Um, Y. (2008b). "Continuous butanol production using suspended and immobilized Clostridium beijerinckii NCIMB 8052 with supplementary butyrate." *Energy and Fuels,* 22(5), 3459-3464.

Lee, Y. D., and Kim, H. S. (1992). "Effect of organic solvents on enzymatic production of cyclodextrins from unliquefied corn starch in an attrition bioreactor." *Biotechnology and Bioengineering*, 39(10), 977-983.

Li, D. M., and Chen, H. Z. (2007). "Fermentation of acetone and butanol coupled with enzymatic hydrolysis of steam exploded cornstalk stover in a membrane reactor." *Guocheng Gongcheng Xuebao/The Chinese Journal of Process Engineering*, 7(6), 1212-1216.

Lienhardt, J., Schripsema, J., Qureshi, N., and Blaschek, H. P. (2002). "Butanol production by Clostridium beijerinckii BA101 in an immobilized cell biofilm reactor: Increase in sugar utilization." *Applied Biochemistry and Biotechnology - Part A Enzyme Engineering and Biotechnology*, 98-100, 591-598.

Lin, Y. L., and Blaschek, H. P. (1983). "Butanol production by a butanol-tolerant strain of Clostridium acetobutylicum in extruded corn broth." *Applied and Environmental Microbiology*, 45(3), 966-973.

Lipnizki F, Hausmanns S, Laufenberg G, Field R, Kunz B. (2000). "Use of pervaporation-bioreactor hybrid processes in biotechnology." *Chem Eng Technol*, 23:569–577.

Liu, J., Wang, M., Wu, M., and Diwekar, U. (2007) "Simulation study of the process for producing butanol from corn fermentation." AIChE Annual Meeting, Conference Proceedings.

Maddox, I. S. (1989). "The acetone-butanol-ethanol fermentation: recent progress in technology." *Biotechnology and Genetic Engineering Reviews*, 7, 189-220.

Maddox, I. S., Steiner, E., Hirsch, S., Wessner, S., Gutierrez, N. A., Gapes, J. R., and Schuster, K. C. (2000). "The cause of "acid crash" and "acidogenic fermentations" during the batch acetone-butanol-ethanol (ABE-) fermentation process." *Journal of Molecular Microbiology and Biotechnology*, 2(1), 95-100.

Marchal R., M. Rebeller and J.P. Vandecasteele, (1984) "Direct bioconversion of alkali-pretreated straw using simultaneous enzymatic hydrolysis and acetone butanol production." *Biotechnol. Lett.* 6, 523–528.

Martinez A., M.E. Rodriguez, M.L. Wells, S.W. York, J.F. Preston and L.O. Ingram, (2001) "Detoxification of dilute acid hydrolysates of lignocellulose with lime." *Biotechnol. Prog.* 17 (2001), 287–293.

McNeil, B., and Kristiansen, B. (1985). "Effect of temperature upon growth rate and solvent production in batch cultures of Clostridium acetobutylicum." *Biotechnology Letters*, 7(7), 499-502.

Meyer, C. L., and Papoutsakis, E. T. (1989). "Continuous and biomass recycle fermentations of Clostridium acetobutylicum - Part 1: ATP supply and demand determines product selectivity." *Bioprocess Engineering*, 4(1), 1-10.

Mitchell WJ. (1998). "Physiology of carbohydrates to solvent conversion by *clostridia*." *Adv Microb Physiol,* 39:31–130.

Monot F, Martin J-R, Petitdemange H, Gay R. (1982). "Acetone and butanol production by *Clostridium acetobutylicum* in a synthetic medium." *Appl Environ Microbiol* 44:1318–1324.

Monot, F., and Engasser, J. M. (1983). "Production of acetone and butanol by batch and continuous culture of Clostridium acetobutylicum under nitrogen limitation." *Biotechnology Letters*, 5(4), 213-218.

Nout, M. J. R., and Bartelt, R. J. (1998). "Attraction of a flying nitidulid (Carpophilus humeralis) to volatiles produced by yeasts grown on sweet corn and a corn-based medium." *Journal of Chemical Ecology*, 24(7), 1217-1239.

Parekh, M., and Blaschek, H. P. (1999). "Butanol production by hypersolvent-producing mutant *Clostridium beijerinckii* BA101 in corn steep water medium containing maltodextrin." *Biotechnology Letters*, 21(1), 45-48.

Parekh, M., Formanek, J., and Blaschek, H. P. (1998). "Development of a cost-effective glucose-corn steep medium for production of butanol by *Clostridium beijerinckii*." *Journal of Industrial Microbiology and Biotechnology*, 21(4-5), 187-191.

Parekh, M., Formanek, J., and Blaschek, H. P. (1999). "Pilot-scale production of butanol by Clostridium beijerinckii BA101 using a low-cost fermentation medium based on corn steep water." *Applied Microbiology and Biotechnology*, 51(2), 152-157.

Pierrot, P., Fick, M., and Engasser, J. M. (1986). "Continuous acetone-butanol fermentation with high productivity by cell ultrafiltration and recycling." *Biotechnology Letters*, 8(4), 253-256.

Qureshi, N. (2006). "Manufacturing butanol from corn fiber." *Industrial Bioprocessing,* 28(5), 2-3.

Qureshi, N., and Blaschek, H. P. (2000a). "Butanol production using Clostridium beijerinckii BA101 hyper-butanol producing mutant strain and recovery by pervaporation." *Applied Biochemistry and Biotechnology - Part A Enzyme Engineering and Biotechnology*, 84-86, 225-235.

Qureshi, N., and Blaschek, H. P. (2000b). "Economics of butanol fermentation using hyper-butanol producing Clostridium beijerinckii BA101." *Food and Bioproducts Processing: Transactions of the Institution of of Chemical Engineers, Part C,* 78(3), 139-144.

Qureshi, N., and Blaschek, H. P. (2000c). "Recovery of butanol from fermentation broth by gas stripping." *Renewable Energy*, 22(4), 557-564.

Qureshi, N., and Blaschek, H. P. (2001). "ABE production from corn: A recent economic evaluation." *Journal of Industrial Microbiology and Biotechnology*, 27(5), 292-297.

Qureshi, N., Annous, B. A., Ezeji, T. C., Karcher, P., and Maddox, I. S. (2005a). "Biofilm reactors for industrial bioconversion process: Employing potential of enhanced reaction rates." *Microbial Cell Factories*, 4:24, 21 pages.

Qureshi, N., Hughes, S., Maddox, I. S., and Cotta, M. A. (2005b). "Energy-efficient recovery of butanol from model solutions and fermentation broth by adsorption." *Bioprocess and Biosystems Engineering*, 27(4), 215-222.

Qureshi, N., Karcher, P., Cotta, M., and Blaschek, H. P. (2004). "High-productivity continuous biofilm reactor for butanol production: Effect of acetate, butyrate, and corn steep liquor on bioreactor performance." *Applied Biochemistry and Biotechnology - Part A Enzyme Engineering and Biotechnology*, 114(1-3), 713-721.

Qureshi, N., Li, X. L., Hughes, S., Saha, B. C., and Cotta, M. A. (2006). "Butanol production from corn fiber xylan using *Clostridium acetobutylicum*." *Biotechnology Progress*, 22(3), 673-680.

Qureshi, N., Meagher, M. M., Huang, J., and Hutkins, R. W. (2001). "Acetone butanol ethanol (ABE) recovery by pervaporation using silicalite-silicone composite membrane from fed-batch reactor of *Clostridium acetobutylicum*." *Journal of Membrane Science*, 187(1-2), 93-102.

Qureshi, N., Saha, B. C., and Cotta, M. A. (2007). "Butanol production from wheat straw hydrolysate using *Clostridium beijerinckii*." *Bioprocess and Biosystems Engineering*, 30(6), 419-427.

Qureshi, N., Ezeji, T. C., Ebener, J., Dien, B. S., Cotta, M. A., and Blaschek, H. P. (2008a). "Butanol production by Clostridium beijerinckii. Part I: Use of acid and enzyme hydrolyzed corn fiber." *Bioresource Technology*, 99(13), 5915-5922.

Qureshi, N., Saha, B. C., Hector, R. E., Hughes, S. R., and Cotta, M. A. (2008b). "Butanol production from wheat straw by simultaneous saccharification and fermentation using Clostridium beijerinckii: Part I-Batch fermentation." *Biomass and Bioenergy*, 32(2), 168-175.

Qureshi, N., Saha, B.C., and Cotta, M.A. (2008c) "Butanol production from wheat straw by simultaneous saccharification and fermentation using Clostridium beijerinckii: Part II-Fed-batch fermentation." *Biomass and Bioenergy* 32: 176-183.

Qureshi, N., Schripsema, J., Lienhardt, J., and Blaschek, H. P. (2000). "Continuous solvent production by Clostridium beijerinckii BA101 immobilized by adsorption onto brick." *World Journal of Microbiology and Biotechnology*, 16(4), 377-382.

Ross, D. (1961) "The acetone-butanol fermentation." *Progr. Indust. Microbiol.* 3, 71–90

Shinto, H., Tashiro, Y., Kobayashi, G., Sekiguchi, T., Hanai, T., Kuriya, Y., Okamoto, M., and Sonomoto, K. (2008). "Kinetic study of substrate dependency for higher butanol production in acetone-butanol-ethanol fermentation." *Process Biochemistry*, 43(12), 1452-1461.

Soni, B. K., K. Das, and T. K. Ghose. (1982). "Bioconversion of agro-wastes into acetone-butanol." *Biotechnol. Lett.* 4:19-22.

Soni, B. K., Soucaille, P., and Goma, G. (1987). "Continuous acetone-butanol fermentation: influence of vitamins on the metabolic activity of Clostridium acetobutylicum." *Applied Microbiology and Biotechnology*, 27(1), 1-5.

Srivastava, A. K., and Volesky, B. (1991). "NADH fluoresence in a fed batch fermentation by *C. acetobutylicum*." *Canadian Journal of Chemical Engineering*, 69(2), 520-526.

Tashiro, Y., Takeda, K., Kobayashi, G., and Sonomoto, K. (2005). "High production of acetone-butanol-ethanol with high cell density culture by cell-recycling and bleeding." *Journal of Biotechnology*, 120(2), 197-206.

Tashiro, Y., Takeda, K., Kobayashi, G., Sonomoto, K., Ishizaki, A., and Yoshino, S. (2004). "High butanol production by Clostridium saccharoperbutylacetonicum N1-4 in fed-batch culture with pH-stat continuous butyric acid and glucose feeding method." *Journal of Bioscience and Bioengineering*, 98(4), 263-268.

Thongsukmak, A., and Sirkar, K. K. (2007) "Removal and recovery of solvents from fermentation broth by pervaporation using liquid membrane." 2007 AIChE Annual Meeting.

Twarog R, Wolfe RS (1962) "Enzymatic phosphorylation of phosphotransbutyrylase." *J Biol Chem* 237:2474–2481.

Vane, L. M. (2008). "Review: Separation technologies for the recovery and dehydration of alcohols from fermentation broths." *Biofuels, Bioproducts and Biorefining*, 2(6), 553-588.

Varga E, Klinke HB, Reczey K, Thomsen AB. (2004). "High solid simultaneous saccharification and fermentation of wet oxidized corn stover to ethanol." *Biotechnol Bioeng* 88:567–574.

Vijjeswarapu, W. K., Chen, W. Y., and Foutch, G. L. (1985). "Performance of a cell recycling continuous fermentation system for the production of n-butanol by Clostridium acetobutylicum." *Biotechnology Bioengineering Symposium*, VOL. 15(15), 471-478.

Volesky B, Szczesny T (1983) "Bacterial conversion of pentoses sugars to acetone and butanol." *Advances Biochem Eng Biotechnol* 27:101–118

Yang, X., and Tsao, G. T. (1995). "Enhanced acetone-butanol fermentation using repeated fed-batch operation coupled with cell recycle by membrane and simultaneous removal of inhibitory products by adsorption." *Biotechnology and Bioengineering*, 47(4), 444-450.

J. Zaldivar, A. Martinez and L.O. Ingram (1999). "Effect of selected aldehydes on the growth and fermentation of ethanologenic *Escherichia coli*." *Biotechnol. Bioeng*. 65. 24–33.

Zhang, M. L., Wei, R. X., Fan, Y. T., Xing, Y., and Hou, H. W. (2007). "Enlargement test studies of bio-hydrogen production using artificial wastewater of corn stalk fermentation lixivium by mixed culture." *Huanjing Kexue/Environmental Science,* 28(8), 1889-1893.

Zverlov, V. V., Berezina, O., Velikodvorskaya, G. A., and Schwarz, W. H. (2006). "Bacterial acetone and butanol production by industrial fermentation in the Soviet Union: Use of hydrolyzed agricultural waste for biorefinery." *Applied Microbiology and Biotechnology,* 71(5), 587-597.

CHAPTER 21

Application of Nanotechnology and Nanomaterials for Bioenergy and Biofuel Production

X.L. Zhang, S. Yan, R.D. Tyagi, R.Y. Surampalli, and Tian C. Zhang

21.1 Introduction

Energy needs have been dramatically increasing since the last century. The problems associated with global warming due to rising atmospheric greenhouse gas (GHG) levels, predicted scarcity of fossil fuels (Srivastava and Prasad, 2000), and economic concerns resulting from the use of non-renewable energy sources, have been the major concerns. Therefore, alternative fuel/energy should be developed to reduce our heavy dependency on fossil fuels. Biofuels/bioenergy, including biodiesel, biogasoline, hydrogen, bioethanol, biobutanol, and biogas, among others , have gained considerable attention due to their inherent merits such as better GHG emission abatement, renewability, long-term sustainability, and clean tail-pipe emission . To create truly sustainable biofuel solutions, many factors need to be considered, such as closed loop production, crop and soil sustainability, invigorate rural economies, sustainable water use, reduced carbon footprint, and market sustainability, among others.

Commercially biofuels are produced from feedstocks such as vegetable oils, animal fats, sugar and starch carbohydrate using processes such as transesterificationand fermentation (Fukuda et al., 2001; Mohammad et al., 2009). There is on-going effort to produce biofuels from nonfood feedstocks such as biomass, biowastes, and non-edible oil seed crops and algae through both biochemical and thermochemical pathways. The second and third generation biofuels also suffer from high production costs and other technical barriers as outlined in Chapter 1. The use of nanotechnology and nanomaterials could be one possible avenue to improve biofuel/bioenergy production efficiency and to reduction the processing cost. Nanotechnology application in biofuel production mainly focuses on 1) reducing the transportation cost of feedstock, 2) breaking down the feedstock more efficiently, and 3) improving biofuel production efficiency. Nano-catalysts have also improved the efficiency of biodiesel production when animal fats and vegetable oils were used as feedstock (Jennifer and Peter, 2006). Besides the direct contributions of nanotechnology to biofuel production, nanotechnology is also applied in many areas related to biofuel production, such as:

- The production and applications of nanoparticle-based catalysts can convert feedstock to biofuel or other value-added oils/lubricants efficiently (Jennifer and Peter, 2006). For example, a US-based company, Nano Chemical Systems Holdings, Inc., uses its patented process to immerse nano-sized molybdenum metal ball bearings to produce high-value oils and lubricants from the byproducts of palm-biodiesel production. The enhanced oils and lubricants give the consumer the advantages of longer machine life from reduced wear and superior performance at high temperatures and pressures. These oils and lubricants are highly biodegradable and have the promise of non-hazardous to the environment. The new oils and lubricants also offer the promise of reduced crank case and other lubricant emissions to improve air quality (AZoNano™.com, 2007).
- Applying the nanotechnology-based coating on large tanks, lines, and other refinery equipment installed in the biofuel facilities offers a clean, environmental friendly solution to issues of thermal insulation and corrosion protection that many facilities face.
- Nanotechnologies have been used to improve the commercial viability of algae bio-fuel. For example, QuantumSphere will be developing a nanocatalyst-based bio-gasification process that uses the nano-metals as catalysts for generating biofuel from wet algae collected from Salton Sea in California (Prakash, 2009). This application is important because algae-farming demands water and high ambient CO_2 concentration in atmosphere for maximal algal growth. Salton Sea is one of the largest and lowest inland seas at 227-feet below sea-level. According to the QuantumSphere technologists, Salton Sea has good algae growth, driven by large agricultural runoff. If the project is successful, the natural anxiety will be go away about the CO_2 limitation and a *water-versus-fuel battle* for algae-driven bio-fuel generation. As another example, researchers at Iowa State University are growing several strains of algae to test nanofarming technology that uses sponge-like mesoporous nanoparticles to extract biofuel oils from algae–without harming the algae crop, resulting in the reduction of the production cost and the generation cycle (softpedia, 2009).
- Nano-scale observations are important to understand how enzymes work. An improvement in resolution of that miniscule amount can mean the difference between seeing where atoms are and understanding how they interact. For example, new, improved-resolution views of a zinc transporter protein deciphered at the U.S. Department of Energy's Brookhaven National Laboratory (2009) provide not just a structure but also a suggested mechanism for how cells sense and regulate zinc. With advances in nanofabrication techniques, the study of fluid dynamics around a nano-object (e.g., enzymes) or in a nano channel is now more accessible experimentally and has become an active research area. For example, individual carbon nanotube transistors of ~2 nm diameter, incorporated into microfluidic channels, has been developed and used to locally sense the change in

electrostatic potential induced by the flow of an ionic solution, functioning as a nanoscale flow sensor (Bourlon et al., 2006).

- Most of the world's largest food and beverage corporations are conducting research and development (R&D) on nano-scale technologies to 1) engineer, process, package and deliver food and nutrients, and 2) reformulate their pesticides at the nano-scale to make them more biologically active. On the other hand, a wide range of current R&D has been conducted, ranging from atomically-modified seeds, nano-sensors for precision agriculture, plants engineered to produce metal nanoparticles, nano-vaccines for farmed fish, nano-barcodes for tracking and controlling food products, and more. All these R&Ds have significantly affected the current and future agriculture practice and biomass production. In nanoforestry, nanotechnology can be used to improve processing of wood-based materials into a myriad of paper and wood products by improving water removal and eliminating rewetting; reducing energy usage in drying; and tagging fibers and flakes (Worbring, 2006). Nanotechnology can also be used for development of the next generation of biomass for building functionality onto lignocellulosic surfaces at the nanoscale for ease in biofuel production.

This chapter is not intended to cover all applications of nanotechnology, but rather focused on recent developments in the field of nanotechnology as applied to bioenergy/biofuel production. This chapter discusses the basic concept of nanotechnology, major applications of nanotechnologies and nanomaterials for biofuel production and storage, current technology limitations, and future prospects. A comprehensive review on some of the topics can be found in forth coming book by Garcia-Martinez (2010).

21.2 Why Nanotechnology?

Concept of Nanotechnology. Nanotechnology is the technique to devise, synthesize, manufacture, and apply the matters with atomic or molecular precision (Zhang et al., 2009). Nanotechnology is the study to create structures and devices at dimensions of 100 nanometers (nm) or smaller (RS/RAEng, 2004). It is well known that small particles provide more surface area than the same materials at a larger size of same weight. The physical properties of carbon element change distinctly in nanoscales. Properties of carbon nanotubes have significant differences with plain carbon materials. Not only is the surface area extensively increased, the tenacity, elasticity, strength, and electricity are also enhanced. While the intensity of the aggregates of carbon nanotubes is 100 times that of steel, the weight of the former is only one-sixth of the latter. It may be considered that magnetic, electrical, optical, mechanical, and chemical properties are different for the material at various dimensions. Recent studies have proven that the nanosized materials have different physical and chemical properties from their macroscale materials. The change in properties of the materials is due to the increased surface area and the quantum effects (RS/RAEng, 2004). The increase in the surface area results in an increase in surface atoms. Surface atoms in nano-size particles have lower coordination numbers compared to macro-particles. Thus, the activity is highly enhanced in nanoscale matters (Brian, 2003). At nanosized substances, quantum effects become non-negligible. For example, silver is not magnetic in the bulk but nanoscale silver showed this property due to subtle electronic interactions. Therefore, the nanotechnology exhibits the overwhelming predominance of a wide range of fields from semiconductors to biotechnology to energy, transportation, medicine, engineering, electronics, agriculture, and consumer products due to unique behaviors and properties.

Why. Nanotechnologies are expected to revolutionize biofuel/bioenergy-related fields for the following reasons:

1. **Advances will require breakthrough technology.** In general, incremental/marginal improvements will not enable biofuel/bioenergy solutions to be relevant commercially. For example, solar cells and batteries are products that have experienced marginal improvements in efficiency or storage capacity, but have not made substantial leaps in either area during the past 10 to 20 years. Therefore, technical advances/breakthroughs are required to advance biofuel/bioenergy solutions to be commercially competitive with existing forms of production and resource allocation. The advancement in nanotechnology may accelerate the pace of change and improvements in these and other industries.
2. **Breakthrough advances in technology will be enabled at the nanoscale.** A great proportion of technical advances will be enabled at the nanoscale (or even the molecular level). Understanding of forces that act at the molecular, atomic and sub-cellular scales are dominating current scientific research. Unique properties of materials arise when the interactions between electrons and photons are controlled with nanoscale precision. Exploiting properties that are not readily attained at the macroscale makes nanotechnologies a far-reaching development and application potential.
3. **Nanotechnology-based innovation will be fostered by entrepreneurial companies.** A substantial portion of the innovation in biofuel/bioenergy industries will be developed by entrepreneurial, early-stage, privately held companies. Historically, these companies have been the source of innovation for developments and new products. The rapid development of personal computers, the internet, and the revolution in telecommunication equipment are examples of this type of innovation. The emergence of nanotechnologies and their applications in biofuel/bioenergy areas should be no different.

21.3 Nanotechnologies as New Concepts/Tools for Biofuel Production

Since the development of nanotechnology, it has been providing us with the capabilities to extend the function of materials, enhance the efficiency of processes, improve applications, and reduce hazard and waste generation. From earlier discussion, it is apparent that there is a considerable challenge to improve the efficiency of biofuel production due to the limitation of conventional technologies. For example, traditional catalysts employed in the production of biofuels have low efficiency/activity and poor selectivity. Common materials that are being employed to aid electron conduction in microbial fuel cell systems to facilitate electricity generation, to, enhance bioenergy storage, and to increase energy density are not efficient enough to be commercially viable. Applications of nanotechnologies and nanomaterials may have significant positive impact on these issues.

21.3.1 Applications of Nanomaterial Catalysts

Nanomaterials have captured a great interest in catalyst application as they help catalysts to become active, effective, efficient, reusable, stable, and pollutant-free. Nanomaterial catalysts consist of nanosized particles that can also be used as nanocarriers. The utilization of nanomaterials as catalysts in biofuel production is mainly in 1) feedstock pretreatment and hydrolysis, 2) storage of products (biogasoline, bioethanol, biogas, biobutanol, biodiesel and hydrogen), and 3) catalytic conversion of animal fat, vegetable oils and microbial lipids into biodiesel.

Nano-Size Particle Catalysts. Nano-sized particle catalysts have started to make contribution in biofuel production especially in biodiesel, hydrogen, and microbial fuel cells. In biodiesel production using soybean oil, researchers employed calcined Mg–Al hydrotalcites with an average particle size of 50 μm as the catalyst (Xie et al., 2005). The rate of conversion from soybean oils to biodiesel was only 67% even with an addition of a large amount of the catalyst (i.e., 7.5 wt% based on the oil weight) and a high mixing speed.

In general, the big size particle catalysts are not efficient and work under strict controlled conditions. According to equation 21.1 (Levenspiel, 1999), it can be seen that if the particle radius (R_p) is reduced, *Th* (Thiele modulus, a smaller value of *Th* will result in a higher utilization rate of catalyst) will reduce as well, and the utilization rate of catalysts will increase. Namely, the smaller the particle size is, the higher the catalyst efficiency is.

$$Th = \frac{R_p}{3}\sqrt{\frac{k}{D_{eff}}} \tag{21.1}$$

where *Th* = Thiele modulus; R_p = the particle radius; k = the pseudo-first order reaction rate constant; and D_{eff} = the effective diffusion coefficient, which is relative to the molecular diffusivity and the pore structure of the material (Carberry, 1976).

Butanol is an advanced n biofuel fuel that has fuel property comparable to gasoline. Currently, many companies (e.g., TetraVitae Bioscience, Inc. and Cobalt Biofuels) are using the nanotechniques to evolve and select for strains of organisms that produce the highest yield of butanol using their natural biochemical pathways.

Some studies examined nanoparticle catalysts application in biodiesel production through transesterification. Biofuel production via transesterification of oils and fats with alcohols of low molecular weight has been developed for a long time; in 1986, the first kinetic studies were performed (Rubio-Arroyo et al., 2009). Rubio-Arroyo et al. (2009) reported that a conversion of 99.05% sunflower oil into biofuel using a combined catalysis between tin oxide supported in mesoporous material and sodium hydroxide, without adding a step to separate the catalyst in the biofuel formation. The action of the heterogeneous catalyst on the oil tends to modify the relation between the aliphatic chain and the double bonds, contributing to the esterification reactions of free fatty acids before the reaction. Nano-γ-Al_2O_3 catalyst particles (< 50 nm) achieved 99.84% conversion with an addition of only 3 wt% catalysts (based on the oil weight). Using nanoparticle-based catalyst (nanocrystalline calcium oxides), a high conversion of 99% was obtained with an addition of 0.6 wt% catalysts (Boz et al., 2009; Reddy et al., 2009). In hydrogen production with chemical reaction, some studies investigated the reaction between aluminum nanoparticles (with an average particles size of 120 nm and the specific surface area of 18 m^2/g) and water (Ivanov et al., 2000). The results showed that hydrogen was produced when the reaction between the ultrafine aluminum and gel-like water, which was formed by adding soluble polymer in water, occurred with the temperature up to 343K. The ultrafine metal powders (UFP) are produced by electrical explosion of wires. UFP have some unique properties that bulk metal doesn't possess, such as a high stability in the air at 300–450°C, a large specific surface area (5–50 m^2/g), and no oxide film on the surface. No passivation on UFP occurs because of the inert gas layer, which was formed during the UFP production process. After desorption of the layer, the activity recovers. The development of nano-scale metal particles enhanced the hydrogen production rate. In Austria, a research group developed a biocatalyst by modifying a natural nanoporous mineral by adding to anaerobic digestion processes. The biogas production was enhanced by 30%; and hydrogen sulfide production was largely reduced (http://www.invenia.es/tech:06_at_atbi_0e1s).

These studies indicated that the high conversion efficiency was achieved due to the abundant pores contained in nano-size particles. It also resulted in substantial reduction of the reaction time. Nano-size catalysts could provide a large amount of pores which assist the intra-particle diffusion. Therefore, even a small quantity of nanoparticle catalysts can provide a high active specific surface area of catalysts, and thus, reduce the dosage of the catalysts.

Catalysts Loading on Nanocarriers. Coating catalysts on nanomaterials is another effective method to enhance the efficiency of catalysts besides improving their activities by generating nanoparticle catalysts. Nanomaterials with various shapes and structures have been employed to support catalysts, such as nanoparticles (surface-attachment) (Matsunaga and Kamiya, 1987; Shinkai et al., 1991; Crumbliss et al., 1992; Kondo et al., 1992), nanofibers (carrying) (Jia et al., 2002; Herricks et al., 2005), nanoporous matrix (entrapping) (Wang et al., 2001), nanotube (adsorption and entrapping) (Mitchell et al., 2002; Besteman et al., 2003; Rege et al., 2003; Yim et al., 2005). Application of nanocoated catalysts for biofuel production includes the pretreatment stage (lignocellulose as feedstock) and the production stage; applications in both stages have been revealed a large impact on the biofuel production efficiency. Nano-structured materials provide a large loading area for catalyst immobilization (Wang, 2006) and increase the diffusion of the substrate molecules to the catalyst, which in turn enhances the production rate (Kim et al., 2004). Immobilizing biocatalysts and/or enzymes on nanomaterials is a significant technique to realize the ease in catalyst recycling, continuous operation, and product purification.

In ethanol production using cornstalks as feedstock, nanotechnology was applied to break down cornstalks to nano-sized particles, thus it greatly reduces the cost of transporting the biomass to fuel production plants, and increases the transportation efficiency (Jennifer and Peter, 2006). In addition, researchers proposed to use continuous high shear force nanomixers and nanodispersers that use variety of turbines for targeting different lignocellulosic biomass for pretreatment of cellulase transgenic biomass (Sticklen, 2009).

Lignocellulose as feedstock has overwhelming advantages due to cost and availability compared to other feedstocks, such as corn and sugars. There are two pathways of producing biofuels: biochemical and thermochemical. In biochemical conversion, there are considerable challenges to obtain fermentable sugar for ethanol fermentation. Traditionally, physical, chemical or biological treatment is used to break down the complex lignocellulosic structure. However, lignin dissolution is often a challenge.

Nano-catalysts, such as enzyme-based nano-particles (Johnson et al. 2009) and sulfonic acid cobalt spinel ferrite magnetic nanoparticles (Hohn et al., 2009), increased the conversion efficiency of lignocellulose. For example, cellulase is an efficient enzyme used to convert pretreated biomass to fermentable sugars for bioethanol production. Yet, the high cost of the enzyme dictates that recyclable enzymes should be developed. Immobilization of cellulases on magnetic nanoparticles ($Fe(OH)_2$) was carried out to apply for the conversion of lignocellulose (Johnson et al., 2009). It was found that 80% of the activity of this enzyme-nanoparticle complex remained intact even after 10 sequential recycles. Acid catalysts to improve lignocellulose conversion to sugars are the most common method in bioethanol production. However, the serious damage to the equipments caused by the use of strong acid is a concern. The use of sulfonic acid cobalt spinel ferrite magnetic nanoparticles as the catalyst in bioethanol production prevented the acid corrosion on the equipments, as well as increased the conversion efficiency of lignocellulose to sugars (Hohn et al., 2009).

Lipases as biocatalysts have achieved a stable yield up to 93% of biodiesel from vegetable oils (Akoh et al., 2007). However, the recycling of lipases to reduce the enzyme cost is difficult. Loading lipases on nanomaterials should be a choice to fulfill the goal of recycling/reuse lipases. Porous silica nano-tubes (NTPS) with pore diameters of 2 nm have been investigated to immobilize enzyme. Lysozyme with an approximate diameter of 3 nm was immobilized on NTPS with a loading of more than 350 mg/g silica (Ding et al., 2004), which was three times higher as compared to most conventional mesoporous silica materials (less than 100 mg/g silica) (Lei et al., 2002; Fan et al., 2003). The loading was completed with adsorption (electrostatic attraction) on the external surface of NTPS and/or entrapping in the openings of NTPS (if the opening size is larger than the enzyme). This study provides a foundation for immobilizing lipases on nanomaterials.

Loading metal catalysts on nanomaterials has been reported for ethanol production. Carbon nanotubes coated with nano-size rhodium and manganese generators have been employed to produce ethanol from carbon monoxide and hydrogen during syngas conversion (Pan et al., 2007). Rh and Mn loaded both inside and outside the carbon tube enhanced dissociation/activation of CO and increased the hydrogenation rate (Figure 21.1). This led to a highly improved ethanol production rate.

Coating the catalysts on nanomaterials has been the most extensive approach to enhance the catalyst efficiency. However, further studies are needed in this area.

21.3.2 Nanomaterial Reaction Media

Gas Storage. Hydrogen storage is a significant step on hydrogen application and usually is considered as part of the hydrogen production process. Two common hydrogen storage technologies, high pressure tanks for gas form and cryogenic vessels for liquid hydrogen, are difficult to use on a daily basis. Hydrogen storage in solid materials provides an alternative path. Metal hydrides (Cohen and Wernick, 1981), chemical hydrides (Kojima et al., 2004), and physisorption (Nijkamp et al., 2001) are the three typical methods. However, metal-hydride systems and chemical hydride systems have low energy efficiencies; they are not completely reversible processes. Physisorption has a very high potential to store hydrogen with high safety, low energy need for adsorption and release of H_2.

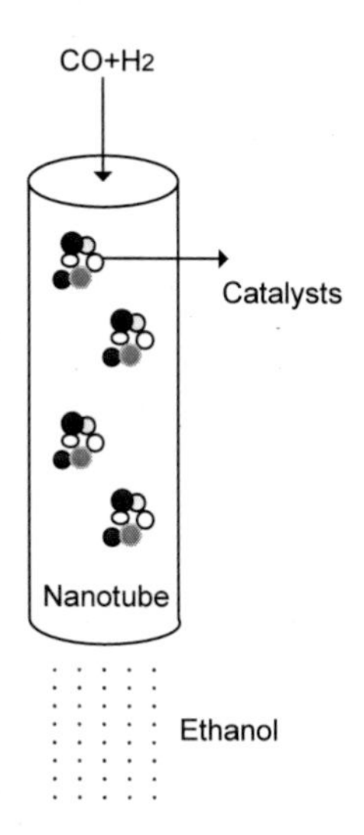

Figure 21.1 Ethanol production by chemical synthesis of syngas (Berger, 2007).

Porous materials especially nanomaterial adsorbents are excellent media for sorption of hydrogen. They can be ultimately applied in a hydrogen-storage system. Currently, most research on hydrogen storage is focused on storing hydrogen in a lightweight, compact manner for mobile applications. A design target for automobile fueling has been set by the U.S. Department of Energy at 6.5% hydrogen by weight (Rosi et al., 2003; Booker and Boysen, 2005). There are several hydrogen carriers, including metal hydrides, synthesized hydrocarbons, carbon nanotubes, ammonia, ammonia borane (H_3B-NH_3), amine borane complexes, imidazolium ionic liquids, doped polymers, and glass microspheres. Of these carriers, nanomaterials hold great potential. For example, carbon nanotubes may hold 50 wt% hydrogen although < 1% storage is practically accepted at cryogenic temperatures (see Table 21.1, Dillon et al., 2002). Single-walled carbon nanotubes (SWCNs) are widely used as a good example of carbon-based nano-structured adsorbent (Anson et al., 2004). Theoretically, by chemisorption alone SWCNs can store hydrogen up to 7.7 wt% as every carbon atom adsorbs one hydrogen atom. Moreover, the hydrogenated SWCNTs' surface still has the potential to increase the capacity of hydrogen storage. Metal-organic frameworks (MOFs) are highly crystalline materials; they have been produced through metal-ions aggregation onto carboxylate rigid organic ligands to form extended frameworks (Nathaniel et al., 2003). MOF compounds (see the hydrogen storage section) can be designed based on the application desire. Up to now, a great number of different MOFs have been synthesized. MOF-5 with a cubic 3-D extended small porous structure can absorb hydrogen up to 4.5 wt% (17.2 hydrogen molecules per formula unit at room temperature, a pressure of 20 bar and 78 ^{o}K (Rosi et al., 2003). Some MOFs have a hydrogen storage capacity up to 10 wt%. Zeolites have three dimensional silicate structures. Depending on the different framework types, zeolites can have a very open microporous structure. Compared to the hydrogen storage capacity of the above two materials, zeolites have the lowest uptake (maximum 2.2 wt%), owing to their relatively low specific surface area. Boron oxide is another option for hydrogen sorption materials (Jhi et al., 2004). Boron oxide is fine powder with sizes between 10 and 100 nm. Boron oxide as an adsorbent has 30% higher T_D (moderate desorption temperature), 110 K but for carbon T_D is 80 K. Apparently, T_D of boron oxide is much nearer to the goal of 200 K.

Nanomaterials also have application in biogas storage (e.g., methanol separation from biogas). Biogas is a very practical fuel in rural areas, but the makeshift covered lagoon digesters (CLD) could not maintain proper temperature for methanogenic activities. Thus, the production would be slowed or almost ceased in winter (Chastain et al., 1999). Therefore, the biogas storage is very important. Traditional storage methods are not practical due to the low efficiency. Researchers are seeking new pathways to solve this problem. Carlos and co-workers investigated the application of the vacuum pressure swing adsorption (VPSA) process for biogas storage (Carlos and Alirio, 2007). Equilibrium and kinetics were employed to evaluate the efficiency of biogas storage by using activated carbon materials, zeolite and metal organic framework solids as adsorbents. Activated carbon materials are favorable to carbon dioxide sorption compared to methane. Zeolite has faster sorption of carbon dioxide than that of methane. However, activated carbon material has a low selectivity, and zeolite faces the problem of recycling. Copper-MOF was found to possess a capacity of 6.6 mol/kg for carbon dioxide sorption (activated carbon: 3.83 mol/kg and zeolite: 3.2 mol/kg) (Simone et al., 2008). Under controlled conditions, selectivity of carbon dioxide or methane can be significantly increased.

On the other hand, during biofuel production processes, it is unavoidable to generate other byproducts, and it is essential to selectively recover biofuel. Nanomaterials exhibited high selectivity for recovery of a desired product.

Table 21.1 Hydrogen storage capacity of different nanomaterials[a].

Material	Density (wt%)	Temperature (K)	Pressure (mPa)
> 10 wt%			
GNFs (Platelet)	53.68	RT	11.35
Li-MWNTs	20	~473–673	0.1
Ammonia Borane[b]	19.6	-	-
K-MWNTs	14	< 313	0.1
GNFs (Tubular)	11.26	RT	11.35
CNFs	~10	RT	10.1
Li/K-GNTs (SWNT)	~10	RT	8–12
GNFs	~10	RT	8–12
Within 1-10 wt%			
SWNTs (Low Purity)	5–10	273	0.04
SWNTs (High Purity)	8.25	80	7.18
CN nanoballs	8	573	0.1
Nano graphite	7.4	RT	1
SWNTs (High Purity + Ti Alloy)	6–7	~300–700	0.07
GNFs	6.5	RT	~12
CNFs	~5	RT	10.1
MWNTs	~5	RT	~10
MOF-5	4.5	78	2
SWNTs (High Purity + Ti Alloy)	3.5–4.5	~300–600	0.07
SWNTs (50% Purity)	4.2	RT	10.1
Li-MWNTs	~2.5	~473–673	0.1
IRMOF-8	2.3	RT	1
SWNT (50% Purity)	~2	RT	Echem
K-MWNTs	~1.8	< 313	0.1
(9,9) Array	1.8	77	10
IRMOF-6	1.1	RT	1
MOF-5	1.0	RT	1
MOF-5	1.0	RT	2

[a] Modified from Zhang et al. (2009). [b] Davis et al., 2009.

Biological Fuel Cells. Recently, considerable studies have been conducted 1) combining the microbial fuel cell (MFC) technology with nanotechnologies and nanomaterials to improve the performance of MFCs (Stolarczyk et al., 2007), and 2) enzymatic biofuel cells with enzyme modified anode and cathode electrodes or enzymes incorporated with electropolymerized films (Brunel et al., 2006; Ramanavicius et al., 2008). To improve MFCs' performance, properties of nanomaterials as both efficient catalysts and reaction media have been utilized. Yan and coworkers (2008) synthesized nanostructured polyaniline (PANI)/ mesoporous TiO_2 composite and employed it as anode in *Escherichia coli* MFCs. The optimal performance of the composite was observed with 30 wt% PANI. This nanostructured composite increased the power density to 1495 mW/m^2, which was almost three times higher than that obtained with conventional electrode in the MFC. Nanomaterials are also used in MFCs to enhance power density due to its high specific surface area and uniform nanopore distribution. Tushar and co-workers employed nanofluids which comprised of platinum nanoparticles anchored to multi-walled nanotubes as an electron mediator and carbon nanotube (CNT)- based electrodes on the MFC technology (Tushar et al., 2008). CNTs as electrodes have the tremendous predominance due to its excellent stability and electronic conductivity, and large surface areas. The power density in this MFC system increased to 2470 mW/m^2. Novel nanoporous membranes have recently been developed for fuel cells and MFCs, such as membranes consisted of polytetrafluoroethylene (PTFE) as the backbone with a nanosized ceramic powder (e.g., Aerosil2001 or Aerosil1301) dispersed in it (Jagur-Grodzinski, 2007). Nanoporous filters (e.g., nylon, cellulose, polycarbonate) were used to replace proton exchange membranes (PEMs) in MFCs (Biffinger et al., 2007).

Enzyme-based biofuel cells provide versatile means to generate electrical power from biomass or biowastes, and to use biological fluids as fuel-sources for the electrical activation of implantable electronic medical devices, or prosthetic aids. Willner et al. (2009) reviewed recent advances for assembling biofuel cells based on integrated, electrically contacted thin film-modified enzyme electrodes. Currently, there are many different methods to electrically connect the enzymes associated with the anodes/cathodes of the biofuel cell elements, such as 1) the reconstitution of apo-enzymes on relay-cofactor monolayers assembled on electrodes, or the crosslinking of cofactor-enzyme affinity complexes assembled on electrodes; 2) the immobilization of enzymes in redox-active hydrogels associated with electrodes; and 3) the use of nano-elements (e.g. carbon nanotubes, carbon nanoparticles) for the electrical contacting of the enzyme electrodes comprising the biofuel cells (Kim et al., 2006; Willner et al., 2008). There are three grand challenges that hamper the development of enzyme-based biofuel cells. First, the interface between the enzyme and the electrode must be characterized and optimized. An ideal enzymatic electrode should be limited by catalyst loading, and not by the transport of electrons between the enzyme and the electrode. The second key challenge is the stability of the bio-electrodes used in the biofuel cell. There is often a trade-off between enzymatic stability and activity, but optimal biofuel cells have to exhibit both high stability and activity. The final challenge is the spatial control of catalyst and substrate distribution at the electrode; mastery of this will be required for practical applications and mass fabrication of biofuel cells.

The unique electronic properties of metal nanoparticles, nanorods, or carbon nanotubes introduce new opportunities for the electrical contacting of redox enzymes with electrodes (Willner et al., 2008). Many studies focused on engineering aspects of designs of anode or cathode electrodes with enzymes and different materials (such as carbon nanoparticles, carbon nanotubes, micro- or nano-catalysts) to enhance the performance of biofuel cells. A fructose oxidizing enzyme, D-fructose dehydrogenase, was used to deposit onto the electrode surface to increase the rate of electron transfer to the anode without the need of a mediator. Another enzyme, laccase from *Trametes* sp., was used to reduce dioxygen at the cathode, completing the circuit. Eliminating the need for mediators may simplify the construction of MFCs as the device can operate under mild conditions. Stolarczyk et al. (2007) studied reduction of dioxygen catalyzed by laccase at the cathode coated with carbon nanotubes or carbon microcrystals. The authors found that the dioxygen reduction started at 0.6 V versus Ag/AgCl, which is close to the formal potential of the laccase used. The carbon nanotubes and nanoparticles present on the electrode provided electrical connectivity between the electrode and the enzyme active sites.

Using nanomaterials to develop biofuel cells is a cutting-edge technique because the nanocoating may enhance the performance of enzyme-based biofuel cells (Ramanavicius et al., 2008). However, many governing factors have been considered and suggested yet without much systematic evaluation and further verification. Factors such as electron and proton transfer resistances can be more overwhelming than the heterogeneous reaction kinetics in limiting the power generation in MFCs. For example, Zhao et al. (2009) studied CNT-supported glucose oxidase (CNT–GOx) in the presence of 1,4-benzoquinone (BQ). They found that the intrinsic Michaelis parameters of the reaction catalyzed by CNT–GOx were very close to those of native GOx. However, the Nafion entrapment of CNT–GOx for an electrode resulted in a much lower activity due to the limited availability of the embedded enzyme. Interestingly, kinetic studies revealed that the biofuel cell employing such an enzyme electrode only generated a power density equivalent to $< 40\%$ of the reaction capability of the enzyme on electrode. In addition, enzymatic bio-fuel cells need delicate optimization of electrode fabrication, which requires pore structure engineering, from macro- to nano-scale. Nano-material synthesis and electrode fabrication require in situ observations and non-intrusive characterizations of the process. Several intriguing in-situ characterization techniques have been demonstrated of their utility, such as fluorescence and polarization, imaging ellipsometry + quartz crystal microbalance + electrochemical technology (e.g., CV, EIS), and microporometry. Moreover, dynamic transient observations need to be followed by studies on mechanistic details in these nanomaterial-based biofuel cells. More research is needed to evaluate the effects of adding nanoparticles as mediators to facilitate the electron transport from the bulk solution to the anode, which may facilitate the growth of the microorganisms that are able to generate soluble redox shuttles (mediators) in the bulk solution. This is important because, in the past, MFCs have been designed and operated as a biofilm process, and the efficiency in generating electricity was dependent on the availability of electrode surface area. To increase power outputs, microorganisms that are not attached to the anode surface (e.g., suspended biomass) need to be enhanced for power generation.

21.4 Limitations of Current Technologies

Nanotechnologies are appealing due to their great potential in improving biofuel production. Applications of nanotechnologies and nanomaterials have significant potential to improve biofuel production. It can improve the production rate, by enhancing reaction kinetics, and may eventually reduce the production cost due to saving in material cost and processing time. However, it also has several limitations. In processes of catalyst coating on nanomaterials, the size of some catalysts is similar or even bigger than the pore of nanomaterials, resulting in the blockage of the channels or pores, and thus, hindering diffusion. For example, in enzyme immobilization processes, usually some of the enzyme molecules embedded in the pores or the inner surface of mesoporous silica, which limits the diffusion of substrate/product to/off the enzyme and results in lower enzyme activity. In addition, the coating process is not fully controllable. Therefore, some of the catalysts are not embedded in the channels while others still stay on the surface, which will be lost during the operation and/or storage stage. Moreover, nanoparticles and nanotubes exhibit mass transfer limitation; their dispersion and recycling are more difficult. The production of nanoparticle catalysts may pose health risks to the worker and users. . The nanocatalysts still have selectivity problems; more often than not, target reaction is enhanced but the affiliated reactions may also be accelerated. The cost is another factor that may go down as the technology becomes matured.

There are issues to be addressed before biofuel cells become competitive in practical applications. Two critical issues are short lifetime and poor power density, both of which are related to enzyme stability, electron transfer rate, and enzyme loading. Recent progress in nanobiocatalysis opens the possibility to improve in these aspects. However, nanomaterials as media also have some drawbacks. In hydrogen storage systems, nanomaterials still cannot largely increase hydrogen storage capacity (the highest hydrogen sorption capacity on nanometarial is around 10 wt% as reported so far). Furthermore, hydrogen desorption still needs a low temperature usually 80 K, but the goal is 200 K. Nanomaterials as anodes in microbial fuel cells are easily clogged by microbes due to their similar pore sizes with microbes, resulting in cell death and extreme reduction of the electrochemical reaction surface.

21.5 Future Research/Development Directions

As stated in section 21.4, there are limitations on nanotechnology application for biofuel production including feedstocks pre-treatment, efficient use of catalysts, bioconversion rate, storage media materials, and so on. To overcome these limitations, further research is needed.

1. Carbon is the major nanomaterial source that is being widely studied. Various other nanomaterials should also be explored.
2. Only a few types of nanomaterials are studied such as nanoparticles, nanotubes (especially single-walled), and nanofibers. Other types of nanomaterials need to be evaluated, like multi-walled nanotubes, which will enhance the nanotube performance.
3. The design of recyclable, highly active and selective heterogeneous catalysts for biofuel production using advanced nanotechnology, synthesis methods and quantum chemical calculations is needed. The chemical fundamentals of catalyst impregnation need to be understood and exploited for simple, rational methods to make different nanoparticles.
4. In today's nano-material and biofuel research, control of materials synthesis and process requires fast, non-intrusive, in-situ characterization over a large sample area. For enzymatic biofuel cell applications, such as in-situ, non-intrusive

observations are highly desirable.

5. Controllable catalyst coating necessary to prevent catalyst loss, catalyst aggregation and catalyst blockage in the channel and improvement of the coating efficiency are important problems for future study. Nanocatalyst stability should be improved otherwise its capability will be lost before being used.
6. Development of nanomaterials with high resistance to ethanol corrosion is necessary.
7. Most of nanomaterial applications for biofuel production are in the laboratory scale; scale-up studies must be conducted for eventual industrial applications.
8. Next-generation analytical tools at the nanoscale (e.g., spectroscopic, scattering and imaging tools that allow for imaging of nanometer sized features in thin liquid layers) are needed to explore the structural changes of nanocatalysts in aqueous environments of biofuel production processes.
9. Application of nanotechnologies may allow us to develop the next generation of biomass, including development of intelligent biomass (e.g., wood, crops, grass, etc.) and biomass-based products with built-in functionality onto lignocellulosic surfaces at the nanoscale that could open new opportunities in the future, such as improve processing of biomass-based materials into biofuels.

21.6 Summary

There are two major pathways to produce biofuels from biomass, namely biochemical and thermochemical. In both pathways, nanotechnology/nonomaterials can play a key role. Nanotechnology/nanomaterials can be applied in some areas, such as biomass pretreatment, enzyme hydrolysis, fermentation, and product separation and recovery during biochemical conversion. In thermochemical conversion, the application involves primarily in the development of different catalysts using nano-materials, improving mass transfer, product separation and recovery Furthermore, other applications of nanomaterials are in biofuel storage, collection and separation as absorbents or electrode design in fuel cells. Applications of nanocatalysts and nanotechnologies can reduce the cost of transportation of raw materials of feedstock, and enhance the feedstock utilization rate and biofuel production rate. Nanomaterial media can facilitate biofuel purification, separation, storage, and increase power outputs. However, nanotechnology application in biofuel production is still at the infant stage. Therefore, further research on new materials, processes and devices with nanotechnologies is needed to overcome the challenges outlined earlier.

21.7 Acknowledgements

The authors are sincerely thankful to the Natural Sciences and Engineering Research Council of Canada (Grants A4984, Canada Research Chair), and INRS-ETE for financial support. The views or opinions expressed in this chapter are those of the authors and should not be construed as opinions of the U.S. Environmental Protection Agency.

21.8 References

Akoh, C.C., Chang, S.W., Lee, G.C., and Shaw, J.F. (2007). "Enzymatic approach to biodiesel production." *J. Agric. Food Chem.*, 55, 8995-9005.

Anson, A., Callejas, M.A., Benito, A.M., Maser, W.K., Izquierdo, M.T., Rubio, B., Jagiello, J., Thommes, M., Parra, J.B., and Martinez, M.T. (2004). "Hydrogen adsorption studies on single wall carbon nanotubes." *Carbon*, 42, 1243-1248.

AZoNano™.com–the A to Z of Nanotechnology (2007.) "Nano chemical systems make green oils and lubricants from palm oil waste by-product." Available at <http://www.azonano.com/nanotechnology%20news.asp> (accessed Oct., 2009).

Berger, M. (2007). *Ethanol production inside carbon nanotubes.* Available at http://www.nanowerk.com/spotlight/spotid=2053.php (accessed June, 2009).

Besteman, K., Lee, J.O., Wiertz, F.G.M., Heering, H.A., and Dekker, C. (2003). "Enzyme-coated carbon nanotubes as single-molecule biosensors." *Nano Lett.*, 3, 727-730.

Biffinger, J.C. Ray, R., Little, B., and Ringeisen, B.R. (2007). "Diversifying biological fuel cell designs by use of nanoporous filters." *Environ. Sci. Technol.*, 41, 1444-1449.

Booker, R., and Boysen, E. (2005). *Nanotechnology for Dummies.* Wile Publishing, Inc. Hoboken, NJ.

Bourlon, B., Wong, J., Mikó, C., Forró, L., and Bockrath, M. (2006). " A nanoscale probe for fluidic and ionic transport." *Nature Nanotechnology*, 2, 104-107.

Boz, N., Degirmenbasi, N., and Kalyon, D.M. (2009). "Conversion of biomass to fuel: Transesterification of vegetable oil to biodiesel using KF loaded nano-[gamma]-Al2O3 as catalyst." *Applied Catalysis B: Environmental,* 89, 590-596.

Brian, F.G.J. (2003). "Nanoparticles in catalysis." *Topics in Catalysis*, 24(1-4), 147-159.

Brunel, L., Tingry, S., Servat, K., Cretin, M., Innocent, C., Jolivalt, C., and Rolland, M. (2006). "Membrane contactors for glucose/O_2 biofuel cell." *Desalination*, 199, 426–428.

Carberry, J. (1976). *Chemical and catalytic reaction engineering*, McGraw-Hill, New York, USA.

Carlos, A.G., and Alirio, E.R. (2007). "Biogas to fuel by vacuum pressure swing adsorption I. behavior of equilibrium and kinetic-based adsorbents." *Ind. Eng. Chem. Res.*, 46, 4595-4605.

Chastain, J.P., Linvill, D.E., and Wolak, F.J. (1999). "On-Farm biogas production and utilization for South Carolina livestock and poultry operations." *Project summary*, Department of Agriculture and Biological Engineering, Clemson University.

Cohen, R.L., and Wernick, J.H. (1981). "Hydrogen Storage Materials: Properties and Possibilities" *Science*, 214, 1081-1087.

Crumbliss, A.L., Perine, S.C., Stonehuerner, J., Tubergen, K.R., Zhao, J., and Henkens, R.W. (1992). "Colloidal gold as a biocompatible immobilization matrix suitable for the fabrication of enzyme electrodes by electrodeposition."

Biotechnol. Bioeng., 40, 483-490.

Davis, B.L., Dixon, D.A., Garner, E.B., Gordon, J.C., Matus, M.H., Scott, B., and Stephens, F.H. (2009). "Efficient regeneration of partially spent ammonia borane fuel." *Angew. Chem. Int. Ed.*, 48, 6812-6816.

Dillon, A.C., Parilla, P.A., Gilbert, K.E.H., Alleman, J.L., Gennett, T., and Heben M.J. (2003). "Hydrogen Storage in Carbon Nanotubes." Presentation given by National Renewable Energy Laboratory and Rochester Institute of Technology at DOE HFCIT Program Review Meeting, DOE Office of Energy Efficiency and Renewable Energy, DOE Office of Science, Division of Materials Science. <http://www.nrel.gov/hydrogen/proj_storage.html> (accessed July 14, 2008).

Ding, H.M., Wen, L.X., and Chen, J.F. (2004). "Porous silica nano-tube as host for enzyme immobilization." *China particuology*, 2(6), 270-273.

Fan, J., Lei, J., Wang, L.M., Yu, C.Z., and Zhao, D.Y. (2003). "Rapid and high-capacity immobilization of enzymes based on mesoporous silica with controlled morphologies." *Chem. Commun.*, 17, 2140-2141.

Fukuda, H., Kondo, A., and Noda H. (2001). "Biodiesel Fuel Production by Transesterification of Oils." *J. Biosci. Bioeng.*, 92, 405-416.

Garcia-Martinez, J. (2010). *Nanotechnology for the Energy Challenge.* Wiley, 440 pgs.

Herricks, T.E., Kim, S.H., Kim, J., Li, D., Kwak, J.H., Grate, J.W., Kim, S.H., and Xia, Y. (2005). "Direct fabrication of enzyme-carrying polymer nanofibers by electrospinning." *J. Mater. Chem.*, 14, 3241-3245.

Hohn, K., Wang, D.H., Pena, L., Ikenberry, M., and Boyle, D. (2009). "Acid Functionalized Nanoparticles for Hydrolysis of Lignocellulosic Feedstocks." *American Society of Agricultural and Biological Engineers* Available at www.asabe.org (accessed June, 2009).

Ivanov, V.G., Gavrilyuk, O.V., Glazkov, O.V., and Safronov, M.N. (2000). "Specific Features of the Reaction between Ultrafine Aluminum and Water in a Combustion Regime." *Comb. Expl.Shock Waves*, 36(2), 213-219.

Jagur-Grodzinski, J. (2007). "Polymeric materials for fuel cells: concise review of recent studies." *Polym. Adv. Technol.*, 18, 785–799.

Jennifer, K., and Peter, V. (2006). "Nanotechnology in agriculture and food production." *Project on Emerging Nanotechnologies*, 4.

Jhi, S.H., Kwon, Y.K., Keith, B., and Gabriel, J.C.P. (2004). "Hydrogen storage by physisorption: beyond carbon." *Solid State Communications*, 129, 769-773.

Jia, H., Zhu, G., Vugrinovich, B., Kataphinan, W., Reneker, D.H., and Wang, P. (2002). "Enzyme-Carrying polymeric nanofibers prepared via electrospinning for use as unique biocatalysts." *Biotechnol. Prog.*, 18, 1027-1032.

Johnson, P.A., Park, H.J., and McConnell J.T. (2009). *Enzyme Immobilization On Magnetic Nanoparticles for Cellulose Hydrolysis.* Available at http://aiche.confex.com/aiche/2009/webprogrampreliminary/Paper170582.ht ml (accessed June, 2009).

Kim, H.J., Kang, B., Kim, M.J., Park, Y.M., Kim, D.K., Lee, J.S., and Lee, K.Y. (2004). "Transesterification of vegetable oil to biodiesel using heterogeneous base catalyst." *Catalysis Today*, 93-95, 315-320.

Kim, J., Jia, H., and Wang, P. (2006). "Challenges in biocatalysis for enzyme-based biofuel cells." *Biotechnology Advances*, 24, 296–308.

Kojima, Y., Suzuki, K.-i., Fukumoto, K., Kawai, Y., Kimbara, M., Nakanishi, H., and Matsumoto, S. (2004). "Development of 10 kW-scale hydrogen generator using chemical hydride." *Journal of Power Sources*, 125, 22-26.

Kondo, A., Murakami, F., and Higashitani, K. (1992). "Circular dichroism studies on conformational changes in protein molecules upon adsorption on ultrafine polystyrene particles." *Biotechnol. Bioeng.*, 40, 889-894.

Lei, C., Shin, Y., Liu, J., and Ackerman, E.J. (2002). "Entrapping enzyme in a functionalized nanoporous support." *J. Am. Chem. Soc.*, 124, 11242-11243.

Levenspiel, O. (1999). *Chemical reaction engineering 3rd ed.* John Wiley and Sons, New York, USA.

Matsunaga, T., and Kamiya, S. (1987). "Use of magnetic particles isolated from magnetotactic bacteria for enzyme immobilization." *Appl. Microbiol. Biotechnol.*, 26, 328-332.

Mitchell, D.T., Lee, S.B., Trofin, L., Li, N., Nevanen, T.K., Suerlund, H., and Martin, C.R. (2002). "Smart nanotubes for bioseparations and biocatalysis." *J. Am. Chem. Soc.*, 124, 11864-11865.

Mohammad, N., Bambang, H.S., Muhammad, A.H., and Anondho, W. (2009). "Biogasoline from Palm Oil by Simultaneous Cracking and hydrogenation Reaction over Nimo/zeolite Catalyst." *World Applied Sciences (Special Issue for Environment)*, 5, 74-79.

Nathaniel, L.R., Juergen, E., Mohamed, E., David, T.V., Jaheon, K., Michael, O., and Omar, M.Y. (2003). "Hydrogen Storage in Microporous Metal-Organic Frameworks." *Science*, 300(5622), 1127-1129.

Nijkamp, M.G., Raaymakers, J.E.M.J., Dille, A.J.V., and Jong, K.P.D. (2001). "Hydrogen storage using physisorption –materials demands." *Appl. Phys. A.*, 72, 619-623.

Pan, X.L., Fan, Z.L., Chen, W., Ding, Y.J., Luo, H.Y., and Bao, X.H. (2007). "Enhanced ethanol production inside carbon-nanotube reactors containing catalytic particles." *Nature Materials*, 6, 507-511.

Prakash, V. (2009). "Nanotechnology to aid the commercial viability of algal bio-fuel production." *About Energy, in the Americas*, April 23, 2009.

Ramanavicius, R., Kausaitea, A., and Ramanavicienea, A., (2008). "Enzymatic biofuel cell based on anode and cathode powered by ethanol." *Biosensors and Bioelectronics*, 24, 761–766.

Reddy, C., Reddy, V., Oshel, R., and Verkade, J.G. (2009). "Conversion of biomass to fuel: Transesterification of vegetable oil to biodiesel using KF loaded nano-g-Al_2O_3 as catalyst." *Energy Fuel*, 20(3), 1310 [S1].

Rege, K., Raravikar, N.R., Kim, D.Y., Schadler, L.S., Ajayan, P.M., and Dordick, J.S. (2003). "Enzyme-polymer-single walled carbon nanotube composites as biocatalytic films." *Nano Lett.*, 3, 829-832.

Rosi, N.L., Eckert, J., Eddaoudi, M., Vodak, D.T., Kim, J., O'Keeffe, M., and Yaghi, O.M. (2003). "Hydrogen storage in microporous metal-organic frameworks." *Science*, 300(5622), 1127-1129.

Royal Society and Royal Academy of Engineering (RS/RAEng). (2004). *Nanoscience and Nanotechnologies: Opportunities and Uncertainties*. London: Royal Society and Royal Academy of Engineering.

Rubio-Arroyo, m.F., Ayona-Argueta,M.A., Poisot, M., and Ramı´rez-Galicia, G. (2009). "Biofuel Obtained from Transesterification by Combined Catalysis." *Energy & Fuels*, 23, 2840–2842.

Shinkai, M., Honda, H., and Kobayashi, T. (1991). "Preparation of fine magnetic particles and application for enzyme immobilization." *Biocat. Biotransform.*, 5, 61-69.

Simone, C., Carlos, A.G., and Alirio, E.R. (2008). "Metal Organic Framework Adsorbent for Biogas Upgrading." *Ind. Eng. Chem. Res.*, 47, 6333-6335.

Softpedia (2009). "Nanotechnology extracts biofuel from algae without killing them." Available at <http://www.softpedia.com/> (accessed Oct., 2009).

Srivastava, A., and Prasad, R. (2000). "Triglycerides based diesel fuels." *Renew. Sustain. Energy Rev.,* 4(1), 11-33.

Sticklen, M. (2009). "Molecular breeding forbiomass and biofuels." Presentation slides available at < http://bioenergy.msu.edu/presentations> (accessed Oct., 2009).

Stolarczyk, K., Nazaruk, E., Rogalski, J., and Bilewicz, B. (2007). "Nanostructured carbon electrodes for laccase-catalyzed oxygen reduction without added mediators." *Electrochimica Acta*, 53(11), 3983-3990.

The U.S. Department of Energy's Brookhaven National Laboratory (2009). "High-res view of zinc transport protein." Available at http://www.firstscience.com (accessed Oct., 2009).

Tushar, S., Reddy, L.M.A., Chandra, T.S., and Ramaprabhu, S. (2008). "Development of carbon nanotubes and nanofluids based microbial fuel cell." *International Journal of Hydrogen Energy*, 33, 6749-6754.

Wang, P. (2006). "Nanoscale biocatalyst systems." *Current Opinion in Biotechnology*, 17, 574-579.

Wang, P., Dai, S., Waezsada, S.D., Tsao, A., and Davison, B.H. (2001). "Enzyme stabilization by covalent binding in nanoporous sol-gel glass for nonaqueous biocatalysis." *Biotechnol. Bioeng.*, 74, 249-255.

Willner, I., Yan, Y.-M.,Willner, B., and Tel-Vered, R. (2008). "Integrated enzyme-based biofuel cells–a review." *Full Cells 09*, 1, 7–24.

Worbring, G. (2006). "The choice is yours." Available at <http://innovationwatch.com/> (accessed Oct., 2009).

Xie, W., Peng H., and Chen, L. (2005). "Calcined Mg–Al hydrotalcites as solid base catalysts for methanolysis of soybean oil." *J.Mol. Catal. A: Chem.*, 246, 24-32.

Yan, Q., Shu-Juan, B., Chang, M.L., Xiao-Qiang, C., Zhi-Song, L., and Jun, G. (2008). "Nanostructured Polyaniline/Titanium Dioxide Composite Anode for Microbial Fuel Cells." *American Chemical Society* (*ACS) Nano,* 2 (1), 113–119.

Yim, T.J., Liu, J., Lu, Y., Kane, R.S., and Dordick, J.S. (2005). "Highly active and stable DNAzyme–carbon nanotube hybrids." *J. Am. Chem. Soc.*, 127, 12200-12201.

Zhang, T.C., Surampalli, R., Lai, K.C.K., Hu, Z., Tyagi, R.D., and Irene Lo (eds.) (2009). *Nanotechnologies for Water Environment Applications*, American Society of Civil Engineers, Virginia, 2009.

Zhao, X., Hongfei Jia, J., Kim, J., and Wang, P. (2009). "Kinetic limitations of a bioelectrochemical electrode using carbon nanotube-attached glucose oxidase for biofuel cells." *Biotechnol. Bioeng.*, published online 7 August 2009 in Wiley InterScience (www.interscience.wiley.com). DOI 10.1002/bit.22496.

Editor Biographies

Dr. Samir Kumar Khanal is an Assistant Professor of Bioengineering in Molecular Biosciences and Bioengineering Department at University of Hawai'i at Mānoa and Collaborating Assistant Professor at Iowa State University. Prior to joining University of Hawai'i, Dr. Khanal was a Research Assistant Professor in the Department of Civil, Construction and Environmental Engineering at Iowa State University, Ames, IA from 2004 to 2007. Dr. Khanal was also a faculty affiliate at Biorenewables Program; Environmental Science, Center for Crops Utilization and Research (CCUR) and Biotechnology Program at Iowa State University. He also received two years of post-doctoral training in Biorenewables at Iowa State University from 2002 to 2004. Dr. Khanal obtained B.S. (First class with honors) in Civil Engineering from Malaviya National Institute of Technology, Jaipur, India, M.S. in Environmental Engineering from Asian Institute of Technology, Bangkok, Thailand, and Ph.D. in Environmental Engineering with focus in Environmental Biotechnology from the Hong Kong University of Science and Technology, Hong Kong in 2002. Dr. Khanal teaches transport phenomena, biomass conversion to biofuel and bioenergy, renewable energy systems, bioprocess engineering, and environmental biotechnology, among others. Dr. Khanal has over 30 refereed journal publications since 2003. Dr. Khanal's research focuses on application of physical, chemical and biological principles for the production of biofuel/bioenergy and valuable bio-based products from agri/forest-residues and biowastes. Dr. Khanal has served as a member of a review panel for the National Science Foundation, and he is a regular proposal reviewer for California Energy Commission, US DOE and the US Department of Agriculture. He is a referee of over 20 different international journals in biotechnology, bioenergy and environmental engineering. Dr. Khanal has two patents from his research at Iowa State University. He is a co-recipient of R&D 100 Award in 2008, 2008 Grand Prize for University Research presented by the American Academy of Environmental Engineers and International Water Association (IWA) Project Innovation Award. Dr. Khanal also recently published a book "Anaerobic Biotechnology for Bioenergy Production: Principles and Application " (Wiley-Blackwell Publishing). Dr. Khanal is a registered professional engineer in the state of Iowa.

Dr. Rao Y. Surampalli, P.E., Dist.M.ASCE, is an Engineer Director with the U.S. Environmental Protection Agency. He received his M.S. and Ph.D. degrees in Environmental Engineering from Oklahoma State University and Iowa State University, respectively. He is an Adjunct Professor at Iowa State University, University of Missouri-Columbia, University of Nebraska-Lincoln, Missouri University of Science and Technology-Rolla, University of Quebec-Sainte Foy, and Tongji University-Shanghai and an Honorary Professor in Sichuan University-Chengdu. He is a recipient of the ASCE National Government Civil Engineer of the

Year Award for 2006, ASCE State-of-the Art of Civil Engineering Award, ASCE Rudolph Hering Medal (twice), ASCE Wesley Horner Medal and ASCE Best Practice Oriented Paper Award. His awards and honors also include NSPE's Founders Gold Medal for 2001, National Federal Engineer of the Year Award for 2001, WEF's Philip Morgan Award, USEPA's Scientific and Technological Achievement Award, EPA Engineer of the Year Award (thrice), American Society of Military Engineer's Hollis Medal, Federal Executive Board's Distinguished Military Service Award, the U.S. Public Health Service's Samuel Lin Award, and Distinguished Engineering Alumnus Awards from Oklahoma State University and Iowa State University. Dr. Surampalli is a Fellow of the American Association for the Advancement of Science, a Distinguished Member of the American Society of Civil Engineers, and a Member of the European Academy of Sciences and Arts. Dr. Surampalli is an Editor of *Water Environment Research* and ASCE's *Practice Periodical of Hazardous, Toxic and Radioactive Waste Management.* He has authored over 400 technical publications, including 9 books, 38 book chapters and 140 peer-reviewed journal articles.

Dr. Tian C. Zhang, P.E., M.ASCE is a Professor in the department of Civil Engineering at the University of Nebraska-Lincoln (UNL). He received his Ph.D. in environmental engineering from the University of Cincinnati in 1994; after being a post-doctor for a few months, he joined the UNL faculty in August 1994. Professor Zhang teaches water/wastewater treatment, biological wastes treatment, senior design, remediation of hazardous wastes, among others. Professor Zhang has over 60 peer-reviewed journal publications since 1994. His research involves water, wastewater, and stormwater treatment and management, remediation of contaminated environments, and detection and control of emerging contaminants in aquatic environments. Professor Zhang is a member of the following professional organizations: American Society of Civil Engineers; American Water Works Association; the Water Environmental Federation; Association of Environmental Engineering and Science Professors; and the International Water Association. He has been the Associate Editor of *Journal of Environmental Engineering* (since 2007), *Practice Periodical of Hazardous, Toxic, and Radioactive Waste Management* (since 2006), and *Water Environment Research* (since 2008). He has been a registered professional engineer in Nebraska since 2000.

Dr. Buddhi P. Lamsal is an Assistant Professor in Food Science and Human Nutrition Department at Iowa State University since 2008. Prior to joining Iowa State University, Dr. Lamsal was a Research Assistant Professor from 2006 to 2008 in Grain Science and Industry Department at Kansas State University. Dr. Lamsal received his BS in Agricultural Engineering from Tamil Nadu Agricultural University, India in 1993; MS in Post-Harvest Technology from the Asian Institute of Technology, Thailand in 1994; and PhD in Food and Bioprocessing from The University of Wisconsin-Madison, WI in 2004. His PhD research was in soluble leaf protein recovery as value-added coproduct from alfalfa forage. He obtained two and half years of post-doctoral training in bioprocessing at Center for Crops Utilization Research (CCUR) at Iowa State University from 2004 to 2006. Dr. Lamsal's research focuses on value-added utilization of coproducts of biofuel industries, biorefinery approach to lignocellulosic biomass processing, and biobased products from biopolymers. Dr. Lamsal teaches food processing and enzyme applications in food and bioprocessing. He is also actively involved in professional societies such as American Association of Cereal Chemists (AACC), Institute of Food Technologists (IFT), and American Society of Agricultural and Biological Engineers (ASABE). Dr. Lamsal is currently an Associate Editor of Cereal Chemistry, and a referee of several food processing/biotechnology related scientific journals. Dr. Lamsal is a registered professional engineer (Agriculture Engineering) in the State of Iowa.

Dr. R. D. Tyagi is an internationally recognized Professor with 'Institut national de la recherché Scientifique – Eau, terre, et environement', (INRS-ETE), University of Québec, Canada. He holds Canada Research Chair on, 'Bioconversion of wastewater and wastewater sludge to value added products.' He conducts research on hazardous/solids waste management, water/wastewater treatment, sludge treatment/ disposal, and bioconversion of wastewater and wastewater sludge into value added products. He has developed the novel technologies of simultaneous sewage sludge digestion and metal leaching, bioconversion of wastewater sludge (biosolids) into *Bacillus thuringiensis* based biopesticides, bioplastics, biofertilisers and biocontrol agents. He is a recipient of the ASCE State-of-the Art of Civil Engineering Award, ASCE Rudolph Hering Medal, ASCE Wesley Horner Medal and ASCE Best Practice Oriented Paper Award. Dr. Tyagi has published/presented over 430 papers in refereed journals, conferences proceedings and is the author of five books, thirty-six book chapters, ten research reports and nine patents.

Dr. C.M. Kao, P.E., M.ASCE is a professor with the Institute of Environmental Engineering at National Sun Yat-Sen University, Kaohsiung, Taiwan. He received his M.S. and Ph.D. degrees in Civil and Environmental Engineering from North Carolina State University in 1989 and 1993, respectively. Prof. Kao has eleven years of experience as a researcher and environmental engineer in contaminated site characterization, soil and groundwater remediation, wastewater treatment, water quality modeling, watershed management, and risk assessment. Currently, he is working on several research projects funded by Taiwan National Science Council, Taiwan Environmental Protection Administration, and Taiwan Petroleum Company. He has organized and served as the conference chair for "The 8th International Conference on Drinking Water Quality Management and Control Technology" and "TEPA-US EPA Bilateral Cooperation Program - Petrochemical Wastewater Treatment Technology and River Basin Management Workshop in Taiwan." Dr. Kao is active in many professional organizations and is a Registered Professional Engineer in the branch of Civil Engineering, a Certified Ground Water Professional, and a Professional Hydrologist in the United States. He also serves on the executive committee of the Taiwan Association of Soil and Groundwater Environmental

Protection and Taiwan Wetland Protection Association. He has received numerous awards including Chinese Engineer Association's "Young Engineer Award," Chinese Institute of Environmental Engineering's "Academic Paper Award" in 2000 and 2003. He is also a Diplomate of the American Academy of Environmental Engineers. Dr. Kao has published more than 80 refereed journal articles.

Index